METALLIZED PLASTICS 2

Fundamental and Applied Aspects

METALLIZED PLASTICS 2

Fundamental and Applied Aspects

Edited by

K. L. Mittal
IBM US Technical Education
Thornwood, New York

SPRINGER SCIENCE+BUSINESS MEDIA, LLC

Library of Congress Cataloging-in-Publication Data

Metallized plastics 2 : fundamental and applied aspects / edited by K.L. Mittal.
p. cm.
"Proceedings of the Second Electrochemical Society Symposium on Metallized Plastics, held May 6-11, 1990, in Montreal, Quebec, Canada"--T.p. verso.
Includes bibliographical references and index.

1. Plastic coatings--Congresses. I. Mittal, K. L., 1945- .
II. Electrochemical Society Symposium on Metallized Plastics (2nd : 1990 : Montréal, Québec)
TP1175.M4M472 1992
668.4'1--dc20 91-44896
CIP

ISBN 978-1-4899-0737-0 ISBN 978-1-4899-0735-6 (eBook)
DOI 10.1007/978-1-4899-0735-6

Proceedings of the Second Electrochemical Society Symposium on Metallized Plastics, held May 6–11, 1990, in Montreal, Quebec, Canada

Originally published by Plenum Press, New York in 1991
Softcover reprint of the hardcover 1st edition 1991

PART IV. ADHESION ASPECTS OF METALLIZED PLASTICS

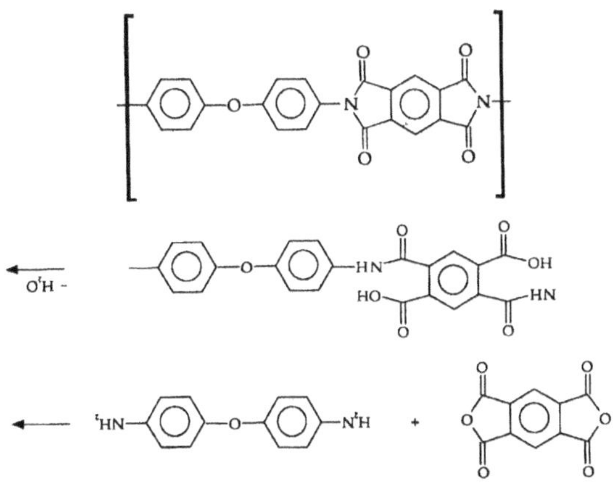

Scheme 1

POLYIMIDE FILM METALLIZATION

The procedure for metallizing Kapton^R film is straightforward. In general the stepwise metallization is accomplished by 1) electrochemical charge injection into the film; 2) oxidation of the film in a copper oxalate solution; 3) thick copper film formation by immersion in electroless copper. A specific procedure for a small scale metallization procedure is given below. The process can be implemented on a large, continuous scale which will be described in a later section and in figure 5.

Aqueous solutions of cupric oxalate, used for oxidatively discharging the reduced film, were prepared by dissolving 2.5g of reagent grade cupric sulfate in 1000ml of deionized water(10mM). To this solution 5.0g of solid potassium oxalate was added to bring the oxalate concentration to 27mM. The final pH was 6.1.

The electroless copper solution used in the metallization process was Shipley CUPOSIT^R CP78, a formaldehyde system which we operated between 40-45C. The solutions were obtained from Shipley Co. of Newton, Mass.. All other process steps in the metallization procedure were operated at ambient temperature.

A small piece of Kapton^R film, approximately 4 cm X 4 cm, was immersed in a 200 ml solution of $[(CH_3)_4N]_2V(edta)$ for 90 seconds and then rinsed in deionized water. The resulting green radical anion film was then immersed in a 200ml solution of 10 mM cupric oxalate for 3 minutes. During the immersion in cupric oxalate the polyimide film acquired a reflective coating of copper metal. The film was rinsed and then immersed in a commercially available solution of electroless copper (Shipley CUPOSIT^R CP78), which was held at a temperature between 40 and 45C for 1 minute.

The adhesion measurements followed IPC (Institute for Interconnection and Packaging Electronic Circuits) test method 650-2.4.9 method A and method C. Copper lines, 1.6 mm wide, were etched in 35 µm thick electroplated copper sheets by $FeCl_3$. The samples were held to a freely rotating wheel, 15.2 cm in diameter, by adhesive tape to insure a constant 90 degree angle of peel during the test. The peel force was measured by an Instron. Crosshead speed was 5 cm/minute and a copper strip 4 to 5 cm in length was peeled.

RESULTS AND DISCUSSION

A cyclic voltammetric response of the thermally cured polymer under 0.1M aqueous $NaClO_4$ is easily demonstrated provided the film is thin enough (figure 1)[10,11]. The pyromellitimide segment undergoes two chemically reversible, one electron transfers. The first reduction is observed at -0.78V (vs Ag/AgCl) and corresponds to the formation of a green radical anion. The second electron transfer at -1.03V results in the formation of a red dianion. These potentials are in general agreement with the voltammetric results of the chemically cyclized film[7].

Electrochemical film reduction can only occur through the combined processes of electron transport and mass transport of charge compensating ions through the film[13]. In principle either can be rate limiting. Factors which can influence mass transport include polymer density, degree of polymer swelling in an electrolyte system, and film thickness. Factors which can affect electron transport are not quite so obvious. In some systems chain mobility, caused by plasticization in solvents that swell, is thought to be crucial for electron self-exchange[14]. Conversely, it is possible that molecular order can also lead to enhanced electron transport by providing a lower activation energy to electron hopping. Disruption of order in such a system can then be expected to impede electron transport.

Although the reduction potentials and visible spectra recorded for chemically and thermally cyclized PI are in general agreement, significant differences are found in the conditions under which these results were obtained. For

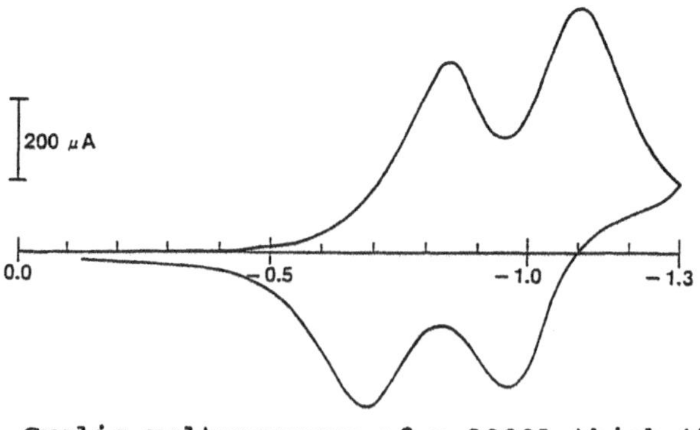

Figure 1 Cyclic voltammogram of a 2000A thick thermally cured polyimide film in 0.1M $NaClO_4$(aq.) at a 0.79cm^2 tin oxide electrode. Scan rate=20 mV/sec; iR compensated.

produce a desired copper thickness over a given surface area. For instance, if a 5000 A thick copper film over a 1 cm^2 surface area were desired, and using in the calculation a copper density of $8.9 g/cm^3$, $5.3X10^{-3}$ g of polymer would have to be reduced. This correlates to a film thickness of 38 µm using a film density of $1.42 g/cm^3$ [6]. Reducing this much film (and necessarily oxidizing the same amount) can require very long periods of time given that the apparent diffusion coefficient for charge transport in redox polymers can be very small. Mazur and Reich[4] have calculated a charge transfer coefficient for chemically cured PI of $8X10^{-10}$ cm^2/sec. Assuming this value for illustration purposes only, reducing 30 µm of Kapton[R] film would require 5 hrs (calculated from the approximate relationship $(D_{ct}t)^{1/2}=d$ where D_{ct} is the charge transfer diffusion coefficient, t is the time and d the film thickness). Therefore for practical purposes only short reduction times and consequently thin copper films on the order of 100 A were prepared in this stoichiometric reaction. Copper films of this thickness deposited on Kapton[R] had very high surface resistivity indicating a very discontinuous metal film. However, it was possible to produce copper films with bulk or near bulk conductivity with long reduction and oxidation times by this method.

A more practical approach to building copper thickness was found in the use of commercial electroless copper solutions to build copper thickness on top of the thin stoichiometric seed layer. Typically the thin copper film was exposed to a formaldehyde based electroless copper solution for 2 minutes which gave a total copper film thickness of approximately 2000 A. This film could then be electroplated to any additional thickness.

While the stoichiometric copper deposition reaction works well with either the potassium or tetramethylammonium salt of the V(II) complex, we normally used the tetramethylammonium salt to avoid the introduction of alkali metal ions into the substrate. We had poor success using larger quaternary ammonium salts because of restricted transport of the larger cations in the polyimide substrate[10].

The entire process for depositing copper films in the range of 2000 A in thickness can be done continuously by using a series of in-line reagent tanks which the Kapton[R] film is wound through. A schematic of a continuous line for this process is shown in figure 5. The Kapton[R] film is wound around water wrapped rollers so that the film rides on a thin film of water as it moves through the apparatus. This is done to avoid abrasion of the copper film. The entire process must be enclosed in a box which provides a nitrogen blanket for the reductant as well as the reduced film. A circulating electrochemical flow cell was used to generate the initial charge of $V(II)edta^{2-}$ as well as to keep this reagent in the V(II) state since during the film reduction V(II) is oxidized to V(III).

In the continuous process, the line speed was adjusted to provide a film residence time in the 0.1M V(II) tank of 1 minute. Since all the other tanks were of the same volume (approximately 4 liters) and configuration, this line speed (ca. 0.6 m/min) gave rise to a 1 minute residence time in the cupric oxalate tank and a total of 2 minutes in electroless copper (figure 5).

The basis of this metallization reaction is the stoichiometric injection of electrochemical charge across the film/solution interface. This makes it readily straightforward to produce an image in the metal deposit. Virtually any water insoluble coating or film which is applied to the polyimide substrate prior to the reduction reaction will serve as a resist since electron transport to the substrate is blocked at the covered surface. One of the easiest approaches to demonstrate this is screen printing an image or test pattern onto the film before reduction. Development of the reduced image in the copper oxalate solution produces a copper image.

One of the more important properties of a metal film deposited onto a polyimide substrate is the adhesion of the metal to the polyimide. Adhesion of films prepared by this process was measured by a 90 degree peel test. A 2000A thick copper film, deposited by the metallization process described above, was electroplated to a thickness of 35 μm and then 1.6 mm wide lines were etched by masking off the lines with plater's tape and then etching the copper with $FeCl_3$. For comparison, copper films on polyimide deposited by sputtering copper over a sputtered chromium oxide[15] adhesion layer, and then electroplating to a thickness of 35 μm, were also tested.

Typical adhesion values of the electrochemically deposited copper ranged from 1 to 1.4 kg/cm while the sputtered material ranged from 1.4 to 1.6 kg/cm. Examination of the fractured interface by ESCA and optical microscopy revealed the presence of large amounts of polyimide on the peeled copper line while no copper was found on the substrate interface. Additionally the substrate interface, after peeling the line, was severely distorted and crazed. These results indicated that a cohesive failure occurred within the polyimide as the mode of adhesion failure. The sputtered material behaved similarly. 1 to 1.4 kg/cm adhesion strength is relatively high for copper to Kapton adhesion. However, when the thick, etched samples were exposed to high humidity conditions, for instance 85C and 85% relative humidity, a dramatic decrease in the adhesion strength occurred. Peel strengths dropped from the 1 to 1.4 kg/cm value to as low as 0.1 kg/cm or lower. The mode of failure of the "wet" samples was different. Only very small amounts of polyimide were found on the fractured copper surface and the substrate interface did not craze or distort. The sputtered material did not undergo such a dramatic reduction in adhesion strength. The low adhesion strength can be reversed by simply drying the films at elevated temperature and low humidity. This also resulted in re-establishing the cohesive failure mechanism associated with the high adhesion strength.

The reversible behavior of adhesion strength suggested to us a mechanical interlocking of the copper and polyimide

- Does the morphology of the polymer surface or the interfacial region have an influence on the nature of the mass transfer of the metal layer?
- Is there a boundary layer existing between polymer and metal with mass transport properties diverging from those of the pure polymer or metal phase?
- What influence does the adhesion at the interfacial layer have on the mass transport?

The answers to these questions are rather complex and today we are far from providing definitive interpretations. In our studies we have simplified these problems to three subjects for investigation:

- Permeability of homo- and copolymers and the influence of morphology on mass transport.
- Adhesion between metal and polymer and the improvement of adhesion by various pretreatment methods.
- Permeation through electroplated polymer films and the influence of metal thickness and polymer surface properties.

PERMEABILITY

The process of permeation through a non-porous polymer membrane is usually explained in terms of the solution diffusion model[7,8]. This model postulates that the permeation occurs in three stages:

- sorption of the gas onto the polymer,
- diffusion through the polymer,
- desorption from the opposite face,

as shown in fig. 2.

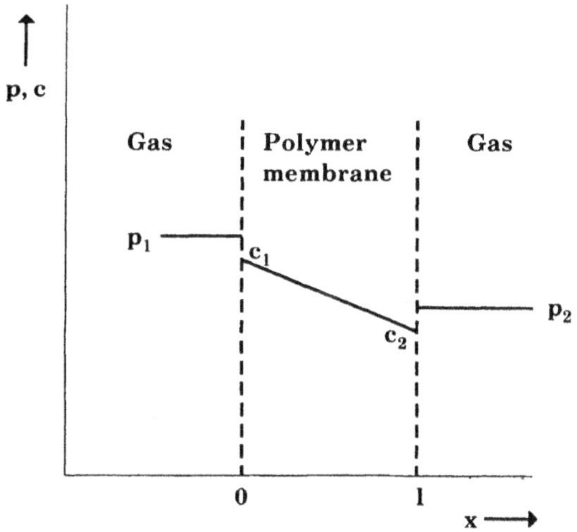

Figure 2. Diffusion process through a polymer membrane.

Equilibrium sorption usually follows Henry's laws[4,7]

$$c = S\,p$$

c = concentration
S = solubility coefficient
p = pressure

The flux J of material diffusing through unit area of membrane with time is given by Fick's first and second laws[4,7].

$$J = -D\frac{\partial c}{\partial x} = -D\frac{\Delta c}{\Delta x}$$

J = flux
D = diffusion coefficient

$$\frac{\partial c}{\partial t} = D\frac{\partial^2 c}{\partial x^2}$$

x = distance
t = time

where in most cases D is a concentration independent diffusion coefficient.

In many practical systems the surface concentration c is not always known and it is conventional to express the flux in terms of the pressure by

$$J = P\,\frac{\Delta p}{l}, \quad \textit{with} \quad P = D\,S,$$

where Δp is the pressure difference and l the membrane thickness. P is the Permeability coefficient.

The temperature dependences of P, D and S usually follow an Arrhenius form[4,7].

$$P = P_0\,e^{-\frac{E_P}{RT}}, \quad D = D_0\,e^{-\frac{E_D}{RT}} \quad \text{and} \quad S = S_0\,e^{-\frac{\Delta H_s}{RT}}$$

The activation energy of the permeability coefficient E_P consists of the activation energy of the diffusion E_D and the heat of sorption ΔH_S

$$E_P = E_D + \Delta H_S$$

The transport of an atom or a small molecule through a polymer depends on its microstructure. The properties of partially crystalline polymers, block or graft copolymers, polymer blends or fiber reinforced composites depend markedly on the position inside the probe, that is the structure of these so-called multiphase systems. We call a material a multiphase system when deviations in density, chemical composition or other characteristic mass quantities, which exceed molecular dimensions, occur along the space coordinate (fig. 3).

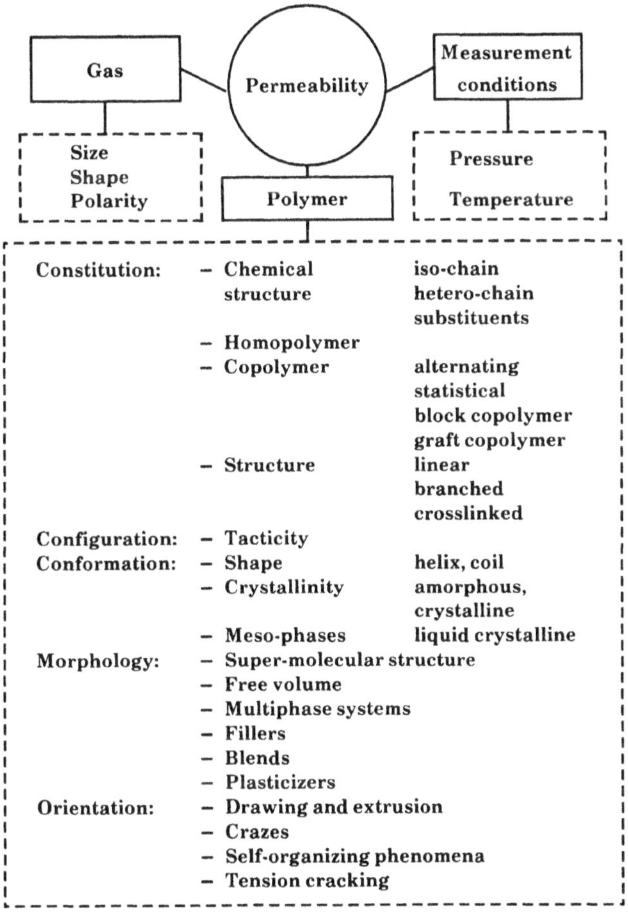

Figure 3. Quantities influencing gas permeability: nature of the permeating gas, the nature of the polymer, and measurement conditions.

Transport rates of gases in multiphase systems suggest that only one of these phases contributes to the mass transport[4,7,9,10]. At least the mass transport coefficients differ by orders of magnitude, so that the other phase(s) act as impermeable barriers. The requirement for diffusion is the interchange of sites in the polymer. As diffusion is an activated process, the mobility increases by increasing temperature. An important concern is how the permeation behavior changes with temperature as the glass transition region is traversed. The extent to which the activation energies of the transport coefficients alters depends on the nature of the permeating gas. Below the glass transition temperature the sorption often does not follow Henry's law. This has usually been described by the dual sorption model because evidently two separate mechanisms are operating in the sorption of gases in glassy polymers (fig. 4)[7].

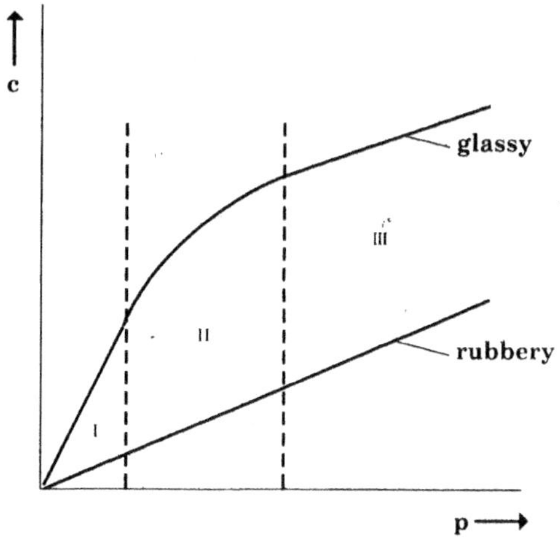

Figure 4. Dual mode sorption. In contrast to a rubbery polymer, for a glassy polymer below T_g the sorption isotherm is a combination of a linear low pressure region (I) and a linear high pressure region (III) which are connected by a non-linear transition range (II).

It is thought that in the glassy state, fixed adsorption sites or holes exist beneath the mobile population at which gas atoms or molecules can be bound. The solubility is, therefore, a combination of Henry's law sorption c_D and a term which follows a Langmuir form c_H

$$c = c_D + c_H = K_D p + \frac{c'_H b p}{1 + bp} \quad \text{and} \quad K_D = S$$

where b is the hole affinity constant which is the quotient given by the rate constants for desorption and adsorption. c'_H is the hole saturation constant, the maximum available adsorption surface, and k_D is the solubility coefficient similar to S. In the low pressure region (I) ($bp \ll 1$) the term $c = (K_D + c'_H b)p$ is valid and after a non-linear transition region (II) the concentration in the high pressure region (III) ($bp \gg 1$) is given by $c = (K_D p + c'_H)$. Study of gas sorption in polymers at temperatures below their T_g and at high pressure enables us to obtain information indirectly about morphology or chain conformation of the polymeric substance. The next figures show some examples of the gas transport properties. In fig. 5 polypropylene (PP) is shown as an example of the influence of the degree of crystallization on the permeability of He, CO_2, O_2 and N_2. As the degree of crystallization is increased and the amount of the non-crystalline phase decreases, the permeation coefficients drop. Permeation investigations were carried out according to the time-lag design. Experiments started with degassed sample, so that time-lag and steady state were obtained from mass transport analysis. Time-lag and steady states refer to D and P, respectively. These measurements were taken at relatively low pressures (up to 1.1 bar)[11]. The Arrhenius plots confirm the solution diffusion model.

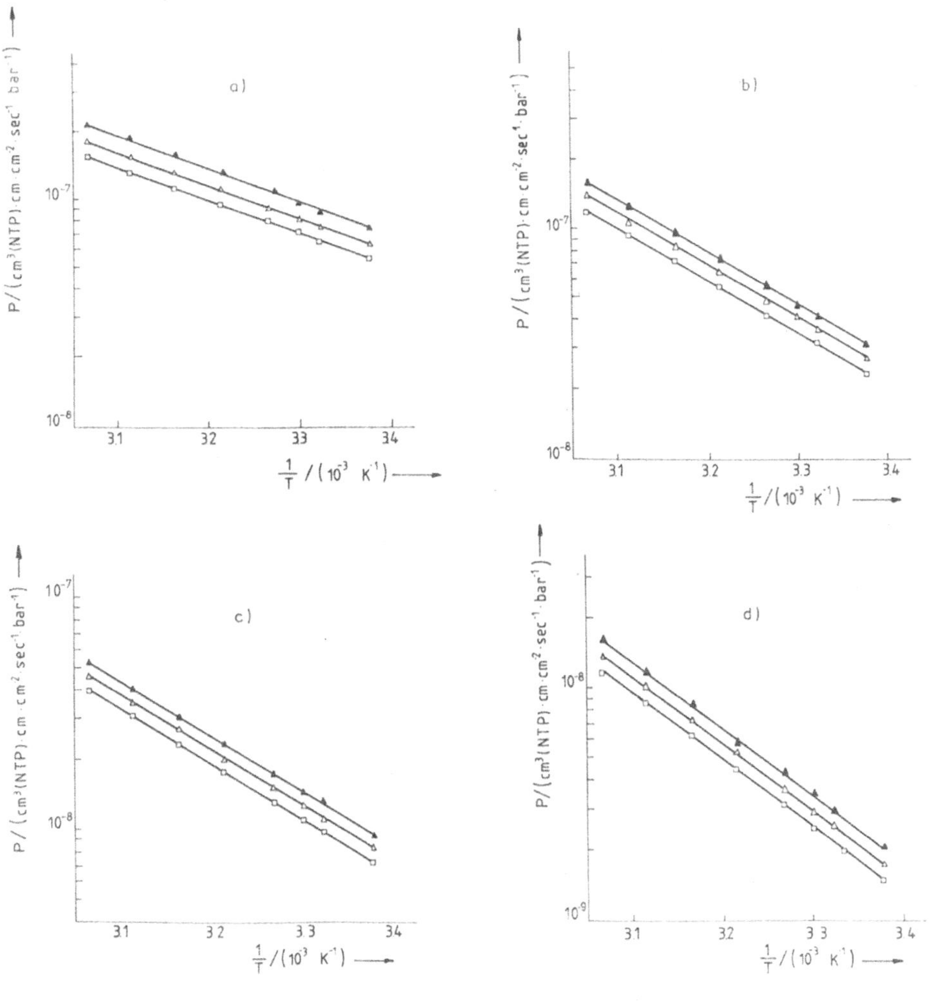

Figure 5. Temperature dependence of permeability coefficient P of a) He, b) CO_2, c) O_2 and d) N_2 through PP of various densities ρ: ▲ ρ = 0.8871 g/cm³, △ ρ = 0.8931 g/cm³, □ ρ = 0.8998 g/cm³.

An important concern is how the activation energy changes with temperature as the glass transition region is traversed. This is shown in fig. 6 for polybutyleneterephthalate (PBTP)[5,12]. Here, too, a decrease of the permeability is shown with increasing degree of crystallization. In the series of the inert gases He, Ne and Ar one can follow the influence of the size of the permeating species. The relatively high permeability coefficients of CO_2 are a consequence of its good solubility in the polymer, which we will not consider in any more detail. In addition we will not discuss why the change in degree of crystallization has different consequences on the different sizes of the inert gases.

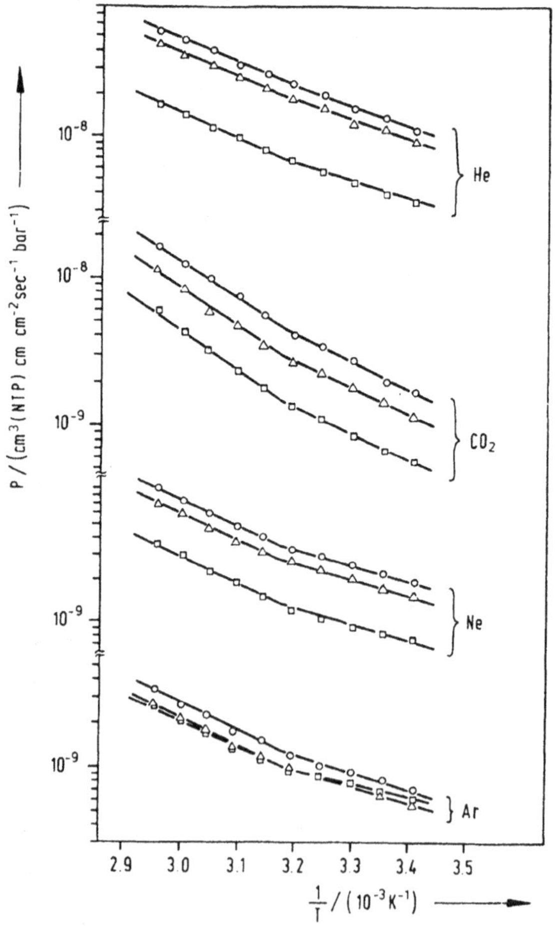

Figure 6. Temperature dependence of the permeability coefficient P of various gases through PBTP of various crystallinities x_{cw}:

○ x_{cw} = 19 %, △ x_{cw} = 37 %, □ x_{cw} = 51 %.

A morphological aspect, which can be discussed by means of the dual sorption theory, is demonstrated in fig.7. Here the solubility of CO_2 in PBTP is shown as a function of the pressure for drawn films[13]. During the drawing process the chain segments are braced and the internal surface increases at which the sorption of gases can occur. This behavior is of technical importance as, for example, many polymer films manufactured for food packing are extruded or biaxially drawn.

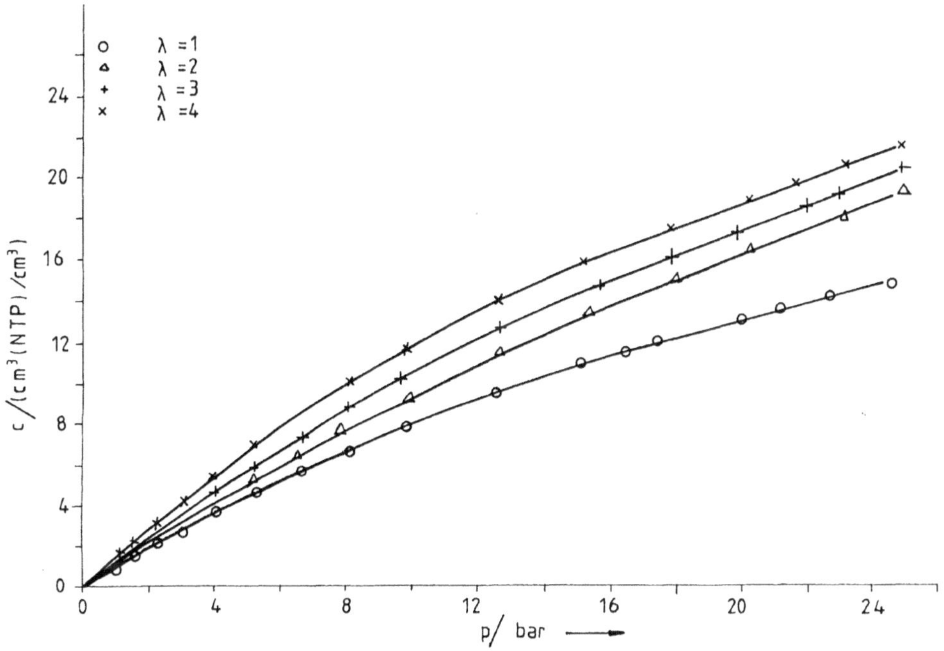

Figure 7. Sorption isotherms of CO_2 in PBTP at T = 298 K as a function of different drawing ratios λ.

Another morphological aspect we will touch on is the permeability of block copolymers. A well-known example of different separation structures, which develop depending on the composition of the block copolymer, is the system styrene/butadiene. The separation structures are shown schematically in fig. 8.

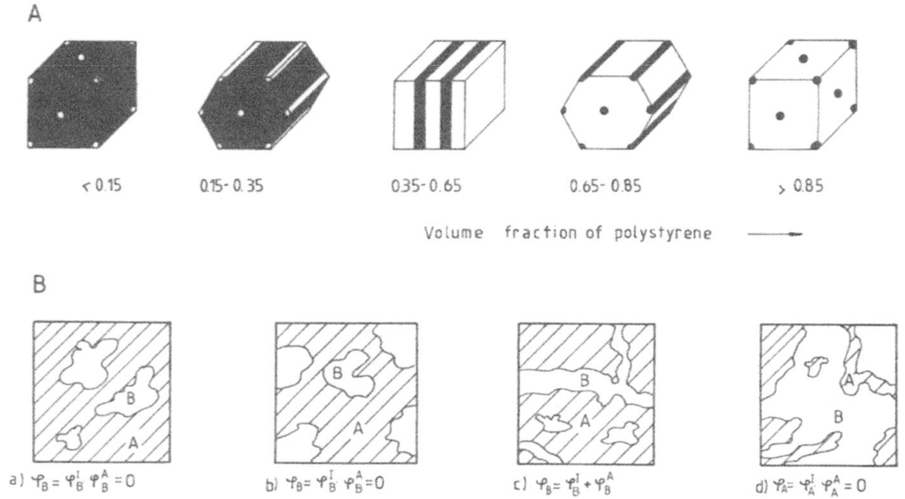

Figure 8. A) Separation structures in the system styrene/butadiene.

B) Separation structures obtained using percolation theory.

ψ_i^I = volume fraction of i, which is isolated

ψ_i^A = volume fraction of i, which is accessible

i = Phase A or B

PERMEABILITY OF ELECTROPLATED POLYMER FILMS

Now we return to the permeation of metallized polymer films. Our investigations are restricted to the determination of the permeability of electroplated polymer films. The investigated polymers were an ABS-copolymer and a PP-homopolymer, the specimen were pretreated by a wet chromic acid etchant and metallized as described above.

As regards the permeability of metal-polymer compound systems, the following questions can be asked:

- What influence on permeability does the surface structure of the polymer have after the etching?
- Does the surface structure influence the precipitation of the metal, or in other words, is the metal film in the first layers disturbed to the extent that small gas atoms or molecules can penetrate it?

First we present the investigations on permeability of electroplated ABS[28] films. The copolymer has a composition of acrylonitrile : butadiene : styrene in the ratio of 23:16:61.

Fig. 14 shows the permeability coefficients of CO_2, O_2 and N_2 as a function of the copper layer thickness. The non-metallized ABS film shows the typical decrease in permeability in the order: $CO_2 > O_2 > N_2$. The permeability of metallized films is nearly constant up to a copper layer thickness of 8 µm. After that the permeation of all three gases decreases considerably. The permeability coefficient of CO_2 drops even below those of O_2 and N_2 so that the ratio of the permeability coefficients reverses.

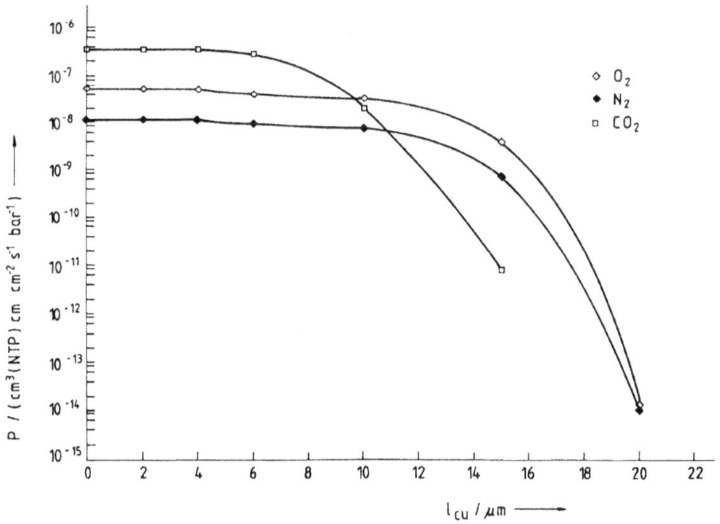

Figure 14. Permeability coefficient P as a function of the copper layer thickness l_{Cu} on ABS.

The mechanism of permeation through an electroplated metal layer has to be fundamentally different from that of the polymer film. It will essentially depend on structural disorders of the metal lattice. The composition of the electroplating baths, which ordinarily contain gloss additives or other components, should be an important factor in this connection. On the other hand, x-ray investigations[29] of the metal layer gave no indication that the chemical pretreatment of the polymer or the polymer surface morphology had any influence on the structure of the metal layer.

To show the influence of etching of polymers on gas permeability, experiments were performed with PP[11,30] as shown in fig. 15.

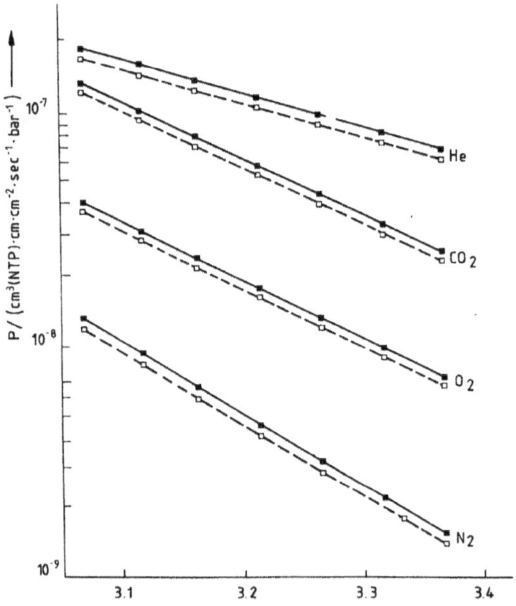

Figure 15. Temperature dependence of the permeation coefficient P. Broken line: permeation coefficient of the unetched PP; continuous line: permeation coefficient of the etched PP.

It turned out that the permeability of the etched polymer film was higher than that of the unetched one. By etching PP, the non-crystalline chain segments are partially decomposed and spherulites or parts of them are removed, as seen in fig. 16. Compared to the unetched sample, the effective distance of diffusion is shortened and the result is that the permeability coefficient increases.

In contrast to ABS, the permeability coefficient of PP drops appreciably after the electroless metallization with a nickel layer of about 0.5 µm, which is summarized in Table I.

MOLECULAR LEVEL METAL AND CERAMIC/POLYMER COMPOSITES: SYNTHESIS OF METAL AND METAL OXIDE CONTAINING POLYIMIDES AND ITS RELEVANCE TO POLYMER METALLIZATION

Ayusman Sen*, Manish Nandi and Jeanine A. Conklin

Department of Chemistry
The Pennsylvania State University
University Park, PA 16802

This paper is concerned with (a) the thermodynamic and kinetic binding sites for Cr(0) in polyimides, and (b) the application of the binding of metals to specific sites on the polyimide backbone to the synthesis of polymer trapped chromium and iron oxide nanoclusters (size < 1.5 nm). In the first part, the organometallic products obtained by the reaction of $Cr(CO)_3(MeCN)_3$ or $Cr(CO)_6$ with a number of compounds that mimic the polyimide backbone were identified. A π-arene complex was formed in each case indicating that Cr(0) prefers to form π-arene complexes (even with electron deficient arene rings) rather than react with the oxygen functionalities present. Furthermore, the tendency to form a π-arene complex increases with increasing electron density on the ring. The first part also describes the synthesis and x-ray crystal structure of a charge-transfer complex of $Cr(CO)_6$ with pyromellitic dianhydride -- one of the building blocks of polyimides. The second part outlines the synthesis of polyimide-trapped metal oxide nanoclusters. $Cr(C_6H_6)_2$ or $Fe_3(CO)_{12}$ was added to polyamic acid solutions. Following thermal curing, polyimide films containing a homogeneous dispersion of nanoclusters (size < 1.5 nm) of Cr_2O_3 or α-Fe(O)(OH) were obtained. Both SEM and SIMS depth profile indicated that the usually observed migration of the dopant to the polymer surface had not occurred.

INTRODUCTION

This paper encompasses the interaction of metals with polyimides. Specifically, we report on (a) the thermodynamic and kinetic binding sites of chromium(0) in polyimides, and (b) the application of the binding of metals to specific sites on the polyimide backbone to the synthesis of the corresponding polymer-trapped metal oxide nanoclusters (size <1.5 nm).

Metallized Plastics 2, Edited by K.L. Mittal
Plenum Press, New York, 1991

The relationship between the macroscopic properties of metal-polymer and semiconductor-polymer interfaces and the nature of the chemical bonds formed at the interface are of fundamental importance, especially to the electronics industry. For example, the very large scale integration of devices (VLSI technology) necessitates the use of multilayered structures at the chip level and for packaging[1]. Such structures are usually fabricated by using alternate metal and insulating layers and thus incorporate metal-insulator interfaces throughout the structure. Hence, good adhesion of the metal to the insulator surface, usually a polymeric material, is of critical importance. To this end, a detailed understanding of the relationship between chemical bonding and adhesion quality will provide new opportunities for optimizing metal-polymer adhesion[2].

Polyimides (e.g., **I**, derived from pyromellitic dianhydride and p,p'-oxydianiline) have emerged as the material of choice as an insulating and pattern-delineating matrix in the area of VLSI technology[3]. Consequently, the

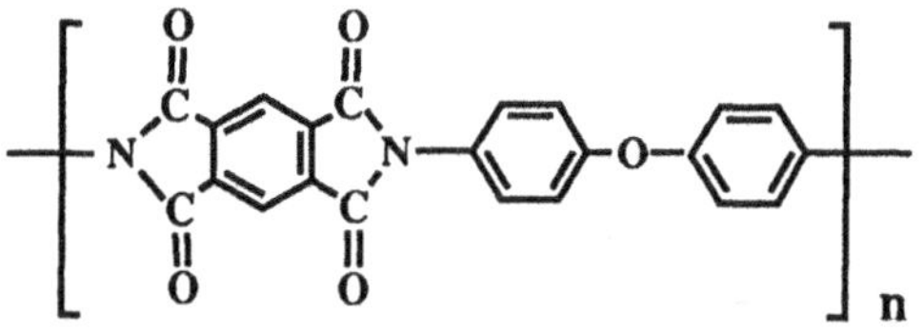

I

problem of adhesion between polyimides and metals is of paramount importance. The adhesion of metals to **I** varies considerably[4]: Cu displays poor adhesion while Cr adheres strongly to the surfaces of **I**. The reason for this difference between Cu and Cr remains unclear. Accordingly,we have examined the interaction of the Cr(0) fragment, $Cr(CO)_3$, with several small molecules that serve as models for **I**. The results, which are summarized below, serve to delineate, for the first time, the relative thermodynamic preference of Cr(0) for the various reactive sites on **I**[5].

The second part of the paper involves the synthesis of polyimide-trapped metal oxide nanoclusters (size <1.5 nm). Ceramic nanoclusters with sizes ranging from <1 to 10nm represent a new class of materials with new properties that are different from those of discrete molecules, as well as bulk solid state materials[6]. Their anticipated applications range from cluster-size dependent catalysis[7] to nonlinear optics[8]. In many of these applications, the ability to form films and fibers is an important requirement. This is most easily achieved by forming composites in which nanoclusters are embedded in a processable polymer matrix. A further advantage of synthesizing such composite is that the polymer matrix will physically prevent the agglomeration of the clusters -- a persistent problem in size-selective cluster synthesis.

Aromatic polyimides have long been the material of choice because of their high thermal stability, chemical resistivity and mechanical strength[9]. It was, therefore, of interest to see if ceramic nanoclusters could be formed in a polyimide matrix. The high glass-transition temperatures (Tg) of polyimides would be expected to further stabilize the nanoclusters by decreasing their mobility, thereby preventing their agglomeration to larger clusters. In our initial studies we have chosen to make chromium and iron oxide clusters because of their expected interesting magnetic and electronic properties[10].

Table I. Bond lengths (Å) and angles (°) with estimated standard deviations in parentheses.

a. Bond lengths (Å).

Cr-C	1.885(4)	C(1)-C(4)	1.466(4)
C-O	1.143(4)	C(2)-C(3)	1.474(5)
O(1)-C(1)	1.188(4)	C(3)-C(4)	1.381(4)
O(2)-C(1)	1.391(4)	C(3)-C(5)	1.377(4)
O(2)-C(2)	1.393(4)	C(4)-C(5^{i})	1.385(4)
O(3)-C(2)	1.184(4)	C(5)-H(5)	0.95(3)

B. Bond angles (°).

C-Cr-C^{ii}	89.3(2)	C(2)-C(3)-C(4)	107.3(3)
C-Cr-C^{iii}	90.7(2)	C(2)-C(3)-C(5)	129.6(3)
Cr-C-O	178.4(3)	C(4)-C(3)-C(5)	123.1(3)
C(1)-O(2)-C(2)	109.5(2)	C(1)-C(4)-C(3)	108.0(3)
O(1)-C(1)-O(2)	120.3(3)	C(1)-C(4)-C(5^{i})	129.3(3)
O(1)-C(1)-C(4)	131.8(3)	C(3)-C(4)-C(5^{i})	122.7(3)
O(2)-C(1)-C(4)	107.6(3)	C(3)-C(4)-C(4^{i})	114.2(3)
O(2)-C(2)-O(3)	121.3(3)	C(3)-C(5)-H(5)	124(2)
O(2)-C(2)-C(3)	107.5(3)	C(4^{i})-C(5)-H(5)	121(2)
O(3)-C(2)-C(3)	131.2(3)		

Symmetry codes: i: -x, 1-y, 1-z; ii: z, x, y; iii: -z, -x, -y.

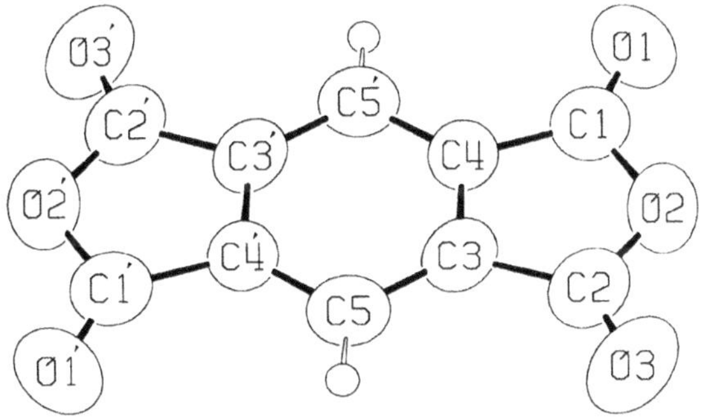

Figure 2. Atom numbering scheme for the pyromellitic dianhydride (PMDA) part.

respectively, above and O(2) 0.056Å below the plane of the carbon atoms thus giving an envelope conformation of the five-membered ring. With one exception the bond distances and angles in PMDA (Table I) are similar to those observed in noninteracting PMDA[21], as well as in most charge-transfer complexes of PMDA with organic donors[18]. The C(1)-C(4) and C(2)-C(3) distances of 1.466(4) and 1.474(5)Å are, however, shorter than in noninteracting PMDA (average: 1.485Å)[21]. Instead, the values are at the low end of those observed with a number of charge-transfer complexes with organic donors (e.g., biphenylene,

1.486Å[18c]; trans-stilbene, 1.472Å[18f]; perylene, 1.464Å[18g]). This bond shortening may be understood in terms of increased contribution of the canonical quinoid structure **II**, brought about by increased electron-density in the PMDA molecule.

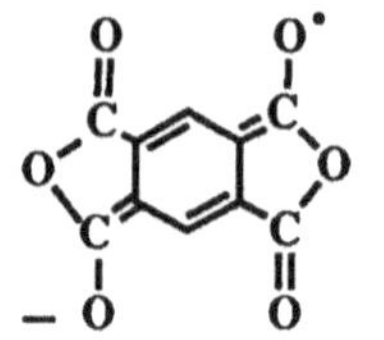

II

Indeed, slight bond alteration can be discerned in the structure of PMDA [compare, C(4)-C(5') versus C(3)-C(5), C(2)-C(3) versus C(1)-C(4), and O(1)-C(1) versus O(3)-C(2)].

The $Cr(CO)_6$-PMDA complex, thus represents the first example of a charge-transfer complex of PMDA with an organometallic species. This observation supports the proposition[17] that charge-transfer from the chromium to the ring derived from PMDA may occur when the former is deposited on PI

The polymers chosen for our studies on the synthesis of polymer-trapped metal oxide nanoclusters have structures depicted in Fig. 3. Upon curing to the fully imidized polymer, both PMDA-ODA and BTDA-ODA formed free-standing transparent golden-yellow films. In the case of the doped polymers, the initial homogeneous solution containing the organometallic precursor and the polyamic acid was green for chromium and wine-red for iron. After curing the chromium containing polymers formed free-standing transparent yellow-brown films. Similarly, the iron containing films were also transparent but a little lighter in color. No discernible changes in flexibility and brittleness of the films were observed on doping.

The presence of dopants in the polymer matrix did not have a dramatic effect on their thermal properties (Table II). Thermogravimetric analysis in N_2 indicated that the polymer decomposition temperature, corresponding to 10% weight loss of the sample, decreased upon doping, with chromium showing a bigger effect than iron. We note that a decrease in polymer decomposition temperature has also been reported for other metal oxide containing polyimides and the results suggest the presence of a metal assisted decomposition pathway for the doped polymers[22]. With respect to glass-transition temperature (Tg), DSC analysis did not reveal a consistent trend. However, since polyimides do not display a sharp glass-transition point the values reported in Table II are of limited accuracy.

The IR spectra of the doped polymers showed no evidence for any chemical interaction between the dopant and the polyimide matrix. The ^{13}C CP/MAS NMR of the chromium containing polymers did not show any change in the ring resonances of the polyimides. Since any metal-arene interaction would result in upfield shifts of these resonances, such an interaction is evidently absent in the cured polymers. The carbonyl resonances are similarly unaffected. These observations, however, should be treated with caution. Since the molar ratios of metal to polymer repeating unit are low (Table II), subtle changes in the polymer may remain undetected by these techniques especially in view of the broadness of the peaks.

The energy dispersive x-ray analysis of the surfaces of the doped polymer films showed intense Cr and Fe peaks on both the atmosphere and the substrate(glass) sides. An x-ray dot mapping of Fe on the atmosphere side of the

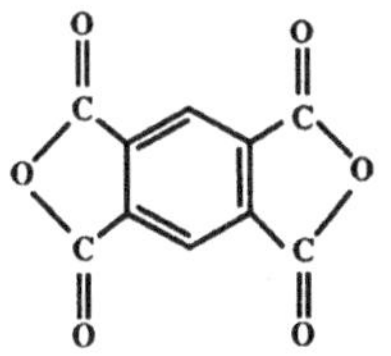

PMDA

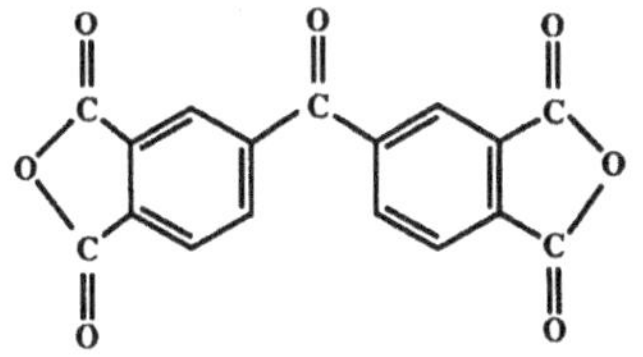

BTDA

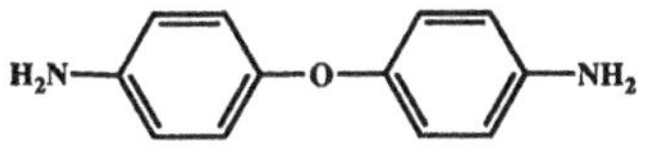

ODA

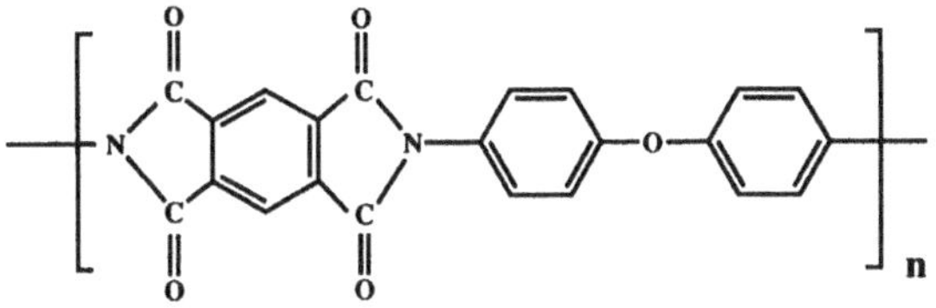

PMDA/ODA

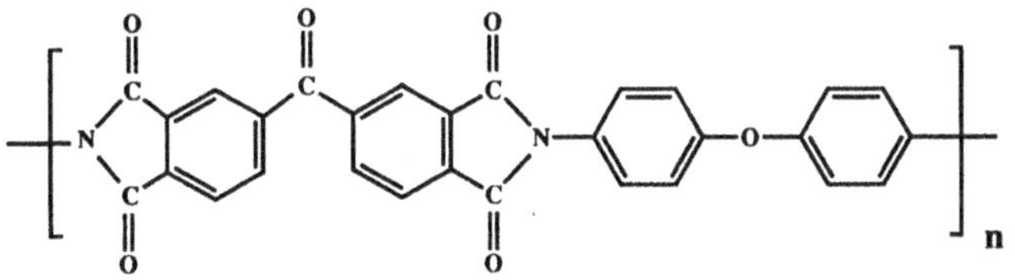

BTDA/ODA

Figure 3. Chemical structures of the monomers and polymers.

Table II. Composition and thermal analysis of the ceramic/polymer composites.

Polymer	Metal, Wt. %	Metal, Mol. %	TGA (°C)[a]	Tg (°C)
PMDA-ODA			551	346
PMDA-ODA/Cr	2.9	20.0	513	333
PMDA-ODA/Fe	1.4	8.9	522	385
BTDA-ODA			545	361
BTDA-ODA/Cr	2.14	18.7	511	332
BTDA-ODA/Fe	0.95	7.7	536	351

[a]10% loss in weight of polymer.

PMDA-ODA/Fe film revealed a uniform distribution of the metal throughout the surface of the films (Fig. 4). A similar homogeneous distribution of Cr was also observed in the PMDA-ODA/Cr film. Although these observations indicate the presence of metal species at the polymer surface, the scanning electron microscopy (SEM) of both the chromium and iron containing polymers showed uniform surfaces with no apparent microstructure, on both the atmosphere and the substrate sides of the films. In contrast to most previous reports on polyimides doped with metal compounds[11], no metal containing aggregates were observed by SEM even at a magnification of 60,000. Clearly, neither the formation of a "metallic surface" nor a significant migration of any metal species to the film surfaces had occurred (also see SIMS results).

The ESR spectrum of the PMDA-ODA/Cr at -120°C showed a single resonance at g = 2.018 which was identified as a chromium(III) species[23]. Further characterization of both the chromium and the iron species present in the PMDA-ODA polymeric matrix was achieved by x-ray photoelectron spectroscopy

Figure 4.
X-ray dot map of PMDA-ODA/Fe.
The dots represent concentrations of iron atoms.

(XPS). XPS of the polymer films obtained at a takeoff angle of 90° for maximum sampling depth showed well-defined Cr (2p) and Fe (2p) on the atmosphere sides of the films. In the case of the PMDA-ODA/Cr, the Cr $2p_{3/2}$ peak was observed at a binding energy of 576.7 eV (Fig. 5). This corresponded to the presence of Cr_2O_3[24]. The O (1s) photopeak at approximately 532 eV also suggested the presence of metal oxides[25]. The Fe ($2p_{3/2}$) photopeak in PMDA-ODA/Fe was at 711.6 eV (Fig. 6) and corresponded to the α–form of Fe(O)(OH)[26]. It should be noted that although the formulation of Cr_2O_3 and α-Fe(O)(OH) is consistent with the XPS data, alternative combinations of oxo and hydroxo ligands cannot be totally ruled out. The XPS of the substrate sides of the films showed no detectable amounts of the metals.

The concentration gradient of the metal species in the BTDA-ODA films was studied by secondary ion mass spectrometry (SIMS). A semi-quantitative depth profiling from the atmosphere side of the films indicated that, for both the chromium and iron containing films, the concentration of the metal species increased slightly on going from the surface to the bulk and then leveled off (Fig. 7 and 8). Thus the combination of XPS and SIMS results indicates a near-homogeneous distribution of metal oxide clusters throughout the polyimide

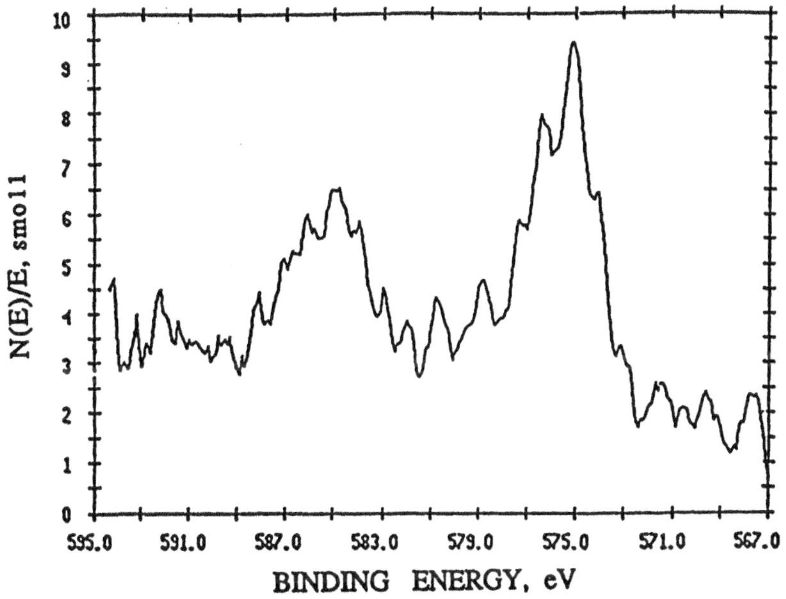

Figure 5. Chromium $2p_{1/2,3/2}$ photopeak region of the atmosphere side of PMDA-ODA/Cr.

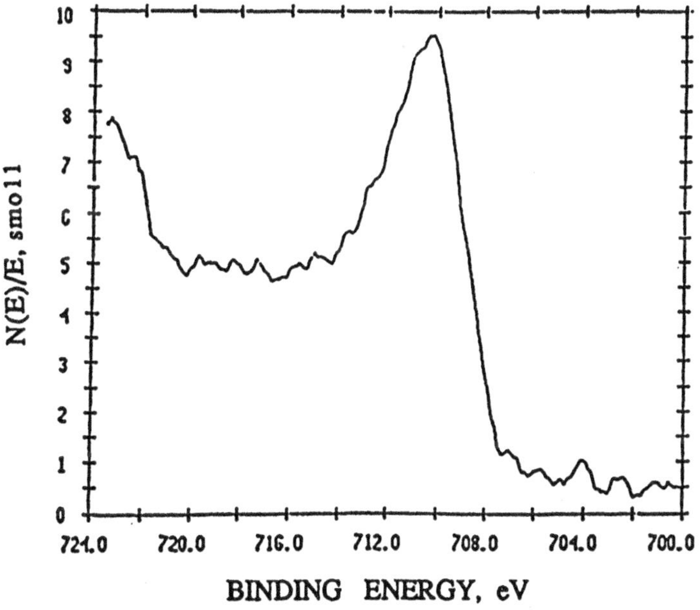

Figure 6. Iron $2p_{1/2,3/2}$ photopeak region of the atmosphere side of PMDA-ODA/Fe.

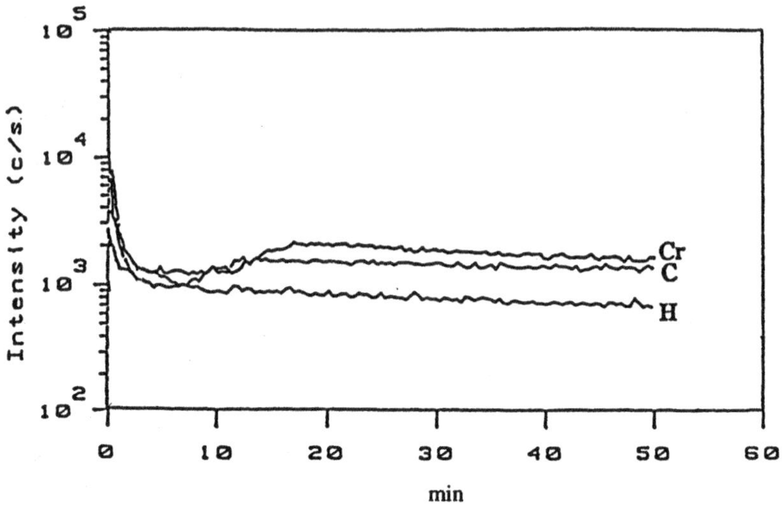

Figure 7. Depth profile of BTDA-ODA/Cr film (C, H and Cr) by secondary ion mass spectrometry (SIMS).

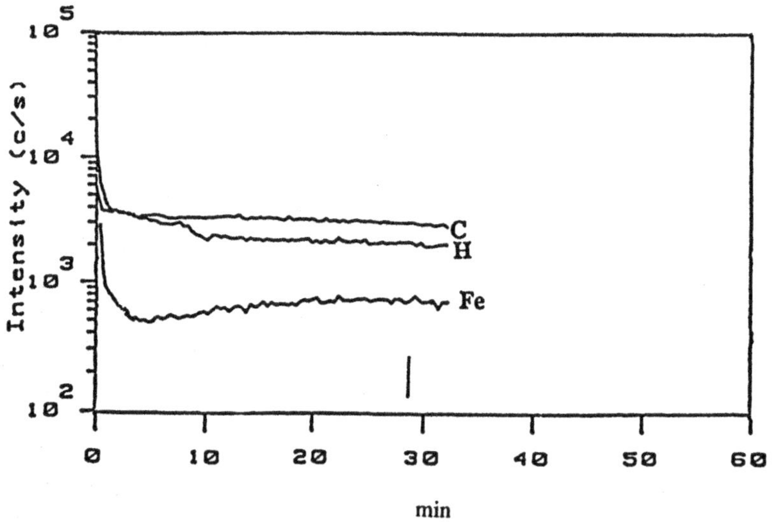

Figure 8. Depth profile of BTDA-ODA/Fe film (C, H and Fe) by secondary ion mass spectrometry (SIMS).

films except near (0-75 Å) the substrate-side surface where no metal was detected (similar narrow metal-free zones near the substrate-side surface have been observed previously[4a] and have been ascribed to zone refining effect brought on by temperature gradients). Our results are in sharp contrast to the previous reports where the migration of the metal species from the bulk to the surface of the polymers was often observed (vide infra)[11].

Finally, in order to probe the internal structure of the polymers, ultramicrotomed cross-sections of the polymeric films were examined by transmission electron microscopy (TEM). The bulk of the films was found to be perfectly homogeneous and no metal containing species were detected. Thus, both SEM and TEM indicate that the size of the metal oxide particles present in the polymeric matrices is less than 1-1.5 nm, the resolution limit of the instruments.

The spectroscopic studies on the polyimide films indicate the formation of homogeneous dispersion of metal oxide (Cr_2O_3 or α-Fe(O)(OH)) nanoclusters (size < 1.5 nm) in the polymers. These composite materials differ in two respects from previously synthesized metal oxide containing polyimides[11]. The latter materials usually exhibit a non-homogeneous distribution of the oxide particles with markedly higher concentration at or near the polymer surface. In addition, large particles with sizes > 100 nm are present. We tentatively ascribe this difference to our use of zero-valent organometallic species (rather than metal salts) as precursors to the metal oxides. The zerovalent organometallic species are readily oxidized[27] and, as a result, lose their π-acid ligands (arene or CO) due to reduced metal to ligand back-bonding[28]. Thus, upon oxidation (e. g., by O_2 in the presence of air) the binding of the metal to donor groups on the polymer backbone (e.g., the carboxylic acid groups in polyamic acid) is expected. In addition, the chromium may also initially bind to the arene rings on the polymer backbone through the well-known arene exchange reaction[29]. We believe that this prebinding of the metal to the polymer backbone prior to the conversion to the oxide prevents (a) the migration of the metal species to the polymer surface, and (b) the agglomeration to large oxide particles.

The source of oxygen for the metal oxides is the water formed during the polyamic acid-polyimide transformation. The evidences for this are as follows. The metal oxides were also formed when the curing was done in a nitrogen atmosphere. For both metals a limiting molar ratio of metal oxide to polymer repeating unit was invariably reached even with higher loadings of organometallic precursors. The limiting value (1: 5 for Cr and 1: 10 for Fe) did not change on going from PMDA/ODA to BTDA/ODA. This is not unexpected since both polyimides would yield one equivalent of water per repeating unit upon curing. A metal oxide to polymer repeating unit ratio that exceeded the above limiting value could only be achieved by deliberate addition of water to the polyamic acid - organometallic precursor mixture *prior* to curing. In this way PMDA-ODA/Cr films containing 9.1% Cr (molar ratio of Cr to PMDA-ODA repeating unit : 1: 1) were obtained. Unfortunately, a serious deterioration of film quality was observed with such high metal oxide loadings.

EXPERIMENTAL

Part 1.

Reagents. All reagents, unless otherwise mentioned, were obtained from Aldrich Chemical Co. $Cr(CO)_6$ was purified by sublimation under reduced pressure. It was stored under N_2 after purification. Acetonitrile was dried by distillation from CaH_2 under N_2 atmosphere. The dry, deoxygenated compound was stored in a glass bottle with a teflon lined cap. 4-methylphthalic anhydride, obtained from Pfaltz & Bauer, Inc. and pyromellitic dianhydride (PMDA), obtained from Aldrich Chemical Co.,were purified by sublimation under reduced pressure and stored under N_2 after purification. Aniline, 4-phenoxyaniline, 4-ethylaniline and 4-*tert*-butylaniline were all purified by distillation from CaH_2 under reduced pressure. After purification they were stored in dark colored glass bottles with teflon lined caps.

Instrumentation. NMR spectra were recorded with Bruker WP-200 (200 MHz) and AM-300 (300 MHz) spectrometers. The chemical shifts for the ^{1}H NMR spectra are reported in parts per million relative to tetramethylsilane at 0.00 ppm. Infrared spectra were recorded with a Perkin-Elmer FT-IR 1710 instrument.

Synthesis of $Cr(MeCN)_3(CO)_3$. The chromium precursor, $Cr(MeCN)_3(CO)_3$, was synthesized from $Cr(CO)_6$ and acetonitrile according to the literature preparation[13]. The compound was obtained as bright yellow, extremely air-sensitive solid which was stored in a N_2 filled glove box.

Synthesis of N-phenyl-4-methylphthalimide. Condensation of 4-methylphthalic anhydride and aniline gave the model imide **2**. Aniline (2 g, 0.02 mol), previously purified, was added dropwise to a solution of 4-methylphthalic anhydride (3.2 g, 0.02 mol) in dry toluene (30 mL) in a 50 mL round bottomed flask under N_2 to form a curdy white solution. It was then refluxed at 90°C for 24 h under inert atmosphere. Upon cooling, a white solid precipitate was obtained. It was separated by filtration and washed thoroughly with dry toluene (3 x 10 mL). The compound was then dried under high vacuum. IR (cm^{-1}, KBr): ν_{CO} 1780, 1731. ^{1}H NMR (THF-d_8 , ppm): 7.75 -7.85 (m, 3H); 7.30 - 7.70 (m, 5H); 2.53 (s, 3H).

Synthesis of N-4-ethylphenyl-4-methylphthalimide. The model imide **3a** was synthesized by the condensation of 4-methylphthalic anhydride and 4-ethylaniline. Previously distilled 4-ethylaniline (0.121 g, 0.001 mol) was added dropwise to a solution of 4-methylphthalic anhydride (0.162 g, 0.001 mol) in dry toluene (30 mL) under N_2. Following the procedure described in the above synthesis, a white solid product was obtained. IR (cm^{-1}, KBr): ν_{CO} 1779, 1710. ^{1}H NMR (THF-d_8 , ppm): 7.70 - 7.85 (m, 3H); 7.30 -7.35 (m, 4H); 2.65 (m, 2H); 2.52 (s, 3H); 1.22 (m, 3H).

Synthesis of N-4-*tert*-butylphenyl-4-methylphthalimide. The imide **3b** was synthesized by the condensation of 4-methylphthalic anhydride (1.62 g, 0.01 mol) and 4-*tert*-butylaniline (1.5 g, 0.01 mol) following the procedure described above. The product was obtained as a white solid. IR (cm^{-1}, KBr): ν_{CO} 1771, 1714. ^{1}H NMR (THF-d_8 , ppm): 7.75 -7.85 (m, 3H); 7.35 -7.65 (m, 4H); 2.53 (s, 3H); 1.36 (s, 9H).

Synthesis of N-4-phenoxyphenyl-4-methylphthalimide. The model imide **4** was obtained by the condensation reaction of 4-methylphthalic anhydride and 4-phenoxyaniline. Previously distilled 4-phenoxyaniline (0.55 g, 0.002 mol) was added dropwise to a solution of 4-methylphthalic anhydride (0.49 g, 0.002 mol) in dry toluene (30 mL) in a round bottomed flask. Following the same procedure as previously described, a white solid was obtained as the product. IR (cm^{-1}, KBr): ν_{CO} 1781, 1720. ^{1}H NMR (THF-d_8 , ppm): 7.75 -7.80 (m, 3H); 7.4 - 7.65 (m, 4H); 7.05 -7.38 (m, 5H); 2.53 (s, 3H).

REFERENCES

1. M. T. Goosey, editor, " Plastics for Electronics," Elsevier Sequoia, New York, 1985.
2. K. L. Mittal and J. R. Susko, editors, "Metallized Plastics 1: Fundamental and Applied Aspects," Plenum Press, New York, 1989.
3.. (a) Reference 1, Chapters 9 and 10. (b) S. D. Senturia, R. A. Miller, D. D. Denton, F. W. Smith, and H. J. Neuhaus in "Proceedings of the Second International Conference on Polyimides," p. 107, Society of Plastics Engineers, Ellenville, NY, 1985.
4. (a) Y.-H. Kim, J. Kim, G. F. Walker, C. Feger, and S. P. Kowalczyk, J. Adhesion Sci. Technol., 2, 95 (1988). (b) J. M. Burkstrand, J. Appl. Phys., 52, 4759 (1981).
5. Preliminary report: M. Nandi and A. Sen, Chem. Mater., 1, 291 (1989).
6. (a) R. P. Andres, R. S. Averback, W. L. Brown, L. E. Brus, W. A. Goddard, A. Kaldor, S. G. Louie, M. Moscovits, P. S. Peercy, S. J. Riley, R. W. Siegel, F. Spaepen, and Y. Wang, J. Mater. Res. 4, 704 (1989) (b) M. L. Steigerwald and L. E. Brus, Annu. Rev. Mater. Sci. 19, 471 (1989). (c) L. E. Brus, J. Phys. Chem., 90, 2555 (1986).
7. (a) A. Kaldor, D. M. Cox, and M. R. Zakin in "Molecular Surface Chemistry: Reactions of Gas-Phase Metal Clusters," Advances in Chemical Physics, I. Prigogine, editor, Vol. 70, Pt. 2, p. 211, Wiley, New York, 1988. (b) R. J. St. Pierre and M. A. El-Sayed, J. Phys. Chem., 91, 763 (1987).
8. (a) T. Kobayashi, editor, "Nonlinear Optics of Organics and Semiconductors," Part II, Springer-Verlag, Berlin, 1989. (b) Y. Wang, N. Herron, W. Mahler, and A. Suna, J. Opt Soc. Am. B. 6, 808 (1989). (c) L.-T. Cheng, N. Herron, and Y. Wang, *J. Appl. Phys.* 66, 3417 (1989). (d) R. K. Jain and R. C. Lind, J. Opt. Soc. Am. 73, 647 (1983).
9. (a) K. L. Mittal, editor, "Polyimides: Synthesis, Characterization and Applications," Vols. 1 and 2, Plenum Press, New York, 1984. (b) M. I. Bessonov, M. M. Koton, V. V. Kudryavtsev, and L. A. Laius, "Polyimides," Consultants Bureau, New York, 1987.
10. A search of the Chemical Abstracts (1977-90) yielded 158 and 2,845 references, respectively, on the electronic and magnetic properties of Cr_2O_3 and Fe_2O_3.
11. Recent references: (a) I. Otsuka, T. Shinada, M. Mitshuboshi, and H. Inone, Jpn. Kokai Tokkyo Koho, JP 63,182,361 (1989), CA: 110116226x. (b) H. Yamamoto, T. Doi, and S. Ozawa, Jpn. Kokai Tokkyo Koho, JP 63,172,741 (1989), CA: 110155731c. (c) G. M. Porta, J. D. Rancourt, and L. T. Taylor, Chem. Mater., 1, 269 (1989). (d) J. D. Rancourt, G. M. Porta, E. S. Moyer, D. G. Madeleine, and L. Taylor, J. Mater. Res., *3*, 996 (1988). (e) R. K. Boggess and L. T. Taylor, J. Polym. Sci., Polym. Chem. Ed., 25, 685 (1987). (f) L. T. Taylor in "Proceedings of the Second International Conference on Polyimides," p. 351, Society of Plastics Engineers, Ellenville, New York, 1985. (g) J. D. Rancourt, R. K. Boggess, and L. T. Taylor, in "Proceedings of the Second International Conference on Polyimides," p. 372, Society of Plastics Engineers, Ellenville, NY, 1985. (h) S. A. Ezzell and L. T. Taylor, Macromolecules, 17 , 1672 (1984). (i) S. A. Ezzell, T. A. Furtsch, E. Khor, and L. T. Taylor, J. Polym. Sci., Poly. Chem. Ed., 21, 865 (1983). (j) T. A. Furtsch, L. T. Taylor, T. W. Fritz, G. Fortner, and E. Khor, J. Polym. Sci., Polym. Chem. Ed., 20, 1287 (1982). (k) E. Khor and L. T. Taylor, Macromolecules, 15, 379 (1982). (l) K. Ohmura, I.

Shibazaki, and T. Kimura, Jpn. Kokai Tokkyo Koho, JP 79,143,462 (1979), CA: 92130214b.
12. Previous report: M. Nandi, J. A. Conklin, L. Salvati, and A. Sen, Chem. Mater., 2, 772 (1990).
13. D. P. Tate, W. R. Knipple, and J. M. Augl, Inorg. Chem., 1, 433 (1962).
14. Reference 9b, chapter 1.5.
15. For a discussion of Cr-arene bond strengths in $(arene)_2Cr$ complexes, see: R. Davis and L. A. P. Kane-Maguire in "Comprehensive Organometallic Chemistry," G. Wilkinson, F. G. A. Stone, and E. W. Abel, editors, Vol. 3, p. 988, Pergamon Press, New York, 1982.
16. G. Van Koten and J. G. Noltes, in reference 15, Vol. 2, Chapter 14.
17. Recent references: (a) B. D. Silverman, Macromolecules, 22, 3768 (1989). (b) M. J. Goldberg, J. G. Clabes, A. Viehbeck, and C. A. Kovac, Polym. Mater. Sci. Eng., 59, 199 (1988). (c) M. J. Goldberg, J. G. Clabes, A. Viehbeck, and C. A. Kovac, Proc. Electrochem. Soc., 88-17,103 (1988). (d) J. G. Clabes, M. J. Goldberg, A. Viehbeck, and C. A. Kovac, J. Vac. Sci. Technol., A6, 985 (1988). (e) M. J. Goldberg, J. G. Clabes, and C. A. Kovac, J. Vac. Sci. Technol., A6, 991 (1988). (f) R. Haight, R. C. White, B. D. Silverman, and P. S. Ho, J. Vac. Sci. Technol., A6, 2188 (1988).
18. (a) I. V. Bulgarovskaya, V. E. Zovodink, and V. M. Vozzhennikov, Acta. Crystallog. Sect. C: Cryst. Struct. Commun., C43, 764, 766 (1987). (b) I. V. Bulgarovskaya, V. K. Bell-Skii, and V. M. Vozzhennikov, Acta. Crystallog. Sect. C: Cryst. Struct. Commun., C43, 768 (1987). (c) J. J. Stezowski, R. D. Stigler, and N. Karl, J. Chem. Phys., 84, 5162 (1986). (d) V. N. Baumer and V. A. Starodub, Synth. Met., 9, 467 (1984). (e) R. Foster, J. Iball, S. N. Scrimgeour, and B. C. Williams, J. Chem. Soc., Perkin Trans. II, 682 (1974) (f) T. Kodoma and S. Kumakura, Bull. Chem. Soc. Jpn., 47, 1081, 2146 (1974). (g) J. C. A. Boeyens and F. H. Herbstein, J. Phys. Chem. 69, 2160 (1965).
19. (a) B. Rees and A. Mitschler, J. Am. Chem. Soc., 98, 7918 (1976). (b) A. Jost, B. Rees, and W. B. Yelon, Acta Crystallogr., B31, 2649 (1975).
20. (a) P. S. Braterman, Struct. Bonding (Berlin), 10, 57 (1972). (b) P. S. Braterman, Struct. Bonding (Berlin), 26, 1 (1976).
21. S. Aravamudhan, U. Haeberlen, H. Irngartinger, and C. Krieger, Molecular Phys., 38, 241 (1979).
22. (a) Reference 11c. (b) E. Khor and L. T. Taylor in "Metal-Containing Polymer Systems," J. E. Sheats, C. E. Carraher, Jr., and C. U. Pittman, Jr., Editors., p. 367, Plenum Press, New York, 1985.
23. A. Abragam and B. Bleaney, "Electron Paramagnetic Resonance of Transition Ions," p. 432, Oxford University Press, London, 1970.
24. J. C. Carver, G. K. Schweitzer, and T. A. Carlson, J. Chem. Phys., 57, 973 (1972).
25. C. D. Wagner, W. M. Riggs, L. E. Davis, J. F. Moulder, and G. E. Muilenberg, editors, "Handbook of X-Ray Photoelectron Spectroscopy," p.42, Perkin-Elmer Physical Electronics Division, Eden Prairie, MN, 1979.
26. N. S. McIntyre, D. G. Zetaruk, Analytical Chem., 49, 1521 (1977).
27. (a) Reference 15, Vol. 3, p. 999. (b) D. F. Shriver and K. H. Whitmire in reference 15, Vol. 4, p. 262.
28. J. P. Collman, L. S. Hegedus, J. R. Norton, and R. G. Finke, "Principles and Applications of Organo-transition Metal Chemistry," p. 43, University Science Books, Mill Valley, CA, 1987.

and stored over 4Å activated molecular sieves. Other alcohols were used as received. Deoxygenation of solvents and solutions was effected by bubbling N_2 directly through the solutions.

Photolysis Reaction of $[PtCl_6]^{2-}$ with Alcohols

In a typical experiment 3.5mL of an alcohol solution that was about 16mM in H_2PtCl_6 was irradiated with 488nm light (~80mW) for times ranging from several minutes to hours. After irradiation, samples were immediately analyzed by ^{195}Pt NMR for platinum species and analytical GC for organic products. Dark controls were also carried through the analysis.

Photolysis Reaction of $PtCl_4$ with Isopropanol

All preparations were carried out under an atmosphere of nitrogen with glassware oven-dried at 105°C. In a typical experiment, 5mL of solution was prepared by dissolving $PtCl_4$ in dry isopropanol (saturated in $PtCl_4$). The solution was irradiated using an argon ion laser (60mW, 488nm) for six hours. After irradiation, samples were immediately analyzed. Organic products were separated by preparative GC with 20% Carbowax 20M 60/80 chromosorb W column at 30°C oven temperature. A dry ice-acetone cooling bath was employed to collect samples which were then characterized by ^1H NMR.

Deposition of Platinum on Mylar

Deposition was carried out onto Mylar sheets (400D, DuPont company) which had been rinsed in acetone and distilled water. The clean sheets were placed onto the watchglass carrier filled with a H_2PtCl_6 alcohol solution (8mM). A laser beam or filtered mercury source was employed to directly irradiate the Mylar sheets for 10-30 mins (focused beam, 450-500mW/cm^2) or 1-2hrs (expanded beam, 60-80mW/cm^2). Samples were then developed in a Na_2PtCl_4 alcohol solution (16mM) in the dark.

Electroless Copper Plating Catalyzed by Photoinduced Deposited Platinum on Mylar

A Mylar sheet was irradiated in a H_2PtCl_6/alcohol solution as described in the previous section. Samples were then placed into an electroless copper plating bath for several minutes, which contained $CuSO_4.5H_2O$ (8.5g/L), formaldehyde (37%,2mL/L) and EDTA dihydrate (57.3g/L), at 65°C adjusted to pH11.2.

Measurements

^{195}Pt NMR spectra were recorded on a Bruker 250-MHz FT-NMR. ^{1}H NMR spectra were recorded on a GE QE-300 NMR spectrometer. Preparative thin layer chromatography was carried out on Silica GEL GF 2000 micron 20x20 cm plates. Preparative gas chromatography was performed on a Varian Aerography 90-P instrument with 20% Carbowax 20M 60/80 chromosorb W column at 30°C oven temperature. A Perkin-Elmer Sigma 3B Dual FID gas chromatograph with a 3% Carbowax 20M 80/100 chromosorb WHP column was used to analyze organic products. Quantitative analysis was accomplished with cyclohexanone as an internal standard.

MECHANISTIC ASPECTS OF HEXACHLOROPLATINATE PHOTOREDUCTION

We have previously reported detailed spectroscopic and chemical studies aimed at elucidating the mechanism by which Pt(IV) is reduced to platinum metal[3,4]. Based on that work, the following reaction scheme appears most plausible.

$$[Pt^{IV}Cl_6]^{2-} + xROH \xrightarrow[(x=1-6)]{xh\nu} [Pt^{IV}Cl_{(6-x)}(OR)_x]^{2-} + xH^+ + xCl^- \quad (1)$$

$$[Pt^{IV}Cl_{(6-x)}(OR)_x]^{2-} \xrightarrow{h\nu} [Pt^{III}Cl_{(6-x)}(OR)_{(x-1)}]^{2-} + {}^{\bullet}OR \quad (2)$$

$$[Pt^{III}Cl_{(6-x)}(OR)_{(x-1)}]^{2-} + {}^{\bullet}OR \xrightarrow[(n=1-4)]{} [Pt^{II}Cl_{(4-n)}(OR)_n]^{2-} + O{=}CR \quad (3)$$

$$[Pt^{II}Cl_4]^{2-} + ROH \xrightarrow{h\nu \text{ or } \Delta} [Pt^{II}Cl_{(4-n)}(OR)_n]^{2-} + H^+ \quad (4)$$

$$[Pt^{II}Cl_{(4-n)}(OR)_n]^{2-} \longrightarrow Pt^0 + O{=}CR + (n-1)ROH + Cl^- \quad (5)$$

tris-isopropoxide which is the photochemically most reactive complex. Thus the primary reaction route based on Equations 1-5 involves x=3 in Equations (2) and (3).

DEVELOPMENT OF A PHOTOLITHOGRAPHIC SCHEME

Although the photoreduction of hexachloroplatinate to platinum metal embodies all the chemistry requisite for a successful photolithographic process, the large induction time prior to the observation of metal formation makes it unsuitable for a direct laser writing process or for any process in which a heavy coating of platinum is required. On the other hand, the photoreduction of tetrachloroplatinate has the necessary speed but, unfortunately, the competing dark reaction channel along with the autocatalytic nature of the reaction means that one will not be able to achieve even moderate spatial resolution. One can, however, achieve a suitable photolithographic process by taking advantage of the photochemical resolution which can be achieved using hexachloroplatinate and the autocatalytic nature of the thermal tetrachloroplatinate reduction. If an aqueous solution containing hexachloroplatinate and alcohol (either isopropanol or ethanol) is irradiated in the presence of a substrate (either Pyrex glass or Mylar has been employed) for a short period of time at 488 nm using an argon ion laser, a small deposit of platinum will be formed. This deposition will only form in the irradiated area and it will not form uniformly in that area. If the laser beam is scanned during the irradiation process, then a series of metallic dots will be formed. This dot formation will be incomplete and not visible to the naked eye. However, such minimal platinum metal deposits can act as catalytic sites for the formation of more platinum metal. Thus, after laser exposure, development of the substrate in the dark in an aqueous solution containing tetrachloroplatinate and alcohol produces a heavy metallic coating. The autocatalytic nature of the process along with the inhibition associated with chloride present in a developing solution assures that metal deposition only occurs in the irradiated area. This type of development will establish a well defined circular area if the substrate has been held stationary during photolysis or a line if the substrate has been moved during the photolysis. The width of the line obtained correlates well with the size of the laser beam, down to 20μm the tightest beam focus tested.

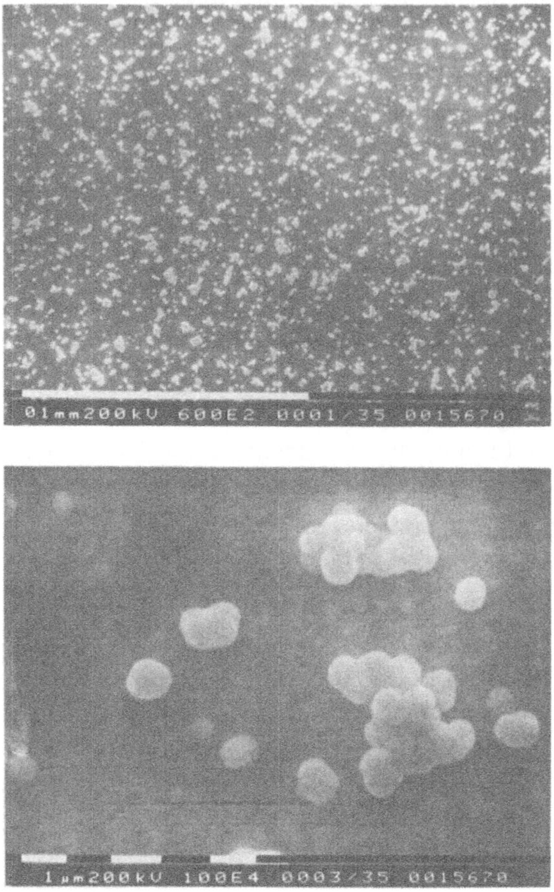

Figure 4. SEM photograph of platinum deposit on Mylar at two different magnifications showing a smooth metallic layer with an overlayer of platinum spheres. The top picture is taken at a magnification of X600 while the bottom is X10,000. Cross sectional analysis indicates a 5μm thickness for the smooth platinum background.

Large area metallization can be obtained by defocusing the laser beam or employing a white light source. Again, a two-step process involving the photodeposition of metallic sites using $[PtCl_6]^{2-}$ followed by development in an alcoholic solution of $[PtCl_4]^{2-}$ is employed. Rates of reaction can be enhanced in this case by employing a UV source. Results achieved using this procedure are illustrated in Figure 4 which provides scanning electronmicrographs of a platinum deposit on Mylar. No pretreatment of the Mylar was found necessary for adhesion. Using this approach, we have platinized areas up to 2 square centimeters; presumably larger areas can also be metallated. As shown in the lower portion of Figure 4, the platinized surface appears very smooth down to submicron resolution. However, surprisingly it is covered with a series of spheres having typical radii ranging from ~1000Å to ~4000Å. Energy dispersive X-ray analysis (EDX) of the smooth film and the spherical particles indicates they consist of the same chemical composition: platinum metal (see Figure 5). A trace contamination of chloride from the initial reagents is observed. (EDX is not sensitive to organic contamination). Cross sectional analysis by SEM of the film shown in Figure 4 indicates a 5μm thickness which is typical of the films investigated

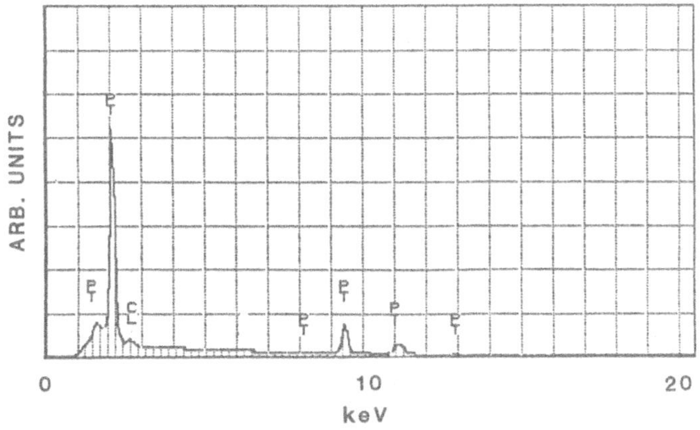

Figure 5. EDX data obtained for the sample shown in Figure 4.

Scotch tape adhesion testing of the metal layer indicated no failure up to 66 N/100mm (3M Tape #250, 90° pull angle), the highest pull strength tested. Four probe resistivity measurements on these films yielded typical resistivities on the order of ~100Ω per centimeter. These high resistance levels suggest film contamination by alcohol or solvent (H_2O). However, such contamination (and the related high resistance) may not be intrinsic to the described process, since as discussed below it is not observed in the case of copper deposition. To date, no effort has been made to minimize the film resistivity. From an application point of view, high resistivity may not represent a major problem in that photolithographic deposition of platinum is of primary interest with respect to metallic mask repair.

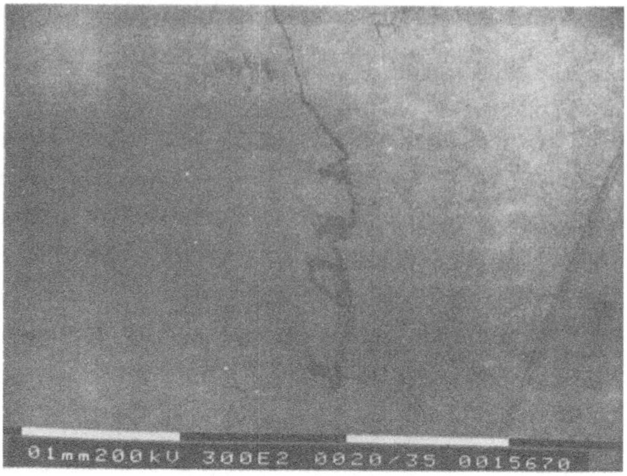

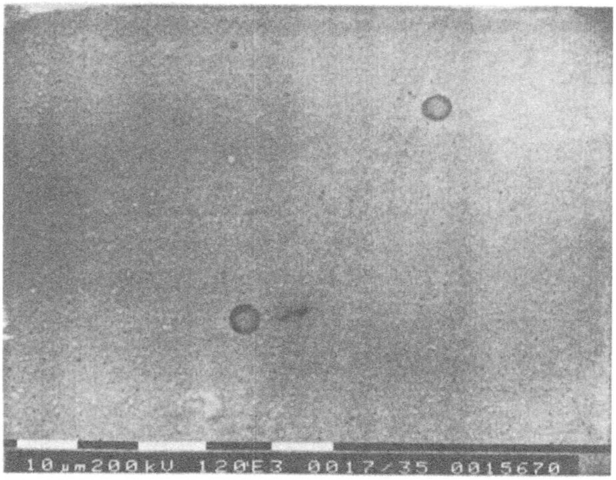

Figure 6. SEM photograph of copper on Mylar at two different magnifications (top: X300, bottom: X1200). The object shown in the bottom photograph is a mechanically induced scratch. . EDX analysis of the bottom of the scratch indicates the presence of platinum. Cross sectional SEM indicates a 1.2μm layer of copper is present.

Copper metal deposition on Mylar can be accomplished by first photochemically depositing a trace amount of platinum metal using a $[PtCl_6]^{2-}$ solution as described above, followed by developing in an electroless copper plating bath (see Experimental Section). Good resolution on a macroscopic scale is observed for this process. No high resolution experiments have been attempted on this system. Typical SEM data for copper plated Mylar are shown in Figure 6. The circular features observed on the otherwise smooth film (lower portion of Figure 6) are either bubbles or craters. EDX analysis of the smooth portion of the film and the circular features shows their composition to be identical, namely copper with a small chloride contaminate (Figure 7). Note that no signal for platinum is observed. The upper portion of Figure 6 shows a mechanically induced scratch. EDX analysis of this area does indicate the presence of platinum.; thus, a limited platinum layer underneath the copper coating is present as is expected. SEM cross section analysis of the film shown in Figure 6 yields a 1.2μm thickness.

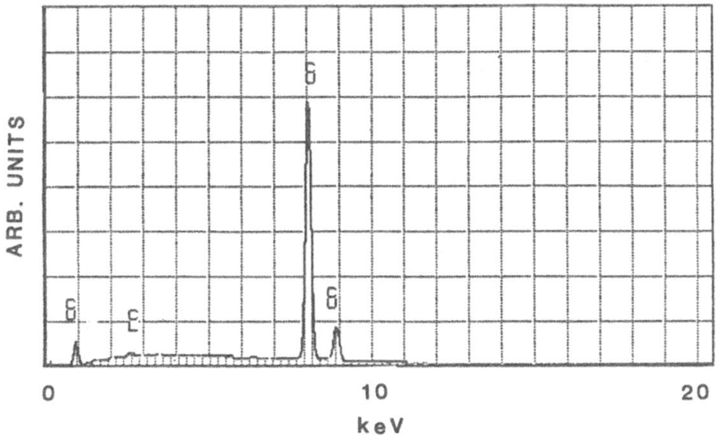

Figure 7. EDX analysis of the sample shown in Figure 6.

The adhesion of the copper overlayer is significantly less than observed for the pure platinum system, failing the lowest strength Scotch tape test employed (44 N/100mm). This result is surprising in light of the good adhesion observed between Mylar and platinum, since the EDX data suggest a similar interface exists in the copper system. The failure may be occurring at the platinum/copper interface, but more likely, the amount of platinum/Mylar interface is too limited in area to provide the expected adhesion.

Resistivity measurements on the copper/Mylar interface show much higher conductivity than observed for the platinum metallized surfaces. Typical resistivities on the order of 10Ω/cm were obtained using a four-point probe. Thus a relatively pure metal deposit is obtained.

CONCLUSIONS

Conducting coatings of platinum or copper with good adhesion can be deposited onto silica or Mylar substrates using a two-step exposure and development process. The initial photochemical step takes advantage of the inhibitory effect of chloride to yield good resolution, limiting metal deposition to illuminated areas. The development step is controlled by the autocatalytic nature of the reduction of $[PtCl_4]^{2-}$ by platinum metal. Photodeposition rates can be controlled by controlling the chloride concentration, the illumination intensity, or the wavelength employed. Strong adhesion of the metal layer to Mylar was obtained, while the conductivity of the coating was only moderate. The latter finding suggests some level of organic contamination in the metallic layer. However, neither interfacial adhesion nor conductivity has been optimized. The nature of the metal - Mylar interface is currently under investigation and it is expected that a better understanding of the chemistry at the interface will lead to improved adhesion. Likewise, a better understanding of the detailed deposition chemistry should provide control over the electrical properties of the surface layer.

ACKNOWLEDGEMENT

The Department of Energy, Office of Basic Energy Sciences is thanked for support of this work under Grant #DE-FG02-85ER13438.

REFERENCES

1 A. Vogler and J. Hlavatsch, Angew. Chem., Int. Ed., 22, 154, (1983).

2 R. E. Cameron and A. B. Bocarsly, J. Am. Chem. Soc., 107, 6116 (1985).

3 A. B. Bocarsly, R. E. Cameron, and Meisheng Zhou, in "Photochemistry and Photophysics of Coordination Compounds," H. Yersin and A. Vogler, Editors, p. 177, Springer Verlag, Berlin, 1987.

4 R. E. Cameron and A. B. Bocarsly, Inorg. Chem., 25, 2910, (1986).

5a. R. E. Cameron, Ph.D. Thesis, (pp 39-40), Princeton University, 1986.

b. F. J. Schwab, Departmental Thesis, Princeton University, 1986.

6 G. A. Shagisultanova, Koord. Chem., *7*, 1527 (1981).
7 I. I. Blinov and G. A. Shagisultanova, Koord. Chem:, *8*, 191 (1982).
8 V. P. Grivin, I. V. Khmelinski and V. F. Plyusnin, J. Photochem Photobiol. A: Chemistry, *51*, 379 (1990).
9 J. Chatt, B. L. Shaw, J. Chem. Soc., 5075 (1962).
10 P. Gröser, Angew. Chem., Int. Ed., *6*, 364 (1967).

METALLIZED INDIUM ISLAND FILMS FOR FLEXIBLE AUTOMOTIVE TRIM

Richard Eisfeller

Davidson Technology Center Textron
Industrial Park
Dover, NH 03820

John S. Judge

John Judge & Assoc., Inc.
11 Magrath Rd.
Durham, NH 03824

Davidson has developed a unique process for metallizing flexible automotive trim to produce a bright metallic reflective appearance while maintaining durability and corrosion resistance. This process involves the vapor deposition of indium films onto polyurethane coated substrates. These films are subsequently etched opening up channels between islands of indium. The films are passivated with a topcoat of sprayed polyurethane which achieves its adhesion through the open channels to the undercoat. These channels are so small (~50 nm) that the film appears continuous. They also allow for flexibility without damage to the metal film.

INTRODUCTION

For many years there have been efforts to produce metallic films of high reflectivity for automotive trim[1]. The corrosion resistance and durability of these films has been less than what is desired. In addition, the deposition of films on flexible substrates without introducing cracking when flexed presents additional difficulties. Davidson has developed a process for the deposition of highly reflective indium films which appears to overcome these previous limitations.[2] This process deposits discontinuous films composed of islands of indium separated by channels of polyurethane. Undercoats and overcoats of sprayed polyurethane films adhere to each the through the channels and encapsulate the indium islands. The microscopic size of the channels leaves the metallic film with the appearance of a continuous reflective surface.

EXPERIMENTAL

Figure 1 is a schematic of the process flow for preparing these indium island films. Parts are prepared for the indium film deposition by spray coating with a polyurethane film. Then the indium is flash

Metallized Plastics 2, Edited by K.L. Mittal
Plenum Press, New York, 1991

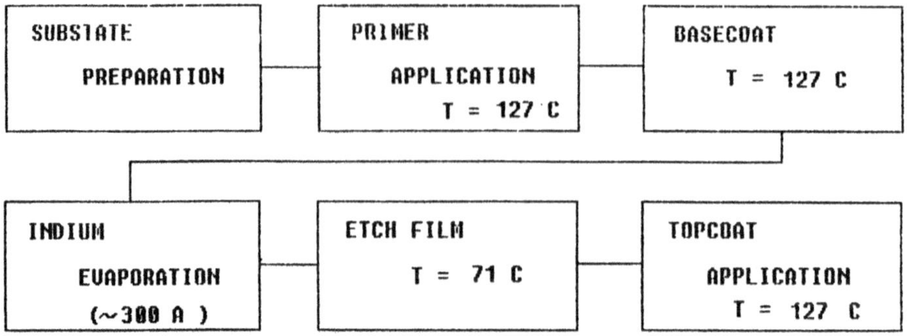

Figure 1. Process flow for preparing bright trim island films.

evaporated from a tungsten boat in an argon atmosphere of $\sim 10^{-3}$ Torr. These films are subsequently etched in NaOH solution of pH 12-12.5. After rinsing and drying a second topcoat of polyurethane is applied. Samples of vacuum deposited indium films were prepared for this investigation without being overcoated. The results of X-ray fluorescence thickness measurements on some of these samples are shown in Table I.

Table I. X-ray Fluorescence Results.

SAMPLE #	THICKNESS(A)	ETCHED
1	197	YES
2	304	NO
3	67	YES
4	158	NO

Actually, #1 and #2 were taken from the same run, as were #3 and #4, but the additional process step of etching was done on #1 and #3. Samples #3 and #4 were produced with half the indium charge in the tungsten boats that samples #1 and #2 had.

Table II contains the structural and property data for a number of samples obtained by varying the process parameters around a central control point. The structural data for all of these films were obtained from measurements taken by scanning electron microscopy. The values of s (the center-to center distance) and λ (the edge-to-edge distance) were calculated using the methods of Underwood.[3] The diffuse reflectance measurements were obtained utilizing a Pacific Scientific Spectrograph II equipped with an integrating sphere. The "scattered reflectance" measurements were obtained utilizing a Labscan Spectral Colorimeter with zero degree illumination and 45 degree circumferential viewing.

RESULTS AND DISCUSSION

Surprisingly the diffuse reflectance measurements, shown in Table I, do not seem to correlate with any of the structural properties measured. These films exhibit various degrees of haziness when viewed at oblique angles and the scattered reflectance measurement has been found to correlate with this visual haziness. One surprising feature of these data is that despite a relatively large variation in metal island size (100 to 300 nm) the percentage of open area does not change much (27 to 45%). Also to the extent that it does vary, the percentage of open area does not correlate with any of the measured variables of the film. It is anticipated that this variable may have a direct bearing on the durability of the adhesion between the topcoat and the substrate. Long term environmental exposures are underway to determine if this is true.

Figure 2 is a photomicrograph taken at 60,000X of sample #2. When this sample is etched, the channels become decidedly wider (Figure 3, sample #1). These figures clearly show indium being distributed as islands separated by channels, presumably of the substrate polyurethane. From these photomicrographs it is apparent that the island sizes in the thicker film are approximately .25 µm. Similar results for sample #3 and #4 show that these island sizes in this thinner film are much smaller - <0.1 µm. Again it is apparent from the SEM photos that the etching widens the channels. Based on these results, two mechanisms of deposition are

Table II. Summary of the Structural and Property Data for Various Indium Island Films.

s* (µm)	λ * (µm)	Ave.diam. (µm)	% open	Diff.Refl. (rel.units)	Scat.Refl. (rel.units)
0.333	0.130	0.206	38.6	41.0	14.2
0.335	0.132	0.203	39.3	52.0	6.5
0.183	0.074	0.109	40.5	37.9	3.4
0.160	0.066	0.094	40.6	38.7	3.3
0.236	0.064	0.173	26.8	43.5	6.5
0.246	0.091	0.155	37.5	49.4	4.8
0.318	0.114	0.206	35.8	52.2	4.4
0.206	0.061	0.142	30.3	38.7	3.3
0.318	0.124	0.193	39.4	50.1	8.8
0.511	0.185	0.323	36.6	43.2	13.9
0.452	0.127	0.325	28.2	49.1	14.0
0.279	0.099	0.180	35.8	49.4	3.6
0.295	0.132	0.163	45.0	43.5	6.2
0.173	0.069	0.104	39.8	43.7	2.8
0.236	0.084	0.150	35.8	50.1	3.8
0.396	0.163	0.234	41.1	45.8	12.2
0.279	0.112	0.168	40.1	48.5	5.5
0.155	0.061	0.091	40.3	43.7	8.3
0.498	0.213	0.284	42.8	46.8	8.3
0.241	0.094	0.147	38.7	48.0	3.7

* s is the average metal island center-to-center distance.
* λ is the average edge-to-edge distance.

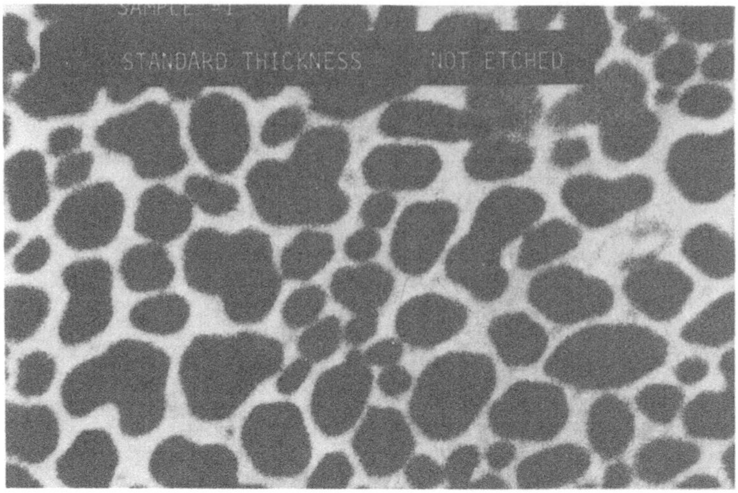

Figure 2. SEM image of unetched indium film at 60,000X.

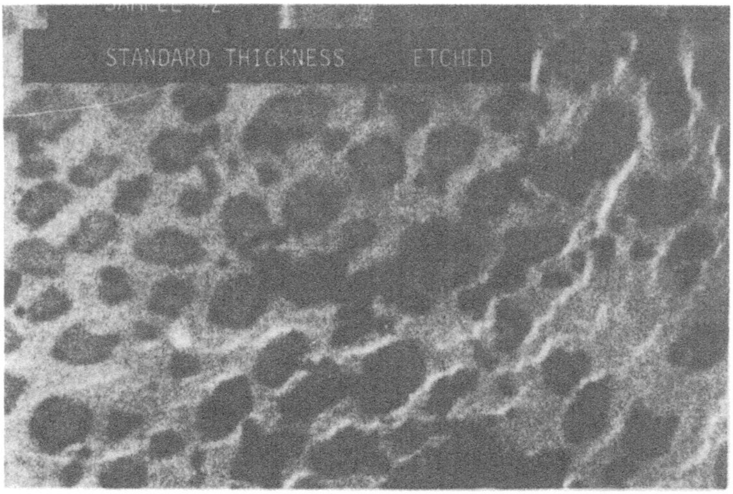

Figure 3. SEM image of etched indium film at 60,000X.

possible - there are more nucleation centers due to different rates of deposition or some sort of coalescence of the islands takes place as deposition proceeds, suggesting that the early nuclei act as if they were in the liquid phase. It is possible that this vacuum deposition of indium onto the basecoat follows the mechanism suggested by Lewis and Anderson[4]; the impinging atom is attracted to the surface by van der Waals forces; as a result the condensate atom loses its velocity component normal to the surface in a short time if the incident kinetic energy is not too high. This physically adsorbed adatom may not be immediately thermally equilibrated and may jump from place to place depending on the binding energy between basecoat and adatom and its kinetic energy parallel to the surface[5]. At high binding energies between adatom and substrate, thermal equilibrium occurs quickly and surface mobility is low. In this case the energy of the metal vapor

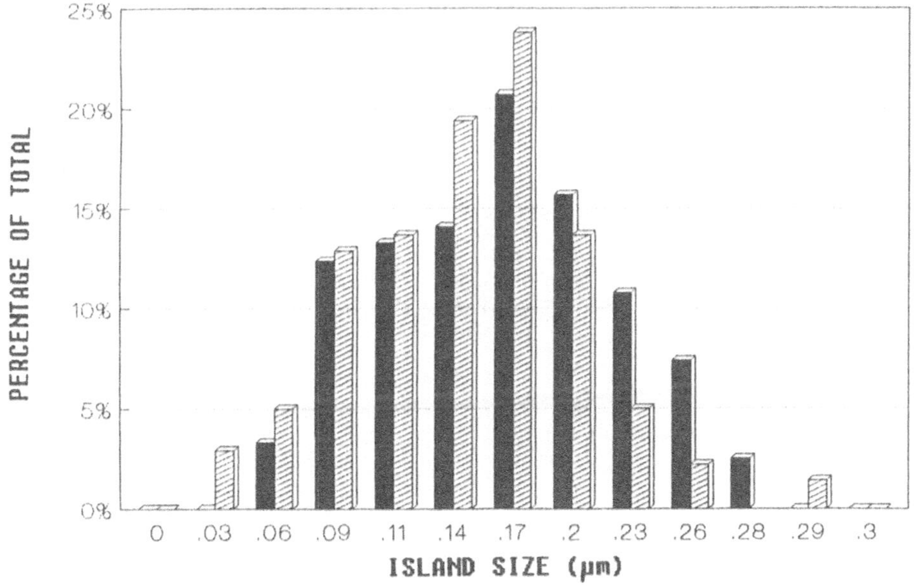

Figure 4. Distribution of island sizes for the thicker film, etched and unetched (solid bars are unetched).

atom is high relative to the non-polar organic basecoat, thus the adatoms remain "hot" and mobile. After a few atoms form metal clusters, they grow into islands by capturing additional adatoms diffusing parallel to the surface. If the melting point of the deposited metal is low enough, the surface mobility allows relatively large island formation. Higher melting point metals commonly associated with vacuum metallizing (e.g. Al, Cr, chromium alloys, stainless steels) form continuous films at thicknesses less than 100 A. As an example, aluminum which melts at 635 ^{o}C, gives a conductive, continuous film at 70 A thickness with 70 % transmission as compared to these indium films which are non-conductive and quite reflective.

Image analyses were subsequently performed on samples #1, #2 and #3 to determine the island size distributions and the coverage or percentage of open area on the samples. It is probable that the adhesion of the polymeric overcoat depends primarily on its adhesion to the substrate in the open channel areas; the ability to measure this open area might be useful in understanding how process variables lead to variations in adhesion. The results for the three samples investigated are shown in Table III.

Table III. Image Analysis Results for Samples 1, 2 and 3.

SAMPLE #	% COVERAGE	AVE. DIAM.(µm)
1	61.42	0.132
2	67.43	0.150
3	40.22	0.0454

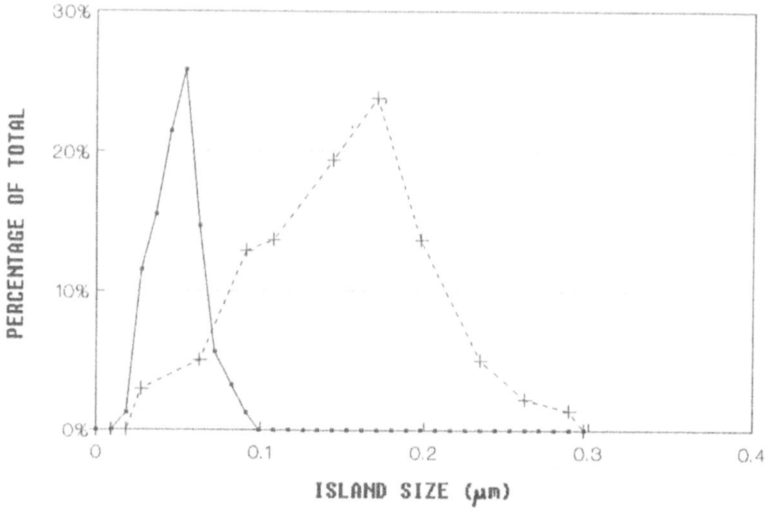

Figure 5. Distribution of island sizes for the thick and thin etched films(---- thick film).

Figure 4 is a plot of the island size distribution, obtained from the SEM micrographs, as a function of the percentage of the total number of islands present for the etched and unetched thicker film. As expected, for the etched film, the distribution shows fewer larger islands and more of a tail in the smaller island region. However, the distribution reaches a maximum at almost the same size for etched and unetched samples but sharpens up for the etched film; it shows almost 25% having the most probable size for the etched sample in comparison to approximately 20% for the unetched film. The affect of etching may be more apparent if the data are viewed as the cumulative percentage of islands with diameter less than a specific value versus the value of that diameter. In that case it is clear that the distribution shifts to lower diameters as etching is performed. In addition to the dependence on etched vs. non-etched condition, the thickness of the etched film plays a role in island diameter. Figure 5 shows the comparative distributions for the thick and thin etched samples. The thinner sample has a much sharper distribution (more uniform) at a much smaller size.

Table I shows the results of the X-ray fluorescence measurements for the amount of indium per unit area of the samples. As expected, these results show that the etched samples are significantly thinner than the unetched ones and that the sample produced with half the charge comes out with almost exactly half the amount of indium. (The values in this table were calculated assuming that these were continuous films.) Calculations on the height of the indium islands, assuming that the indium islands are spherical segments and by utilizing the island size data and the X-ray fluorescence data, island height is estimated at approximately 409 A for the unetched film, and 300 A for the etched film.

The effect of island size on scattered reflectance is shown in Figure 6. The regression line shown in the figure indicates an increase from about 5 to 15 as the island size increases from 100 to 300 nm. Thus the films become more hazy as the island size increases. A remarkable correlation of the average island edge-to-edge distance with island size is shown in Figure 7. The edge-to-edge distance remains about half the island

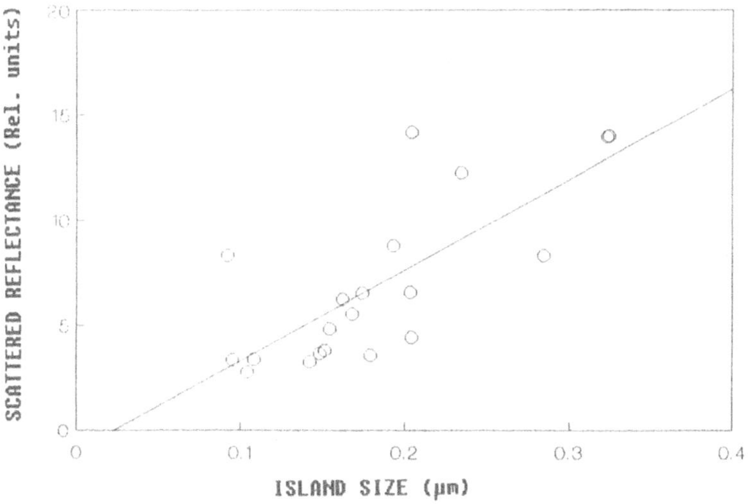

Figure 6. Dependence of scattered reflectance on average indium island size.

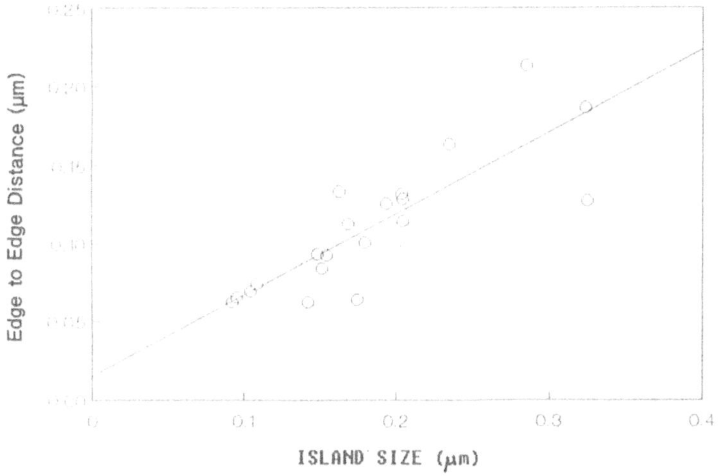

Figure 7. Dependence of the edge-to-edge distance on indium island size.

diameter for all of these specimens. Apparently as the island sizes and the distance between them begin to approach the wavelengths of visible light more scattering occurs, contributing to the reflectance and the haziness observed. The fact that the edg-to-edge distance is dependent on the island size suggests that the islands were close packed to begin with, i. e. before etching; thus, larger sizes initially could lead to larger spaces after etching.

CONCLUSIONS

We have shown that discontinuous island films are formed by vacuum evaporation of indium onto polyurethane substrates. These films are bright and reflective giving the appearance of continuous metal films, but unlike continuous metal films they are flexible. Thicker films form larger islands and subsequent etching of the films reduces the size of the islands and opens the channels between the islands. As island sizes begin to approach the wavelengths of visible light, some scattering begins to take place leading to a hazy appearance.

REFERENCES

1. D. F. Gentle, Electroplat. Metal Finish., 24(2), 12 (1971).
2. R. Eisfeller, Proc. 27th Tech. Conf. Soc. of Vacuum Coaters, 98 (1984).
3. E. E. Underwood, "Quantitative Stereology," Addison-Wesley, 1960.
4. B. Lewis and J. Anderson, "Nucleation and Growth of Thin Films," p 56, Academic Press, New York (1978).
5. G. Kovacs and P. Vincent, J. Colloid Interface Sci., 90, 335 (1983).

METALLIZATION OF PLASTICS BY PHYSICAL VAPOR DEPOSITION TECHNIQUES

C. T. Wan, K. A. Taylor,
D. L. Chambers and G. T. Susi

Xytorr Corporation
3650 Parkway Lane
Hilliard, Ohio 43026, USA

Common physical vapor deposition techniques, such as vacuum evaporation and magnetically enhanced sputtering as used in the metallization of common plastics, are described and compared for different applications. Typical applications range from the metallization of various types of plastics used for trophies, automobile dashboards and compact discs to the deposition of cermet coatings onto aerospace parts fabricated with fiber impregnated polyimide composites or other moldable high temperature polymeric materials.

INTRODUCTION

Physical vapor deposition (PVD) refers to a variety of coating techniques commonly used in a vacuum or a partial vacuum of gases for the deposition of metallic, ceramic or cermet coatings. These coatings are deposited from a vapor onto substrates by applying thermal or mechanical energy to the material to be deposited. This paper describes the application of PVD techniques such as the vacuum evaporation and the sputtering process with emphasis on the use of magnetically enhanced sputtering in high technology applications. Typical applications include (1) decorative coatings for automobile parts and jewelry, (2) reflective coatings on compact and laser discs, (3) coating on web for electronic components and circuitry, solar cells and radiation control, food packaging, (4) electromagnetic interference (EMI) coating for rejection of high frequency noise and (5) coatings for wear, corrosion and erosion resistances.

Metallization Process and Plastics[1]

A typical metallization process for plastic materials consists of applying three layers of coatings: (1) The first layer is a lacquer-type organic coating, referred to as a base-coat whose primary function is to produce a smooth surface so that the metal coating will have maximum luster. It is also used to enhance the adhesion between the metal coating and the substrate. Since plastic material is known to outgas in the vacuum environment, the base-coat

Metallized Plastics 2, Edited by K.L. Mittal
Plenum Press, New York, 1991

also serves to seal the surface of the substrate. (2) The metal coating deposited by the PVD technique is usually aluminum to render the parts highly reflective appearance. (3) The second organic coating applied over the metal coating is referred to as the top-coat and its main function is to protect the metal coating from adverse chemical reactions and physical damage. In some applications, the top-coat is not employed when the metal coating is adequate for the intended wear applications.

The durability of coatings is of prime importance. One of the main factors governing durability is adhesion. Oxygen active metals tend to form chemical bonds with the plastic substrate and are strongly adherent, whereas those metals that form only physical bonds may easily be removed. One of the simplest methods of qualifying adhesion[2] is to apply a short length of adhesive tape to the surface of the coating. When the tape is removed, the coating is either wholly removed, partially removed, or not removed at all from the substrate. This method is only qualitative and gives no indication of the relative magnitudes of adhesion strength. It is, however, generally adequate for most industrial applications.

Among the most commonly used plastics, ABS (Acrylonitrile-Butadiene-Styrene) is commonly used in automobile interior components, housewares and toys. Acrylic thermoplastics are known for their crystal clarity. Clear acrylic plastic has a light transmittance of 92% and an index of refraction of 1.49, which is high enough for use in lenses and other optical applications. Lately, it has found application as disc material coated with aluminum for the laser discs.

Polycarbonates are amorphous thermoplastics which offer superior impact and heat resistance. They are used in automotive applications such as bumpers and interior dashboards. Electronic applications include connectors, terminal blocks, meter covers, business machine housings, computer disc packs, and compact discs.

Cellulosics are synthetic plastics made from naturally occurring polymers referred to as acetate, butyrate and ethyl cellulose. They are routinely used in the manufacture of costume jewelry, novelties and components for household appliances.

A very important application for the metallization of plastic is the metallization of flexible polyester web material. Common applications of metallized web include UV and gas barrier coatings for food packaging and decorative coatings for gift-wrap material. Coated web is widely used for controlling lighting and solar radiation. Aluminum coated web is also used in the manufacture of capacitors. One of the most recent developments is the deposition of microwave receptor films for use in microwave-penetrable packaging structures. Molecules in the metal coatings absorb microwave energy and become agitated thereby creating radiant heat which causes food surfaces to become crisp or brown similar to a pan frying.

Polyimide is a moldable thermosetting polymer characterized by dimensional stability and high strength at elevated temperatures. Polyimides[3] are of interest for use as dielectric thin films in multi-level structures in microelectronics applications where strong and stable metal/polymer interfaces are required. Ohuchi and Freilich[4] have shown that titanium is an effective reaction layer which prevents metal diffusion into the organic phase. LeGoues et. al.[5] have shown that the microstructure of these interfaces is strongly related to the tendency of the metal to react with the polyimide at the interface.

Chromium was shown by Haight et. al.[6] and Dunn and Grant[7] to form a strong chemical bond with the polyimide surface and thus it is widely used as an adhesion promoter between metals and polyimide substrates. Moldable fiber impregnated polyimides are currently used to fabricate high temperature radomes, compressor vanes and other structural parts for aerospace applications that require wear and erosion resistant surfaces for successful operation over long periods of time. For this application, a chromium intermediate layer is required to promote adhesion of the ceramic or cermet coating. These types of corrosion and erosion resistant coatings are routinely applied by the magnetically enhanced sputter-deposition technique.

METALLIZATION BY EVAPORATION

Evaporation Process

Vacuum metallization is a general term referring to the evaporation and subsequent condensation of a metal onto a substrate in a vacuum environment. The metal utilized in most industrial processes for coating plastics is aluminum. However, other materials such as copper, gold and thermally stable compounds also evaporate in vacuum and are used in special applications.

Metallization by evaporation is accomplished by evaporating the coating material from a heated source in a vacuum environment. The heated source can be a filament, boat or crucible made of refractory metal or an intermetallic material. The heating energy can be derived from electrical resistance-heating, RF induction-heating or electron beam heating. The configuration of the various sources is shown in Figure 1.

A simple evaporation process starts with attaching a precise amount of aluminum to each filament and heating it at a programmed rate to liquify the aluminum to wet the surface of the tungsten wire. A further increase in the filament temperature then results in evaporation. For continuous evaporation over a long period of time, the aluminum is wire-fed or pellet-fed to either an intermetallic boat or an RF induction-heated crucible. The requirement of precise control of the coating thickness in some applications is satisfied by using electron beam heated crucibles.

The relationship between the coating thickness and the position of the evaporation source to the substrate is the same for both RF induction-heated crucible and boat type sources because both of them are small area sources. Uniformity of the deposited coating thickness across the width of a planar substrate is achieved by moving the substrate across a series of evaporating sources equally spaced in a line perpendicular to the direction of substrate travel. Uniformity, assuming all sources are equally powered, is a function of the spacing (L) between the sources and the distance (h) below the substrate. The coating thickness profile for a small area source is determined using basic emission laws[8] which state that

$$d/d_o = \frac{1}{(1 + (r/h)^2)^2} \qquad (1)$$

where d/d_o is the ratio of the film thickness (d) at any point on a substrate to the thickness (d_o) at a point directly above the source and r is the radial distance from the center of the source.

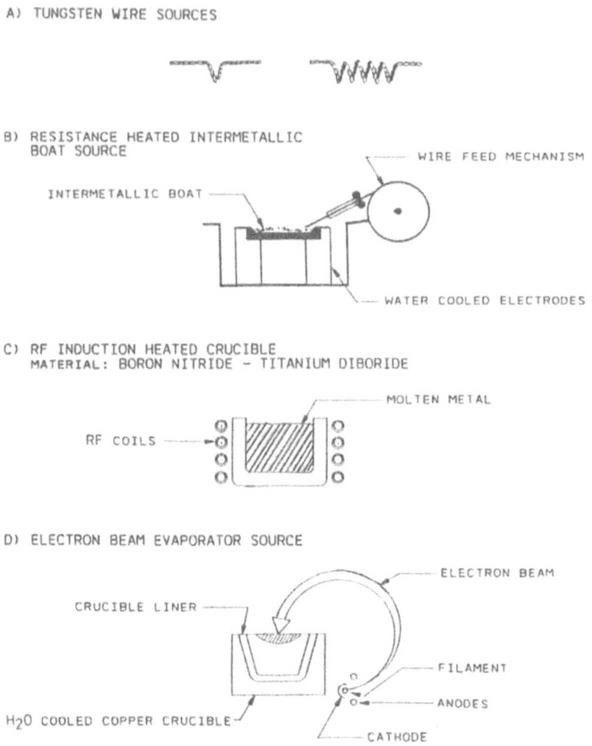

Figure 1. Typical source configurations for metal evaporation.

The thickness of the coating at any point from four sources on a line can be approximated by summing the contribution from each source in the form of a thickness ratio, as derived from Equation (1) that affects the selected point as follows:

$$d/d_o = \left[\frac{1}{(1 + (r/h)^2)^2}\right]_0 + \left[\frac{1}{(1 + ((L-r)/h)^2)^2}\right]_1 + \left[\frac{1}{(1 + ((L+r)/h)^2)^2}\right]_{-1} + \left[\frac{1}{(1 + ((2L-r)/h)^2)^2}\right]_2 \quad (2)$$

Subscripts refer to the locations of sources placed along a line coinciding with the x-axis of Figure 2 at locations designated as 0, 1, −1 and 2. Figure 2 shows calculated thickness profiles and the percent of non-uniformity with respect to the ratio (L/h) where (L) is the distance between sources. Optimum uniformity is approached when L/h = 0.8. Uniformity is dependent not only on the contributions from adjacent sources, but also on the contributions received from sources on each side beyond the adjacent pair. The (d/d_o) profile between the end pair of sources decreases rapidly, as the fourth term in the preceding equation is not included in the calculation. The decrease continues at a more rapid rate at points beyond the last pair as both the second and fourth terms are not included. Grazzini et.

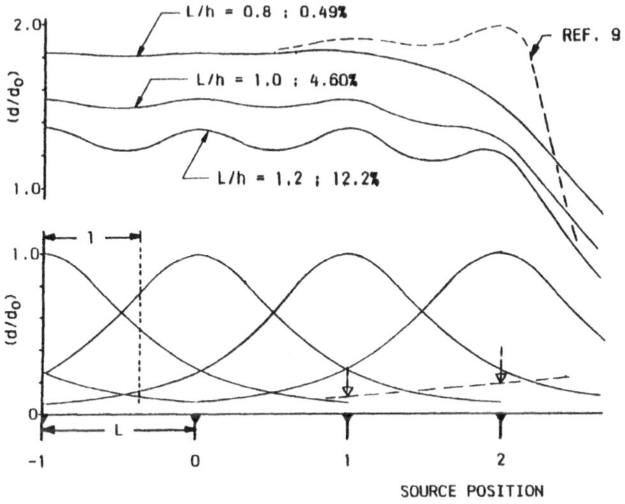

Figure 2. Deposited coating thickness uniformity obtained with in-line multiple sources.

al.[9] have shown that end effects can be minimized by decreasing the distance (h) between the end source and the substrate.

The combined work of Hertz, Knudsen and Langmuir as summarized by Glang[10] resulted in the following equation for the mass evaporation rate (G) from solid surfaces as expressed in (g/cm^2/sec)

$$G = 7.759\ (M/T)^{0.5}(p'-p) \qquad (3)$$

where

M = Molecular weight of the material being evaporated
T = Evaporating temperature (K)
p'= Vapor pressure of the evaporant (pascal) at evaporant temperature (T)
p = Chamber pressure (pascal).

Metallization by the vacuum evaporation process has developed into a powerful technology. Its use is expanding in areas where electroplating was once the standard process.

Web Coating Process

In metallization of flexible plastic web, the coating material is evaporated onto the web as it passes over multiple evaporation sources evenly spaced on a line normal to the direction of web travel. The spacings between the sources and the distance between the sources and the web are dictated by Equation (2) for a specified uniformity. The entire process is carried out in vacuum environment. A schematic of a basic roll-to-roll system is shown in Figure 3. Systems of this type accept plastic web in roll form ranging in diameter from 0.4 to 1.3 meter and in widths from 0.5 to 2.2 meters. The web travels from the unwind shaft and is guided around a chilled capstan roll where the metallic coating is applied and then onto the wind-up core at velocities up to 8 m/sec resulting in production rates of many thousands of square meters per hour.

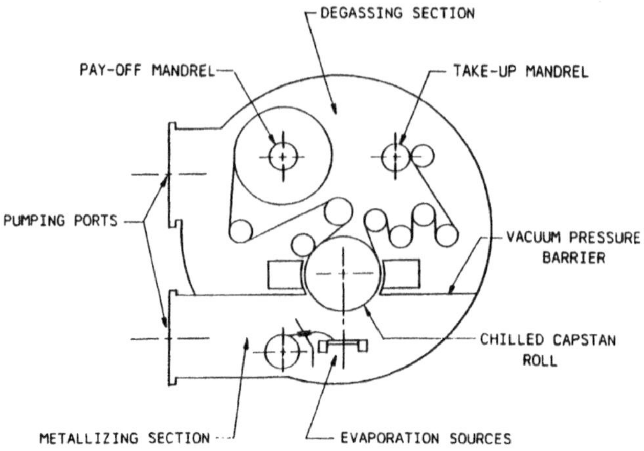

Figure 3. Schematic of a roll-to-roll high vacuum coater for paper and plastic films.

High production rate of web materials such as cellulose acetate and cellophane that have high rates of outgassing in vacuum is achieved by exposing only the small section receiving the metal coating to high vacuum conditions. The bulk of the material to be coated is separated from the coating zone by a pressure barrier in two zones which are pumped separately by a pumping system designed to pump high gas loads. The two small zones as shown in Figure 3 are located on each side of the capstan roll. They form regions of intermediate pressure and act as dynamic seals which prevent the reactive outgassing contaminants from passing into the coating zone and adversely affecting the coating process.

The most common method for depositing metallic coatings onto continuously moving web material involves direct evaporation from intermetallic sources. These sources can be either resistance-heated boats or RF induction-heated crucibles. Boat sources, used primarily to evaporate aluminum, typically have long shallow cavities and they must be used in conjunction with wire-feed mechanism to replenish the source material during long duration evaporation cycles. The physical properties and operating procedures necessary for successful boat operation are described by Parent[11]. RF induction-heated crucibles are hollow intermetallic cylinders which contain several kilograms of evaporant in chunk or billet form. One advantage of induction-heating is that the input energy is coupled directly into the metal evaporant charge; it is unnecessary to heat the crucible hotter than the vaporization temperature required to evaporate the charge. Vaporization temperatures obtained by induction-heating are such that films of aluminum, copper and chromium are readily deposited.

The thickness of the coating at any point across the width of the web can be estimated by[10]

$$w = 9.8 \times 10^{-3} G(t/s)(d/d_o) \qquad (4)$$

where

w = Thickness in μm
t = Residence time that the web is exposed to the evaporant (sec.)

s = Density of the evaporant material (g/cm^3)
d/d_o = The geometrical factor described by Equation (2).

Film thicknesses as a function of the evaporant temperature and the velocity of the traveling web calculated using Equation (4) are shown in Figure 4. These curves show that small changes in evaporant temperature will have a profound effect on the thickness of the deposited film. A small change in the temperature of a single source in the multiple source array will significantly affect the uniformity of the film thickness and a light or a dark band will occur in the direction of web travel.

One of the most significant factors in the vacuum metallization process is the control of the substrate temperature during coating cycle. Heat is transfered to the web by the radiant energy from the evaporation sources and the latent heat of condensation of the evaporant flux. This is particularly important in web coating process because the thin web material is very heat sensitive. For this reason, web coating systems require chilled capstan rolls to support the traveling web and remove the heat energy influx to the web. The capstan roll is jacketed for chilled fluid passage and is designed for a maximum rate of heat transfer to prevent excessive temperature rise in the web material.

Experience has shown that with precise control of the vacuum level, adequate pumping capacity for the removal of outgassing constituents, strategic location of the evaporant sources, and the proper substrate temperature, a high quality metallized product can be produced.

High coating production rates can be achieved by using electron beam sources in place of the tungsten filaments or boats. For example, deposition rates for aluminum evaporated from a 5-cm diameter crucible heated by an electron beam range from 1.1 to 7.95 μm/min with power inputs ranging from 4 to 22 kW. Deposition rates of approximately 60 μm/min have been achieved with a 12.7 cm diameter crucible having a power input of 40 kW.

Roll-to-roll coaters, as shown in Figure 3, are capable of producing 12 μm thick PET web material with an aluminum coating

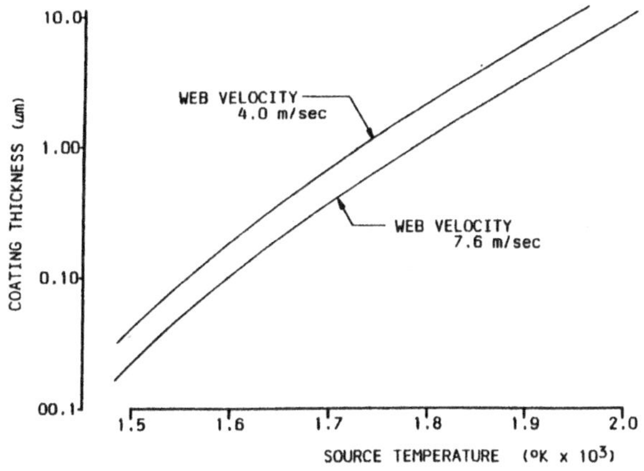

Figure 4. Calculated coating thickness as a function of source temperature and web velocity for aluminum.

approximately 130 Å thick (2 ohms/square) at an annual production rate of 450 tons, based on a 60 cm diameter roll of plastic web, 100 cm wide and 3,500 working hours per year[12].

The sputtering process is also used in coating web where oxide or nitride coatings, and multilayer coatings are deposited onto the web that is traveling at speeds up to 30 meters per minute.

Advantages and Disadvantages of Evaporation

The advantages of evaporation are:

(1) Deposition rate is very high.
(2) Equipment cost is relatively low compared to sputtering.
(3) Process is simple.
(4) Process uses simple basic forms of the source material.
(5) Large quantity of parts can be coated in a single batch.

The disadvantages of evaporation are:

(1) Limited number of materials can be deposited.
(2) The process is relatively hard to control.
(3) The process is basically a batch process.
(4) Replacement cost is high for the evaporating sources.

METALLIZATION BY SPUTTERING

Sputtering is one of the oldest method in the PVD category. It was first observed as far back as 1852 when an Englishman named Grove, noticed that in a normal glow discharge, cathode material was deposited on the surrounding surfaces. Immediately after its discovery, sputtering was widely used to deposit metals such as gold and silver for decorative and optical applications. It was the only method available at that time for the deposition of thin films. The discovery of vacuum evaporation some 50 years later caused sputtering to be discarded by workers in the thin film field because the coatings were found to be highly contaminated. Sputtering technology developed in recent years has not only significantly reduced the level of contamination in the sputter-deposited coating but also pushed the technology to a level surpassing evaporation technology in depositing more complex coatings.

Uniform coatings over large area substrates such as large plates of architectural glass and flexible polyester web material require substrate motion perpendicular to the cathode width. Thickness control then depends on the distribution of the magnetic field for magnetically enhanced sputtering. Variation in the magnetic field at points along the cathode length will result in changes in the erosion or sputtering rate of the source material along the cathode length. Therefore, the coating thickness distribution will vary along the width of the substrate. Most of the erosion takes place in the "race track" area defined by the magnetic field on the cathode surface. As the erosion proceeds, the bombarding ions are focused into an increasingly narrow region resulting in a V-shaped erosion profile. The end of source material usefulness occurs when the erosion groove penetrates to the backing plate.

Typical deposition rates for coatings of aluminum and copper deposited by magnetically enhanced sputtering from a target source

12.7 cm wide x 43.2 cm long are approximately 2.1 to 3.84 µm/min, respectively, with an applied power of 13 kW.

With all the improvements in deposition rates and flexibility in coating material, and shape of substrate, sputtering is becoming more attractive as a high volume production process. It is very versatile in terms of the range of materials that can be deposited. All conductive materials are deposited with ease. The absence of fractionation and dissociation effects and the ability to synthesize compound films by reactive sputtering allow the preparation of many complex alloys, carbides, nitrides, borides, and oxides, many of which are difficult or impossible to deposit by other means. Sputtering is a very stable process which is capable of depositing uniform coatings for long period of time. These unique characteristics make the sputtering process cost effective in the high volume production. Products can be in large-area-sheet form, three-dimensional form, or flexible web form. Multilayer coatings requiring precise control of the structure and stoichiometry can also be deposited on glass substrates or flexible polyester web material in roll-to-roll vacuum coating systems. Sputtering is also used to coat three-dimensional objects by placing two cathodes opposing each other. Sputtering has replaced metallization by evaporation in many applications.

Sputtering Mechanism[13,14,15]

Sputtering is a phenomenon in which atoms are ejected from the surface of a material being bombarded by ions of sufficient energy. Atoms are ejected from the surface of the target material during ion bombardment in a manner similar to that of an array of billiard balls which have been struck by the cue ball. The end result of the ejection of a massive number of atoms is the erosion of the target surface or source material. In other words, the target material is sputtered. The sputtered atoms travel to any nearby surface or substrate forming a coating. Atoms of almost any material can be transferred in this manner resulting in the ability to construct coatings atom-by-atom. Sputtering is not an evaporation process, therefore, the source material does not have to be heated to its boiling point; this makes it a low temperature deposition process compared to vacuum evaporation.

The number of atoms sputtered by each ion is defined as the yield of the material and is dependent on the energy, the angle of incidence and the mass of the ion, and the type of material being sputtered. For instance, the yield of chromium is 1.3 for an argon ion with an energy of 600 eV and striking the cathode at 90° angle. This means that 1.3 chromium atoms are sputtered by one argon ion which has a kinetic energy of 600 eV. Yield curves for different materials with respect to ions and energies are available in current literature [16-19]. Yield data are used to develop an understanding of the sputtering process and to estimate the sputtering rates of different materials for a given set of conditions. The ions used most frequently in sputtering applications are inert gas ions because of their non-reactive property. Argon is the most common gas used because of its large mass and relatively low cost.

The basic sputtering process, prior to the development of the magnetically enhanced cathode, was diode sputtering as shown in Figure 5. Inert gas (usually argon) bled into the vacuum chamber and present in the region between the anode and cathode is ionized to generate electron-ion pairs. Ionization occurs when a potential is applied between the two electrodes and a luminous gaseous discharge is

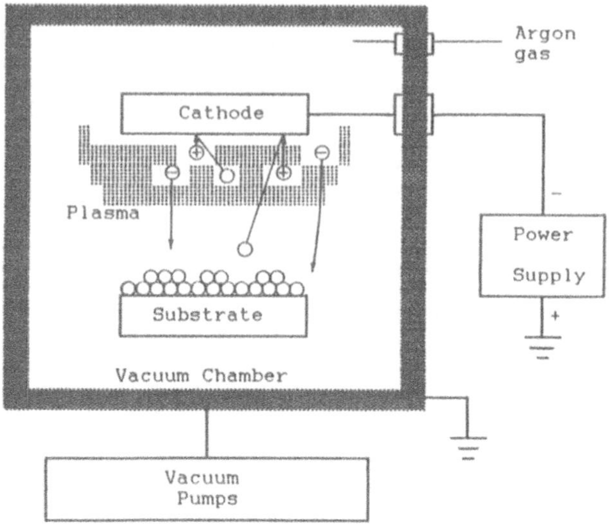

Figure 5. Schematic of basic diode sputtering configuration.

ignited. Electrons are attracted to the anode and the ions are attracted to the cathode. Due to the large mass and energy of the gas ion impacting on the cathode target surface, sputtering of the cathode material occurs. In diode sputtering, the discharge is classified as an abnormal glow discharge which is characterized by low current and high voltage at pressures higher than 2.67 Pa. In the early days of process development, the maximum power that could be applied to a given size cathode was relatively low and power density was usually less than 0.5 W/cm^2 resulting in a relatively slow process.

Magnetically Enhanced Sputtering

The development of the magnetron cathodes[20,21] changed sputtering from a low rate deposition process used primarily in R&D and limited production to a high rate deposition process suitable for many production applications. Planar magnetron cathode uses magnetic field to trap secondary electrons emitted from the cathode surface. Figure 6 shows a diagram of the cathode construction and the general direction of the magnetic field lines. Electrons are trapped in the region close to the cathode surface. In this region, electrons are compressed to a high concentration by the magnetic field and therefore, the mean free path between the electron and the gas atom is greatly reduced. Hence, more ionizing collisions take place. In addition to the above, electrons residence time in this region is much longer due to the entrapment, enabling them to make more ionizing collisions before they are lost to the chamber walls. The result of high concentration, shorter mean free paths and longer residence time of the electrons is a massive increase in ion density available to bombard the cathode. Due to high ion density, the discharge current increases dramatically compared to the diode sputtering. Hence, the cathode operates at much higher power density and high sputtering rate is achieved. Substrate temperature during deposition is also reduced because the substrate is subject to significantly less secondary electron bombardment which is the result of electron entrapment by the magnetic field.

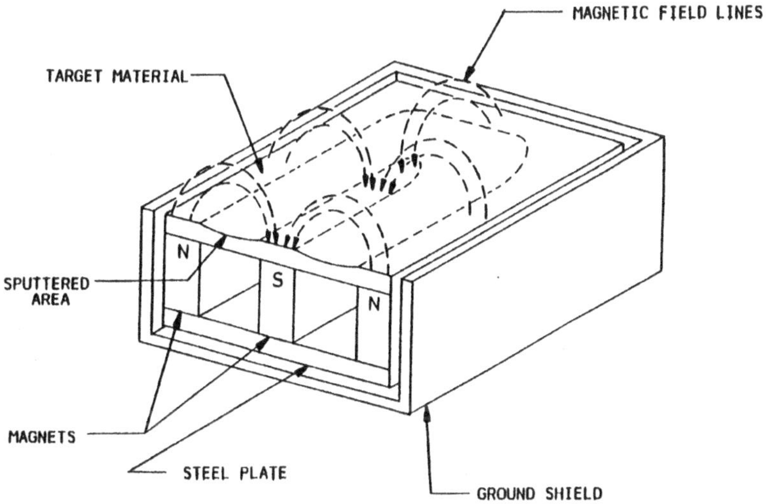

Figure 6. Schematic of magnetically enhanced cathode structure.

Sputtering Process Parameters

Process parameters which control the resulting film properties are:

* Sputtering pressure and gas flow rate
* Sputtering voltage, current and power
* Deposition or sputtering rate
* Cathode-substrate distance
* Substrate temperature.

The sputtering pressure determines the characteristics of the glow discharge and therefore, the current-voltage relationship of the sputtering process. Higher pressures will increase the sputtering current for the same voltage which results in increased sputtering power. It is important to determine the proper argon pressure to achieve the desired coating properties. Vossen and Cuomo[22] describe the major sources of gaseous impurities in the sputtering system as residual gases (such as H_2O) remaining after the initial pumpdown, wall desorption due to bombardment, occluded gases in the sputtering source material, and backstreaming of pump fluids. The presence of oxygen in the sputtering chamber significantly reduces the sputtering rate for most materials. Jones et. al.[23] have shown the sputtering rate of SiO_2 to decrease by approximately 40% with added oxygen compared to sputtering without oxygen. To reduce contamination from the residual gas, it is desirable to use the highest possible gas flow rate so that impurities from the chamber walls can be swept out of the sputtering chamber. In preparing compound films like oxides or nitrides, a reactive gas such as oxygen or nitrogen is mixed in with argon so that the sputtered atoms will react to form metal oxide or nitride coatings. In a reactive sputtering process, it is important that the pumping system is designed for maximum pumping speed in order to achieve the maximum gas flow rate.[24]

The current and voltage during sputtering determine the glow discharge characteristics and the sputtering power and therefore, the sputtering rate. These parameters are generally monitored during sputtering to alert the operator to any process changes which can result in irregularities in the thickness and microstructure of the resultant coating. Deposition rate is directly proportional to the sputtering rate and is an important factor in the formation and properties of the coating. It is also very important in high volume production operations because it determines the production rate, equipment costs, and manufacturing cost.

Cathode-to-substrate distance determines the deposition rate for a given sputtering power. An increase in the cathode-to-substrate distance results in a decrease in the deposition rate. In order to deposit a uniform coating over a large area, the cathode length must be greater than the width of the substrate by an amount determined by the cathode-to-substrate distance. The cathode-to-substrate distance should be minimized in order to take advantage of the increased deposition rate. However, when the substrate is close to the cathode, higher substrate temperature may result. There is a minimum distance below which melting or deformation of substrate will occur when low temperature materials such as plastics are being coated. Another significant effect of shorter substrate-to-cathode distance is that when the distance is less than the minimum, the substrate passing by the cathode will impede the flow of argon gas causing the sputtering conditions to change drastically and the process may become unstable. This may cause coating uniformity problems if not well controlled.

In diode sputtering, the coating and the substrate are constantly bombarded by secondary electrons emitted from the cathode unless the substrate is shielded or negatively biased. Substrate temperatures during diode sputtering range from 300°C to 400°C depending on the power input to the cathode. However, in the magnetically enhanced sputtering process, secondary electrons are mostly trapped in the magnetic field close to the cathode, the substrate temperature is lower and ranges from 90°C to 150° C. Electron bombardment of the substrate is advantageous in many applications, but for heat sensitive materials such as organic polymers, the heating of the substrate due to this type of electron bombardment must be reduced. The process characteristics of high deposition rate with lower substrate temperature make the magnetically enhanced sputtering process a viable process for coating heat sensitive materials such as polyester web.

Morphology and Process Parameters

Substrate temperature affects both the sticking coefficient (defined as the ratio of the number of atoms adhering to the substrate to the number of atoms striking the substrate) and the coating structure. During the sputtering process, argon atoms frequently become entrapped in the coating. For some applications, this may be as high as two atomic percent. Argon entrapment decreases as the substrate temperature is increased because the sticking coefficient is reduced. A schematic representation of the influence of substrate temperature and argon working pressure on the microstructure of metallic coatings as derived by Thornton[25] is shown in Figure 7. T is the substrate temperature and T_m is the melting point of the coating material in absolute degrees. The T/T_m dependence of the coating morphology shown in Figure 7 results from surface diffusion processes, similar to bulk diffusion, which indicates recrystallization.

Pressure dependence is the result of collisions between the sputtered atoms and the argon atoms at elevated pressures. Coatings in the Zone T transition region have dense fibrous structures with smooth, highly reflective surfaces and are desired for most decorative applications. Coatings in the Zone 1 region have a structure consisting of tapered crystals separated by open voided boundaries.[26] This structure is caused by high partial pressures of argon during sputtering, a rough finish on the substrate surface, or a surface at a large angle relative to the coating flux. The extreme Zone 1 structures result in poor reflecting surfaces. Zone 2 is characterized by columnar dome-shaped grains separated by distinct dense intercrystalline boundaries and Zone 3 consists of equiaxed grains resulting from bulk diffusion processes. The melting points of metal and cermet coatings relative to the temperature of plastic

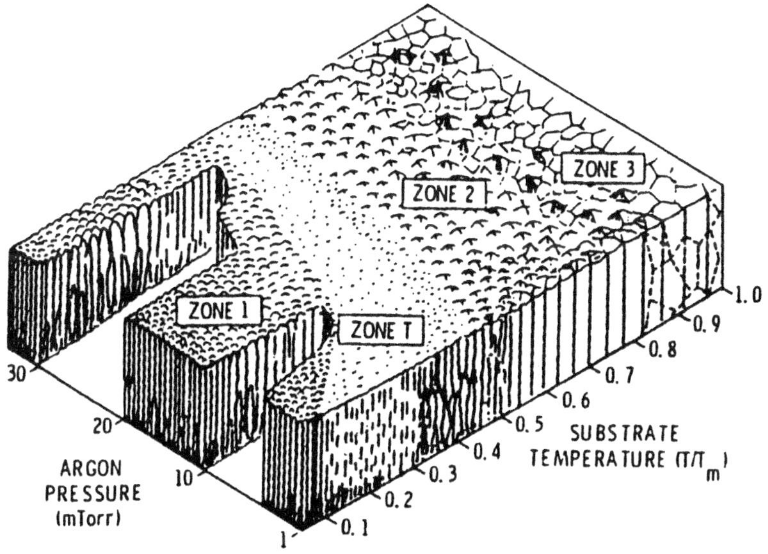

Figure 7. Schematic representation of the influence of substrate temperature and argon pressure on the structure of metal coatings.

substrates during coating are such that the characteristics of Zone T coatings is readily achieved when argon pressures is less than 0.67 Pa.

Coefficients of thermal expansion of plastics are greater than those of most metals by a factor of 10. Thornton and Hoffman[26] have shown that in many applications, the temperature differences between the conditions of deposition and application result in compressive thermal stresses that can exceed the yield strength of the coating material and are capable of fracturing relatively strong coating-to-substrate bonds. This typically occurs when metal coatings are deposited to thicknesses greater than 0.2 µm. Conversely, Chambers et. al.[27] have shown that boride coatings approximately 37.5 µm thick deposited on chromium intermediate coatings which have been deposited on fiber impregnated polyimide substrates had excellent adhesion and resistance to erosive particles suggesting that the density of coatings in a state of compressive stress is increased so that the occurrence of coating defects is minimized.

Advantages and Disadvantages

The advantages of magnetically enhanced sputtering are:

1. Multilayer and alloy films can easily be deposited.
2. Refractory materials can be deposited.
3. Oxide and nitride films can be readily deposited by reactive sputtering.
4. Good adhesion is readily obtained.
5. Thickness uniformity over large areas is easily obtained.
6. Deposition rates for most metals are greater than 2,000 angstroms per minute.
7. Large quantities of parts can be produced in a cost-effective manner.

The disadvantages are:

1. The source material generally must be in sheet form.
2. The substrate must be cooled unless low deposition rates or short runs are acceptable.
3. Capital equipment cost is higher relative to other coating processes.

SUMMARY

The development of physical vapor deposition technology has resulted in high volume production processes for the deposition of high quality coatings. The metallization of plastic materials by vacuum evaporation has replaced the electroplating process in many applications.

The magnetically enhanced sputtering cathode assembly has proven to be a cost-effective source. A wide range of materials can be deposited over large areas at rates acceptable for high volume production. This process is used in coating three dimensional parts and flexible web. In recent years, its use has expanded to the aerospace industry for depositing corrosion and erosion resistance coatings on airplane jet engine and rocket engine polyimide composite parts.

ACKNOWLEDGMENT

The authors would like to express their appreciation to Mr. George E. Stephens of Galileo Vacuum Systems, Somers, CT for providing information relative to current flexible web coating technology.

REFERENCES

1 W. J. Miller in "Decorating Plastics," J. M. Margolis, Editor, p. 77, Hanser Publishers, New York, 1986.
2 Papers in "Metallized Plastics 1: Fundamental and Applied Aspects," K. L. Mittal and J. R. Susko, Editors, Plenum Press, New York, 1989.
3 K. L. Mittal, Editor, "Polyimides: Synthesis, Characterization

and Applications," Vols. 1 and 2, Plenum Press, New York, 1984.
4 F. S. Ohuchi and S. C. Freilich, J. Vac. Sci. Technol. A6, 1004 (1988).
5 F. K. LeGoues, B. D. Silverman, and P. S. Ho, J. Vac. Sci. Technol, A6, 2200 (1988).
6 R. Haight, R. C. White, B. D. Silverman and P. S. Ho, J. Vac. Sci. Techno. A6, 2188 (1988).
7 D. S. Dunn and J. L. Grant, J. Vac. Sci. Technol. A7, 253 (1989).
8 R. Glang in "Handbook of Thin Film Technology," L. I. Maissel, and R. Glang, Editors, p. 1-56, McGraw-Hill, New York, 1970.
9 F. Grazzini, G. Tonino, and N. Merlini in "Paper Film and Foil Converter", Maclean Hunter Publishing Co., Chicago, IL, February, 1980.
10 R. Glang, see Reference 8, p. 1-27.
11 E. D. Parent in "Proceedings 21st Annual Technical Conference", Society of Vacuum Coaters, p. 8, 1978.
12 Web coating system specification of Officine Galeleo, High Vacuum Technology Division, Florence, Itali, 1986.
13 E. Kay in "Advances in Electronics and Electron Physics," Vol. 17, p.245, 1962.
14 G. K. Wehner and G. S. Anderson, see Reference 8, p. 3-1.
15 J. L. Vossen and J. J. Cuomo in "Thin Film Processes," J. L. Vossen and W. Kern, Editors, p. 11, Academic Press, New York, 1978.
16 G. K. Wehner, Phys. Rev. 108, 35 (1957).
17 G. K. Wehner, Phys. Rev. 112, 1120 (1958).
18 N. Laegreid and G. K. Wehner, J. Appl. Phys. 32, 365 (1961).
19 D. Rosenberg and G. K. Wehner, J. Appl. Phys. 33, 1842 (1962).
20 J. A. Thornton and A. S. Penfold, see Reference 15, p. 75.
21 R. K. Waits, see Reference 15, p. 131.
22 J. L. Vossen and J. J. Cuomo, see Reference 15, p. 47.
23 R. E. Jones, H. F. Winters and L. I. Maissel, J. Vac. Sci. Technol. 5, 84 (1968).
24 A. S. Penfold in "Proceedings of the 29th Annual Technical Conference," Society of Vacuum Coaters, p. 381, 1986.
25 J. A. Thornton, J. Vac. Sci. Technol. 11, 666 (1974).
26 J. A. Thornton and D. W. Hoffman in "Proceedings 22nd Annual Technical Conference", Society of Vacuum Coaters, p.5, 1979.
27 D. L. Chambers, C. T. Wan, K. A. Taylor and G. T. Susi, J. Vac. Sci. Technol. A8, 2113 (1990).

ION BEAM ASSISTED METALLIZATION OF PLASTICS

Th. Flottmann and W. Lohmann

Akzo Research Laboratories Obernburg
Postfach
D-8753 Obernburg/Main, West Germany

Ion beam techniques suitable for generating electrical conductivity inside or on polymers include (a) ion implantation leading to intrinsic conductivity, (b) ion beam mixing of pre-deposited metallic layers, and (c) ion beam assisted deposition of metallic films. These processes are reviewed with regard to basic principles, main parameters, and typical examples. Possibilities and limitations of each method are also presented.

INTRODUCTION

The industrial use of ion beam methods, which originally was restricted to semiconductor processing, is to an increasing extent finding other applications because of their specific, sometimes unique properties. Omitting sideline activities, the major fields for this technology are:

- Improvement of <u>tribological</u> (friction, wear) and <u>corrosion behaviour</u> of metallic materials, and

- Generation of <u>electrical conductivity</u> inside insulators (plastics, ceramics) or <u>metallization</u> of such materials.

Only the latter field will be treated in this paper. For a review of the tribological applications see, for example, ref. 1.

Unlike other physical surface treatments, the direct injection of ions (<u>ion implantation</u>) (see fig. 1) leads to an intrinsically conductive surface via a carbonization or a doping mechanism (cf. Section 1). Although only graphite-like conductivity is achieved, there are distinct advantages, because the typical adhesion problem occurring at the metal-insulator interface does not arise.

Two options are available in using ion beams for "real metallization" of insulating materials. <u>Ion beam mixing</u> (IBM) (see fig. 2) can improve the adhesion of a thin, pre-deposited metallic layer applied, e. g., by physical vapor deposition (PVD)-methods. The effect is an interface broadening induced by the IBM process. A

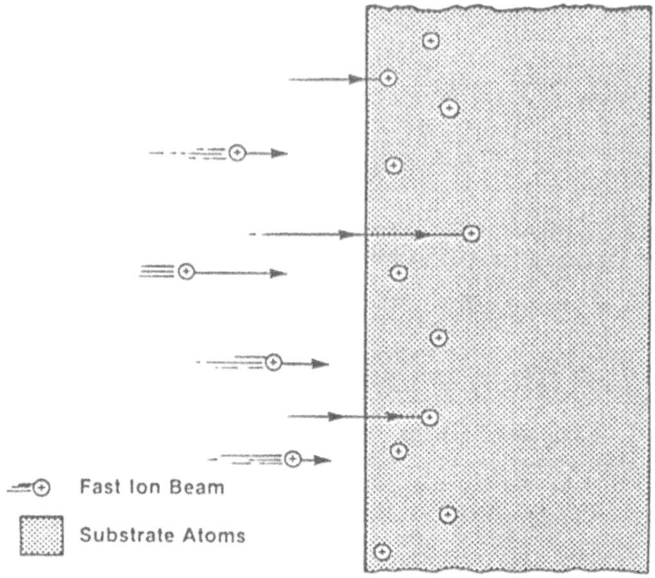

Figure 1. Principle of ion implantation

special IBM method combines the reduction of metallic precursors by ion beam lithography and subsequent interface mixing. More details of such IBM techniques are given in Section 2.

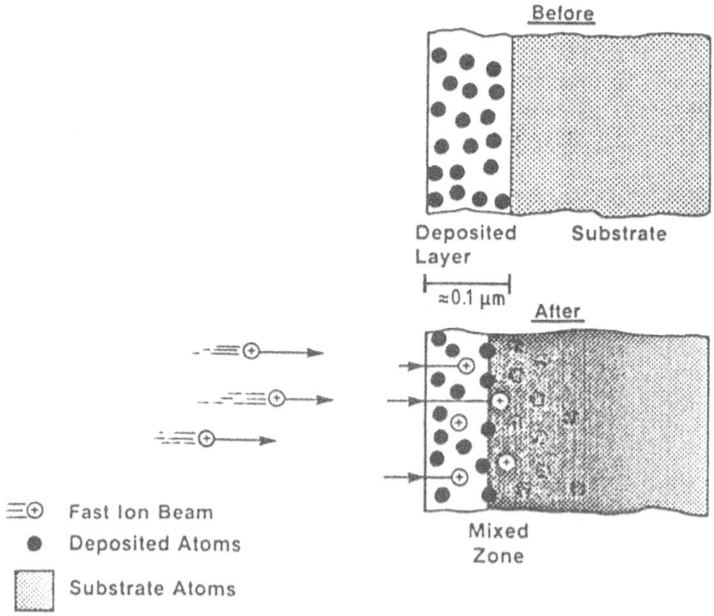

Figure 2. Principle of ion beam mixing

One major drawback of IBM, the thickness limitation of the metallic starting layer due to the ion beam range, is overcome by the ion beam assisted deposition (IBAD) (see fig. 3) of metal layers onto an insulating surface. This method combines PVD and ion beam treatment, resulting in a growth of the metal deposit under simultaneous ion bombardment. As will be discussed in Section 3, an enhanced adhesion between metal and insulator is also observed in this case.

The ion beams required for the processes mentioned above are supplied either by ion implanters as used in the semiconductor industry or by ion beam sources. The choice is governed by the preferred ion beam process. Since a sufficient ion range (> 0.1 μm) and relatively high ion doses (up to 10^{16} - $10^{17}/cm^2$) are essential to both ion implantation and IBM, medium-energy, high-current ion implanters are best suited for this purpose. Modern equipment can provide a mass-separated beam up to a current of 10 mA at energies between 5 and 200 keV. The achievement of even higher ion beam currents is possible since a mass-analyzed beam is often not necessary. This also holds for IBAD processes for which non mass-separated, low energy (< 50 keV) ions at high current (100 mA) from Freeman- or Kaufman-type ion sources can be used. A set-up for ion beam assisted metallization is schematically shown in fig. 4.

In the following sections, the principles of the three metallization methods will be explained and typical examples will be given. Further examples can be found in ref. 2.

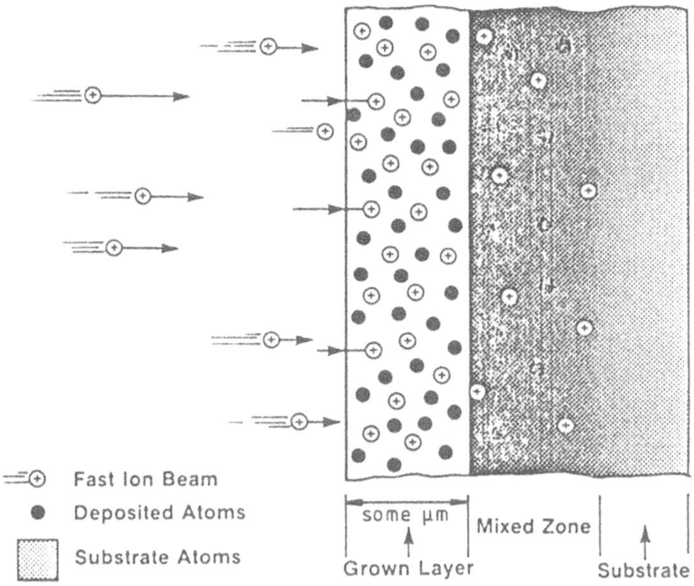

Figure 3. Principle of ion beam assisted deposition

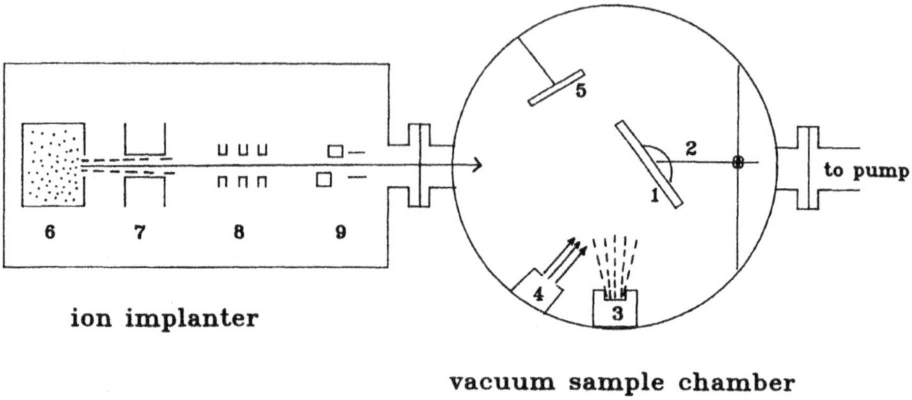

Figure 4. Schematic set-up of equipment for ion beam assisted metallizaton; 1 = sample holder, 2 = sample manipulator, 3 = evaporation device, 4 = ion source, 5 = sputter target, 6 = ion source, 7 = mass separator, 8 = collimator, 9 = x-y-scan.

1. ION IMPLANTATION

When a fast ion beam strikes the surface of a material, ions penetrate deep into the substrate. The range of penetration depends on their energy. Ion energies in the keV-MeV range correspond to ion ranges from about 100 nm to several μm. Before coming to rest, the ions lose their energy by nuclear collisions and electronic excitations (calculations can be performed, e. g., with the help of ref. 3). The share of both forms of energy deposition depends on ion species, ion energy, and the substrate material. In general, heavy ions experience more nuclear collisions near the end of their track than do light ions. At high energies (>> 100 keV), electronic energy loss prevails for most ion species.

There are two mechanisms by which ion implantation can induce conductivity in polymers: carbonization and doping.

Nuclear collisions of ions with polymer molecules can lead to a destruction of the polymer backbone molecules (radiation damage) [4,5]. Volatile fractions are expelled from the surface and a carbon-enriched surface layer is left behind. Such a layer shows graphitic conductivity.

Electronic interactions of ions with the polymer also cause radiation damage. They give rise to chain scission and cross-linking [4,5]. In addition, they can result in a doping effect, similar to that in semiconductors. Depending on the chemical nature of the polymer matrix, the implanted ions induce mobile charges that contribute to the conductivity of the polymer surface layer. The achievable conductivity is in general lower than that of graphite. However, semiconductor properties can be achieved in this way, see, for example, ref. 6.

The typical ion doses needed to achieve conductivity range from about 10^{14} to $10^{17}/cm^2$. The ion species used are mostly noble gases (for carbonization) or chemically reactive ions (for doping).

Example 1

In fig. 5, the effect of ion implantation on the surface conductivity of Kapton^R PMDA-ODA polyimide is shown[7]. Ar ions at 190 kV acceleration voltage damage the polyimide surface mostly by nuclear collisions. The polymer is degraded and carbonizes with increasing ion dose. As can be seen, the specific resistivity of the surface layer decreases by several orders of magnitude. At doses higher than $10^{16}/cm^2$, the saturation conductivity of graphitic

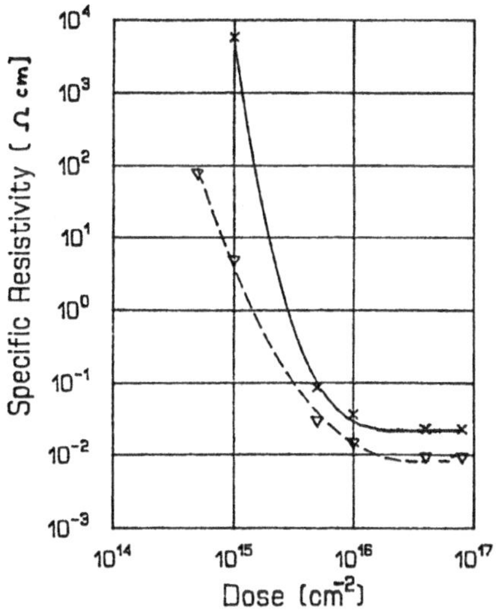

Figure 5. Specific resistivity of ion implanted Kapton polyimide as a function of Ar^+ dose[7]. U = 190 kV, I = 3.7 $\mu A/cm^2$. (x) implanted, (∇) annealed.

material is reached. Systematic experiments[7] with different ion species, ion currents and energies have shown that the absorbed power density in the polyimide is the most relevant parameter for the induced specific conductivity. Post annealing in vacuum at 350 °C further lowers the resistivity by a small amount, probably due to defect healing in the polymer. However, exposure of the irradiated polymer to normal environment again gives rise to higher resistivities (after some months).

Example 2

Fig. 6 gives an example of a doping effect by ion implantation[8]. Polyphenylenesulfide had been implanted with different ion species at high energies in the MeV range. Fluorine and iodine ions clearly induce an increase of the specific PPS conductivity by several orders of magnitude, whereas lithium ions do not seem to have a large effect. With iodine ions, conductivity can be achieved already at relatively low doses (< $10^{15}/cm^2$). The cited conductivities are lower than that of graphite. The authors of the cited paper conclude that the ion induced conductivity increases primarily with the electronic energy loss density of the ions.

Example 3

Amorphous polyetheretherketone (PEEK) films have been implanted with a variety of ions (He, N, F, As, Xe and I) in the energy range 50 keV to 32 MeV[9]. At the low end of this range, the light ions (He and N) show a different dependence of the electrical conductivity of the PEEK on dose than to the heavy ions. The authors of the cited paper attribute the higher effeciency of the heavy ions in enhancing the conductivity to their higher electronic stopping powers. Implantations in the MeV energy range yielded a surface layer on the PEEK with a high conductivity (up to 2.5 $(\Omega\ cm)^{-1}$).

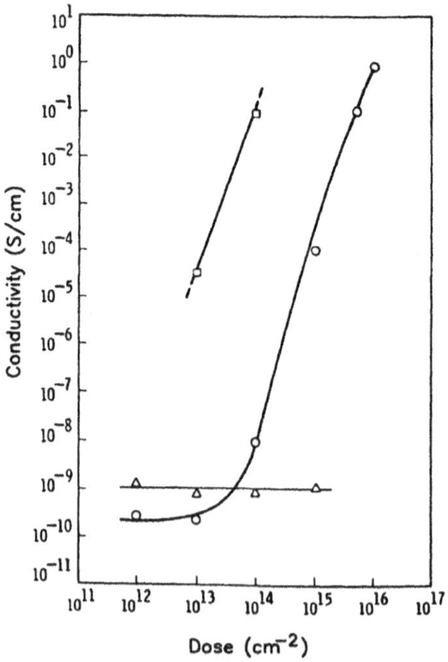

Figure 6. Specific conductivity of ion implanted polyphenylenesulfide as a function of ion dose[8]. (O) 5.6 MeV F, (Δ) 0.32 MeV Li, (□) 50 MeV I

The advantages of ion implantation in polymers can be summarized as follows: The doping depth can be controlled through the ion energy. There is no problem of adhesion of the metallized layers, as they form an integral part of the polymer surface and no mechanically weak interface is present. When using masks or focused ion beams (direct writing), a spatial structuring of the surface conductivity is possible.

There are also disadvantages: The specific conductivity which can be induced is relatively low. Metallic conductivity has not been achieved yet. There are problems with the stability of the modified polymer surface layers.

2. ION BEAM MIXING

To achieve metallic conductivity on polymers, a metallic layer has to be deposited onto the surface. In the ion beam mixing technique, this layer is bombarded with fast ions which penetrate to the metal/polymer interface. At the interface, they deposit energy in form of nuclear collisions or electronic excitation. This can cause a broadening of the interface and also induce chemical interactions between metal atoms and polymer molecules. In both cases, an increased adhesion of the metal film may result.

The metal films are typically deposited in the same vacuum equipment by PVD methods. Since the mixing ions have to reach the interface, the thickness of these pre-deposited films has to be adjusted to the ion range. Typical thicknesses are in the order of several hundred nanometers to a few micrometers for ion beams of 100 keV to a few MeV energies. The ion species used are normally noble gases. For IBM, ion doses may be chosen at least one order of magnitude lower than for ion implantation, i. e. 10^{13} - $10^{16}/cm^2$.

The conductivity of thin metal films generally depends on the specific PVD process mode which is used for deposition (physical properties of metal films are given, for example, in ref. 10). In most cases, it is below the bulk value of the metal, but higher than graphitic conductivity. These metal films can serve well as a basis for electrochemical plating. After thus increasing their thickness to several micrometers, their conductivity may approach that of bulk metal.

Example 1

Fig. 7 shows the effect of IBM on thin Ti films on Kapton^R PMDA-ODA polyimide [11]. The evaporated Ti films have been irradiated with He-ions at various energies and doses. Whereas evaporation alone results in poor adhesion, a clear improvement is seen after ion irradiation. The maximum in adhesion is observed for a dose of about $10^{14}/cm^2$. With Ar^+ions, this effect could not be found, the polyimide surface carbonized. Fig. 8 shows the specific energy loss of He^+ and Ar^+ions for comparison. The values were obtained from ref. 12. As can be seen, the electronic energy loss is the dominant mechanism for He^+ions. For Ar^+ions, the nuclear collisions contribute to a much larger extent. Also, the total deposited energy density is larger and more radiation damage is expected. From this it can be speculated that the adhesion enhancement of the Ti films is induced by an electronic excitation process.

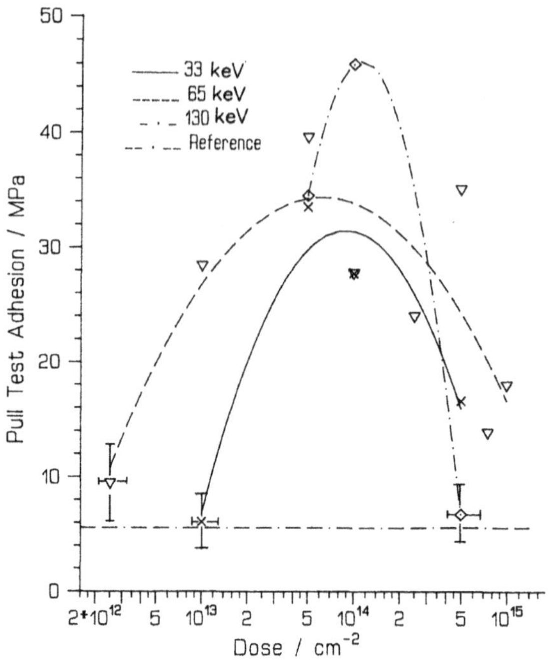

Figure 7. Adhesion of ion beam mixed Ti film on Kapton polyimide as a function of He^+ dose[11]. Ti film thickness: 80 nm.

Example 2

The adhesion of 30 nm Ni films on glassy carbon (GC) was examined as a function of 170 keV Kr^+ implantation at doses of $5 \cdot 10^{15}$, $1 \cdot 10^{16}$ cm^{-2} and $5 \cdot 10^{16}$ cm^{-2} through a 7 nm Si intermediate layer[13]. Auger electron spectroscopy depth profiling showed that the as-deposited Ni/Si/GC specimens consisted of a rather sharp Ni/GC interface with the Si dispersed into both the Ni and the GC. Kr^+ implantation caused the Si intermediate layer to be mixed extensively into both the Ni film and the GC substrate. Also, the GC and Ni were intermixed. X-ray photoelectron spectroscopy revealed different types of chemical bonding (Ni-Si, Si-C and Si-O) which were extensively intermixed after ion implantation. The adhesion of the films was examined using a scratch test. The as-deposited films were removed at forces less than 1 N. In contrast, with increasing implantation dose, the critical force for film removal was increased up to 45 N at $5 \cdot 10^{16}$ cm^{-2}. This adhesion increase was primarily attributed to the graded interfacial bonding produced by the ion implantation.

Example 3

Spatial structuring of ion beam mixed metal films is achieved with masking techniques. Fig. 9 presents an example[14]. A metal film evaporated onto a substrate is bombarded through a mask. Only the ion irradiated structures adhere well to the substrate, whereas the other parts can be removed afterwards, e. g. by a wet etching process. A thickening of the adherent metal layers is possible with electrochemical plating processes.

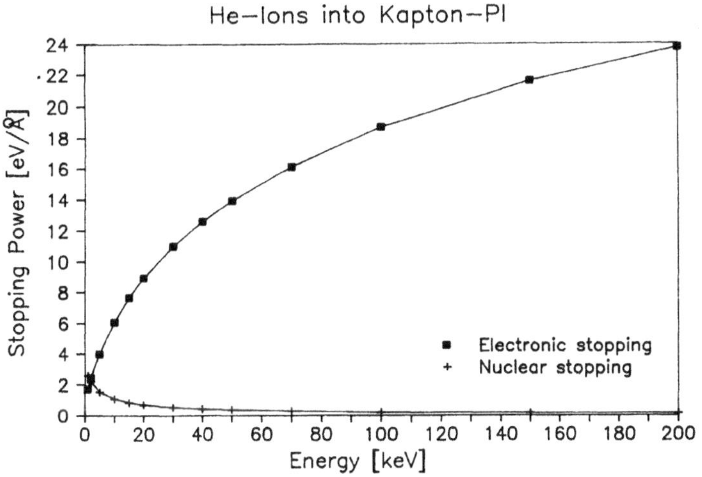

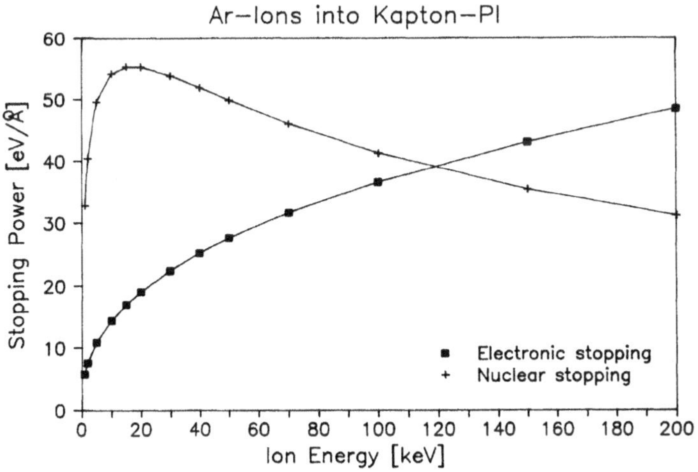

Figure 8. Specific energy loss of He and Ar ions in Kapton polyimide as a function of ion energy[12].

Example 4

Another variant of the ion beam mixing technique which combines with wet chemical processing is shown schematically in Fig. 10. First, a metalorganic compound is deposited from solution on a polymer substrate, e. g., by spin coating. This precursor film is irradiated through a mask with ions. The ions act in a twofold way: firstly, they dissociate the precursor compound. The organic residue evaporates or can be washed off with a solvent. In addition, they promote adhesion of the metal to the polymer by interface mixing. If the resulting thin film contains a noble metal, this can be used as a catalyst for subsequent Cu plating in an electroless plating bath [15]. Eskildsen and Sorensen have demonstrated this first with a Pd-acetate system[16].

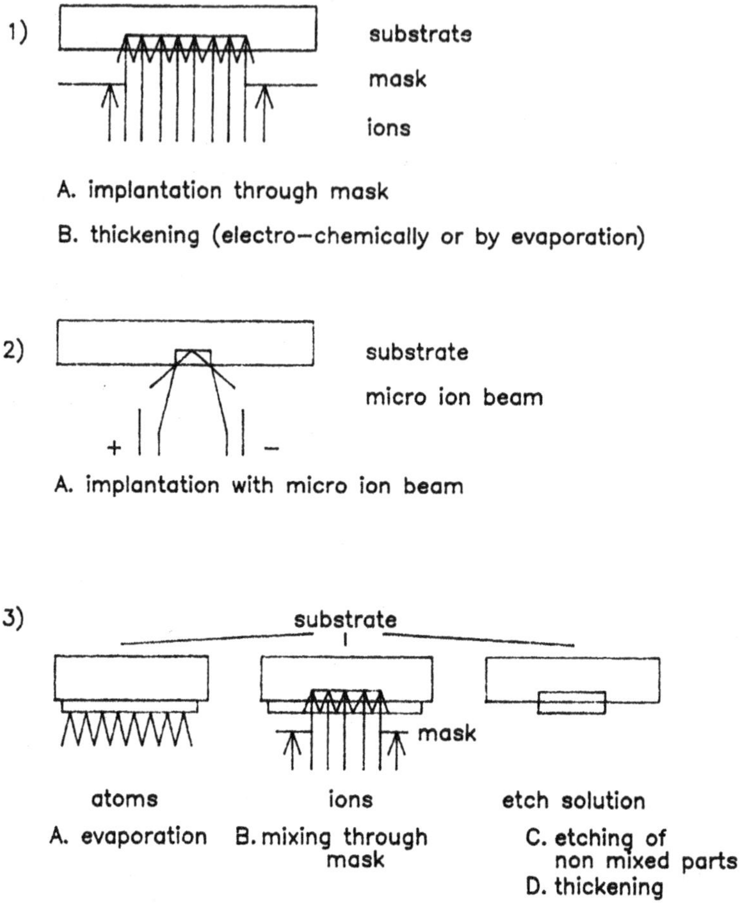

Figure 9. Schematic representation of different ion beam processes for the generation of structured metallization on plastics[14].

The advantages of ion beam mixing techniques include: Lower doses can be used in comparison to ion implantation. Spatial structuring of the metal films is possible. Compatibility with wet chemical processing is given.

On the other hand, radiation damage of the ion irradiated polymer surface can occur. Because of the requirement that the ions penetrate to the interface, only thin metal films can be deposited. These can be made thicker, however, by electrochemical methods.

For completeness, it should be mentioned that ion pre-bombardment before PVD metallization can for some polymers be equally effective in adhesion enhancement as ion beam mixing[17]. This does not set a limit to the film thickness.

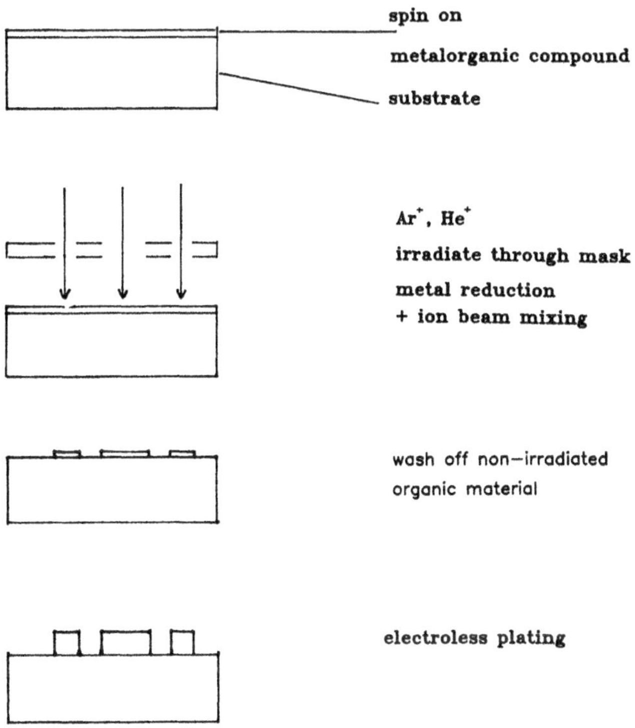

Figure 10. Schematic of ion beam catalysis technique for electroless plating[15].

3. ION BEAM ASSISTED DEPOSITION

In this deposition technique, metal film deposition and ion bombardment are carried out simultaneously. The metal is again vaporized by PVD (thermal or electron beam evaporation, sputtering). At the same time, ions from an ion source irradiate the growing metal film from a different position in the same vacuum chamber. When the first atomic layers of the metal are deposited onto the polymer surface, the arriving ions perform the interface mixing, as in the case of the (static) ion beam mixing technique. After the film has grown thicker, the bombarding ions no longer penetrate to the interface. Continuing ion irradiation now affects the growth of the metal thin film. Depending on ion species, ion energy, and ion/atom arrival ratio, microstructural film parameters such as crystal structure, inherent stress and density can be influenced. In principle, this allows the growth of dense and stress-free metal films. Film thicknesses are not limited in this deposition technique and layers several μm thick can be produced. Noble gas ions from high current ion sources are used in order to achieve high deposition rates. Broad beam Freeman ion sources (5 - 50 keV) and Kaufman ion sources (< 5 keV) provide currents ranging from a few mA to several hundred mA. The ion/atom arrival ratios can typically be chosen from 1 : 10 to 1 : 1000.

Example 1

Ion beam assisted deposition of copper has been performed on silicon substrates. The effect of argon ion bombardment on the structure and properties of 5 μm thick copper films was studied[18] (No such study has been carried out on a plastic substrate, to our knowledge). Primary deposition variables were ion flux, ion energy and substrate temperature. The ion flux was varied to produce an equivalent energy flux of 0.2 - 100 eV/atom. The copper films were deposited by electron beam evaporation and simultaneous irradiation from a Kaufman source under 22 ^{o} incidence. At 600 eV ion energy, the resistivity of films deposited between 60 ^{o} and 103 ^{o}C rose steeply with ion flux, reaching values > 12μΩcm at high flux levels. At 230 ^{o}C the resistivity is much lower, approaching bulk values at low ion flux. In contrast, films deposited at 62 and 125 eV argon energy showed only a small increase in resistivity from 2 to 2.6μΩcm. On silicon substrates exposed to 600 eV ions, a substantial decrease in tensile stress was seen with increased ion flux. At 600 eV, also a large decrease in crystallite size occurs with increasing ion flux, whereas at 62 eV only little change in size was observed.

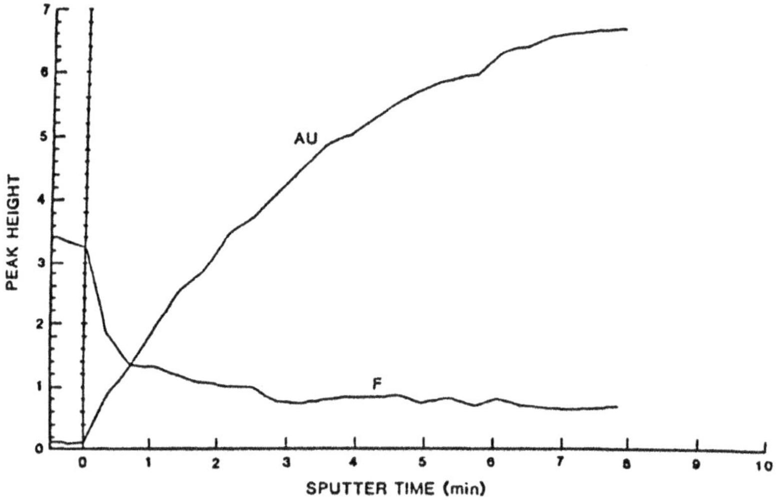

Figure 11. ESCA depth profile of ion beam mixed Au/Teflon interface[19].

Example 2

Metallization of fluoropolymers such as Teflon is difficult, because the polymer surface is chemically inert. However, with ion beam assisted deposition technique, an adherent metallization layer can be obtained. For example, Cu, Ag and Au metals have been evaporated onto Teflon with simultaneous irradiation using 400 eV

Ar^+ions[19]. Compared to non-irradiated specimens, the adhesion of the metal films was good, as evaluated with the Scotch Tape test and a scratch test.

Fig. 11 shows an ESCA sputter depth profile from the metallized layer which was removed by a peel test[19]. The Fluorine peak at the surface indicates Teflon remaining on the Au layer. As sputtering is increased, the F signal drops and the Au signal increases. This indicates the strong intermixing of Teflon and Au due to the ion beam.

For the IBAD method, the advantages can be summarized as: The thickness of the metallization films is not limited. Low ion energies and doses reduce the radiation damage of the polymer. The metal film microstructure can be influenced.

On the other hand, this method involves a rather complicated vacuum process.

SUMMARY AND OUTLOOK

The three basic ion beam assisted techniques for the metallization of polymers have been described. The principles have been explained and typical examples have been given. We have seen that a whole spectrum of conductivities, ranging from semiconductor-like to metallic properties, can be achieved. Well adherent metallic films on polymer surfaces can be produced with these techniques, even on polymers which can hardly or not at all be metallized by conventional methods. The metal films are generally thin (from several nm to few μm), but can be further thickened by wet chemical plating processes such as electroplating or electroless plating. When using masks or focused ion beams, spatial structuring of the metallization layer with high resolution is possible.

Table I summarizes the most important advantages and disadvantages of the ion beam assisted metallization methods.

Table I. Evaluation of Ion Beam Assisted Metallization Techniques

advantages	disadvantages
- combination of any metal with any substrate material	- complicated vacuum process
- good adhesion on smooth surfaces	- line-of-sight process
- low temperature process	- small thickness of coating
- spatial structuring possible	- radiation damage
- compatible with wet chemical processing	
- non-hazardous to environment	

Ion beam processing is a relatively young technology. It is a very versatile technique, as it can be used, in principle, to coat any polymer surface with any metallic element. So, many new applications are still possible.

However, fundamental research is still necessary to investigate the basic mechanisms of adhesion promotion. In particular, it is essential to understand:

1) the physical and chemical interactions of ions with plastics, and

2) the physical and chemical interactions at ion irradiated metal-polymer interfaces.

ACKNOWLEDGEMENT

This work was partly funded by the German Bundesministerium für Forschung und Technologie under contract No. 13 N 5420/0.

REFERENCES

1. J. K. Hirvonen, Mater.Sci.Eng. A 116, 167 (1989)
2a. M.S.Dresselhaus, B. Wassermann, and G.E. Wnek, Mat.Res.Soc.Symp.Proc. 27, 413 (1984)
2b. T. Venkatesan, S.R. Forrest, M.L. Kaplan, P.H. Schmidt,C.A. Murray, W.L. Brown, B.J. Wilkens, R.F. Roberts, L. Rupp, Jr., and H. Schonhorn, J. Appl. Phys. 56, 2778 (1984)
2c. J.E.E. Baglin, G.J. Clark, and J. Böttiger, in "Thin Films and Interfaces II," J.E.E. Baglin, D.S. Campbell, W.K. Chu, Editors, p. 179, Elsevier, New York, 1984
2d. I.H. Loh, R.W. Oliver, and P. Sioshansi, Nucl. Instrum. Methods B 34, 337 (1988)
3. J.F. Ziegler, J.P. Biersack, and U. Littmark, "The Stopping and Range of Ions in Solids," Vol. 1, Pergamon Press, New York (1984)
4. G. Marletta, Nucl. Instrum. Methods B 46, 295 (1990)
5. P. A. Ingemarsson, M. P. Keane, and U. Gelius, J. Appl. Phys. 66, 3548 (1989)
6. N. Koshida, Y. Suzuki, and T. Aoyama, Nucl. Instrum. Methods B 37, 708 (1989)
7. W.Lohmann, and T.Flottmann, unpublished data (1987)
8. J. Bartko, B.O. Hall, and K.F. Schoch, J. Appl. Phys. 59, 1111 (1986)
9. C.J. Bedell, C.J. Sofield, L.B. Bridwell, and I.M. Brown, J. Appl. Phys. 67, 1736 (1990)
10. R.F. Bunshah, Editor, "Deposition Technologies for Films and Coatings," Noyes Publications, New Jersey (1982)
11. W.Lohmann, and T.Flottmann, unpublished data (1989)
12. J.P. Biersack, in "Ion Beam Modification of Insulators," P.Mazzoldi, G.W.Arnold, Editors,p. 648, Elsevier, Amsterdam (1987)
13. A.A. Galuska, Applied Surface Science 40, 41 (1989)
14. G.K. Wolf, in "Stromlose Beschichtung," DGO-VDI-TZ Editor, p. 164, E.G. Leuze Verlag, Saulgau (1988)
15. T. Flottmann, A. Tulke, E. Esper, and W. Lohmann, Mater. Res. Soc. Symp. Proc. 167, 309 (1990)

16. S.S. Eskildsen, and G. Sorensen, Appl. Phys. Lett. 46, 1101 (1985)
17. C.-A. Chang, J.E.E. Baglin, A.G. Schrott, and K.C. Lin, Appl. Phys. Lett 51, 103 (1987)
18. R.A. Roy, J.J. Cuomo, and D.S. Yee, J. Vac. Sci. Technol. A6, 1621 (1988)
19. I.H. Loh, and J.K. Hirvonen, Mater. Res. Soc. Symp. Proc. 108, 241 (1988)

PLASMA INDUCED DEPOSITION OF COPPER FILMS ON POLYMER SURFACES

R. Padiyath, M. David, and S.V. Babu

Department of Chemical Engineering
Clarkson University
Potsdam, NY 13699

Thin copper films have been deposited on Kapton® and silicon substrates by reducing films of copper formate in a hydrogen RF discharge. Auger electron spectroscopy (AES) analysis and X-ray diffractometry (XRD) have been used to characterize the films. The oxygen content of the as-deposited copper films has been reduced to about 1% by annealing in a hydrogen ambient at about 300°C. However, only the upper 20-25 nm thick surface layers of the films were found to be rich (>90%) in copper. The remainder of the film was contaminated with carbon and traces of iron. Process conditions that may minimize the carbon content in the bulk of the film are discussed. A new laser ablation technique appears to be more promising.

INTRODUCTION

Metallization of insulating surfaces is commonly achieved by electroless plating[1]. For instance, copper plating of high aspect ratio through-holes in multi-layer printed circuit boards as well as lithographically defined conductor channels on the surface of a laminate is achieved by electroless plating. At present it is the method of choice for high aspect ratio through-hole plating since uniformly thick deposits are produced. Before plating can be initiated in an electroless bath, the substrate surface must be catalytically activated, usually by adsorbing palladium on the surface. The overall plating process has many process steps and requires a high degree of process control since satisfactory conductor characteristics are obtained in only a narrow range of process parameters. Furthermore, since copper electroless plating is autocatalytic, the plating process is extremely sensitive to contamination Finally, electroless plating deposition rates are very small. Typical rates are only 100 to 250 μm per hour and depositing the necessary thickness of copper in a through-hole may need as much as 24 hours.

Copper/polyimide multilayer structures for high performance thin film packages have been developed[2]. Since copper is superior to aluminum in many respects, it is proposed as a replacement for aluminum as an interconnection material in multilevel metallization schemes for future generation of sub-micron devices. However, deposition and patterning of copper with a high degree of reliability and reproducibility on

Metallized Plastics 2, Edited by K.L. Mittal
Plenum Press, New York, 1991

dielectric substrates is still a challenging task and its implementation faces many hurdles. Since the common polymeric materials cannot withstand high temperatures, deposition techniques involving elevated temperatures such as, chemical vapor deposition (CVD)[3], laser induced chemical vapor deposition (LCVD)[4], cannot be used for deposition over such surfaces.

This paper presents results of some of the preliminary experiments that explored a novel plasma enhanced reduction technique for the deposition of high quality copper films of thickness in the range of 50 to 100 nm. Several polymeric and dielectric substrates including polystyrene, Kapton®, polytetrafluoroethylene (PTEE), as well as silicon and glass have been used as substrates. The technique consists of preparation ot a thin film of copper formate[5,6] or copper chloride precursor by spin-coating followed by reduction of the precursor to metallic copper in a hydrogen RF plasma. Since the initial reduction process is carried out at a low substrate temperature (20°C to 30°C), temperature sensitive polymeric materials can be metallized. However, some limitations are imposed by a subsequent anneal step. The efficiency of the process and hence the product quality is dependent on the nature of the precursor and the spin-coating process. This technique is especially suitable for large area deposition of very thin films. The thickness of metallic films that can be obtained by this process is limited by the extent of plasma-solid interaction, but this can be further increased by carrying out a subsequent electro- or electroless deposition. A new laser ablation technique in which a copper formate target is reactively ablated to form copper films appears more promising and is described briefly later.

The as-deposited copper films were annealed in a hydrogen ambient at 300°C in the PlasmaTherm deposition chamber. The annealing temperature limits the choice of substrates to Kapton and silicon, even though films have been prepared on the other substrates. The remainder of the paper discusses the results obtained with Kapton and silicon substrates.

The plasma reduction technique described here is somewhat similar to the metalorganic deposition (MOD) process described by Mantese et al[7]. In the MOD (not to be confused with Metalorganic Chemical Vapor Deposition (MOCVD), a thin film of metalorganic precursor obtained by spin-coating is pyrolyzed to form a film of metal, metal-oxide or other compounds. Superconducting films of Y-Ba-Cu-O were also prepared by this method.

EXPERIMENTAL METHODS

A PlasmaTherm plasma reactor (model 730/740) and a multi-port reaction chamber were used in the deposition studies. In the PlasmaTherm reactor, which was described earlier[8], either of the two parallel plate electrodes could be powered. A 600 W generator supplied the RF power at 13.56 MHz through an automatic matching network. The electrodes were cooled by an ethylene glycol-water mixture. The deposition chamber was pumped by a Roots blower backed by a mechanical pump to a base pressure of about 1 mTorr. A capacitance manometer monitored the chamber pressure which was regulated by an exhaust valve controller. The flow rate of hydrogen was controlled by a mass flow controller (MKS 5850E). A schematic of the PlasmaTherm plasma reactor is shown in figure 1.

An additional, smaller plasma reactor (fig. 2) fabricated from a six-way stainless steel cross was also used in the deposition studies. This reactor uses the control panel and the pumping units of the PlasmaTherm system. The gas entered through one of the ports and left through another. The RF power was coupled into the discharge volume through a U-shaped stainless steel tube (tube o.d. 1/8"), through which the cooling

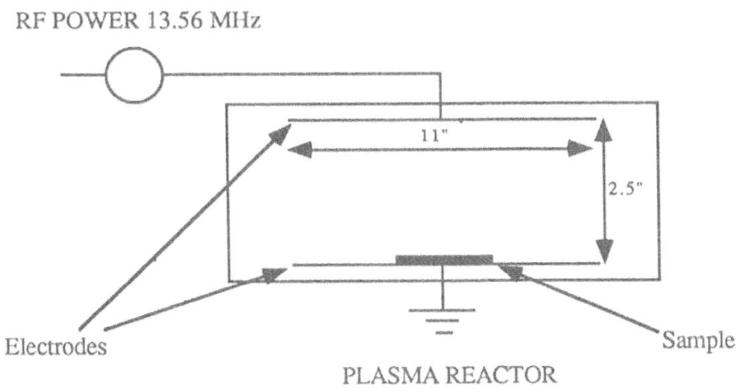

Figure 1. Schematic of the PlasmaTherm plasma reactor.

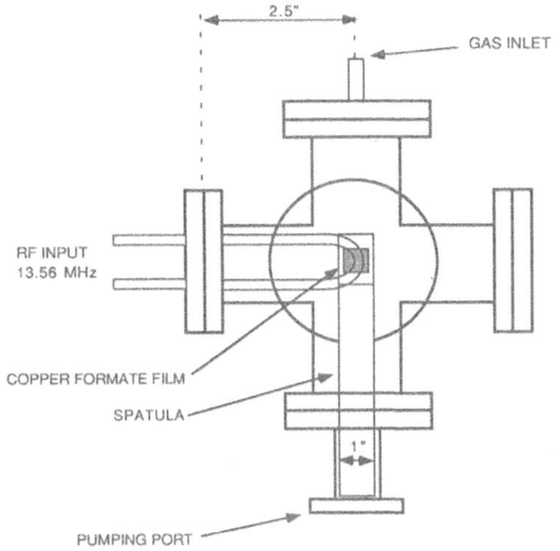

Figure 2. Schematic of the multi-port plasma reactor.

liquid was circulated. A much higher power density compared to the PlasmaTherm reactor could be maintained in the chamber due to the reduced discharge volume.

The polymeric substrates used in the present study did not have good wettability[9]. Hence, they were pre-treated in an RF plasma prior to the spin coating process. Table I lists the plasma and flow conditions used for pre-treatment of the various substrates. It was found that the plasma treatment improved the wettability and the copper formate solution could be applied easily. The pre-treatment of PTFE films with the hydrogen plasma improved its wettability, whereas exposing to an oxygen plasma produced a totally non-wettable surface. Thin uniform films of copper formate were prepared on a plasma pre-treated substrate by first placing a few drops of saturated aqueous copper formate solution on it and turning the substrate at a slow speed to spread the solution. An infrared heating lamp was used for evaporating the water from the spreading

Table I. Process Conditions Used for the Pre-treatment of Various Substrates.

Substrate Material	Plasma conditions				
	Gas(es)	Pressure (Torr)	Flow rate (sccm)	Power (Watts)	Time (min)
Polystyrene	oxygen	1	100	100	2
Kapton®	oxygen	1	100	100	2
PTFE	hydrogen	1	100	100	2
Silicon	CF_4+O_2	1	100	100	5
Glass	CF_4+O_2	1	100	100	5

solution. The spinning speed was increased once most of the water evaporated. The entire coating process took about a minute.

The copper formate coated substrates were then placed on the powered electrode and reduced to copper in a low pressure hydrogen plasma discharge. Typical process conditions were 0.2 to 2 Torr pressure, 100 sccm H_2 flow rate and 200-600 Watts of RF power. The substrates were left in the plasma discharge for about 15 minutes. The copper films were annealed at 300°C in a hydrogen atmosphere to improve the crystallinity and reduce the resistivity. This limited the Auger analysis to films obtained on Kapton® and silicon substrates only.

An optical microscope (Olympus model PME) was used for observing the deposit morphology. Crystallinity of the films was measured by an X-ray diffractometer (Siemens model D500). Elemental analysis and depth profiling of the deposition product was performed by Auger electron spectroscopy (Perkin-Elmer PHI 600 equipped with a 04-303 differentially pumped ion gun). The electron gun was operated at a voltage of 3 kV and 100 nA to obtain the Auger survey and the ion gun was operated at 2 kV and 3 µA for sputtering. The copper films sputtered at a rate of about 9 nm/min under these conditions.

RESULTS AND DISCUSSION

The optical micrographs of the copper films deposited in hydrogen plasma are shown in figures 3a, 3b, and 3c. The deposited copper films have morphological features similar to those of the starting spin coated films. Therefore, the spin coating process is critical and the presence of impurities or particulates affects the final morphology of the films. While the films obtained from aqueous copper formate solution were smooth, copper chloride films contained needle like structures, similar in shape to $CuCl_2 \cdot 2H_2O$ crystals. However, the morphology of the copper chloride films was greatly improved when ethanol was used as the solvent instead of water, but it was still not as good as that of copper formate films. Thus the choice of the solvent is critical in determining the final film morphology.

Films obtained by plasma reduction of copper formate as well as

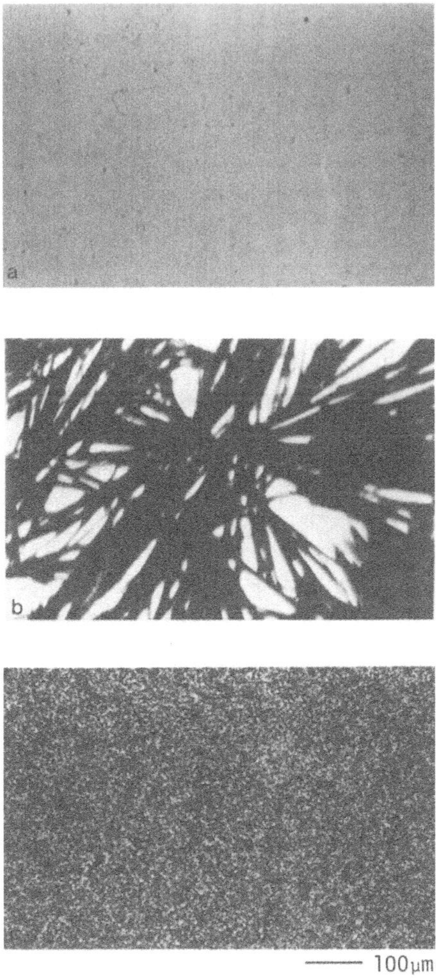

Figure 3. Copper films obtained by plasma reduction: (a) from copper formate. The dark dots are from particulates in the film. (b) from copper (II) chloride with water as solvent showing needle like structure. (c) from copper (II) chloride with ethanol as solvent.

copper chloride showed low resistivity (~10-15 μΩ), with films prepared in the smaller plasma reactor having the highest conductivity. Even though the film morphology was comparatively poor, films obtained with copper chloride precursor showed comparable resistivity. The copper films obtained by the reduction of copper formate on all the pre-treated polymeric substrates passed the "Scotch tape test", but those obtained on silicon substrates and with copper chloride showed comparatively poor adhesion.

The X-ray diffractogram of copper formate films reduced in a hydrogen plasma is shown in figure 4. The characteristic peaks of copper formate are absent indicating that the reduction was complete. However, the X-ray diffractogram of the film obtained by the reduction of copper

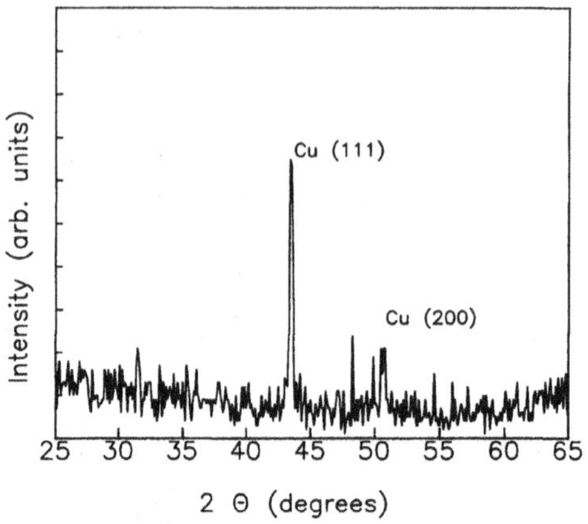

Figure 4. X-ray diffractogram of an annealed copper film obtained by hydrogen plasma reduction of copper formate.

chloride showed peaks of copper (I) chloride indicating that the reduction was not complete. The peak intensities increased upon annealing.

Due to the unsatisfactory morphology, poor adhesion of $CuCl_2$ films and the comparatively higher resistivity ot the films made in the PlasmaTherm chamber, only the copper films prepared from copper formate in the small chamber were analyzed by Auger spectroscopy. The Auger depth profile of a copper film on Kapton® substrate is shown in figure 5. The thickness of the film was about 75 nm, as measured by a Dektak profilometer. Only the surface (20-25 nm) layers are rich in copper (~92%). The film contained negligible amount of oxygen throughout the depth of the film. Thus, while the reduction of copper formate appears to be complete, the final product is contaminated with carbon at greater depths and iron sputtered from the stainless steel electrode. The high concentration of oxygen on the surface suggests post-annealing surface oxidation of the films. Iron contamination can be eliminated by either using a

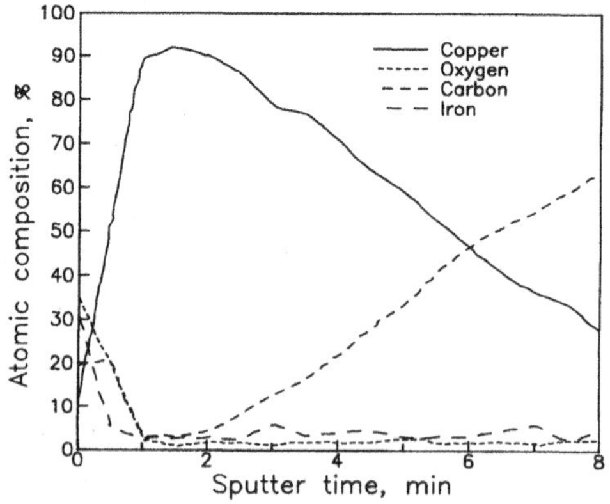

Figure 5. AES depth profile of an annealed copper film obtained by hydrogen plasma reduction ot copper formate on Kapton substrate.

microwave discharge or reducing the film downstream from the plasma discharge.

In order to establish that the carbon in the copper film originated from the copper formate and not from the Kapton® substrate, a copper film obtained by reduction of copper formate on a silicon substrate was analyzed. The Auger profile of this film was found to be essentially similar to that obtained with Kapton® as the substrate. This shows that the removal of carbon by hydrogen was not as efficient as the removal of oxygen.

In a hydrogen plasms, dissociation of copper formate may proceed by reaction with hydrogen, as shown below:

$$Cu(HCOO)_2 + H_2,\ H \rightarrow Cu + C + H_2O + CO_2 \qquad (1)$$

$$mC + nH \rightarrow C_mH_n \qquad (2)$$

The large amount of carbon present at greater depths in the films, as evidenced by the Auger profile, indicates that the second reaction does not completely volatilize the carbon. Under the conditions used, the occurrence of this reaction appears to be limited to a depth of 20-25 nm from the surface due to either slow hydrogen diffusion into the film or slow product hydrocarbon diffusion out of the film. Thus, it may be possible to reduce the carbon content in the bulk of the films further either by longer reaction times or by using a higher substrate temperature. The increased temperature should facilitate hydrogen diffusion into the film as well as removal of product hydrocarbon from the bulk of the film. Also, a microwave discharge containing a higher concentration of reactive hydrogen radicals should enhance diffusion fluxes to the substrate and improve film composition.

Conventional lithographic techniques may be used to obtain patterned copper films by this process. However, since copper formate is soluble in water, a photoresist that can be developed in a non-aqueous solvent has to be used. Thus a negative photoresist like Riston® may be used for wide lines. For finer line patterns, dry processing of the exposed photoresist may be necessary.

Presently, a new laser induced deposition technique for deposition of thin copper films is under development[10]. This process may also be used to metallize polymers since the substrate is kept at room temperature. In this technique, copper is deposited by reactively ablating a copper formate target with a partially focused excimer laser beam (248 nm). The laser beam is focused onto a thick (1.5 mm) solid film of copper formate on a glass slide. A thin film of copper is obtained on a substrate placed in close proximity to the target film. The films are then annealed at 200°C for about 30 minutes in a hydrogen plasma. AES depth profile of a copper film obtained by this technique is shown in figure 6. The copper content of the film is high and essentially constant throughout the entire depth. However, the film was contaminated with small amounts of carbon and oxygen.

CONCLUSIONS

A new technique for low temperature deposition of copper films is described. It consists of preparation and subsequent reduction of a thin film of a copper containing precursor material in a hydrogen RF plasma. The hydrogen plasma is found to be efficient in the reduction of copper formate. The morphological features of the films obtained by this technique are dependent upon the spin-coating process. The films are

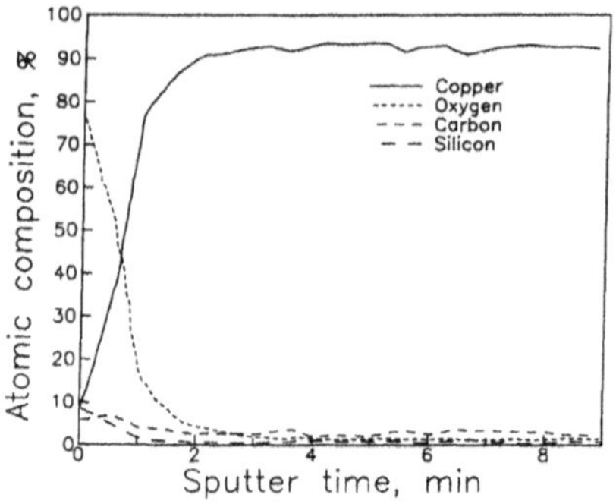

Figure 6. AES depth profile of copper film obtained by laser induced reactive ablation of copper formate.

rich in copper (>90%) only at the surface (20-25 nm) with carbon present at greater depths.

ACKNOWLEDGMENT

The authors wish to thank Dr. Lois Walsh, Rome Air Development Center, Griffiss AFB, NY, for the Auger analysis of the copper samples.

REFERENCES

1. J. F. D'Amico, M. A. DeAngelo, and F. R. McLarnon, J. Electrochem. Soc., 132, 2330 (1985); D. P. Seraphim, R. Lasky, and C. Y. Li, "Principles of Electronic Packaging" McGraw-Hill, 1989.
2. R. J. Jensen, C. P. Cummings, and V. Harshadrai, IEEE Trans., CHMT-7, 384 (1984).
3. A. Sherman, "Chemical Vapor Deposition for Microelectronics", Noyes Publications, (1987).
4. F. A. Houle, C. R. Jones, T. Baum, C. Pico, and C. A. Kovac, Appl. Phys. Lett., 46, 204 (1985).
5. M. Ritz, S. V. Babu, V. Srinivasan, and R. C. Patel, in "Photon, Beam and Plasma Stimulated Processes at Surfaces", V. M. Donelly, I. P. Herman, and M. Hirose (editors), Materials Research Society, 75, 433 (1987).
6. A. Gupta and R. Jagannathan, Appl. Phys. Lett., 51, 2254 (1987).
7. J. V. Mantese, A. L. Micheli, A. H. Hamdi, and R. W. Vest, MRS Bulletin, 14(10), 48 (1989).
8. M. David, S. V. Babu, and D. H. Rasmussen, AIChE J., 36, 871 (1990).
9. S. Wu, "Polymer Interface and Adhesion", Marcel Dekker, New York (1982).
10. S. V. Babu, R. Padiyath, M. David, and L. Walsh, AIChE J., submitted for publication, (1990).

METALLIZATION OF POLYTETRAFLUOROETHYLENE (PTFE) BY MEANS OF PLASMA-ENHANCED CHEMICAL VAPOUR DEPOSITION

H.Meyer, R.Schulz, H.Suhr[*], C.Haag[*], K.Horn[+] and A.M.Bradshaw[+]

Schering AG, Dept. GT-FO, D-1000 Berlin 65, Germany
[*]Dept. of Organic Chemistry, University of Tübingen, D-7400 Tübingen, Germany
[+]Fritz-Haber-Institut der Max-Planck-Gesellschaft, D-1000 Berlin 33, Germany

A process has been developed for coating PTFE with copper. It consists of a plasma pretreatment with reactive gases such as O_2 or CF_4/O_2 followed by a plasma deposition of a thin metallic layer of Pd, Pt, Au, or Cu and subsequent electroless and electrochemical copper plating. The coatings show excellent adhesion properties, limited only by the cohesive strength of the substrate. The PTFE surfaces have been studied with scanning electron microscopy (SEM), and x-ray photoelectron spectroscopy (XPS) after different steps of the pretreatment. SEM reveals directly the roughening of the PTFE surface by the plasma etching step. XPS results for Pd-activated surfaces point to a chemical interaction between Pd and the fluorine atoms of the PTFE. This new process of metallization is a simple technique, avoids hazardous chemicals, requires no decontamination and can equally well be applied to other polymers.

INTRODUCTION

Good dielectric properties combined with high thermal stability render fluoropolymers and, in particular, polytetrafluoroethylene (PTFE) very attractive for applications in

Metallized Plastics 2, Edited by K.L. Mittal
Plenum Press, New York, 1991

electronic devices. However, metallization of PTFE is difficult because it is hydrophobic and chemically very inert. The introduction of functional groups by means of chemical reactions in order to improve adhesion has so far required the use of highly aggressive reagents. There is thus an obvious need for a more convenient metallization process. Adhesion between a metal and a polymer can, however, result from other effects[1], for example surface roughening which not only increases the specific surface area, but may also generate a specific surface morphology, e.g. cavities, which provide an anchoring of the metal layer to the substrate. Adhesion may also occur because of the existence of an intermediate layer, i.e. the replacement of the discontinuity at the phase boundary by an intermediate zone which results from interpenetration of the phases; it may be characterized by a concentration gradient. Interdiffusion or implantation may generate this kind of adhesion-promoting layer.

In order to enhance adhesion one may aim at making use of all three effects simultaneously. In the case of PTFE, which is polycrystalline, fluorine atoms are bound to a carbon skeleton. The typical features of this polymer are high stability of the C-F bond (bond energy 486 $kJmol^{-1}$, compared with 415 $kJmol^{-1}$ for a C-H and 365 $kJmol^{-1}$ for a C-C bond). In contrast to a hydrocarbon bond, the carbon atoms are electron-deficient in the C-F bond because of the high electronegativity of fluorine[2]. Also, the electron lone pairs on the F atoms have a very low polarizability, and the impossibility of expansion of the fluorine octet weakens the intermolecular forces. The weakness of these forces is also shown by the very low coefficient of friction of PTFE and its poor wettability.

In order to promote chemical bonding between the coatings and the PTFE substrate, one has to use highly reactive chemicals with strongly reducing properties. Alkali metals, being strongly electropositive, have been traditionally used, for example in the form of sodium-ammonia and sodium-naphthalene complexes[3]. In order to avoid these chemical approaches and their undesirable side effects, e.g. contamination, environmental problems and inconvenient working procedures, attempts have been made to devise physical processes which achieve the same result. Thus improved processes for the metallization of PTFE make use of sputtering [4,5], irradiation with high energy ion beams[6,7] or with low energy argon ion beams[8]. An overview of such techniques for fluoropolymers has been given by Siperko and Thomas[9]. In the present paper we evaluate a new method of physical pre-treatment: plasma-enhanced chemical vapor deposition (PECVD)[10] for promoting the adhesion of metallic coatings.

Table I. Adhesion of copper strips to PTFE after various pre-treatments. All samples were subsequently treated in a reductive copper bath and thickened by electroplating to 10 μm copper.

Pre-treatment		Peel strength (Nmm^{-1})
Aqueous Pd solution		0
Ar-sputter etch,	aqueous Pd	0.1
O_2/Ar-etching,	aqueous Pd	0.2
O_2/CF_4-etching,	aqueous Pd	0.3
O_2/Ar-etching,	PECVD of Pd	1-2
O_2/CF_4-etching,	PECVD of Pd	4-5

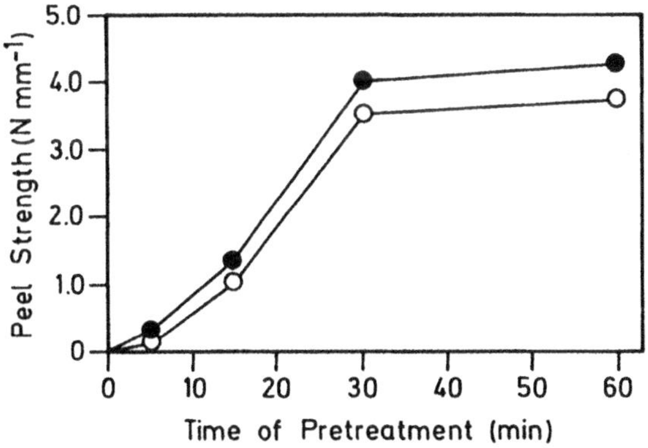

Figure 1. Peel strength of Cu to PTFE vs. time of CF_4/O_2 pre-treatment; maximum (filled circles) and minimum (open circles) values of the peel strength are shown. The samples were subsequently activated by plasma deposited Pd, metallized by a reductive copper bath, followed by electrochemical copper plating.

EXPERIMENTAL

The experiments have been carried out in a parallel plate reactor (Plasma Lab 80 by Plasma Technology Ltd, Great Britain) using PTFE foils of 0.8 mm thickness. These were placed on the lower water-cooled electrode, which was connected via a matching network to a 13.56 MHz rf generator. The pressure of the gases was monitored by MKS gas flow controllers; the gases entered the reactor through the upper, grounded electrode. For the PECVD experiments, argon was fed through a glass evaporator containing the organo-metallic precursor. The temperature of the evaporator was controlled by a thermostat. The following compounds have been used as precursors: $Pd(\pi-C_3H_5)(\pi-C_5H_5)$, $Pt(CH_3)_3(\pi-C_5H_5)$, $Au(CH_3)_2(C_5H_7O_2)$, and $Cu(C_5F_6HO_2)_2$. The preparation of these compounds and the necessary parameters for the plasma-enhanced CVD of metallic layers from the precursors have been published before[11-14]. Here we deal exclusively with the deposition of Pd, in view of its importance as the initial step in electroless copper deposition.

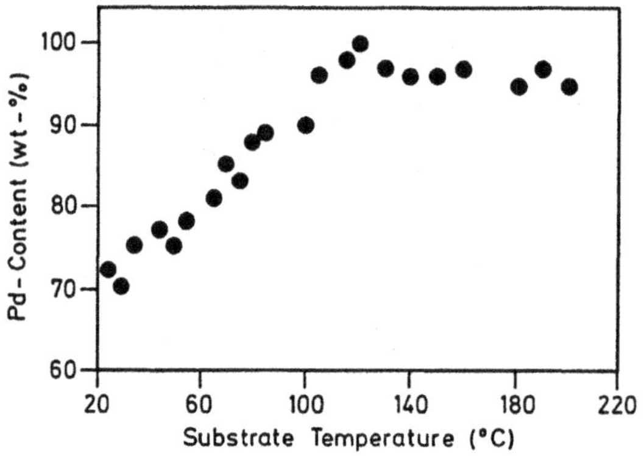

Figure 2. Palladium content in the films deposited on PTFE at various substrate temperatures.

In order to measure adhesion by a peel test (90°), an aluminum mask was placed over the substrates for the preparation of metal strips. Samples which were exposed only to the plasma etch treatment were subsequently activated with the proprietary NEOGANTH™ process[15] (aqueous solution of a Pd complex followed by chemical reduction), treated in a reductive copper bath and then electroplated to a thickness of 10 µm in an acid copper bath. After electroplating, all samples were annealed for 24 hours at 323 K or were exposed to air at room temperature for several days prior to the peel test. The SEM studies were made in a Hitachi 4000 X scanning electron microscope at beam energies of 5 keV. Surface analytical measurements were carried out in a Surface Science Instruments photoelectron spectrometer, equipped with a flood gun for charge compensation.

RESULTS AND DISCUSSION

The results of the peel tests for samples subjected to various pre-treatments are given in Table I. It appears that the adhesion of copper to PTFE can be significantly enhanced by plasma etching pre-treatment. While untreated PTFE cannot be activated by the aqueous Pd solution and coated in the reductive copper bath, with plasma-etched materials such coating is possible. This improvement is probably due largely to a significant increase in surface area.

There is a further drastic improvement in adhesion when plasma etching is followed by plasma deposition of a thin metallic layer. Figure 1 shows that there is then a correlation between the duration of plasma etching and adhesion strength. It has been reported in the literature that plasma treatment of polymers in reactive gases leads to a certain roughening under prolonged treatment, whereas shorter exposures favor the effects of chemical surface modification[16]. The best results were obtained with a 30 minute pre-treatment at 0.8 W/cm^2 and a bias of -600 volts, followed by the deposition of a Pd layer at 0.4 W/cm^2 at a bias of - 430 volts for 5 minutes. SIMS spectra of the back of peeled-off copper strips showed that the metal film was covered with a thin PTFE layer. When such peeled-off strips were dissolved in nitric acid, a very thin white deposit remained. With respect to the use of copper-coated PTFE in electronics it is significant that the adhesion of copper to PTFE was only reduced by 5 to 10 % after dipping the samples into a tin solder bath at 540 K for 20 seconds.

Obviously, the composition of the deposits depends on experimental parameters. An increase of substrate temperature, power density or negative bias reduces the carbon content of the Pd films. The effect of substrate temperature on the purity of the deposited Pd layer is shown in Figure 2. The best adhesion values for the specimens subjected to the peel tests were obtained with metallic layers of about 100 Å thickness. A further increase of metal thickness caused a reduction in adhesion strength. It was not possible to distinguish between thin film growth or island growth of these metal layers. Independent studies of thicker Pd coatings clearly indicate continuous films rather than island growth[17].

In order to obtain information about the adhesion mechanism of the metal layer on the PTFE surface, the morphology and surface chemistry of the pre-treated PTFE surface were investigated.Consider the scanning electron micrographs shown

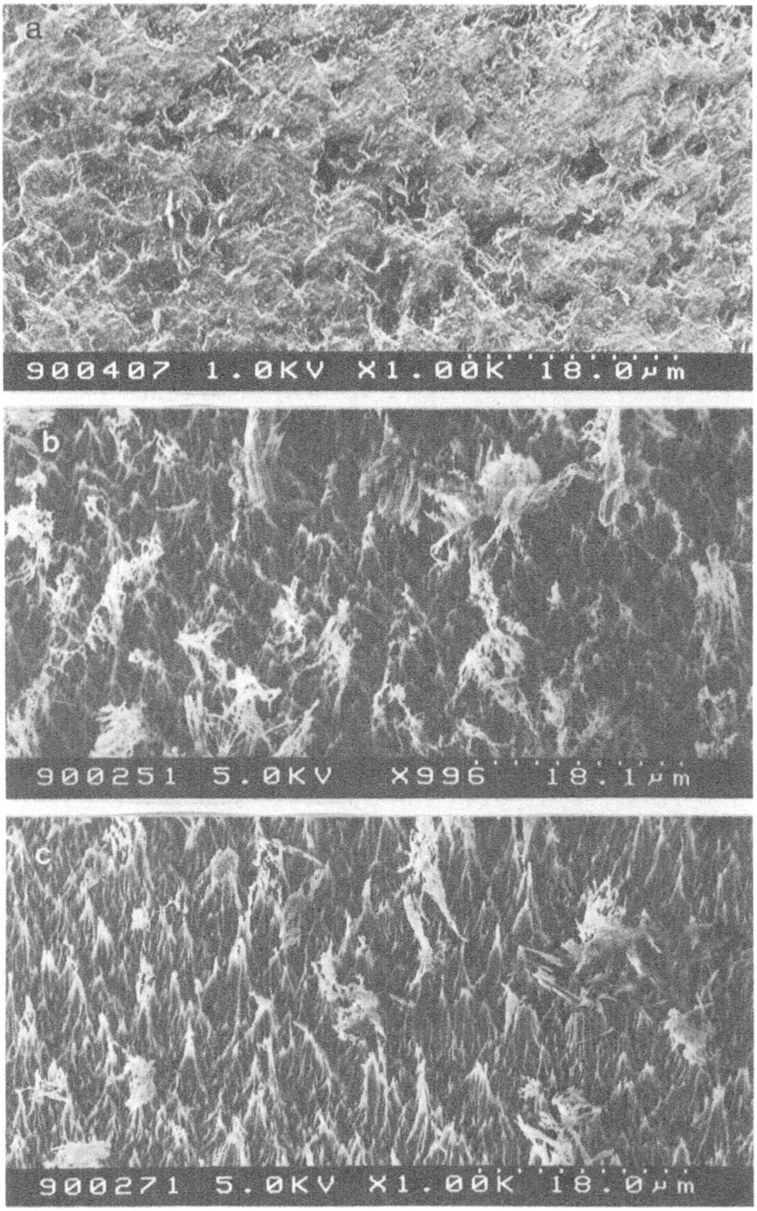

Figure 3. a) SEM micrographs of PTFE surfaces, magnification 1,000 x (a) as received, (b) CF_4/O_2 plasma etched, (c) CF_4/O_2 plasma etched followed by PECVD of Pd.

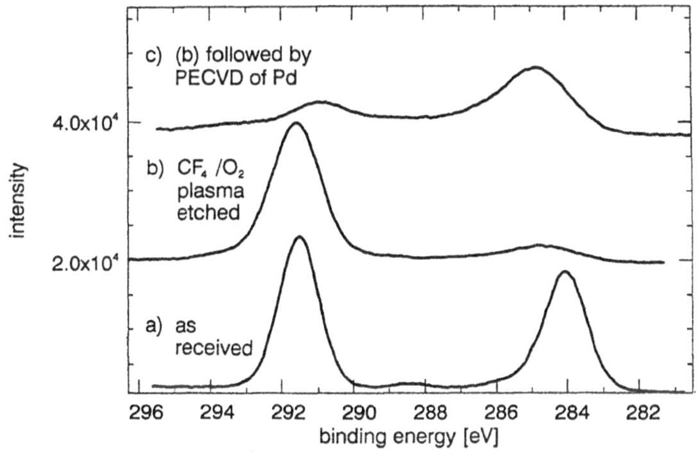

Figure 4. Carbon 1s photoelectron spectra of PTFE surfaces after different pre-treatments as indicated. Data were recorded with Al Kα radiation.

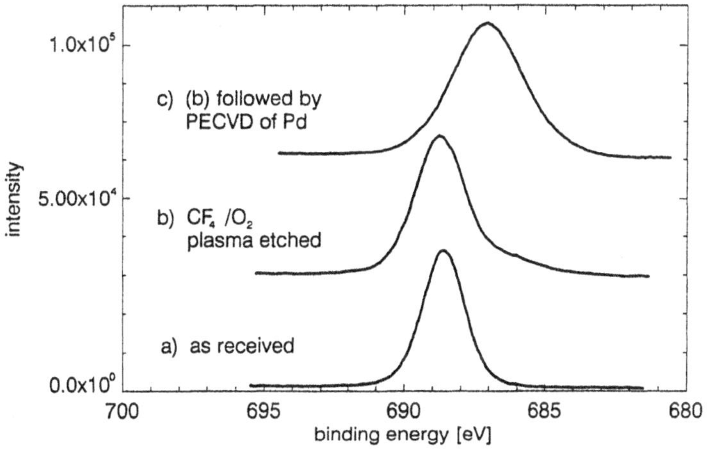

Figure 5. Fluorine 1s photoelectron spectra after identical pre-treatment as in Figure 4.

in Figure 3. They demonstrate gross changes in the surface structure after different pre-treatment steps. Obviously, reactive ion etching with CF_4/O_2 results in a pronounced roughening of the PTFE surface. As expected, PECVD of Pd does not enhance the roughness any further. The enlarged surface probably contributes to some extent to adhesion, as indicated by the data obtained from the peel test presented in Table I.

The high values of copper peel strength measured with samples pre-treated with PECVD of palladium prior to the subsequent plating process suggest that there is some kind of chemical interaction between the metal and PTFE. Therefore the PTFE surface was studied by XPS after different pre-treatment steps. In Figure 4 carbon 1s spectra of the PTFE surfaces are shown. The plasma etching process leads to an efficient cleaning of the surface from adhering organic contamination as is evident from the almost complete removal of the C 1s signal at about 284 eV binding energy, leaving only the peak at 292.5 eV which arises from the $-CF_2-$ units of PTFE. After Pd deposition by PECVD using $Pd(\pi-C_3H_5)(\pi-C_5H_5)$ as precursor, additional C 1s peaks appear in the region around 285 eV. These are probably due to carbon atoms of hydrocarbon fragments formed by the decomposition of the Pd organic compound. It is difficult to make a more specific assignment, given the possibility of many lines due to different hydrocarbon

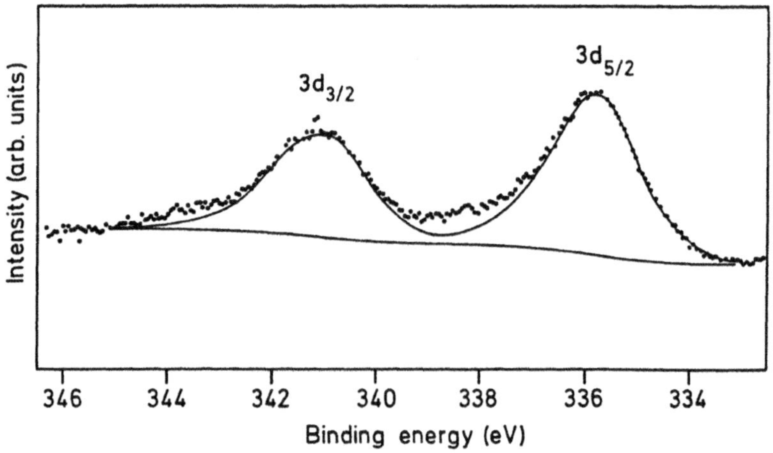

Figure 6. Palladium 3d photoelectron spectra of Pd deposited onto PTFE by PECVD.

species in this region. The fluorine 1 s spectra, shown in Figure 5 indicate that plasma etching has some influence also on the fluorine atoms. Here, only the PECVD of Pd affects this signal. A shift towards lower binding energy connected with a certain broadening of the signal can be observed. This chemical shift implies an interaction with a more electropositive element than carbon[18]. If this element is Pd, this should be observable in the Pd 3d spectrum, shown in Figure 6. The Pd 3d spectrum exhibits a line indicative of metallic Pd by its pronounced asymmetry towards higher binding energies. However, this asymmetry is much larger than in the reference spectrum of metallic Pd[19], and thus suggests the existence of a shifted line. This is consistent with an interaction of Pd and fluorine. Throughout these pre-treatment steps an increase in oxygen content of the surface could not be observed.

The surface-sensitive photoemission experiments are compatible with the peel tests which indicate that enlargement of the surface contributes to adhesion, but does not constitute the dominant effect. Plasma etched PTFE which was activated by sputter coating or vapor deposition of Pd showed no further improvement in adhesion. The observed enhancement of adhesion caused by the PECVD process thus seems to be a specific property of the plasma deposition treatment. It is likely that ions are driven into the polymer surface. Due to the high bias voltage their energy is apparently sufficient to break bonds in the substrate and to form new bonds. It is not clear, however, whether the plasma-induced dissociation of the organometallic precursor molecules imparts sufficient kinetic energy for them to penetrate the PTFE to any extent. The peel-off experiment shows cohesive fracture of PTFE, since adhesion strength on the order of 4 N/mm seems rather high with regard to the tensile elastic modulus of PTFE which is 410 N/mm^2 [20].

CONCLUSIONS

By combining a plasma etch pre-treatment with the plasma-enhanced deposition of a very thin metal film, followed by conventional plating techniques, excellent adhesion of metal coatings to PTFE is achieved. The process can also be applied successfully to other engineering polymers. In our view, this method offers several advantages which renders it attractive for manufacturing processes: it is a simple technique, based on standard plasma etching technology used in the electronics industry; and there is no contamination by other metal ions or inorganic species, only the desired metal is deposited. Finally, there are no strict "line-of-sight" limitations as in some other vacuum techniques, i.e. holes, steps, etc. are also metallized.

REFERENCES

1. K.L.Mittal, J. Vac. Sci. Technol. 13, 19 (1976).
2. F.A.Cotton and G.Wilkinson, "Advanced Inorganic Chemistry", 4th edition, Wiley Interscience, New York, 1980.
3. W.Goldie, "Metallic Coatings of Plastics" vol. 2, p.303-323,Electrochemical Publ. Hatch End, Middlesex,Great Britain,1969.
4. R.Michael, D.Stulik, J. Vac. Sci. Technol., 1861 (1986).
5. D.R.Wheeler and S.V.Pepper, J. Vac. Sci. Technol. 20, 442 (1982).

6. J.S.Sovey, J. Vac. Sci. Technol. 16, 813 (1979).
7. J.E.E.Baglin, G.J.Clark, and J.Boettiger, Mater. Res. Soc. Symp. Proc. 25, 179 (1984).
8. C.Chang, J.E.E.Baglin, A.G.Schrott, and K.C.Lin, Appl. Phys. Lett. 51, 103 (1987).
9. L.M.Siperko and R.R.Thomas J. Adhesion. Sci. Technol. 3, 157 (1989).
10. C.Haag and H.Suhr, Appl. Phys. A47, 199 (1988).
11. E.Feurer and H.Suhr, Thin Solid Films 157, 81 (1988).
12. E.Feurer, S.Krauss, and H.Suhr, J. Vac. Sci. Technol. A7, S 2799 (1989).
13. E.Feurer and H.Suhr, Appl. Phys. A44, 171 (1987).
14. C.Oehr and H.Suhr, Appl.Phys., A45, 154 (1988).
15. NEOGANTH ™, Technical Bulletin Schering AG, Berlin, Germany.
16. J.Friedrich, H.Wittrich, and J.Gähde, Acta Polymerica 31, 59 (1980).
17. H.Suhr, A.Etspüler, E.Feurer, and C.Oehr, Plasma. Chem. Plasma Proc. 8, 9 (1988).
18. M.Chtaib, E.M.Roberfroid, Y.Novis, J.J.Pireaux, R.Caudano, P.Lutgen, and G.Feyder, J. Vac. Sci. Technol. A7, 3233 (1989).
19. C.D.Wagner, W.M.Riggs, L.E.Davis, J.F.Moulder, and G.E.Mullenberg, "Handbook of X-ray Photoelectron Spectroscopy", Perkin-Elmer Corp. Eden Prairie, Minnesota (1978).
20. H.Batzer, Editor, "Polymere Werkstoffe", p. 216, Thieme Verlag, Stuttgart, 1984.

METALLIZATION OF PLASTICS VIA

LOW TEMPERATURE ARC VAPOR DEPOSITION (LTAVD)

K.R. Narendrnath and D. Mager

Vapor Technologies, Inc.
2100 Central Avenue
Boulder, Colorado 80301

Low temperature arc vapor deposition (LTAVD) is a new addition to the field of cathodic arc plasma physical vapor deposition processes. Almost any metal, alloy, and intermetallic compound as well as many ceramics and fully reacted metals can be deposited utilizing LTAVD. With its ability to deposit coatings at room temperature, LTAVD has been utilized on a commercial basis in applying coatings to plastics to address electrical, optical, thermal, corrosion/ oxidation, mechanical and decorative requirements. Plastics which have been coated include: polyimide, polyetherimide, fluorocarbons (Teflon), ABS, polyurethane, polypropylene, polyethylene, PVC, polyester, nylon, phenolics, polycarbonate, epoxies and other plastics and composites. LTAVD is a dry process and hence is environmentally safe.

INTRODUCTION

Metallization of a limited number of plastics has been approached through "wet" methods such as electroless- or electro-plating. "Dry", in-vacuo techniques such as thermal evaporation and sputtering have been used to deposit very thin layers of a few nanometers to several micrometers on a broader group of plastics.[1] A combination of both techniques has been used to build up the coating thickness.

Many plastics, e.g., polyimides (PI) and polyetherimides (PEI), are good insulators with high thermal stability, and, thus, are excellent candidates for a variety of applications; however, they also exhibit a chemical inertness which results in poor coating adhesion. The approach to enhance adhesion[2] has been either through methods resulting in rougher substrate surfaces or through methods designed to alter the surface chemistry or both.

Recently, cathodic arc plasma deposition (CAPD) processes[3,4] have been explored as a means of metallizing plastics. In the present paper we detail a CAPD process which can metallize a variety of plastics, even at ambient temperature. The resulting coating microstructures can be dense and can exhibit high adhesion strengths. Results of metallization of several plastics with strong coating adhesion are presented. The

Metallized Plastics 2, Edited by K.L. Mittal
Plenum Press, New York, 1991

patented process[5], termed the Low Temperature Arc Vapor Deposition (LTAVD) process, has been developed to the point of commercial viability and is being used in the U.S.A. and abroad for a variety of applications.

LOW TEMPERATURE ARC VAPOR DEPOSITION PROCESS

Low temperature arc vapor deposition is a physical vapor deposition process that can be used to deposit coatings of any conducting materials such as metals, alloys, intermetallic compounds, and many semiconducting materials. It can also be used to synthesize fully reacted metals and ceramics. Substrates can range from low deflection temperature plastics, fiber and paper to any high temperature materials such as metals, glass, and ceramics. Substrates of complex geometries including the insides of pipes can be coated. Because of the nature of the process, the composition transfer from source to the target can be achieved with high fidelity. The distinctive feature of the LTAVD process is its ability to deposit thick coatings even around 25°C (or at higher temperatures, if desired).

The necessary hardware for operating an LTAVD system, as shown in Fig. 1, consists of three units: (a) a vacuum chamber housing typically a non-planar consumable electrode source surrounded by the substrates to be coated; (b) an evacuation system capable of achieving a vacuum in the range 10^{-2} to 10^{-8} torr (dictated by the application); and (c) a proprietary DC power source that provides control over the arc traverse on the electrode.

In essence, the process consists of generating a low voltage, high current arc in vacuum, unlike in the normal atmosphere where several hundred volts may be required.[6] The arc traverses the surface of the target, Fig.1, and sublimes it into a high energy plasma.

Fig.2 is an instantaneous, magnified cross-sectional view of the arc travel process on the cathode. A cathode spot and a microcrater left behind in its wake are shown on the cathode surface. The products of erosion of interest to LTAVD, as indicated, are a large flux of ions along with some metal vapor and molten metal droplets. The relative extents of the various components of the erosion product are a function of the cathode material, arc current, chamber atmosphere, etc. The ion plasma content can approach 100%.[7]

The high fraction of the generated ion flux is responsible for the distinct advantages of the LTAVD process in particular, and the CAPD processes in general. The characteristics of the ions are listed below:

(a) The ions possess high kinetic energies, up to 100eV[7] and, therefore, move with considerable velocities estimated[8] to be as high as 10^4 m/sec.

(b) The ions can exist in several different charge states: Copper ions can be present[9] as Cu^{1+}, Cu^{2+} and Cu^{3+}; molybdenum ions can exhibit up to five different charge states[10] and so on.

(c) Except for a small fraction, most of the ions tend to travel away from the cathode. Although this would imply line-of-sight coatings, good non-line-of-sight coatings are possible since the plasma can wrap around even complex contours. The electrode, acting as a locus of points rather than as a point source, also assists in the line-of-sight coatings.

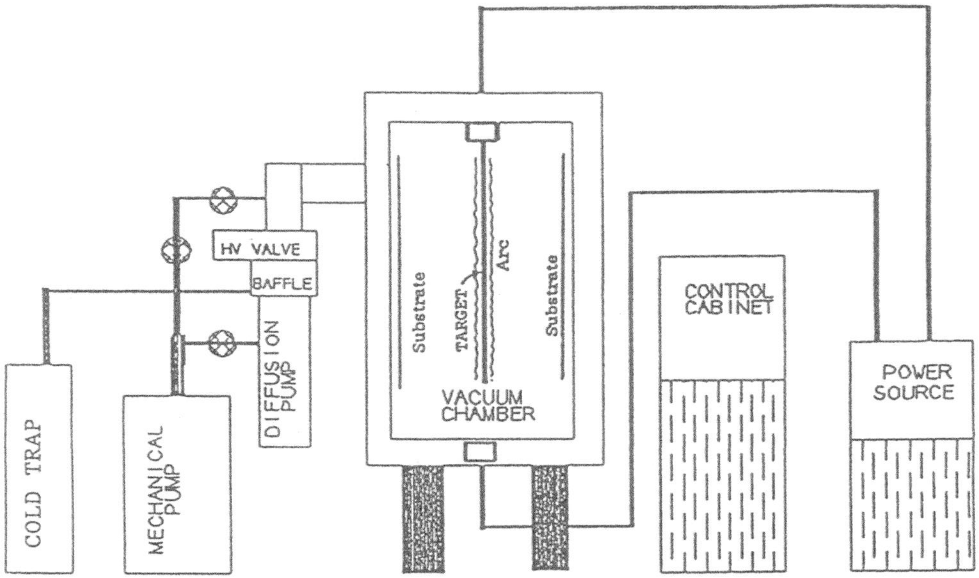

Fig.1 Schematic of the hardware for the Low Temperature Arc Vapor Deposition (LTAVD) process

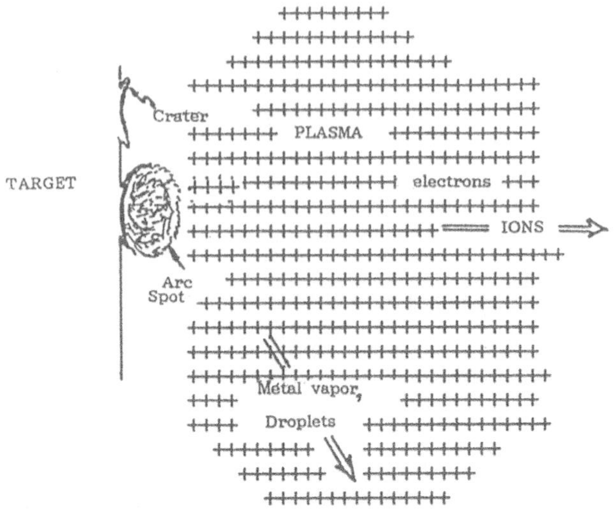

Fig. 2 Physical process underlying the Low Temperature Arc Vapor Deposition (LTAVD) process.

These characteristics provide, at the production level, a generous handle to the operator to manipulate the ion flux and obtain desired coatings on a host of substrates. For instance,

(a) The ions can be deposited as pure element, or, because of their high instability, can be reacted with deliberately injected reactive gases to deposit compounds. This has been used to produce coatings such as tantalum carbide, chromium nitride, titanium dioxide, etc. for tribological, decorative and optical applications, respectively.

(b) Since cathodic erosion occurs rapidly as a flash and since the ions have high velocities, the cathode surface composition can be transferred with good fidelity onto the substrate; stoichiometry or composition can, therefore, remain essentially the same.

(c) The substrate can be biased negatively where feasible to increase the momentum of the ions resulting in higher reaction temperatures, better coating microstructures, and enhanced adhesion.

(d) A range of deposition rates, from a few tens of nanometers to several micrometers per minute, can be obtained depending on the desired coating and its application.

(e) The coating deposition rate, as shown in Fig. 3, is seen to approximate proportionally to d^{-1}, rather than to d^{-2}, relationship for the d values indicated, where d is the cathode-substrate separation distance. Therefore, by a proper choice of cathode-to-substrate distance, uniform coatings can be deposited on complex substrates with steps or similar non-planarities.

(f) Further, since the plasma has the ability to conform to complex contours, non-planar substrates can receive full coating coverages.

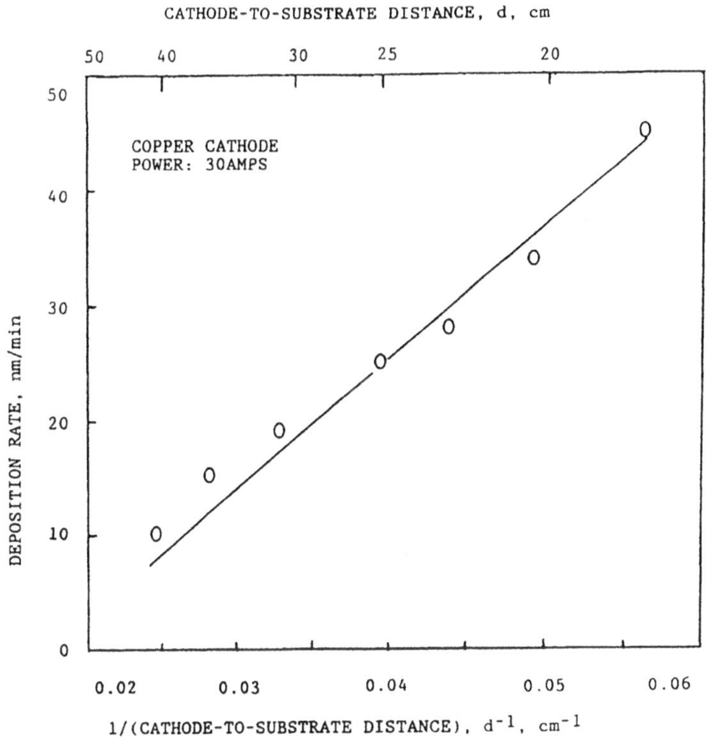

Fig. 3 Coating deposition rate as a function of d^{-1}

(g) A distinct feature, as mentioned earlier, is the ability to coat at ambient temperature. On plastic substrates such as ABS Cycoloy (Trademark by GE), with a deflection temperature as low as 65°C, the final temperature monitored continuously on the surface after an hour of copper deposition had not exceeded 40°C.

Steady state temperature measurements made on the surface of a 2mm thick polyurethane foam which completely surrounds the target during nickel deposition through LTAVD is presented in Fig. 4 as a function of input power. No heat sinks are used in this experiment. It is noted that the substrate temperature can be limited to about 80°C with an input power of 1.5kW even after 20 hours of continuous coating.

COATING PROPERTIES

The characteristics typical to the LTAVD process mentioned in the preceding section result, in practice, in desired aspects such as a dense coating with a non-columnar microstructure and good adhesion strength.

Coating Microstructure

Relative to deposition by thermal evaporation (where the process of material deposition is one of condensation of neutral atoms and subsequent growth and rearrangement) in ion-assisted deposition, other mechanisms are responsible for film growth[11]. The ions, bombarding the condensed matter, drive the matter deeper into the substrate. Simultaneously, a sputtering action can be in progress resulting in the removal of some of the loosely bonded deposited matter. The available micropores are also filled up by the rearranging atoms. One would expect, therefore, that the effect of the large flux of energetic ions emanating from the cathode surface and bombarding the substrate would be reflected in an altered coating microstructure. Such ion-induced modifications have in fact been studied before[12] and it has been clearly recognized that such modifications lead to film densification.

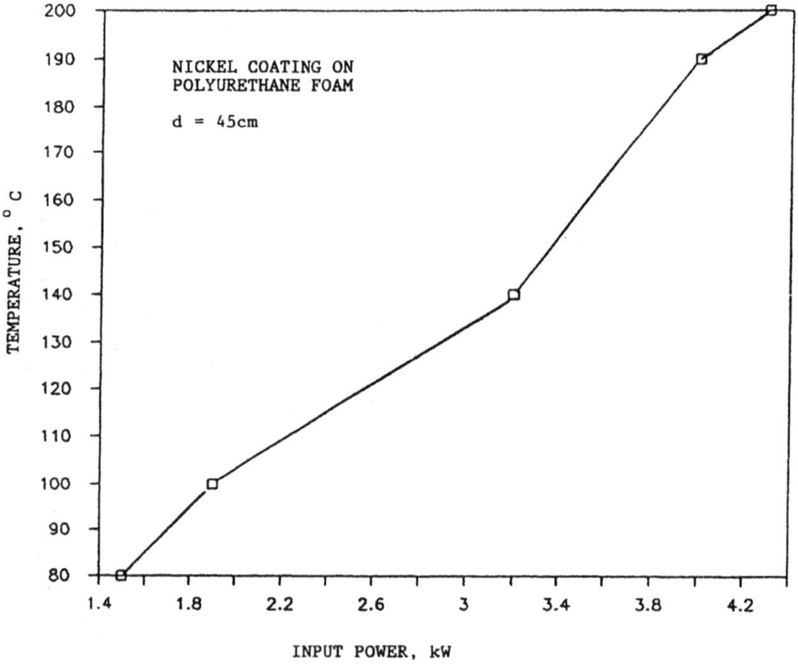

Fig. 4 Steady state temperature of polyurethane foam (after 20 hours) as a function of input power during Nickel deposition.

Supercomputer simulation studies modelling vectored momentum in atom and ion collisions and resultant film densification[11] could correctly predict the observed increase in the density of ZrO_2 films by approximately 15% over the thermally evaporated coatings, with only a 30% ion flux. Similarly, in modelling titanium deposition, even a 20% Ti^+ ion-to-Ti atom arrival ratio and an average kinetic energy of 50eV/Ti^+ ion maximum densification of the coating is achieved[13].

The above findings are consistent with our observations: coating densities of thick films obtained using the LTAVD process have been up to 95% of theoretical density.

Coating Adhesion

Adhesion[14,15], being an important aspect to maintain the integrity of the coating with the substrate, has been studied by several investigators, e.g. Helmersson, et al[16]; the studies attest to the excellent adhesion of the coatings deposited by CAPD in general. The enhanced adhesion can be attributed, in part, to the high energy of the incident ions which promote the various aspects involved in adhesion such as minimization of voids at the interface as mentioned earlier, interatomic bonding, and local diffusion at the coating-substrate interface.

It is noted here that adhesion is also affected by surface cleaning, surface modification (mechanical, chemical and plasma) and surface handling procedures. Experiments to qualitatively understand the relative contributions of the LTAVD process and the different cleaning procedures on a variety of substrates are in progress.

Adhesion strengths obtained through LTAVD processes on plastic substrates have been high; typical values are presented in Table I. All adhesion strengths were obtained by the direct pull-off technique[17] using a Sebastian I (supplied by Quad Group, Spokane, WA) adhesion tester. The ability to measure adhesion was often limited by the bulk substrate failure before the coating came off the substrate.

Most polymeric substrates experimented upon with the LTAVD process have exhibited adherent coatings with the exception of friable plastics loaded with soft materials such as talc, and plastics which have volatile surfaces due to the incorporation of plasticizers and other compounds.

Over 25μm of conductive metals are routinely coated onto aerospace composites of epoxy or polyimide loaded with Kevlar (Trademark of DuPont), carbon-carbon or graphite to provide for EMI/RFI shielding of aircraft avionics. LTAVD provides excellent adhesion in these exotic applications where the parts are subjected to extreme vibration and windshear in addition to the normal degrading environmental exposure of sun, rain, abrasives, and atmospheric corrosives.

As can be seen from Table I, LTAVD can be used to coat plastics with different metals, alloys, compounds or compacted mixtures of metal powders. For example, polyester is coated with over 8μm of platinum to provide for an opaque medical x-ray image. 75μm of nickel are routinely coated onto a polyurethane foam of 97% porosity to provide for high surface area catalysis. Titanium nitride, with the color and appearance of gold, is coated onto polycarbonates for decorative purposes. The coating thickness can range from a few tens of nanometers to several micrometers.

The success of LTAVD in coating plastics is probably due to the high ion energies resulting in good adhesion as well as the low temperatures at which the substrates can be maintained without any cooling. The low temperature coating ability also allows materials of extremely different thermal expansion coefficients to be mated at or near room temperature. For example, thin polyester films of thermal expansion coefficient of 72μm/m/K have been coated with 25μm thick films of silicon (thermal expansion coefficient of 4.67μm/m/K) without curling, wrinkling or delamination.

Table I. Metallized Substrates, Coatings and Adhesion Strengths via Low Temperature Arc Vapor Deposition.

SUBSTRATE	COATING	THICKNESS µm	ADHESION# MPa(psi)
1. ABS	Cu	1.1	36.4(5280)
2. PPS	Ni	2.0	34.7(5040)
3. PEI	TiN	1.1	48.2(7000)
4. Polyphthalamide	TiN	0.4	41.5(6020
5. Polyimide	Ni	1.9	35.4(5140)
6. Polycarbonate (Lexan)	Cr	0.1	11.4(1650)
7. Polycarbonate	W-Ag	0.06	27.6(4000)
8. PTFE (Teflon)	NiCr	0.6	4.7(680)
9. PEEK	Cu	8.1	18.9(2740)
10. Polyester (Mineral-filled)	Cu	1.1	10.4(1512)
11. Polyester	Pt	8.0	*
12. Polyurethane foam	Ni	75.0	*
13. Epoxy (with Kevlar)	Cu	25.0	*

Cleaning procedure: Typical cleaning procedures involved cleaning with a mild soap solution, thorough rinsing with running water followed by rinsing with double distilled water, drying, sonicating in non-aggressive organic solvents (e.g., Isopropyl alcohol) and, if necessary, by a Freon (Trademark of DuPont) dip prior to oven drying.

\# Adhesion strengths measured by direct pull-off technique[17] using Sebastian I adhesion tester.
* Adhesion strengths not measured.
Lexan: Trademark of G.E.
Teflon, Kevlar: Trademark of DuPont

Acronyms:
ABS: Acrylonitrile-Butadiene-Styrene; PEEK: Polyethyl ether ketone
PEI: Polyetherimide; PPS: Polyphenylene sulfide;
PTFE: Polytetrafluoroethylene.

Microdroplets

One product inherent to cathodic erosion, as indicated in Fig. 2, are microdroplets (also termed "macros"). The macros can range in size from a fraction of a micrometer to several micrometers. Macros carry almost no charge and are pure metals in composition when deposited from elemental cathodes. The percentage of macros varies, being higher for lower melting materials. Although the macros of the order of a micrometer in size do not affect applications such as wear resistance, EMI/RFI shielding, etc., they would effectively eliminate optical and sub-micron range electronic applications.

Macros, however, can be controlled or minimized either at the source or at the substrate by controlling different operating parameters.[18] Further, since macros are essentially neutral particles and the plasma is highly charged, it is possible to effectively separate the two by using curvilinear electro-optics so that only the plasma reaches the substrate surface. Although this would compromise the deposition rates to some extent, it would yield extremely smooth and uniform coatings. In fact, several investigators have reported[19,20,21] obtaining extremely smooth, optical grade coatings of pure metals, oxides, and nitrides via CAPD, through filtered arc evaporation.

SUMMARY

A new physical vapor deposition technique, called Low Temperature Arc Vapor Deposition (LTAVD), has been introduced, its process and advantages are detailed, and some commercial applications are mentioned. The process consists of the erosion and subsequent generation of plasma from the source electrode by a traveling arc spot of intense temperature. The plasma is characterized by ions of high kinetic energy and multiple charge states and, therefore, is highly reactive. It can be condensed pure, or reacted with deliberately injected reactive gases to deposit as a compound. An important feature, unique to LTAVD, is its ability to deposit coatings even at ambient temperature. This feature makes available the advantages of the arc vapor deposition process to the metallization of a wide variety of plastics with low deflection temperatures. LTAVD coatings have been used in decorative, industrial and military applications. Since LTAVD is a dry process, carried out in vacuum and with no toxic gases, it is environmentally safe. The process and its characteristics are summarized in Table II.

The LTAVD process in particular, and CAPD processes, in general, are still in their nascent stages. Many aspects of the process are not clearly understood[4] and hence provide much scope for research, both fundamental and applied.

Table II. Characteristics of the LTAVD Process

Coating energy/ion	Up to 100eV
Ionization percentage	Up to 100%
Coating density	Up to 95%
Coating adhesion	Can be $\gtrsim$70MPa ($\gtrsim$10,000 psi)
Energy efficiency	E.g., up to 40g Ti/kWhr
Source shape	Non-planar
Source utilization	Up to 90%
Source-Substrate distance	1cm-200cm
Chamber pressure	10^{-2} - 10^{-8} torr
Chamber atmosphere	Vacuum, Inert, or reactive gas
Safety	Environmentally safe, dry process

ACKNOWLEDGEMENTS

The authors thank Mr. Michael Walsh, President and CEO of VTI, for permission to publish this paper. Thanks are also due to Dr. Kash Mittal of IBM, Thornwood, NY for his interest. Substrate surface cleaning prior to deposition, and coating thickness and adhesion measurements after deposition were carried out by Ms. Chun Zhao.

REFERENCES

1. K.L. Mittal and J.R. Susko, Eds., "Metallized Plastics 1:Fundamental and Applied Aspects", Plenum Press, New York 1989.
2. K.L. Mittal, J. Vac. Sci. Technol., 13(1), 19 (1976).
3. H. Randhawa, Thin Solid Films, 167, 175 (1988).
4. P.A. Lindfors, W.A. Mularie and G.K. Wehner, Surface Coatings Technol., 29, 275 (1986).
5. E. Pinkhasov, U.S. Patent 4,351,855, Sept.28, 1982.
6. J.E. Daalder, J. Phys. D:Appl. Phys., 10, 2225 (1977).
7. C.W. Kimblin, J. Appl.Phys., 44, 3074 (1973).
8. W.M. de Cock and J.E. Daalder, Proc. VII Intl. Symp. on Discharges and Electrical Insulation, Novosibirsk, USSR, 288-292 (1976).
9. H.C. Miller, J.Phys.D:Appl.Phys., 12, 1293 (1979).
10. V.M. Lunev, V.D. Orcharvenko and V.M. Khoroshikh, Sov. Phys.-Tech. Phys. 22, 858 (1977).
11. K-H. Muller, J. Appl. Phys., 59 (8), 2803 (1986).
12. M.E. Harper, J.J. Cuomo and H.R. Kaufman, J. Vac.Sci. Technol., 21, 737 (1982).
13. P.J. Martin, D.R. McKenzie, R.P. Netterfield, P. Swift, S.W. Filipczuk, K-H. Muller, C.G. Pacey and B. James, Thin Solid Films, 153, 91 (1987).
14. H.K. Pulker, A.J. Perry and R. Berger, Surface Technol., 14 25 (1981).
15. K.L. Mittal, J. Adhesion Sci. Technol., 1(3), 247 (1987).
16. U. Helmersson, B.O. Johansson and J.-E. Sundgren, J. Vac. Sci. Technol., A3, 308, (1985).
17. K.L. Mittal, Electrocomp.Sci. Technol., 3, 21 (1976).
18. P.A. Lindfors, Proc. Conf. on the Application of Ion plating and Implanation to Materials, 3-5 June, 1985, Atlanta, Georgia, R.F. Hochman, Ed. ASM, p161.
19. I.I. Aksenov, V.A. Belous, V.G. Padalka, and V.M. Khoroshikh, Sov. J. Plasma Phys., 4, 425 (1978).
20. H. Randhawa, J. Vac. Sci. Technol., A7,2346 (1989).
21. P.J. Martin, R.P. Netterfield and T.J. Kinder, paper presented at Intl. Conf. on Metallurgical Coatings/Intl. Conf. on Thin Films, San Diego, California, 1989.

MAGNETRON SPUTTERED TiN HARD COATINGS ON PLASTICS

O. Knotek and F. Löffler

Aachen University of Technology,
Templergraben 55, W-5100 Aachen, FRG

Advances in processing methods in the film technology field due to the introduction of the PVD technique, and especially of magnetron sputtering, allow hard coatings to be deposited. Given a suitable choice of process parameters, techniques of this kind also permit reduction of process temperatures, making it possible to coat temperature-sensitive materials such as plastics.

In the present study, magnetron sputtering in the DC reactive mode was used to deposit TiN hard coatings on various plastics representing the three main groups: thermoplastics, elastomers and duroplastics.

Suitable analytical techniques had to be developed in order to investigate this new compound system. Investigations described in the paper include scratch tests for adhesion measurement, microhardness tests, scanning electron microscopy (SEM) and optical microscopy, stylus-disc tribometer tests, X-ray structure analysis, fatigue tests, corrosion tests, surface roughness analysis and measurements of the electrical conductivity of the coating. A secondary objective of the study was to establish the relevance of different test methods to various substrate materials.

INTRODUCTION

Magnetron sputtering is one of a number of processes, including vapor deposition, electroplating and electroless deposition, which can be used to deposit metallic coatings on plastics[1]. Magnetron sputtering is very flexible in terms of the coating materials which can be applied, and allows deposition of extremely homogeneous coatings. Extensive experience in the electroplating and vapor deposition coating of plastics proved useful in the present study[2-5].

Owing to the large number of coating parameters, substrate materials and analytical methods included in the study, the present paper can illustrate only the most important findings. TiN was selected as the coating material because its high hardness (approximately 3 000 HV 0.05) imposes special requirements on the coating process, making it particularly suitable for demonstrating the potential of sputtering technology. It is, however, equally possible to deposit a purely metallic coating (e.g. copper, titanium or chromium).

Metallized Plastics 2, Edited by K.L. Mittal
Plenum Press, New York, 1991

EXPERIMENTAL

Coating Process

Coatings were deposited on the high-power sputtering principle, using the Leybold-Heraeus "Z400 Universal Laboratory System". The airlock enabled ancillary coating times to be shortened considerably, with less frequent run-up torques and hence reduced pump loads. The coating technique is described elsewhere[6].

To deposit TiN coatings, a Ti target was sputtered in a nitrogen reactive gas and argon inert gas atmosphere. The DC mode was used. The coating thickness is proportional to the coating time, and ranged from 0.2 μm at a sputtering time of 5 minutes to 2.7 μm at a sputtering time of 60 minutes.

The plastic substrates coated in the study were glass-fibre reinforced plastic (GFRP), carbon-fibre reinforced plastic (CFRP), polytetrafluoroethylene (PTFE), polymethacrylate (PMMA – filled with glass whiskers), transparent PMMA, polypropylene (PP), polystyrene (PS), polyvinyl chloride (PVC) and polyurethanes (PUR). The base material for the fibre-reinforced plastics was an epoxy resin.

Preparation of the Plastic Substrates

Plastic specimens required pretreatment before the coating process. The primary aim was to ensure that the plastic was entirely free of contamination which might otherwise lead to unacceptable characteristics, e.g. poor adhesion, a rough surface and lack of reproducibility.

All specimens, irrespective of the type of plastic employed, were cleaned for five minutes in an alcohol-filled ultrasonic bath. Given the prescribed short bath residence time, previous tests had shown alcohol to be the only medium which did not damage the plastic. Ultrasonic treatment caused no unacceptable heating due to internal oscillations in the plastic.

In order to evaluate the influence of a chemically-activated plastic surface on coating adhesion and structure, a number of specimens were treated with specially-selected chemicals. PTFE, PP and PS specimens were immersed for one minute in tetrachloromethylene, and CFRP, GFRP, PMMA, PUR and PVC specimens in concentrated sulphuric acid. In both cases, immersion was followed by repeated rinsing with water and a five-minute ultrasonic treatment in the alcohol bath. This final stage was intended to freeze the etching process initiated by chemical treatment.

An alternative method of activating plastic surfaces is the formation of active radicals through the agency of metal ions formed in the sputtering process[7]. The metal ion reacts with oxygen. This reaction is particularly rapid in organic media. In order for the reaction process to take place, however, the plastic must previously have been enriched with oxygen. Apart from chemical composition, the prime determinants for the oxidation rate are the diffusibility and solubility of the oxygen in the polymer. Since the diffusion rate of amorphous plastic is significantly higher than that of the semicrystalline material, the polymers PS, PVC and PMMA were selected for this form of treatment. Specimens were irradiated by a 20 cm distant ultra-violet lamp (100 W) for one hour, to initiate the above-mentioned reaction.

Ion etching is another alternative for pretreating plastic surfaces. The sputtering polarity is reversed, causing ionic bombardment of the substrate. This cleaning technique raises the temperature of the plastic, and may lead to unacceptable heating.

Evaluation Methods

Various test methods are available for evaluating hard coatings[8]. These methods are, however, designed primarily for use with metallic substrate materials, and their applicability to plastic substrates is limited. They include microhardness measurement, the scratch test, roughness analysis, X-ray microstructural analysis, the stylus-disc tribometer test and examination under optical and electron microscopes.

Other coating appraisal methods have been developed to supplement these established techniques. Of particular interest are tests determining resistance to liquids, fatigue tests and electrical conductivity measurements. A number of inferences on film structure can be drawn from electrical conductivity measurements. If a layer is discontinuous or if there are a large number of cracks in the coating, the conductivity will decrease. Taking various intermediate variables into account, a low electrical resistance (i.e. a high electrical conductivity) indicates a dense, homogeneous coating structure.

Especially interesting in this context is the fact that coating thickness is inversely proportional to the measured electrical resistance. In glass-fibre reinforced epoxy resin, for example, a resistance of 100 Ω was measured after a coating time of 15 minutes, whereas the resistance after 5 minutes was 320 Ω. At a constant coating rate, the product of resistance and coating thickness thus remains roughly the same, demonstrating the significance of this relationship. In general, it may be stated that electrical conductivity measurements on metallized plastics constitute a suitable test method.

Various methods are available for measuring the adhesion of metallic coatings to plastics[9-10]. The scratch test is frequently used on superhard coatings such as TiN[8]. A diamond cone under a predefined load scratches the surface of the specimen, causing the coating to detach from the substrate material at the so-called critical load.

RESULTS AND DISCUSSION

X-ray analysis of titanium nitride coatings on carbon-fibre reinforced epoxy resin revealed that coatings deposited by different processes were largely identical in their composition. *Fig. 1* shows the relative intensity of reflection at different angles of reflection together with the associated orientations.

A change in orientation from the (200) direction to the (111) direction is evident. This may be attributed to the growth-induced stresses in the sputtered films. The peaks for the titanium nitride film deposited without bias voltage or heating occur at slightly higher reflection angles, with a lower (200) peak. This may be explained by the fact that the higher temperatures produced by heating and by application of a bias voltage permit partial relief of the induced stresses, with an associated shift towards the distribution for the powder material.

Film structure and adhesion (the critical load in the scratch test is between 2 N and 20 N) are more heavily dependent on the coating process than are coating composition and lattice structure.

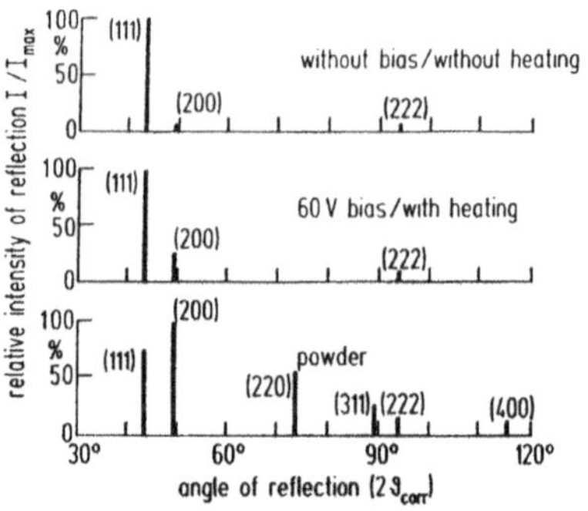

Fig. 1. X-ray Intensity Distribution of TiN as a Coating on a CFRP Substrate and in Powder Form (according to ASTM catalogue)

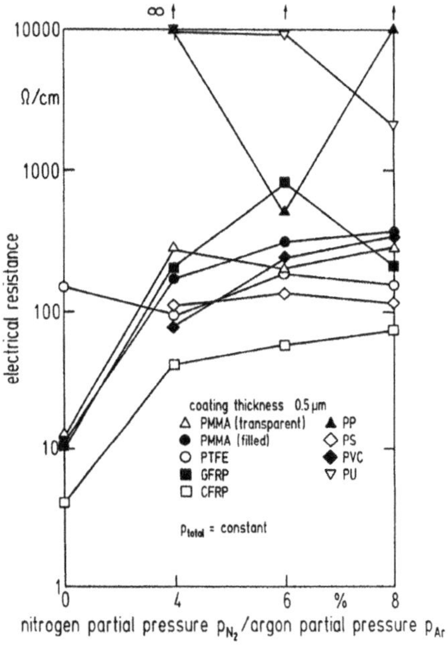

Fig. 2. Electrical Resistance as a Function of the Partial Pressure Ratio

The following observations rely largely on the electrical resistance criterion, which has proved its usefulness for evaluation[8]. The optical quality and adhesion of coatings were generally found to correlate well with high electrical conductivity (below 10 Ω/cm).

*Fig. 2 i*ndicates that electrical resistance rises with increasing nitrogen partial pressure. The influence of the various specific resistivities of titanium and titanium nitride is virtually negligible, reaching a maximum factor of two. The fluctuations by as much as a factor of 20 indicated in the figure may be attributed to the considerably greater toughness of titanium as opposed to titanium nitride, with possibly superior

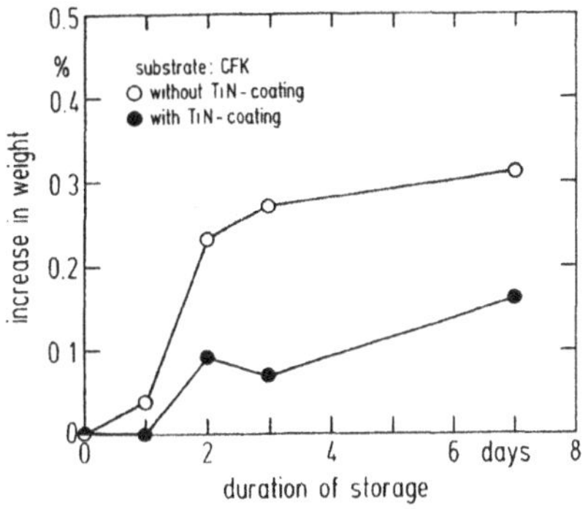

Fig. 3. Weight Increase after Immersion in Water

adhesion and continuity of the film structure. For example, some microcracking is observable on the surface of the TiN coating, entailing a decrease in electrical conductivity.

It is also apparent that coatings on the most temperature-sensitive materials, such as PP and PUR foils, were subject to cracking, with resulting interruption of conductivity.

Fig. 3 shows the extent to which a coating protects carbon-fibre reinforced plastic substrates against sorption of water. Weight increase commences a day later than on the uncoated specimen, and is also considerably lower during the subsequent phases. The occurrence of any water absorption at all in the coated specimen is explained by the absence of a resin surface layer at the edges of the original specimens. As a result of the cutting process, severed fibres project from specimens, leaving a particularly rough edge which cannot be covered completely. Absorption of water by the plastics also creates stresses which inevitably cause further cracking in the surface zones, permitting additional water absorption up to the saturation point[7].

Chemical pretreatment results in higher conductivity on both PMMA types, while ionic etching produces higher conductivity only in the filled thermoplastics, as shown in *Fig. 4*. This may be attributed to the higher thermal resistance of the filled thermoplastic, which allows the improved properties produced by ionic etching to show up particularly clearly at higher temperatures.

The manufacturing process used to produce the plastic specimen had an adverse effect. The plate had been stretched in a preferred direction to achieve greater uniformity and lower roughness. The stresses induced by this process were reversibly relieved by heating, causing the specimen to contract in a preferred direction. *Fig. 5* documents this process. The inherently homogeneous film has contracted in a preferred direction, owing to the substantial deformation forces, and has arranged itself in folds on the substrate material.

This rolling effect induced by the substrate manufacturing process is also evident in *Fig. 6*, which indicates a tenfold increase in electrical resistance after various forms of pretreatment. The explanation is to be found in the deformation of the substrate material produced by higher loading of the thermoplastic specimen, inducing cracks which considerably reduce the homogeneity of the film. One hour of

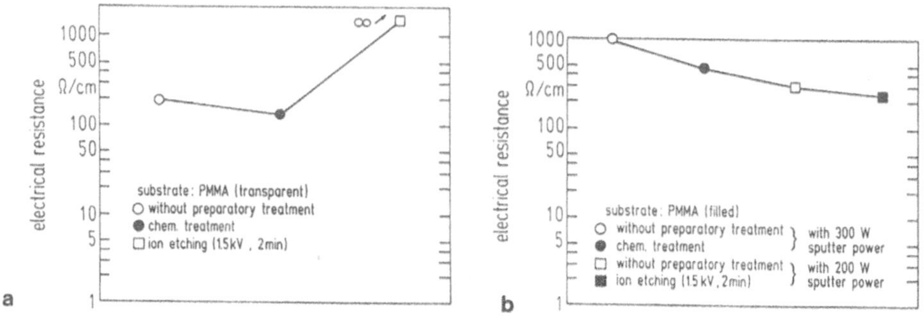

Fig. 4. Electrical Resistance of the TiN Film as a Function of Pretreatment a) transparent plastic b) filled plastic

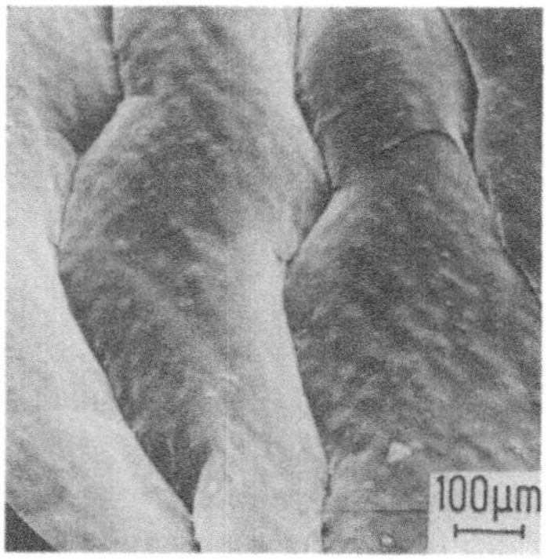

Fig. 5. SEM Image of a TiN Film on Polystyrene (60 V bias, 5 minutes coating time)

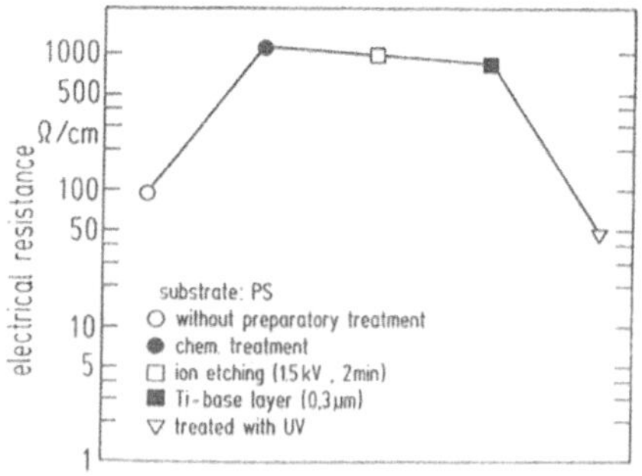

Fig. 6. Electrical Resistance of the TiN Film as a Function of Pretreatment (300 W power, 60 V bias)

ultra-violet radiation had a distinctly positive influence on the conductivity and hence on the structure and adhesion of the film.

The PVC specimen proved particularly sensitive to chemical pretreatment. It swelled during the coating process, with blistering throughout the substrate. This observation is confirmed by the high electrical resistance indicated in *Fig. 7*. The graph shows the favourable effects of ultra-violet radiation or ion etching. The surface of the PVC specimen etched prior to coating appears extremely tough. Even the scratch shown in *Fig. 8*, produced at a load of 40 N, was not associated with any visible spalling. Plastic deformation induced only ultra-fine adjacent cracks, without severe interruption of film coherence.

As also indicated in *Fig. 7*, a titanium base layer affects the increase in electrical conductivity. A good, dense titanium film may be inferred from the extremely favourable electrical resistance value. Further examination of the film indicates, however, that the titanium nitride coating does not completely cover the titanium base layer, since the process parameters for deposition on plastic substrates used in this instance are not suitable for coating metallic substrates.

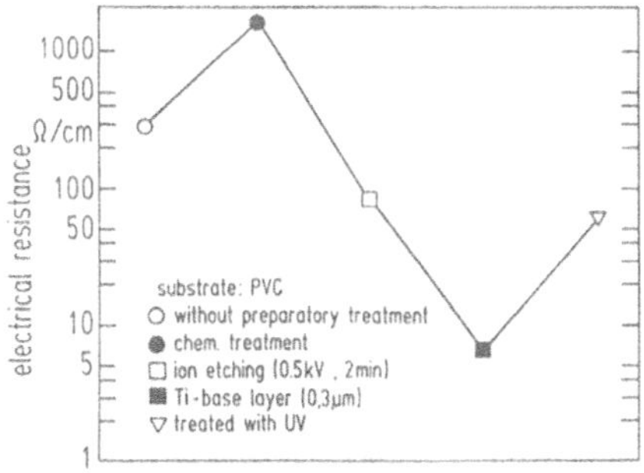

Fig. 7: Electrical Resistance of the TiN Coating as a Function of Pretreatment (300 W power, 60 V bias)

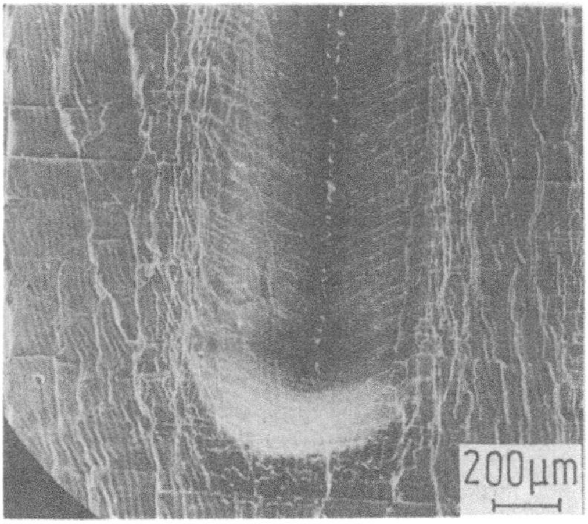

Fig. 8: SEM Image of a TiN Surface Showing a Scratch (300 W, 5 minutes coating time, etched prior to coating, PVC substrate)

CONCLUSION

The duroplastics coated in the study were carbon-fibre and glass-fibre reinforced epoxy resins. The titanium nitride coatings exhibited a wide range of good adhesion, homogeneity and water resistance properties. Coatings were deposited well and reproducibly over a period of one hour. Ion etching, chemical pretreatment, bias voltage and substrate heating produced additional improvements in these coating properties.

Coatings on an elastomer, represented by polyurethane foil, showed a distinct improvement in the usual tendency of the plastic to absorb distilled water. Coating quality was improved by a slightly superstoichiometric TiN coating.

A scatter of coating properties was evident in the case of the thermoplastics. Polytetrafluoroethylene coatings exhibited a homogeneous structure with a short coating time. Application of a bias voltage produced an increase in the adhesion and in the electrical conductivity of polymethylmethacrylate coatings with short coating times. No improvement was observable when sputter power was increased. Conversely, an increase in sputter power on polypropylene resulted in higher conductivity of the coating, still further enhanced by increasing bias voltage. Ultra-violet activation was advantageous for coatings on polystyrene and polyvinylchloride.

REFERENCES

1 G. Siegle, in "Haus der Technik – Vortragsveröffentlichung", K. Giesen, Editor, Vol. 276, Vulkan Verlag, Essen, 1971.

2 K. Heyman, W. Roedel and G. Woldt, in "Metallische Überzüge auf Kunststoffen", H. Wiegand and H. Speckhardt, Editors, p. 48, Carl Hanser Verlag, Munich, 1966.

3 G. Müller, "Galvanisation von Kunststoffen", E. G. Leuze-Verlag, Saulgau, 1966.

4 H. Jörder, Textil Praxis, 12, 917 (1957).

5 R.E. Hurley and E. W. Williams, Thin Solid Films, 92, 99 – 106 (1982).

6 R. F. Bunshah, Editor, "Deposition Technologies for Films and Coatings", Noyes Publications, New Jersey, 1982.

7 B. Dolezel, "Die Beständigkeit von Kunststoffen und Gummi", Carl Hanser Verlag, Munich/Vienna, 1978.

8 L. I. Maissel and R. Glang, Editors, "Handbook of Thin Film Technology", McGraw-Hill, 1970.

9 R. E. Van de Leest,Thin Solid Films, 124, 334 – 341 (1985).

10 K. L. Mittal, Editor, "Adhesion Measurement of Thin Films, Thick Films and Bulk Coatings", ASTM, STP 640, Philadelphia, PA, 1978.

DIFFUSION AND GETTERING SIMULATIONS OF ION IMPLANTED COPPER IN POLYIMIDE.

J. H. Das and J. E. Morris

Department of Electrical Engineering,
T.J. Watson School of Engineering, Applied Science and Technology, State University of New York
Binghamton, New York 13902-6000.

Recently we have observed gettering of ion-implanted copper in polyimide films. In this paper we have modeled the thermal process by incorporating the experimentally obtained dual activation energies into conventional diffusion models. The lower activation energy (small diffusant/substrate interaction) is associated with free atomic diffusion, while the larger activation energy (large diffusant/diffusant interaction) corresponds to diffusants with their movements restricted by clustering within the interaction distance of similar species. Computer simulations of the process validate the observed gettering mechanism via clustering. Experimental high temperature diffusion results are also in agreement with the simulation of the subsequent diffusion of clusters.

INTRODUCTION

In recent years, polymers have found increasing use in integrated circuits and printed circuit board applications due to their thermal stability and usefulness as insulators and protective layers. Metal/polymer structures have become a particularly important area of investigation. There is, however, a lack of proper understanding of interdiffusion mechanisms at metal/polymer interfaces.

Recent studies on metal/polymer interfaces have provided only rather undefined or even elusive diffusion parameters, specially the activation energy, frequency factor, jump length, correlation factor, etc. The results are notably different for the various techniques used to investigate such properties.

One RBS (Rutherford Back Scattering) study of diffusion of Cu films deposited on Dupont Kapton H® found that diffusion was thermally activated with an activation energy of about 0.5 eV when the films were heated in air. Similar films heated in vacuum exhibited a rather constant diffusion coefficient independent of temperature[1].

Separate TEM (Transmission Electron Microscopic) studies have revealed the formation of small Cu clusters at the Cu/polyimide interface and Fickian type diffusion in the polymer[2].

A recent direct measurement of Cu diffusion into polyimide by means of low energy ion sputtering in combination with a radiotracer technique reported a two part penetration profile into the polymer: an initial non-Fickian steep drop in the diffusion activity and a Fickian tail into the polymer. The former was attributed to cluster formation via surface diffusion of the Cu tracer during deposition[3].

Metallized Plastics 2, Edited by K.L. Mittal
Plenum Press, New York, 1991

In all of the above, the Cu was deposited as a surface film. With ion implanted Gaussian profiles of Cu in polyimide we have observed a three-part activity of the implanted Cu ions, namely: (1) a low temperature (below 460 K) diffusion with an activation energy of 0.41 eV, (2) a high temperature (above 525 K) activation energy of 1.9 eV, attributed to cluster motion, and (3) an intermediate range of temperatures over which self-gettering was observed, in which the half width of the profile decreased and the peak concentration increased[4-6].

A summary of these results is shown in Figure 1. By comparison it can be concluded that our low temperature results are in fair agreement with other results obtained by various techniques, while there are large discrepancies in the high temperature results. It is believed that the cluster diffusion is only observable at the low concentrations resulting from the ion implantation technique.

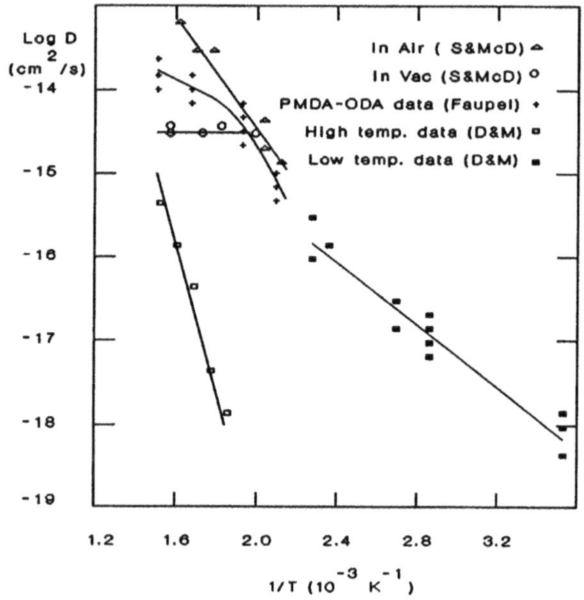

Figure 1. Arrhenius plots of the Cu diffusion constant in polyimide. Data replotted from Shanker and MacDonald[1], Faupel et al[3] and Das and Morris[6]. The D&M high temperature data are attributed to clusters.

Clustering at the film surfaces has been attributed to the relatively weak chemical interaction between Cu and polyimide, in contrast to the significantly higher Cu/Cu interaction energy. In the case described below, with a lack of abundance of Cu atoms (dilute dose), clustering was rather unexpected.

Although the mechanism responsible for the gettering is still not completely understood, there is a second order relaxation process over the gettering temperature range. The polyimide phase transition which would account for such a phenomenon could also initiate clustering by a change in bulk and/or interfacial energies[6].

The gettering process has also been investigated through computer simulations of diffusion where we have included the Cu/Cu interaction in the diffusion model, thus allowing copper atoms to be trapped by other atoms/clusters within the interaction or capture distance. Simulation results validate the gettering process under the above assumptions and reveal several interesting dependences of the process itself upon the initial implant profiles and doses.

High temperature diffusion was simulated by allowing individual atoms to migrate freely (smaller activation energy) while clusters were treated as units. The migration of such units was allowed at a reduced frequency (higher activation energy) in a conventional diffusion model. Simulation results are in good agreement with the experiments.

THEORY

Diffusion

An atom in the lattice, with its normal thermal vibration frequency $\nu \sim 10^{13}$ Hz about the equilibrium site, strikes the potential barrier E_m, ν times per second. However, an atom needs an adjacent vacancy for a successful jump. The frequency of jumping f_m to a neighboring lattice site with Z nearest neighbors is given by

$$f_m = Z.\nu.\exp-\frac{E_a}{kT} \quad . \tag{1}$$

where $E_a = E_m + E_v$ is the activation energy of diffusion and E_v is the formation energy of a vacancy[7].

From macroscopic diffusion, we have a net flow of atoms across a plane per second dN/dt, provided there exists a concentration gradient dC/dx in the region around that plane. dN/dt can be expressed as

$$\frac{1}{L^2}\frac{dN}{dt} = -D\frac{dC}{dx} . \tag{2}$$

With a cross-sectional area L^2 and d as the nearest neighbor distance, a three dimensional diffusion coefficient D is defined as

$$D = \frac{f_m d^2}{6} . \tag{3}$$

Equation (2) is known as Fick's Law of Diffusion. Several features about these expressions are of particular interest as they provide a means of measurement of the diffusion of atoms in the direction favored by the concentration gradient and hence of the activation energy of diffusion.

The Distribution Profile

The implant profile of copper can be approximated as Gaussian with a peak concentration $N_p(t)$ at range R_p and a half-width W(t). The solution to the diffusion equation is particularly simple provided the diffusion constant D is concentration independent[8]. The concentration N(x,t) at a depth x after a diffusion time t is expressed as

$$N(x,t) = N_p(t).\exp\frac{(R_p - x)^2}{2.W(t)^2} . \tag{4}$$

Also $N_p(t)$ and W(t) are related by

$$N_p(t)=\frac{0.399Q}{W(t)}. \tag{5}$$

where Q is the total concentration per unit area.

The Rutherford Backscattering Spectroscopy (RBS) profile 1 of Figure 2 shows an as-implanted distribution of 220 keV ion implanted Cu in polyimide. Profile 2 of Figure 2 shows the gettering effect with an increase in $N_p(t)$ and a decrease in W(t) with time, The profile has the appearance of a distorted Gaussian. Profile 3 of Figure 2 shows a typical distribution at 623 K, with a higher W(t) and a lower peak than profile 2, showing a subsequent high temperature diffusion.

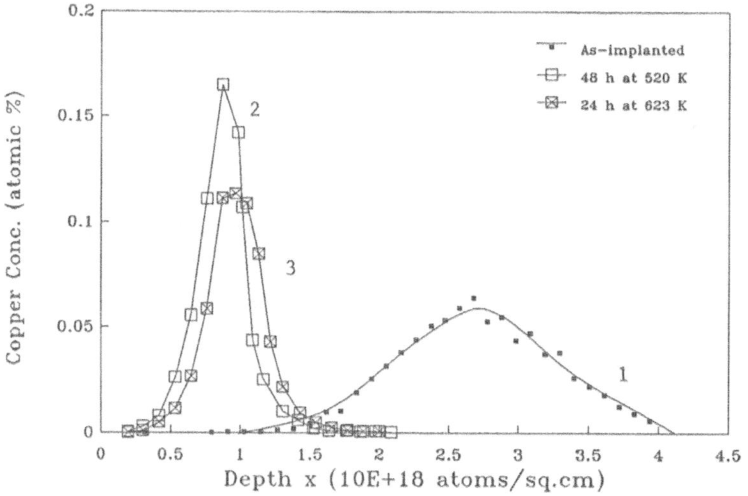

Figure 2. Concentration of Cu in VN-type Kapton as a function of depth and thermal treatment [experimental].

For a Gaussian distribution of free atoms, half-widths $W(t_1)$ and $W(t_2)$ provide a measure of the diffusion constant, given by

$$D=\frac{W(t_2)^2-W(t_1)^2}{2(t_2-t_1)} \tag{6}$$

which implies that the half-width increases with time due to diffusion of the atoms. Consequently the peak concentration decreases in order to conserve the total number of atoms in the Gaussian as shown by Equation (5), and can similarly provide the means to determine D.

Application of Equation(6) to profiles 2 and 3 of Figure 2 gives rise to a negative value of D. The simple diffusion represented by Equations (5) and (6) is not an adequate representation of the migration processes.

Initial Distribution

Due to the randomness of the distribution process, a certain number of the as-implanted atoms are expected to be implanted within interaction distances even for dilute dosage. Therefore, the initial distribution is expected to contain both free atoms and atoms in clusters. Since the number of available sites in a lattice is fixed, one can expect the number of atoms in clusters to increase with $N_p(0)$. Figure 3 shows normalized as-implanted profiles of the remaining free atoms $N_{free}(x,0)$ in the simulated lattice. As the peak concentration C_C (given by $N_p(0)/N_Y$, where N_Y is the total number of lattice points in the Y direction of the lattice) is increased the profiles begin to distort as clustering begins to take effect (as is visibly evident from the dips in the profiles.) We have observed an exponential $N_{free}(R_p,0)$ decrease with increasing C_0.

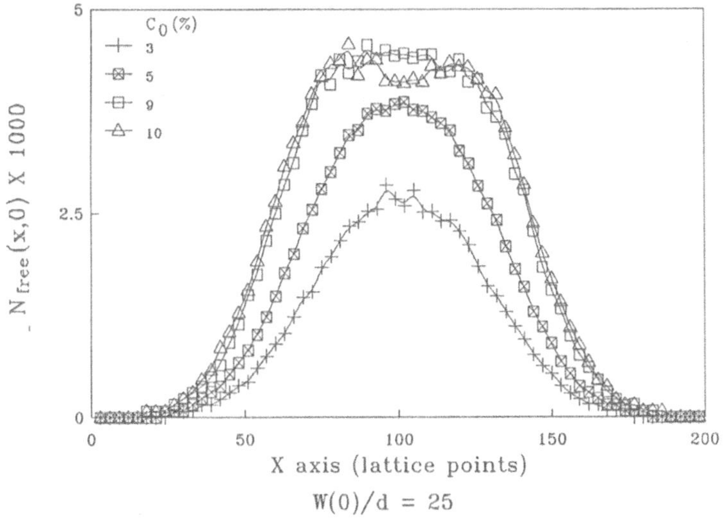

Figure 3. Initial distribution of free atoms in the lattice for constant normalized profile half-width $W(0)/d=25$ [Simulation]. The central dip is due to atoms in clusters which are not included in N_{free}.

Simulation Model

We have simulated the experimental situation in the model in the following fashion: If a diffusant has no other diffusant within its interaction distance, it is regarded as rather loosely bound to the lattice and requires less energy for a jump. This state corresponds to the relatively small 0.41 eV activation energy of diffusion. On the other hand, an atom within interaction distance of another atom/cluster is considered to be "stuck" due to the large interaction energy, and further migration of atoms in this state requires the migration of the cluster as a whole with a much larger activation energy. This state corresponds to the 1.9 eV value observed for clusters. Note that by formally considering the higher activation energy to be associated with the individual atoms within the cluster, one retains the possibility of the individual atom subsequently escaping from the cluster.

To investigate the gettering process, it has been simulated up to 90 nearest neighbour jumps of free atoms with the clusters fixed, assuming a negligible cluster migration during such a small period of simulation and a moderate thermal activation in the temperature range of gettering.

The high temperature diffusion was investigated by increasing the probability of such cluster diffusion to correspond to the 1.9 eV activation energy. The process was simulated for 500 nearest neighbour jumps of the free atoms.

THE SIMULATION

Recently, there has been increasing interest in the physical mechanisms governing the geometrical structure in diffusion limited aggregation(DLA)[9-10]. Several growth mechanisms have been proposed[11-13].

The model presented here is a variant of the Eden one[14]. In this model all particles were distributed throughout the lattice at the beginning of the process to simulate an implant profile given by Equation (4) with varying Np and W(0). These particles walk randomly until they visit a site adjacent to another particle and get "trapped". Edges of such clusters tend to grow more rapidly than other perimeter sites because perimeter sites near the center are "shadowed"[13].

The 15×10^6 lattice was chosen to be a 3D simple cubic structure, with the number of lattice sites varying to accommodate the dose and half width requirements.

Figure 4a shows a cross-section of such a simulated initial distribution with C_0=8% and W(0)/d=35 showing mostly free atoms with a few initial clusters, whereas in Figures 4b and 4c the same lattice is seen after time $40t_{nn}$ and $120t_{nn}$ respectively, where t_{nn} is the required time per nearest neighbor jump. Clustering is evident and the number of free atoms clearly decreases with time as atoms which are not trapped diffuse away.

SIMULATION RESULTS

Gettering process

Growth mechanism. In order to explain the simulated gettering process, a time dependent gettering rate, G(t), is defined as a measure of the growth in the distribution peak, compared to its initial value $N_p(0)$, as

$$G(t) = \frac{N_{max}(t)}{N_p(0)} \tag{7}$$

where the total number of atoms at the peak at time t, $N_{max}(t)$, analogous to $N_p(t)$, which can be expected at or near R_p, can be expressed as

$$N_{max}(t) = N_{mfree}(t) + N_{mclust}(t) \tag{8}$$

where $N_{mfree}(t)$ and $N_{mclust}(t)$ are the number of free atoms and numbers of atoms in clusters at the peak.

Since the cluster movements are restricted by a much larger energy requirement, for the purposes of the simulations, which were limited between 50 and 90 nearest neighbor jumps of free atoms, clusters were considered to be fixed in the lattice.

During each random jump a fraction of $N_{mfree}(t)$, denoted by $N_{fc}(t)$, is lost to clustering. Hence $N_{mclust}(t)$ of Equation (8) is expected to increase with time as the sizes and the numbers of such clusters increase in this region. Another fraction of these atoms will continue to migrate

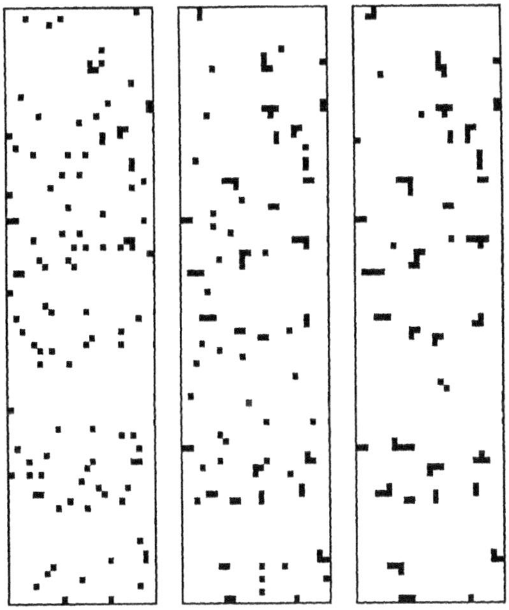

Figure 4. A cross section of the lattice xy plane showing progressive clustering.

as free atoms according to concentration gradients of their distributions. Hence, $N_{mfree}(t)$ at an incremental time t+1 can be expressed as

$$N_{mfree}(t+1) = N_{mfree}(t) - N_{fc}(t) + [N_{Din} - N_{Dout}(t)] \tag{9}$$

Where $N_{Din}(t)$ and $N_{Dout}(t)$ are the atoms diffusing in the and out of the peak region, respectively.

Activity at the peak: $N_{max}(t)$. Figure 5a shows the time variation of activity G(t) (Equation 7) for implants with various implant dosages.

Several interesting features can be observed from these results: There exists a primary growth process with a higher rate of growth in G(t), which is followed by a secondary growth process with a much lower growth rate. Considerable fluctuation of the G(t) magnitude is seen at dilute dosages which is seen to be decreasing with the increasing C_0. The magnitude of both primary and secondary growth decreases with increasing C_0.

When the quantity $[N_{Din}(t) - N_{Dout}(t)]$ in Equation (9) is positive, there is a growth in the peak, due to diffusion of the free atoms surrounding the gettering region.

Activity at the peak: $N_{mfree}(t)$. Figure 5b shows the variation of $N_{mfree}(t)/N_p(0)$ with time for various initial atomic distributions. According to Equation (9) the number of these atoms decreases with time due to ongoing clustering and diffusion of the free atoms. The loss rate is seen to increase with C_0 which is expected; $N_p(0)$ affects the peak concentration and hence the clustering at the peak, while the W(0) variation has negligible effect on the rate of loss there.

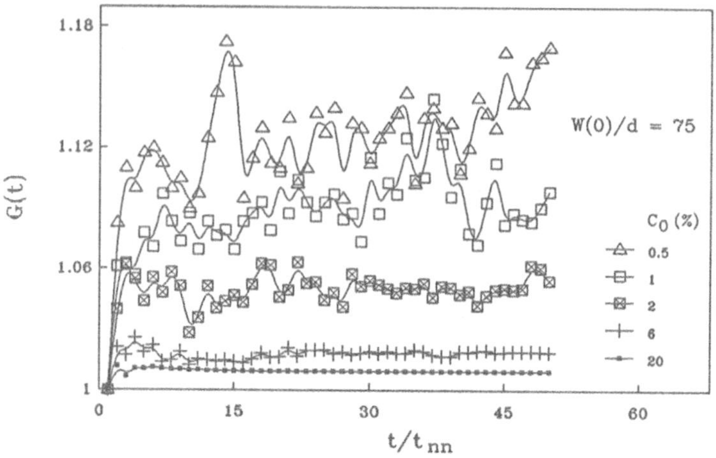

(a)

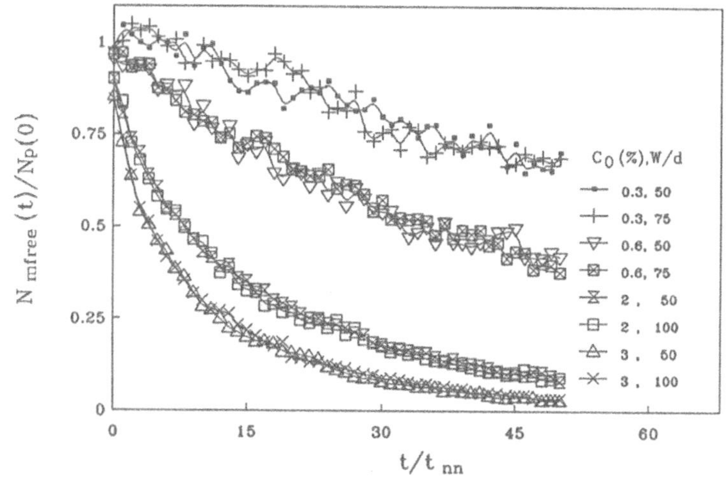

(b)

Figure 5. Time variation of (a) G(t) and (b) $N_{mfree}(t)/N_p(0)$ showing the effects of cluster growth. The reduction in $N_{mfree}(t)$ reflects the diffusion and clustering of the free atoms.

Primary growth. The primary contributor to the growth is attributed mostly to the localized depletion of free atoms created by as-implanted atoms in clusters. Due to these depleted regions the concentration gradient of free atoms around them favors diffusion in the inward direction resulting in further growth of the existing clusters and the formation of new clusters since all atoms are essentially "trap" sites. Such diffusion causes a rapid increase in G(t). As the localized concentration gradient vanishes, the rate of migration decreases, resulting in a slowdown in the overall growth process. The magnitude of such growth is also seen to decrease with increasing dosage, since increased dose represents an increased number of atoms in the initial clusters leaving fewer free atoms in the peak region. In addition the increased number of atoms in the lattice due to the higher dose also reduces the lifetime of the available free atoms, since all atoms are essentially "trap" sites.

Secondary growth. The secondary growth mechanism can be attributed to the atoms in the entire distribution. Due to the longer time duration of the process, atoms far away from the initial peak region have the opportunity to diffuse toward the peak region and contribute to the gettering process.

In the case of C_0=20% in Figure 5a, the gettering is seen to be negligible due to a much reduced concentration of free atoms in the initial distribution and an abundance of "trap" sites. The activity under the peak region ceases within a short time, showing a complete clustering of atoms at the peak.

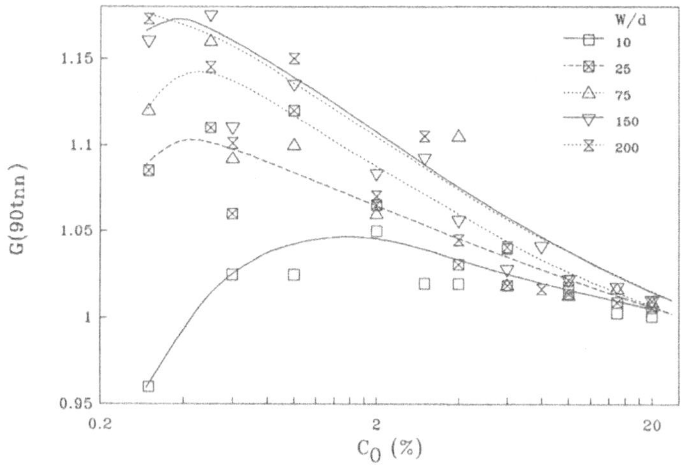

Figure 6. The variation of gettering with C_o showing competing trends.

For a given C_0, a smaller W(0)/d also represents an increased concentration gradient at the edge of the distribution. At lower C_0, a

G(t) versus implant-profiles. In order for an increase in the G(t) magnitude to occur, an abundant supply of free atoms near the gettering region is required to provide a large $N_{Din}(t)$ for gettering. Longer lifetimes of the free atoms surrounding the gettering region are also essential in order to increase the gettering, which requires fewer clusters surrounding the gettering region. Hence there is an optimum distribution profile suitable for a maximum gettering effect.

W(0) variation. Figure 6 shows the variation of $G(90t_{nn})$ (as defined by Equation 7) with C_0 for various W(0)/d. At lower C_0 (below 1%), a sharp drop in G(t) can be seen with decreasing W(0)/d. According to Equation (5), at these low concentrations, and for small W(0)/d cases, there are relatively fewer atoms in the lattice.

lack of "trap" sites (atoms) and the large concentration gradient favors conventional diffusion of free atoms and the overall gettering effect is diminished. In one case, with C_0=0.3%, G(t) falls below 1, again showing dominant conventional diffusion of free atoms.

At higher C_0 (above 1%), with increasing W(0)/d, the distribution spreads out and the contribution of the atoms at and near the edge of the distribution becomes less significant and the effect of W(0)/d variation on G(t) is seen to be diminishing.

N_p variation. Figure 6 also shows the effect of the variation of G(t) with C_0 and hence N_p. N_p influences both the concentration gradient at the edges and the concentrations at the peak of the distribution. At larger W(0)/d and dilute dosages, initial clustering is small providing an abundance of free atoms with a longer lifetime, favoring large activity at the peak region, resulting in an increased G(t) in these moderate dosages. A further increase in C_0 still implies more "trap" sites but the density of initial clusterings begins to dominate the process, resulting in a reduction of the number of free atoms and a reduced lifetime of such free atoms. Consequently, a decrease in G(t) is seen with further increases in C_0.

High Temperature Diffusion

In contrast to ideal cases of gettering (clusters immobilized) of figure 5a, figures 7a and 7b show the time variation of the peak of the distribution for various cluster diffusion probabilities with C_0=1 and W(0)/d=25 and 50, respectively.

In all cases the primary growth process is evident, which can be attributed to the fast and brief nature of the primary growth mechanism. Subsequent variation can be attributed to the combined effects of the secondary growth process and the cluster diffusion away from the peak region. The cluster diffusion is seen to reduce G(t) and the rate of reduction is seen to increase with increasing probability of cluster movement. A comparative study of figures 7a and 7b reveals that diffusion is enhanced in all cases with the reduced half width providing larger concentration gradients at the edges of the distribution.

At smaller migration probability, N_{max} is seen to increase in a similar fashion to that of the ideal cases of figure 4a. Simulation results are in general agreement with the experimental gettering results at moderate temperature.

Figure 8 shows the as-implanted distribution profile and the distribution profile after a simulated cluster diffusion with a 0.01 cluster diffusion probability. The process was simulated for 500 nearest neighbor jumps of free atoms. In this early stage of gettering the peak of the profile is seen to increase together with a decrease in its half-width.

DISCUSSION

Due to the randomness of the implant process some atoms are found to be in clusters even at a dilute implant dose, creating a void in the free atom distribution profile. Such a distribution is favorable for further clustering through diffusion.

The lack of movement of the clusters and their further growth in numbers and sizes results in an accumulation of total atoms in excess of the initial distribution at or near its peak resulting in the further gettering of free atoms.

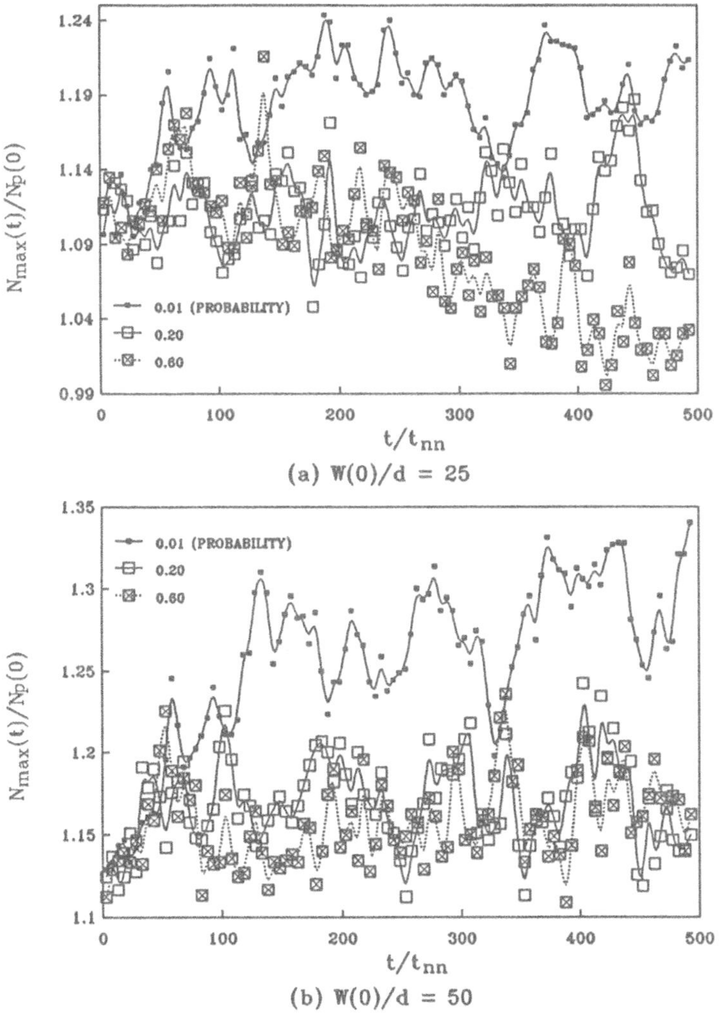

Figure 7. Time variation of $G(t)=N_{max}(t)/N_p(0)$ showing the effects of cluster diffusion. Out-diffusion of clusters dominates the inward diffusion of free atoms for narrow implants and/or high temperatures (high diffusion probability.)

The random diffusion process was simulated by scanning each XY plane of the lattice back and forth sequentially. The XY planes were scanned in an odd/even fashion producing a "combing effect" on the lattice to enhance randomness of the process.

A comparative study of distribution profiles of Figure 3 for C_0=10% and 9%, respectively, reveals that for relatively larger dosages (C_0 greater than 5%) there exists a certain distribution profile for which the initial clustering will be most favorable for the secondary growth

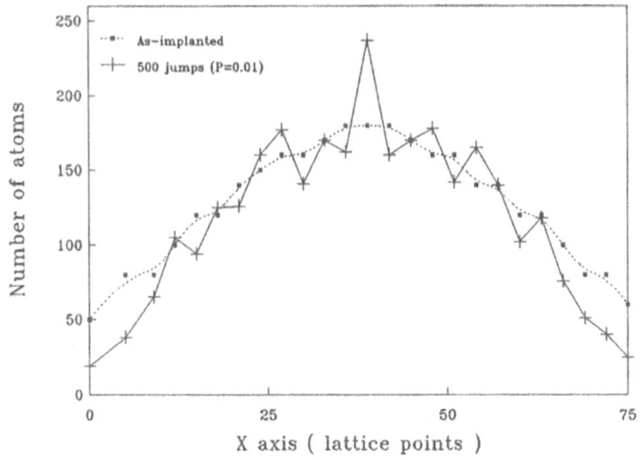

Figure 8. Simulation of thermal diffusion of atoms with a Gaussian distribution, showing the overall tendency to sharpen the profile.

process, and one can in fact expect a secondary G(t) peak at that distribution. Such a G(t) peak can be expected to be much smaller than the primary peak shown in Figure 6 located around C_0=1%. The magnitude of this peak is expected to be reduced by the reduced lifetimes of the free atoms at such high concentrations. The G($90t_{nn}$) plot failed to record such a peak due to (1) the longer time requirements of secondary gettering, and (2) the expectation of the magnitude of such a gain to be below the accuracy of our simulation methods.

There exists a minimum requirement on Q below which gettering cannot be observed.

The secondary growth rate was found to be dependent upon the concentration gradients at the edges of the distribution and also upon the cluster distributions in the initial distribution, while the primary growth mechanism was dominated by localized depletion of free atoms produced by initial clusters.

Though the process was simulated for the ion-implanted case, the simulation is a gross approximation of the process: (1) Boundary errors were large for extreme cases, e.g. small doses and also large W(0)/d points. (2) The choice of lattice, though apparently large, is a gross approximation to an infinite lattice. (3) The randomness of the process was simulated in a sequential fashion. (4) Due to limitation of the lattice size, the best lattice for the simulations was only 80 planes thick compared to about $20x10^3$ planes of 1 mm^2 window of the RBS profiles.

In Figures 5 and 7 large fluctuations in G(t) can be noted, particularly at very dilute dosages. Such fluctuations are also attributed to the limitations of the simulation where an infinite lattice was represented by a finite lattice.

CONCLUSIONS

The diffusion and clustering of ion-implanted metal ions in a polymer have been simulated by computer for comparisons with experiments on copper in polyimide.

Overall simulation results are satisfactory and consistent with the experimental observation of gettering and high temperature diffusion. The simulations adequately describe the gettering process via clustering of adjacent atoms due to strong inter-atomic interactions, and a much smaller cluster migration rate than that of free atoms due to large activation energy of cluster motion.

The simulation results predict a fast primary growth in peak height which is attributed to localized depletion of free atoms due to initial clustering of the implanted atoms at and near the peak region. The rate and magnitude of such a growth is strongly dependent upon the initial dosages and profiles of the implants. The primary growth is subsequently followed by a much smaller secondary growth rate which is attributed to the free atoms in the entire distribution contributing to gettering.

The high temperature diffusion of clusters formed by large interactions, requiring a higher activation energy for subsequent migration, is also in agreement with high temperature diffusion results. The simulation shows a gettering peak which is attributed to the primary growth mechanism prior to the subsequent diffusion of clusters at these higher temperatures.

REFERENCES

1 K. Shanker and J. R. MacDonald, J. Vac. Sci. Technol. A5, 2894 (1987).

2. F. K. LeGoues, B. D. Silverman and P. S. Ho, J. Vac. Sci. Technol., A6, 2200 (1988).

3. F. Faupel, D. Gupta, B. D. Silverman and P. S. Ho, Appl.Phys. Lett. 55, 357 (1989).

4. J. E. Morris and J. H. Das, Technical Digest, Intl. Conf. VLSI & CAD Korea, 534 (1989).

5. J. E. Morris and J. H. Das, Proc. Second Intl. Conf. Solid State & Integrated Circuit Technol. China, (1989).

6. J. H. Das and J. E. Morris, J. Appl. Phys. 66, 5816 (1989).

7. C. A. Wert and R. M. Thomson, "Physics of Solids", p54, McGraw-Hill Book Company, 1964.

8. B. L. Crowder, J. F. Ziegler, F. F. Morehead and G.W. Cole, "Ion Implantation in Semiconductors and Other Materials", p267, Plenum Press, New York, 1973.

9. X. R. Wang, Y. Shapir and M. Rubinstein, J. Phys. A11, L507 (1989).

10. T. Nagatani, Phys. Rev. A38, 2632 (1988).

11. H. E. Stanley, J. Phys. A10, L211 (1977).

12. S. R. Forrest and T. A. Witten, Jr., J. Phys. A12, L109 (1979)

13. T. A. Witten,Jr. and L. M. Sander, Phys. Rev. Lett. 47, 1400 (1981).

14. M. Eden, in "Proc. Fourth Berkeley Symposium on Mathematical Statistics and Probability", J. Neyman, editor, p223, University of California Press, 1961.

PART II. INVESTIGATION OF INTERFACIAL INTERACTIONS

COMPARISON BETWEEN THE INTERFACIAL CHEMISTRY OF METALLIZED POLYIMIDES AND POLYIMIDE FILMS ON BULK METAL SUBSTRATES

M. Grunze, A. Killinger, C. Thümmler, C. Hahn* and T. Strunskus*

Lehrstuhl für Angewandte Physikalische Chemie am Physikalisch-Chemischen Institut
Universität Heidelberg
6900 Heidelberg, F.R.G.
*Laboratory for Surface Science and Technology and Department of Physics
University of Maine, Orono, ME 04469, USA

In this short overview the pertinent differences in the chemical interaction in differently prepared polyimide/metal interfaces are summarized. Considered are the cases of metals (Cr, Cu) deposited onto cured polyimide surfaces, interfaces prepared by spin coating polyamic acid onto a bulk substrate and subsequent thermal imidization, vapor deposited polyamic acid on bulk substrates (Cu, Ag, Si, Cr) with subsequent thermal imidization, and deposition of polyamic acid in the form of mono- and multilayers by Langmuir-Blodgett (LB) techniques onto a Si (100) surface. New FTIR and XPS data are discussed for copper deposition onto polyimide.

INTRODUCTION

Aromatic linear polyimides (PI) are a class of organic polymers with favourable mechanical and dielectric properties for applications in electronic devices which can easily be processed into planar films. Originally polyimides were introduced in Japan as an interlevel dielectric for multilevel metal transistors[1]. Today they are used routinely in VLSI devices as multilevel dielectric insulation[2-4]. There are also several other applications which have emerged in the last 10-15 years. Because of their high absorption coefficient for α-particles they are used as protective coatings in charge sensitive memory devices. Their mechanical properties and high temperature stability is utilized in composite materials for aerospace uses[5], and they are considered as high temperature adhesives[6] as well as lubricants in spacecraft applications[7]. More recent applications of some polyimides take advantage of their crystallinity which makes them applicable as orientating layers for LCD devices, or, when crystallization is reduced by adjustment of the curing procedure, as light guide materials[8].

However, reliable adhesion, particularly in humid environments, remains to be a major issue of concern in the application of PI's. Only recently spectroscopic results on polyimide/metal interfaces have become available which can shed some light on the chemical and physical parameters determining the adhesion bond between polyimide and metallic, oxidic, or semiconducting substrates.

Experiments to determine the chemical interaction between polyimide and metals have concentrated on poly [N,N' -(phenoxyphenyl)-pyromellitimide] (PMDA-ODA) and related model compounds. The two commonly used methods to prepare PMDA-ODA

Metallized Plastics 2, Edited by K.L. Mittal
Plenum Press, New York, 1991

polyimide films are spin coating (SC) and vapor deposition polymerization (VDP). They differ in the way the film is applied to the substrate. VDP is a solventless technique in which the monomers PMDA and ODA are codeposited by evaporation onto the substrate where they react to form polyamic acid. Spin coating (SC) requires that the polymer precursor polyamic acid be applied in a polar solvent, typically N-methyl pyrrolidone (NMP). The interfacial chemistry and adhesion is directly influenced by the way in which the interface is formed. Therefore, variations in the preparation of the polymer and/or metal film are expected to lead to polyimide/metal interfaces with different physicochemical properties.

In the following the results obtained for the interfacial chemistry of metals deposited onto cured polyimide, and polyimide films deposited by different techniques onto metals and silicon will be compared.

SUBMONOLAYERS AND MONOLAYERS OF METALS DEPOSITED ONTO CURED POLYIMIDE SURFACES

Polyimide substrates were prepared by spin coating polyamic acid onto a substrate followed by solvent extraction, imidization and curing at temperatures exceeding 250°C. The metals were evaporated at room temperature under ultrahigh vacuum conditions onto the polyimide films and their chemical interaction was followed by x-ray photoelectron spectroscopy (XPS)[9-12,14,15,18,20], x-ray absorption fine structure (NEXAFS) measurement[10,16], high resolution electron energy loss spectroscopy (HREELS)[13], infrared reflection absorption spectroscopy (IRAS)[17,37], and recently also by scanning electron tunneling microscopy (STM)[18].

The first systematic XPS study on the interaction of evaporated metals with polyimide surfaces was published in 1984 by Chou and Tang[9]. They studied monolayers and submonolayers of Cr, Ni, Cu and Ag on freshly cured (T=623 K, 30 min) polyimide substrates. They concluded that Cr and Ni reacted with the pendant oxygen in the substrate, whereas Cu and Ag did not form chemical bonds. They also presented a simple thermodynamic model to predict whether a given metal will form a chemical bond with the polymer during metallization at room temperature. According to this model Al, Mg, Mn, Sn, Ti and V should react with the pendent oxygen in polyimide.

Later experiments indicated that Cr [10], Ti[11], Al[13,14] react with polyimide surfaces at coverages around and exceeding one monolayer by formation of metal oxides, carbides and nitrides. In the case of Cr[10,15,16,20] and Al[14] spectroscopic data suggest that the initial interaction between the metal and the polymer involves a charge transfer with the PMDA part of the polymer. Clabes et al.[10] compared their XPS data for submonolayer coverages of Cr on PI with chemically and electrochemically reduced polyimide films. Reduction of PI involves electron transfer into the lowest unoccupied orbital (LUMO) of the PI which, according to calculations performed by Haight et al.[15], has its highest amplitude on the phenyl carbon atom next to the carbonyl group and the carbonyl oxygen atom in the PMDA part of the polymer. Clabes et al. interpret their data by postulating that the Cr atom initially interacts with two carbonyl oxygen atoms of two parallel PI-chains by charge donation into the LUMO of the PMDA parts of the PI macromolecules (Fig. 1a). A coordination with two adjacent carbonyl ligands belonging to different monomer units thus allows the chromium atom to reach the stable Cr^{2+} configuration, which, according to Clabes et al., is not possible through interaction with a single PMDA unit. An initial interaction between the metal (Cr) and the polyimide carbonyl groups was previously also concluded by Jordan et al.[16,20] from detailed XPS and NEXAFS studies.

Supported by their "ab initio" calculations Haight et al.[15] interpreted their XPS data by a chromium/PMDA charge transfer complex where the most stable coordination site for the Cr-atom is above the phenyl ring of the PMDA part of the molecule, analogous to a metal-π arene complex. This bonding configuration is shown in Fig. 1b. The chemical

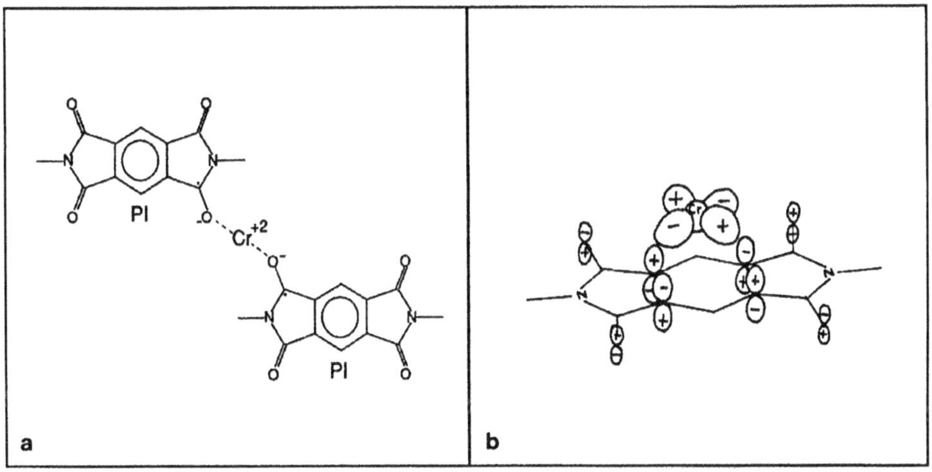

Figure 1. (a)Model for the initial interaction of a chromium atom with polyimide according to ref. 10. One chromium atom interacts with the carbonyl oxygen atoms of two adjacent polyimide chains.

(b)Model of the initial interaction of the chromium $d_{x,y}$ orbitals with the LUMO's of the PMDA part of the polyimide, analogous to a metal π-arene complex according to ref. 15.

shifts in the C1s, O1s and N1s spectra upon chromium deposition were explained by theoretical calculations based on this model without invoking formation of covalent or ionic bonds. Support for this model was suggested by the fact that the changes in the polyimide C1s, O1s, and N1s spectra are qualitatively the same irrespective of whether chromium or copper is deposited. As an example, we replotted the C1s spectrum of polyimide after deposition of 0.012 nm Cr and 0.1 nm Cu in Fig. 2. In both cases the predominant changes are the reduction of the carbonyl C1s intensity (E_B~288 eV) and the PMDA phenyl carbon intensity (E_B~285 eV).

A direct comparison of the initial interaction of Cr and Cu with spin-coated polyimide films was published by Dunn and Grant[17] using infrared reflection absorption spectroscopy. Metallization of a 15 nm thick polyimide film with a 1 nm film of chromium or copper was done in a vacuum system at a pressure of 3×10^{-7} Torr during evaporation. After metallization the films were exposed to laboratory air as they were transferred into the infrared spectrometer. In their studies chromium and copper caused similar changes in the polyimide spectra, in particular in the carbonyl stretching band and the out-of-plane C-N-bending vibration of the $(O{=}C)_2$-N group. No evidence for complex formation between the metals and the aromatic rings in the polymer chains as suggested in ref. 15 was observed. Dunn and Grant concluded that for both metals the carbonyl groups on the polymer surface are the primary reaction sites.

Recent experiments by Mack et al.[18] on copper deposition on polyimide films prepared by *vapor deposition*, however, suggest that copper interacts primarily via the imide part of the molecule because the XPS N1s emission was comparatively more strongly attenuated than the C1s and O1s emissions when copper was deposited. We repeated the IRAS experiment published by Dunn and Grant[17] and the XPS measurement by Mack et al.[18] for copper evaporation on polyimide films produced by vapor deposition to establish a direct comparison between the two techniques. The polyimide film preparation, the copper deposition and the XPS and IR-analysis (in a single reflection mode) was done in a UHV-system without exposing the sample to air in between[37].

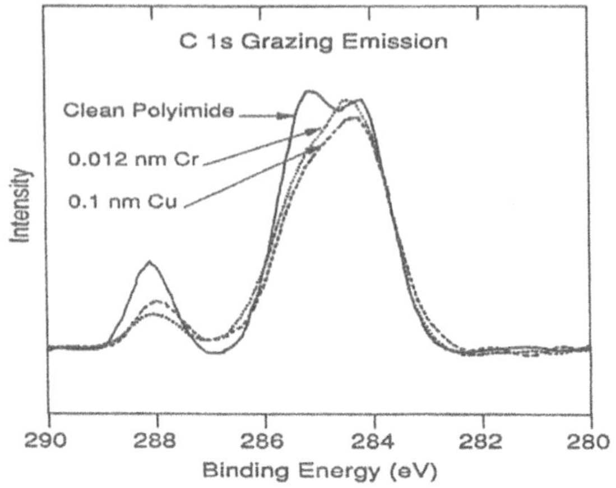

Figure 2. C1s spectra of clean polyimide, and after 0.012 nm Cr and 0.1 nm Cu deposition. Note the similarity in the spectral changes induced by Cr and Cu. Replotted from data presented in ref. 15.

In Fig. 3 we compare the FTIRAS spectra of the cured polyimide film (a) with the difference spectra between the original polyimide film and the film after the first copper deposition (0.1 nm) (b), the second copper deposition (total of 0.2 nm) (c), and heating the film after the second deposition to 570 K for 15 min. (d). The spectra displayed have been background subtracted. From a qualitative point of view our IRAS results for the 0.1 nm deposition of copper are identical to those of Dunn and Grant, i.e. attenuation of the carbonyl stretching band, attenuation of the imide related bands at 1383 and 1119 cm^{-1}, and the ether C-O-C stretch at 1248 cm^{-1}, but no obvious effect on the ODA ring stretching mode around 1500 cm^{-1}. Following the second deposit, however, (Fig. 3c) a clear attenuation of the ODA ring stretching mode around 1500 cm^{-1} is observed. Heating the 0.2 nm deposit of copper further enhances the effect seen in spectrum 3c, in particular for the imide related modes and the ODA ring mode (Fig. 3d).

The XPS data, which are not shown here, are identical in their spectral changes to those published by other groups[15,18]. However, the detailed changes in the O1s, C1s and N1s spectra upon copper deposition depended on the evaporation rate of the copper. Variations were observed in the degree of attenuation of the C1s, O1s and N1s signal relative to the Cu 2p signal, as well as in the appearance of low binding energy shoulders on the N1s and O1s emission[42]. Despite these irreproducibilities in the C1s and O1s bands, in all experiments the Cu 2p data ruled out Cu^{2+} formation. Also, we always observed a strong reduction of the O1s shake up intensity and no changes in the C1s shake up intensity. The latter observation is strong evidence for the disturbance or elimination of conjugation in the imide ring. As discussed by us in ref. 34, this result could be explained by an opening of the imide ring and the resultant non-planarity of the PMDA phenyl ring with the carbonyl groups which would eliminate charge conjugation between the imide and phenyl ring.

The IRAS and XPS data would be consistent with a model where the initial interaction of copper with the fivefold planar imide ring of polyimide leads to ring opening and, considering the disproportionate loss in the C1s and O1s XPS emission observed in some experiments, even to elimination of one CO group. In this picture, the decrease in absorption in the IRAS data would be explained by polymer bond breaking. However, chemical intuition would argue against such a strong reaction, and we can not unambiguously identify any reaction products in the infrared or photoemission data. It is therefore also possible to explain the spectral changes by a weak, but localized chemical interaction between the copper atoms (or clusters) and the imide ring of the polymer. The oscillator strength of the carbonyl stretching mode and the imide related vibrations will be effected through the interaction of the dynamic dipole with the screening charge located on the copper atoms or clusters and by the interaction of the effective dipole with the local electric field. As with the "surface selection rules" governing dipole excitation processes at surfaces, we would expect an image doubling of perpendicular dipole components and a cancellation of parallel ones, i.e. the specific adsorption site of the copper atoms or small clusters would determine whether an enhancement or attenuation of a specific adsorption band is observed. The details of the applicability of the surface selection rule for molecules interacting with small particles depend, however, on the geometry and size of the metal spheres and the refractive index[43]. Since these parameters are not known for the Cu/PI system, we can not unambiguously identify whether physical or chemical causes lead to the reduction in the intensity of the vibrational bands in the polyimide spectra. We note that the spectral changes in the XPS data can also be interpreted by changes in final state screening of the core holes in the presence of a metallic deposit, alternatively to postulating chemical reactions between the polymer and the metal.

Copper deposition of more than 0.1 nm then leads to copper cluster formation (as evident from the Cu 2p data) and adsorption over the ODA part of the molecule. Heating the copper film then decreases all polyimide bands in proportion to their intensity in the clean polyimide and leads to copper migration into the polymer. At this stage also irreversible CuO formation is observed in the XPS data.

The results discussed above, taken on vapor deposited polyimide films without exposing the sample to air, are qualitatively identical to those published by other groups for copper deposition on spun-on polyimide films and vapor deposited polyimide films. The mechanism we propose is consistent with results obtained previously. At this stage we can not distinguish between cleavage of the imide ring with possible elimination of CO or a weak, but localized interaction of copper with polyimide in the initial interaction stage. However, the FTIRAS results combined with an XPS analysis described above *exclude* complex formation and copper oxide formation as major reactions for submonolayer copper deposits[42].

COMPARISON BETWEEN COPPER METALLIZATION LAYERS ON POLYIMIDE AND POLYIMIDE FILMS ON COPPER

Polyimide on *bulk copper* and copper metallization layers on *polyimide* are well studied model systems to demonstrate the differences in interfacial chemistry as a function of preparation method. In the following we will summarize the results for the PI/copper interfaces formed by (i) copper film deposition on cured polyimide, and (ii) spin coating of the polymer precursor polyamic acid (PAA) onto a copper film.

Kim and coworkers[19] measured the adhesion strength by 90° peel tests for (i) and (ii). They found that in case (ii) adhesion was significantly enhanced as compared to copper deposited onto cured polyimide (i) and they attributed this to the difference in interfacial chemistry, i.e. chemical reaction between polyamic acid and bulk copper (ii) as compared to copper atoms or clusters interacting with cured polyimide. The difference in interfacial chemistry was evident in cross sectional TEM observations[19]. In the case of a sputter deposited copper film on a cured polyimide film, a sharp boundary was observed; whereas in the case of a polyimide/copper interface prepared by spin coating polyamic acid and subsequent imidization, cuprous oxide (Cu_2O) particles were found in the polymer matrix. That copper oxide particles are distributed over a thickness of ~500 nm was reported by Burrell et al.[21] for spun-on polyimide films on copper. In their XPS and IR-reflection absorption measurements they attributed the observed oxidative degradation and chemical modification of the thick polyimide films to copper oxide particle formation. Relatively thick polyamic acid films were spun onto copper substrate by Shih et al.[22]. They followed the imidization and curing by sheet resistance, parallel plate capacitance measurements, FTIR and x-ray photoemission. Evidence for the interaction of copper with polyamic acid and for copper oxidation controlled by the supply of oxygen to the interface was obtained from FTIR, and cross-sectional TEM micrographs. Shih et al. reported that the amount of cuprous oxide found in the polymer matrix was significantly less when curing was performed in a reducing gas atmosphere as compared to nitrogen or vacuum curing. They concluded that Cu-ions are dissolved in the polyamic acid/solvent layer in the initial curing stage and subsequently diffuse into the polymer matrix.

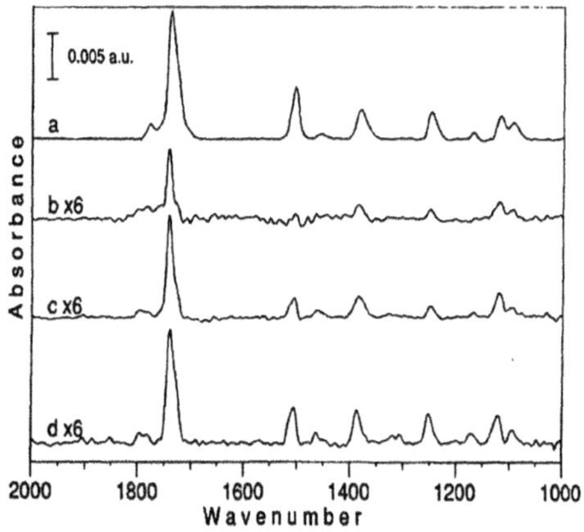

Figure 3. (a) FTIRAS spectra obtained in a single reflection geometry from a vapor deposited polyimide film cured for 30 minutes at 573 K.

(b) Difference spectrum with respect to 3a after deposition of a submonolayer of copper (0.1nm) under ultrahigh vacuum conditions.

(c) Difference spectrum with respect to 3a after deposition of a total of 0.2 nm copper on the PI film.

(d) Difference spectrum with respect to 3a after heating the 0.2 nm copper film to 573 K for 15 min.

To evaluate the role of the solvent (NMP) in the formation of cuprous oxide particles in the polymer matrix during imidization, Kowalczyk et al.[23] studied the PI/Cu interface prepared by vapor deposition of PMDA and ODA on Cu by TEM. As shown in Fig. 4, no copper oxide particles are found in the polymer matrix of the vapor deposited film. However, if prior to imidization a drop of the solvent NMP is applied to a vapor deposited polyamic acid film, copper oxide particle formation in the polymer matrix can be

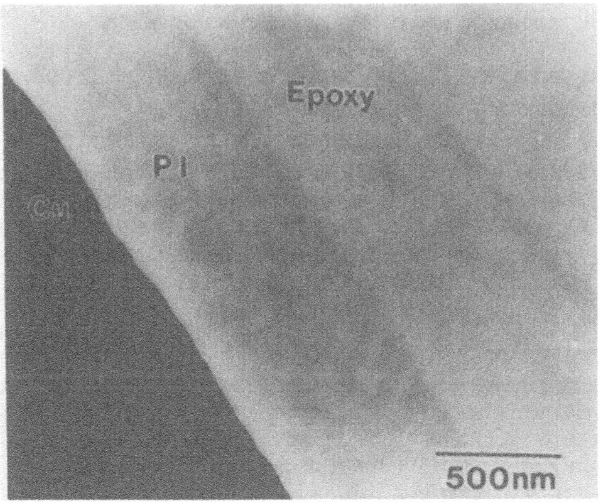

Figure 4. TEM micrograph of a copper/polyimide interface prepared by vapor deposition of the polymer. No evidence for copper cluster migration into the polymer matrix is found. With permission, from ref. 23.

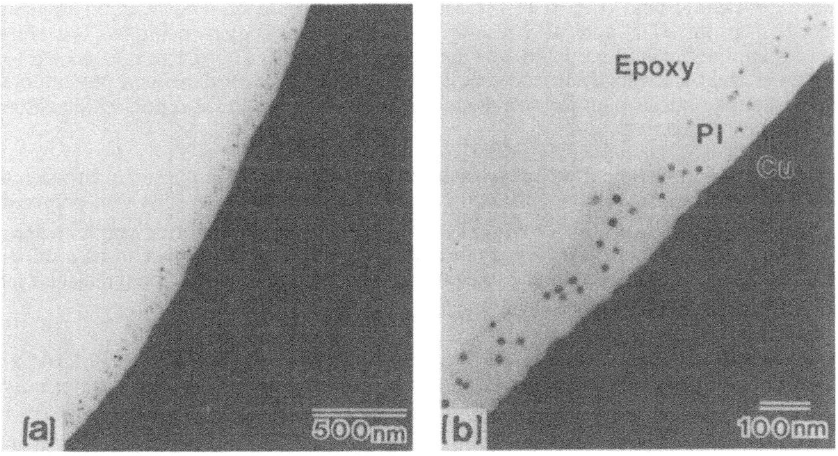

Figure 5. Formation of Cu_2O particles in the polyimide matrix after applying a drop of NMP to the polyimide film prepared by vapor deposition. With permission, from ref. 23.

(a)Magnification: ca. 3×10^9 (b)Magnification: ca. 1×10^9

observed by TEM[23] as shown in Fig. 5. The results of Kowalczyk et al. clearly demonstrate that the solvent provides mobility for the copper ions formed at the interface to diffuse into the polymer matrix and precipitate as cuprous oxide.

Results on vapor deposited polyimide films on copper revealed that it is not possible to produce polyimide films of thickness less than 4 nm [24,25]. According to Kowalczyk et al.[26] this is due to incomplete imidization at the interface caused by carboxylate formation, whereas we believe[25] that this is a consequence of fragmentation and loss of functional groups of the PMDA and ODA molecules at the interface. Both, the strong initial reaction of PMDA and ODA or of PAA with the copper surface, can provide an explanation for the dissolution of the copper surface and eventual cuprous oxide formation in the polymer matrix facilitated by the solvent.

VAPOR DEPOSITED FILMS AS MODEL SYSTEMS FOR SPIN-COATED POLYIMIDE FILMS ON BULK SUBSTRATES

The above comparison for differently prepared polyimide/copper interfaces raises the question as to what extent vapor deposited polyimide films could provide a suitable model to study polyimide/substrate interfaces of practical relevance. Curing and imidization of PAA dissolved in NMP has been investigated by Brekner and Feger[27]. They demonstrated that complexation of PAA by NMP has a marked influence on imidization. Only at temperatures high enough for decomplexation to occur (T~380 K) can the highly reactive free groups react to imide or, in a reverse direction, to anhydride. Decomplexation precedes imidization when NMP is not in excess. This implies that for molecules at the interface the decomplexed acid groups can interact with the substrate, possibly in a competitive reaction to imidization. That ultrathin films (d< 4nm) of PMDA and ODA codeposited onto silver[30], copper[25] or gold[34] react instantaneously to form the polymer precursor was shown previously by XPS measurements. Recent FTIR experiments reveal that the reaction between PMDA and ODA in the condensed phase starts at 300 K and that at room temperature predominantly an amine carboxylate salt (Fig. 6, III) is present in the film[37]. Polyamic acid (Fig. 6 IV) is formed at about 375 K. The reaction sequence deduced from our FTIR and XPS results is schematically shown in Fig. 6. Therefore, solventless polyamic acid prepared by vapor deposition onto a substrate is expected to show the same chemical reactivity and reactions as occurring with dissolved polyamic acid at temperatures high enough for conversion of the carboxylate salt to polyamic acid and decomplexation, respectively.

For spectroscopic methods to be applied to analyze the polymer/substrate interface the polymer films have to be sufficiently thin. To prepare sufficiently thin polymer films by spin coating was only successful in a few cases. Russat[28] reported results where polyamic acid dissolved in NMP was spun onto gold resulting in polyimide film thicknesses ranging from 1.3 to 2.9 nm after imidization and curing. However, the interfacial reaction on gold could not be unambiguously established[28].

Deposition of polyamic acid by coevaporation of the anhydride (PMDA) and the diamine (ODA) can provide sufficiently thin polyamic acid and, after curing, polyimide films to study the interfacial reaction with x-ray photoelectron spectroscopy[24-26,29-34], near edge x-ray absorption fine structure (NEXAFS)[26] and infrared reflection absorption spectroscopy (IRAS)[33,34,37]. This vapor deposition preparation method to study the interfacial reactions has been applied by our group to copper[24,25], silver[29-31] and gold[34] substrates and by Kowalczyk et al. to Cu, Cr and Si surfaces[26]. The first description of the vapor phase preparation of polyimide films was given by Salem et al.[36] for thick PI films (d > 10 µm). Application of UHV surface studies to such films was first described in reference 29.

A detailed study of monomer interaction (PMDA and ODA) and adsorption on clean silver surfaces[30] showed that both molecules undergo partial fragmentation upon room temperature adsorption. Spectroscopic analysis of vapor deposited polyamic acid indicates that the reactive sites for the interaction with a silver substrate, as well as for evaporated gold atoms and clusters, are the amine carboxylate groups in the molecule[34]. Evaporated gold atoms and clusters preferentially interact via an electron transfer from the metal to the carboxylate part of the molecule[34]. The interaction with a bulk silver surface is believed to involve silver carboxylate formation in either a monodentate or bidentate bonding configuration of the PMDA part of the polymer chain. This model for the interfacial chemical bond was derived from XPS and IR reflection absorption experiments for the ultra-thin polyimide films obtained after curing the polyamic acid layers[30,34]. Such a salt-like ionic surface bond would explain the sensitivity of the interfacial bond towards humidity, i.e it suggests that hydrolysis of the interfacial bond is a likely cause for interfacial failure.

There are strong similarities in the XPS and NEXAFS data for the case of polyimide vapor deposited onto Cr surfaces and Cr evaporated onto cured polyimide surfaces[26] suggesting that the same reactions occur in these differently prepared laminar structures. With the highly reactive chromium, carbide, nitride and oxide species are formed at the interface when the PAA/Cr interface is annealed. Due to the variety of the chemical species present it was not possible to identify the nature of the chemical bond between polymer and metal.

The results for a 0.4 nm thin vapor deposited PMDA/ODA film on a Si(111)-7x7 surface (the notation 7x7 refers to the clean surface reconstruction) show that three chemical states are induced by the interaction with the polymer precursor. This is evident in the Si 2p XPS data and can be explained by more than one geometric adsorption site, more than one functional group being involved in the bond formation, or a stepwise interaction pathway[26]. The C1s data show a very strong attenuation in the carbonyl C1s intensity suggesting that the carboxylate groups and the amide carbonyl groups are involved in the interaction with the substrate. Kowalczyk et al.[26] clearly demonstrated in their work that covalent bonds to the silicon substrate are formed in the polymer precursor state.

LANGMUIR-BLODGETT (LB) DEPOSITION OF POLYIMIDE FILMS

Kakimoto et al.[38] reported the first successful preparation of mono- and multilayer films of polyimide by depositing the alkylamine salt of polyamic acid by the LB technique and subsequently removing the long alkyl chain by thermal or chemical treatment. This technique was also used by Sotobayashi et al.[39] to deposit polyimide on highly oriented pyrolytic graphite. The PI monolayers formed a regular array of highly oriented polymer chains as evident in the STM image[39] and the NEXAFS measurements by Schedel-Niedrig et al.[40]. We performed a comparative XPS and NEXAFS study on the orientation and interfacial chemistry [41] of 1,3 and 5 monolayers of polyamic acid deposited by LB techniques in the form of the octadecyl-ammonium salt and octadecyl-ester and on the resulting polyimide films following thermal treatments. The substrate was a Si (100) wafer etched with HF directly before deposition. For the alkylamine salt of polyamic acid and the resulting polyimide, films of thickness equal to or exceeding three monolayers NEXAFS measurements indicate a preferential orientation of the polymer chains with respect to the surface plane.

In the context of this short summary we would like to point out the difference in the composition of the silicon surface when the polymer was deposited as the alkylammonium salt or the ester. Fig. 7 compares the Si 2p XPS data for a) 5 layers of the alkylammonium salt, b) 5 layers of the ester, c) difference between the salt and the ester films, d) the imidized polyamic acid alkylammonium salt film, e) the imidized ester film and f) the difference spectrum d-e. The oxidation states +1, +2, +3 and +4 of silicon show photoemission at approximately 1, 2, 3 and 4 eV higher electron binding energy than the Si(0) emission at ~99 eV. The stronger emission from higher oxidation states of silicon com-

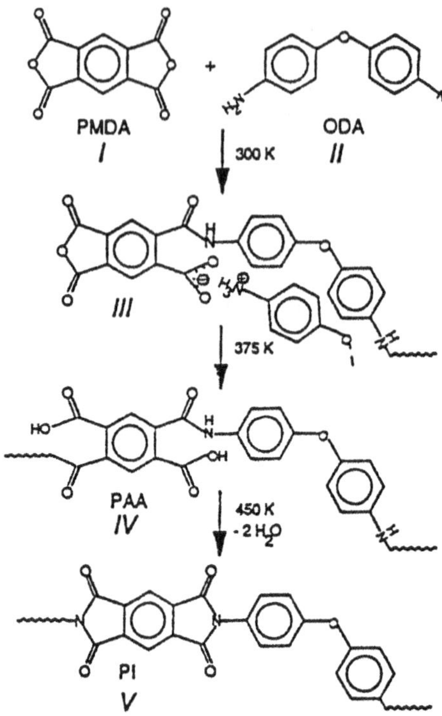

Figure 6. Reaction sequence between PMDA and ODA in the condensed phase to polyimide.

pared to the corresponding Si(0) emission as well as the presence of the higher oxidation states in spectrum 7a and in the difference spectrum 7c for the alkylammonium salt demonstrate that for the initial LB deposit the alkylammonium salt of polyamic acid reacts more strongly with the silicon substrate. After imidization, both samples exhibit about the same intensity ratio between the Si(0) and Si(+4) photoemission bands. There is, however, a difference in line shape of the Si(0) emission. The lower binding energy shoulder in the difference spectrum 7f could indicate that some of the silicon surface atoms are in an electron-rich environment, i.e. partially reduced state, at the interface after the ammonium salt of polyamic acid has reacted to polyimide. From our spectra we can not distinguish between silicon oxide species or silicon in different bonding environments with the polymer. We also note that for the same initial deposit (5 layers) the resulting octadecyl ester film must be thicker due to the stronger attenuation of the silicon substrate emission. This could be due to a better homogeneity of the film, or a different packing density.

The C1s, N1s and O1s spectra reveal in agreement with the above observations that the resulting polyimide films from the two deposits have different compositions in the interface layer when monolayers are deposited. Despite these chemical differences at the interface, both polyimide precursors can be used to grow thick polyimide films by the Langmuir-Blodgett deposition. This example complements the previous comparison between dissolved and solventless polyamic acid with respect to the observation that the interfacial chemistry between the polymer and the substrate depends on the method of preparation. In particular the LB technique offers a well controlled experimental procedure to study these effects further.

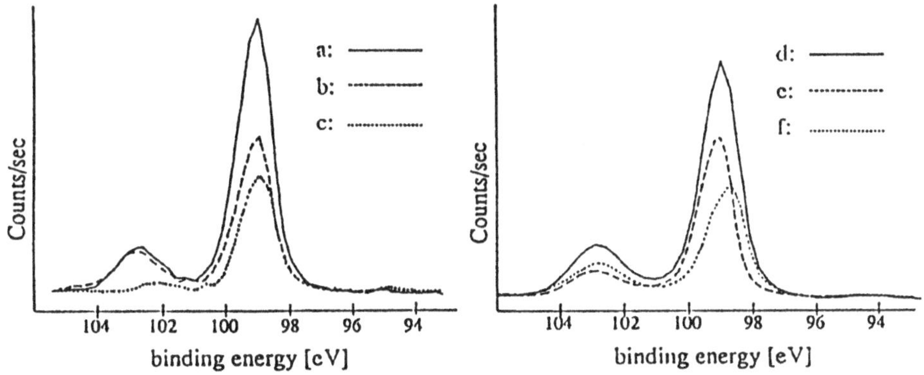

Figure 7. Si 2p XPS data for LB deposited polyamic acid (onto a freshly etched Si (100) surface) and the resulting polyimide films (after thermal treatment for 30 minutes at 250°C)

(a) 5 layers of the alkylammonium salt

(b) 5 layers of the ester

(c) difference spectrum a-b

(d) the imidized polyamic acid alkylammonium salt film

(e) the imidized ester film

(f) difference spectrum d-e

The results summarized so far refer to studies related to the interfacial bond between polyimide films and the inorganic substrates. Typically delamination between polyimide and the substrate is, however, not caused by adhesional failure, but rather by cohesive failure in the polymer itself. Although the effects of adhesion promoters (typically aminosilanes) in the formation of a strong interfacial bond have not been studied to our knowledge by spectroscopic techniques for polyimide laminar structures, it is informative to recall the results obtained for a reorientation of the polymer chains between the polymer/metal interface and the polymer bulk.

XPS and IR data on silver led to the conclusion that at the surface the polymer chains are oriented away from the surface plane. This geometry is induced by the chemical bond to the surface[30,33]. Small angle x-ray scattering data by Russel[35], however, prove that for thick polyimide films (d~ 75 μm) the polyimide polymer chains in spun-on films are oriented parallel to the surface plane. Such an orientation of the polymer chains with respect to the substrate was confirmed by IR absorption reflection studies[33] and NEXAFS measurements by our group for vapor deposited polyimide films (d>8nm) on copper and gold, respectively. At present it is not known at what distance away from the surface a transition from a nearly perpendicular (induced by the polymer/substrate bond) to a parallel orientation relative to substrate plane of the polymer chains occurs, but it is tempting to speculate that such an orientational change in the polymer is related to the locus of failure in the case of cohesive delamination.

CONCLUSION

With their high temperature stability and easy processibility polyimide based polymers became an integral component in microelectronic device technology. Problems associated with adhesion in laminar structures have been mainly overcome by a careful control of process conditions, yet understanding of the interfacial chemistry is still rudimentary and controversial. The application of spectroscopic techniques to identify the chemical composition and bonding at the interface has been established and such techniques are now routinely used in basic and applied research efforts. The combination of FTIRAS and XPS can provide insight into the interfacial chemistry between polyimide and metals as demonstrated for copper deposition onto polyimide surfaces. However, delamination typically occurs by cohesive failure in the polymer. To identify the locus and chemical and/or physical mechanism of cohesive failure remains a challenge for future research.

ACKNOWLEDGEMENTS

Financial support for some of the authors´ work summarized here was obtained from the Office of Naval Research , IBM, and the Bundesministerium für Forschung und Technologie through grant No. TK 03240 . We thank Dr. W. Schrepp (BASF AG, Ludwigshafen) for providing the LB deposited polyamic acid films. Stimulating discussions with S. Kowalczyk, Ch. Wöll, G. Hähner and W.N. Unertl are gratefully acknowledged.

Note added in proof:
An excellent overview on the adhesion of polyimides to metal and ceramic surfaces including a summary on adhesion theories and adhesion measurements, by L.P. Buchwalter, came to the attention of the author. This review is published in J. Adhesion Sci. Technol., 4 , 697-722 (1990).

REFERENCES

1 K. Sato, S. Harada, A. Saiki, T.Kimura, T. Okubo, and K. Mukai, IEEE Trans. on Parts, Hybrids&Packaging, PHP-9 (3), 175 (1973).
2 K.L. Mittal, editor , "Polyimides-Synthesis, Characterization and Applications", Vols. 1&2, Plenum Press, New York (1984).
3 R. Jaccodine, K.A. Jackson, and R.C. Sundahl, editors,"Electronic Packaging Materials Science III", Mater. Res. Soc. Symp. Proc., 108, (1988).
4 D.P. Seraphim, R. Lasky, and Che-Yu Lin, editors, "Principles of Electronic Packaging" McGraw Hill, New York (1989).
5 L.M. Poveromo, in ref. 2, p. 1003.
6 A.K. St.Clair and T.L. St.Clair, in ref. 2, p. 977.
7 R.L. Fusaro, in ref. 2, p. 1053.
8 R. Reuter, H. Franke and C. Feger, Applied Optics, 27, 4565 (1988) and C. Feger, R. Reuter and H. Franke ,in, "Polymers in Information Storage Technology", K.L. Mittal, editor, pp. 227-233, Plenum Press, New York (1989)
9 N.J. Chou and C.H. Tang, J.Vac.Sci. Technol. A2, 751 (1984)
10 J.G. Clabes, M.J. Goldberg, A.Viehbeck and C.A. Kovac, J.Vac.Sci.Technol. A6, 985 (1988)
11 F.S. Ohuchi and S.C. Freilich, J.Vac.Sci.Technol, A4, 1039 (1984)
12 N.J.DiNardo, in "Metallized Plastics 1: Fundamental and Applied Aspects" K.L. Mittal and J.R. Susko, editors, Plenum Press, New York (1989)
13 J.J.Pireaux, M. Vermeersch, N.Degosserie, C. Grégoire, Y.Novis, M.Chtaib,and R. Caudano in "Adhesion and Friction" M. Grunze and H.J. Kreuzer, editors, p. 53, Springer Series in Surface Sciences, 17, (1989).
14 Lj. Atanasoska, S.G.Anderson, H.M. Meyer, Z. Lin and J.H. Weaver, J.Vac.Sci.Technol. A5, 3325 (1987).
15 R. Haight, R.C. White, B.D. Silverman, and P.S.Ho, J.Vac.Sci.Technol., A6(4) 2188 (1988).

16 J.L.Jordan-Sweet, C.A.Kovac, M.J. Goldberg, and J.F.Morar, J.Chem.Phys., 89, 2482 (1988).
17 D.S.Dunn and J.L. Grant, J.Vac.Sci.Technol., A7 , 253 (1989).
18 R.G. Mack, E. Grossman and W.N. Unertl, J.Vac.Sci.Technol., in press.
19 Y.H.Kim, G.F Walker, J.Kim and J. Park, J. Adhesion Sci. Technol., 1 , 33 (1987).
20 J.L. Jordan, C.A. Kovac, J.F. Morar, and R.A. Pollak, Phys.Rev.B,36 , 1369 (1987).
21 M.C. Burrell, P.J. Codella, J.A. Fontana, J.J. Chera, M.D. MacConnell, J.Vac-Technol., A7 , 53 (1989).
22 D.Y.Shih, J. Parasczak, N. Klymko, R. Flitsch, S. Nunes, J. Lewis, C. Yang, C. Cataldo, R. McGouey, W. Graham, R. Serino, and E. Galligan, J.Vac.Sci.Technol., A7 (3) 1402 (1989).
23 S.P.Kowalczyk, Y.H. Kim, G.F. Walker, and J.Kim, Appl. Phys. Lett. 52 , 37 (1988).
24 M. Grunze, J.P. Baxter, C.W. Kong, R.N. Lamb, W.N. Unertl, and C.R. Brundle Vapor phase growth of polyimide films, Paper presented at the American Vacuum Society National Symposium in Anaheim, November 1987.
25 R.N. Lamb, M. Grunze, J. Baxter, C.W. Kong and W.N. Unertl, in "Adhesion and Friction", M. Grunze and H.J. Kreuzer, editors, p. 36, Springer Series in SurfaceSciences, 17, (1989).
26 S.P. Kowalczyk and J.L. Jordan-Sweet, Chemistry of Materials, 1, 592 (1989).
27 M.J. Brekner and C.Feger, J.Polym. Sci., Part A: Polym. Chem., 25, 2005-2020 (1987).
28 J.Russat, Surface Interface Anal., 11, 414 (1988).
29 M. Grunze and R.N. Lamb, Chem. Phys.Lett. 133 (1987).
30 M. Grunze and R.N. Lamb, Surface Sci., 204 (1988) 183
31 M. Grunze and R.N. Lamb, J.Vac.Sci.Technol A5, 1685 (1987).
32 R.N. Lamb, J. Baxter, M. Grunze, C.W. Kong and W.N. Unertl, Langmuir 4 , 249 (1988).
33 M. Grunze, W.N. Unertl, S. Gnanarajan and J. French, Mater. Res. Soc. Symp. Proc., 108 , 189 (1988).
34 T. Strunskus, M. Grunze, and S. Gnanarajan , Proceedings of the International Symposium on the Metallization of Polymers, Montreal, ACS series, in press (1990).
35 T.P. Russel J. Polym. Sci.; Polym. Phys. Ed., 22, 1105 (1984).
36 J.R. Salem, R.O. Sequeda, J. Duran, W.Y. Lee and R.M. Yang, J. Vac. Sci. Technol. A 4 , 369 (1986)
37 C. Hahn, T. Strunskus and M. Grunze, J. Electr. Spectr. Rel. Phen. in press (1990).
38 M. Kakimoto, M. Suzuki, T. Konishi, Y.Imai, M.Iwamoto, and T.Hino, Chem. Phys. Lett., 823 (1986).
39 H. Sotobayashi,T.Schilling and B.Tesche, Langmuir, 6, 1246 (1990).
40 T. Schedel-Niedrig, H. Sotobayashi, B. Tesche, A. Ortega-Villamil and A. Bradshaw, Surface Sci., in press (1990).
41 G. Hähner, Ch. Wöll, A. Killinger. C. Mainka and M. Grunze, in preparation
42 T. Strunskus, C. Hahn, D. Frenkel and M. Grunze, J. Vac. Sci. Technol., submitted
43 W. Jaisli, H. Kuhlman and G. Schultz- Eckloff, Surface Sci. 118 L269 (1982).

CHEMICAL INTERACTIONS AT POLYMER-METAL INTERFACES OF INTEREST TO MICROELECTRONICS

S. Pignataro

Dipartimento di Scienze Chimiche, Università di Catania
Viale A. Doria 6, 95125 Catania, ITALY

Two types of polymer-metal interfaces of interest to microelectronics have been considered.

The first is that between a commercial encapsulating resin and an air exposed metal surface. This system has been studied by moulding the resins to the metal. The joint has been mechanically broken and the chemical species at the fracture locus have been characterized by means of ESCA and SIMS. This study has been made on several types of epoxy and siloxane resins bonded to Ag, Cu, Ni surfaces. Diffusion of some of the resin constituents to the joint interface as well as various interfacial chemical reactions have been followed with time, temperature and environmental conditions. In some cases migration or reaction kinetics could be quantified.

The second system reported here involves polyamic acid/metal or a polyimide/metal case.

The samples were prepared by spin coating very dilute polyamic acid solutions on various air exposed metal substrates. Such films are thin enough to allow the direct observation of the interfaces by ESCA. The samples were then cured in vacuum in the analysis chamber of an ESCA spectrometer in order to follow the formation of polyimide and the ESCA analysis was then repeated. It is shown that the interfacial chemistry of these apparently more simple systems is very complex and depends upon several factors.

INTRODUCTION

For the development of high-tech industry it is mandatory to understand the interfacial chemistry of solid systems obtained by the contact of metallic and polymeric partners, regardless of which of the two surfaces acts as the substrate.

In this paper I would like to discuss the interfacial chemistry of two entirely different systems of the above type. This in order to show

Metallized Plastics 2, Edited by K.L. Mittal
Plenum Press, New York, 1991

and summarize most of the types of phenomena which may occur in connection with the complexity of the involved partners.

The first is very complex dealing with polymer containing resin on air exposed metal surface. The second deals with an apparently more simple polyamic acid/metal or a polyimide/metal system.

In all cases the experiments were made simulating plant conditions.

It will be shown that the interfacial chemistry has a high degree of complexity and, therefore, in tailoring such interfaces several factors should be considered. The phenomena and reactions reported refer to polymer on metal systems. However, their study and understanding may be very useful also for the development of the metallized plastics field, since it is heavily conditioned by the chemistry at the interfaces which in many respects is similar to that presented here.

POLYMER RESIN-METAL CASE

In the last few years a number of papers have been published[1-4] from our laboratory dealing with the interfacial chemistry of polymer resin-metal systems. Such a system is the one generally found in a microelectronic device, where a moulding resin is used to encapsulate the silicon chip. The experimental set up consisted of moulding a resin to a metal frame (see fig.1); subsequently the resin-metal joint was mechanically broken, the adhesion strength was measured with a pull test machine, and the newly obtained failure surfaces were analysed by ESCA or SIMS. In these studies different metal substrates and different resins were used.

In particular the resins were commercial epoxidic or siloxanic products supplied by Hysol and Dow Corning respectively, and the metals were Ag, Ni, Cu or Ni plated copper.

The procedures used in making the joint were those used in the industrial process of microelectronic device manufacturing. In particular the moulding conditions used were 180 °C, 70 kg/cm^2.

What we learned from these experiments on the interfacial chemistry of this very complex system is that a variety of solid state chemistry processes are operative there. In particular, of the various resin constituents some were found to be confined in the bulk of the resin itself;

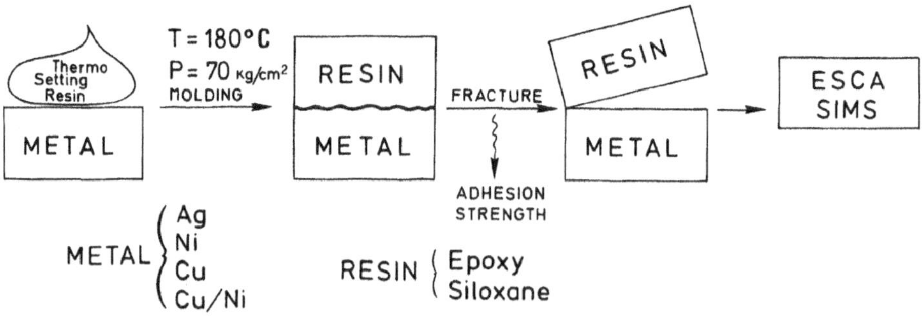

Figure 1 Experimental procedure used for the study of the interfacial chemistry in polymer resin-metal systems.

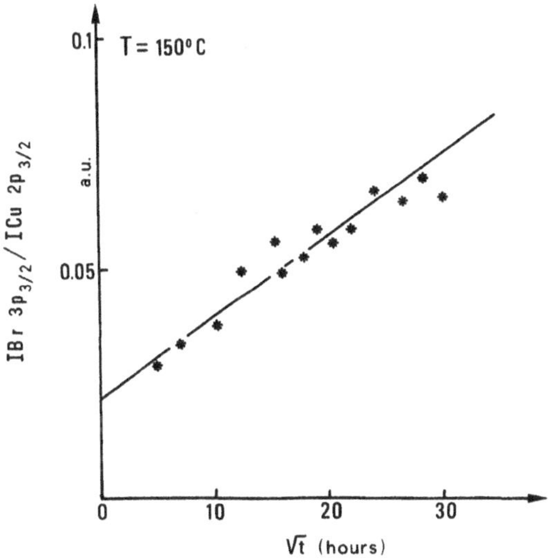

Figure 2 Time dependence of the concentration of the bromine coming from the antiflame additive of the resin at the resin-metal interface of a microelectronic device. The rate law observed is in agreement with a diffusion controlled mechanism of migration (ref.3).

others were found to be homogeneously distributed between the bulk of the resin and the interface; some others were found to segregate to the interfacial zone and some constituents were in a chemical state different from that found in the bulk. Also some trace elements present as additives in the metal substrate were found to migrate to the interfacial zone. The migration phenomena were found to depend both on the chemical nature of the metal and the resin. Time and moulding temperature as well as ambient atmosphere were found to affect the interfacial chemistry. In the case of an epoxy resin, containing bromine as bromobisphenol A (TBBPA) an antiflame retardant, moulded to a copper surface the kinetics of bromine migration to the metal surface could be followed and the diffusion controlled mechanism substantiated (see fig.2). The migration phenomenon obeys a simple rate law of the form: $Y=A\sqrt{X}+B$, where X is the time of storage, Y is the Br concentration (%), A is the slope and B the value of Y when the storage time is zero. In addition TBBPA was found to retard the oxidation of copper substrate with storage time at high temperature, in agreement with the well known[5] antioxidant power of the phenolic moieties of the TBBPA. Fig. 3 shows this phenomenon. Interpenetration of the "metal" phase into the "resin" phase has been also found. The morphology of the interfacial zone was obviously affected. It is important to note here that such phenomenon is the same as found[6] in the apparently more simple system of a copper surface spin coated with polyamic acid.

Most importantly, some of the observed effects can be interpreted in terms of competition between the various migrating species and/or reactions in the interfacial zone of the resin-metal system. Thus it becomes very important to understand the interplay of each of the constituents of the system considered in order to provide the best degree of reliability and the best performance of the joint.

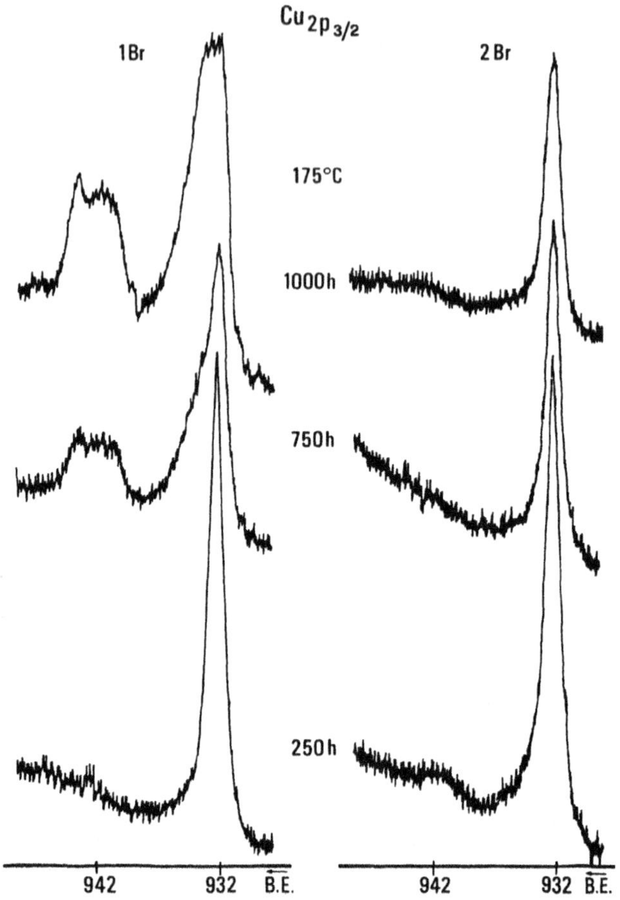

Figure 3 Cu $2p_{3/2}$ ESCA peaks at the resin-copper joint stored at 180°C for various times. The resins used were formulated with X and 2X content of the TBBPA flame retardant (1Br and 2Br in the figure). The oxidation of the copper moulded to a resin with a higher bromine (from TBBPA) content is clearly retarded (ref.4).

Particularly enlightening in this respect is the case of the nickel plated copper substrate coated with the above mentioned epoxy resin containing the bromobisphenol A as anti-flame retardant. Fig.4 shows the SIMS images of Cu^+, Ni^+, Br^-, O^- on the "metal" failure surface obtained with the procedure described in fig. 1.

Copper is found to migrate through the nickel and to react with the brominated additive of the resin which, in turn, migrates to the interfacial zone in order to react with the metallic system. Bromine is found mainly where copper migrates to the top of Ni.

The above described phenomena have been connected to failure cases observed in commercial electronic devices (see fig. 5) manufactured with the above chemicals. In this figure the bromine relative content at the metal side of the resin-metal interface is reported against the percent failure of the device. The percent failure increases abruptly above a threshold of bromine content.

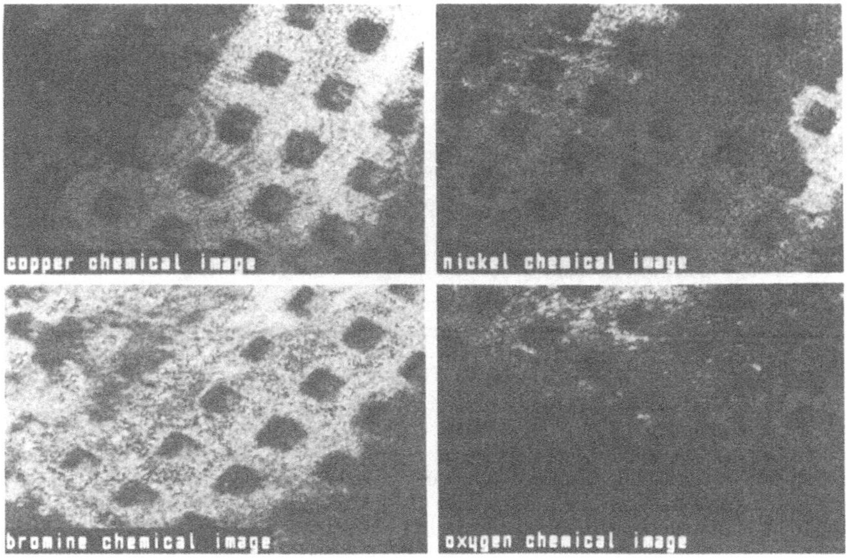

Figure 4 SIMS images of Cu^+, Ni^+, Br^-, O^- at the metal failure surface of the resin-nickel plated copper joint broken after 500 hours of storage at 200°C. Copper migrates through the nickel layer reaching the interface zone with the resin and reacts with the antiflame retardant containing Br in the resin itself (ref.4).

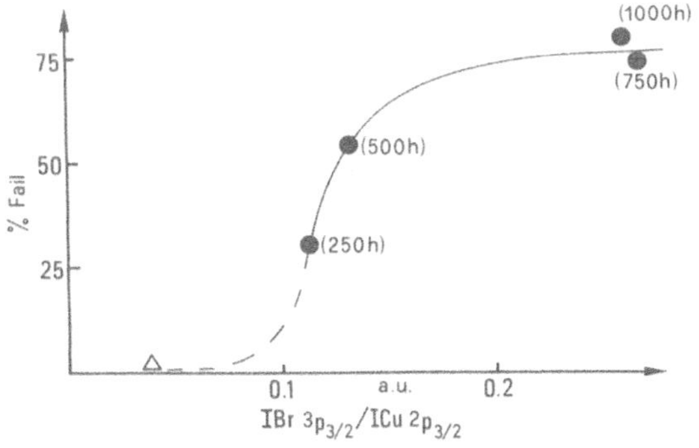

Figure 5 Threshold dependence of the percent of failure of a commercial electronic device from the concentration of bromine (from TBBPA) at the resin-copper joint interface (ref.3). The number of tested devices was 20.

POLYAMIC ACID-METAL AND POLYIMIDE-METAL CASES

In this example BTDA-ODA polyamic acid (PAA) solutions in N-methylpyrrolidone from DuPont de Nemours (Pyralin 2555) were used. Very dilute solutions of the commercial product (about 0.3% in weight) were deposited by spin coating on various air exposed metal surfaces. The PAA films are thin enough (thickness 25±5 Å estimated by means of angle resolved ESCA measurements) to allow the direct observation of the interfaces by means of ESCA. The samples were then cured in vacuum in the analysis chamber of an ESCA spectrometer in order to follow the formation of polyimide (PI) and the ESCA analysis was then repeated[7,8].

The knowledge of these types of interfaces is most important again for microelectronics industry and several papers[9-12] have recently reported on this subject, however the majority of papers deal with the metal-on-polymer systems and only a few are devoted to the polymer-on-metal types.

Fig. 6 shows the situation for the Si/SiOx native substrate studied through the C 1s ESCA peaks. ESCA peak analysis was performed by means of computer program of gaussian peak fitting (with inelastic background subtraction). It is to note in particular that in fig.6 and followings the main C 1s component of the cured BTDA-ODA has been resolved in two peaks (B.E. 285.0 and 286.0 eV) with a 14:10 ratio. Same results are obtained by resolving this peak in two components having a 12:12 ratio[8]. The spectrum

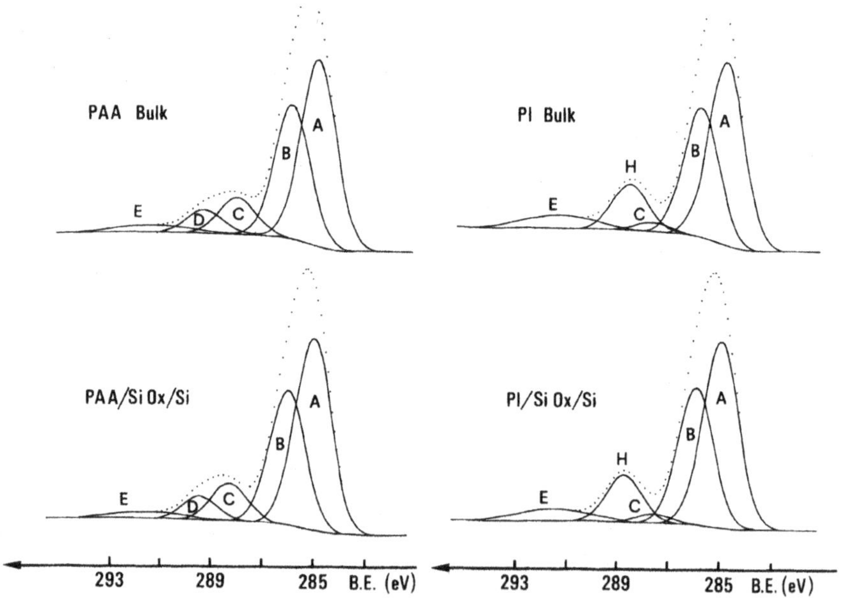

Figure 6 ESCA C 1s spectra of the polyamic acid and polyimide interfaces with air exposed Si substrate in comparison with the corresponding spectra of the bulk of the polymers. These examples are typical of systems where the interfacial chemistry is practically absent. The assignment of the bands is as follows: A= C-C and C-H bonds; B= C-O and C-N bonds and six aromatic carbons of the BTDA unit; C= amidic and benzophenonic carbonyls; D= COOH groups; E= shake up; H= imidic carbonyls.

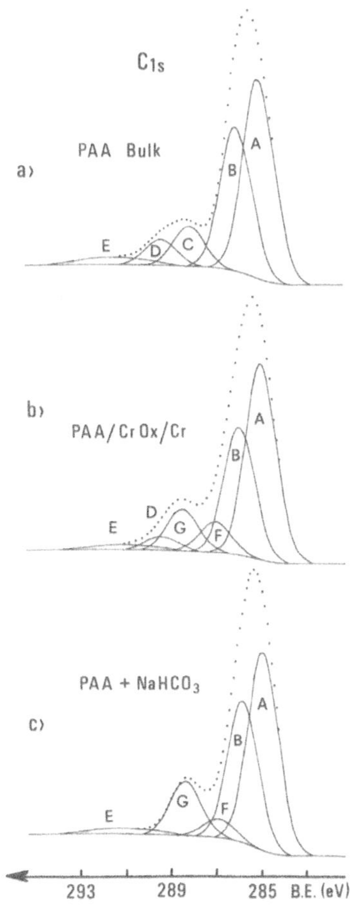

Figure 7 C 1s ESCA peak of the polyamic acid interface with air exposed Cr substrate, in comparison with the corresponding spectrum of the bulk of the polymer. The assignment of the bands is as follows: A= C–C and C–H bonds; B= C–O and C–N bonds and six aromatic carbons of the BTDA unit; C= amidic and benzophenonic carbonyls; D= COOH groups; E= shake up; F= aromatic carbon of the BTDA unit modified by salt formation; G= carboxylate groups and amidic, imidic and benzophenonic carbonyls. The similarity with the C 1s spectrum of a polyamic acid sample with added $NaHCO_3$ shows that a reaction with Na salt formation occurs at the interface.

of the PAA film spin coated on this substrate is very similar to that of a reference thick PAA layer (few µm) and spectrum of the polyimide film obtained by the above described curing is very similar to that of a reference thick (few µm) PI layer.

This is indicative of a very simple and clear situation in which PAA does not appreciably interact with the substrate; the imidization proceeds neatly with formation of a polyimide layer on top of the inorganic layers; the interaction of the polymer and inorganic layers is again very poor. The interfacial chemistry in this system is, therefore, practically absent and this is in agreement with the well known poor adhesion of the Si/SiOx substrate with the polyimide.

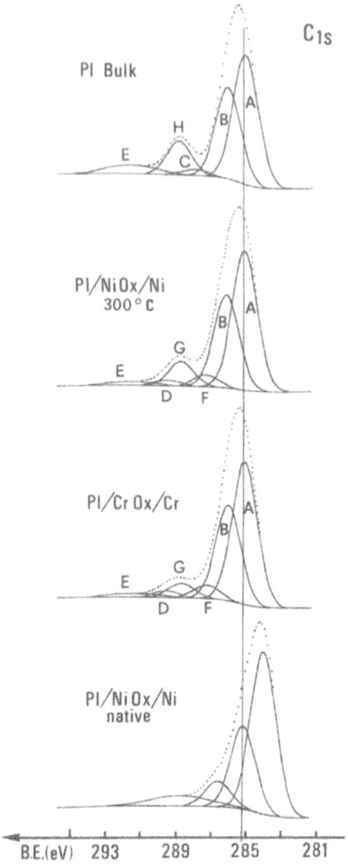

Figure 8 C 1s ESCA peak of some metal/metal oxide substrates on top of which polyamic acid has been cured in vacuum at 300°C in comparison with the corresponding spectrum of the bulk of the polymer. The assignment of the bands is as follows: A= C-C and C-H bonds; B= C-O and C-N bonds and six aromatic carbons of the BTDA unit; C= amidic and benzophenonic carbonyls; D= COOH groups; E= shake up; F= aromatic carbon of the BTDA unit modified by salt formation; G= carboxylate groups and amidic, imidic and benzophenonic carbonyls; H= imidic carbonyls. The imidization results in different interfaces depending upon the chemical nature of the inorganic substrate.

If the substrate is Cr/CrOx or Ni/NiOx, important interfacial chemical processes can be observed. Fig. 7 shows again the C 1s energy region of the ESCA spectra of the PAA film spin coated on the Cr/CrOx native substrate together with the same spectrum for a reference PAA thick sample. In the same figure the C 1s band of PAA thick film, obtained through the reaction of PAA solution in N-methylpyrrolidone with an aqueous solution of $NaHCO_3$ is also reported (7c).

The difference between the spectra reported in fig. 7a (PAA bulk) and 7b (PAA/CrOx/Cr sample) shows that PAA interacts with the inorganic substrate, while the similarity of the spectra in fig. 7b and 7c shows that the main process is the reaction of the acidic sites of the PAA with inorganic ions. The situation is very similar also for other substrates studied in our laboratory such as a) Ni/NiOx native substrate; b) substrate obtained by heating Ni/NiOx in air for 30 minutes at 300°C; or for the Cu/CuOx native substrate reported by other authors[6].

The interfacial chemistry becomes very complex and starts to resemble that described in the interaction of the resin with metal system, reported in the previous section of this paper, when the imidization process of the PAA on the above substrate is considered. Fig. 8 shows the C 1s energy region of the ESCA spectra of some metal/metal ox substrates on top of which the above PAA film was cured in vacuum at 300°C in the preparation chamber of an ESCA spectrometer and the same spectrum for a reference thick (few µm) PI sample.

The spectra show that the imidization does not proceed in the simple way observed for the Si/SiOx substrate, and in particular results in different interfaces for the various inorganic layers employed. The C 1s spectra are indeed all different from each other and are different from that of the reference PI sample. In the case of the Ni/NiOx native substrate a dramatic decomposition of the polymer layer may be inferred from the above spectra.

CONCLUSIONS

The interfacial chemistry of polymer-metal systems is, in general, very complex and in tailoring such interfaces several factors have to be considered. These range from trivial factors such as the purity of the materials used and the cleanliness of the preparation apparatus, to more fundamental factors such as the physico-chemical ways chosen to bring in contact the metallic and polymeric species; the chemical nature of the species involved per se and the surface pretreatments or the presence of additives; the temperatures used; the interdiffusional capabilities of the species involved and the thickness of eventual barriers; the possibility of catalytic effects, and so on.

In particular, we demonstrated that the interfacial chemistry of an apparently simple PAA-metal or a PI-metal system depends in a dramatic way upon the chemical Ni species used in the specimen. This fact is of great importance and suggests new philosophies in the field of metallized plastics.

Let us consider, for example, only the ways of bringing in contact the metal and the polymer and among these one of the emerging methodology: the sputter deposition of metals on polymer. The above findings suggest that in the deposition of a given metal by sputtering on a given polymeric substrate the physico-chemical characteristics of the joint such as adhesion, optical parameters, etc., will depend upon the relative percentage of positive, negative, atomic neutral and clusters species as well as the energy of the sputter deposited particles. These various metallic species may drive different chemical reaction pathways giving different molecular products showing different physico-chemical behaviour.

ACKNOWLEDGMENTS

CNR grants (P.F. "Materiali e dispositivi per l'elettronica a stato solido"; P.F. "Chimica Fine II") are gratefully acknowledged.

REFERENCES

1. A.Torrisi, G.Marletta, O.Puglisi and S.Pignataro, Surface Interface Anal., 5, 161 (1983).
2. S.Pignataro, A.Torrisi and G.Ferla, Surface Interface Anal., 7, 129 (1985).
3. A.Torrisi and S.Pignataro, Surface Interface Anal., 9, 441 (1986).
4. A.Torrisi, A.Cavallaro, A.Perniciaro, G.Ferla and S.Pignataro, Mater. Res. Soc. Symp. Proc., 108, 365 (1988).
5. J.Pospisil in "Advances in Polymer Science", K.Dusek editor, vol.36, p.69, Springer Verlag, Berlin, 1980.
6. Y.-H.Kim, J.Kim, G.F.Walker, C.Feger and S.P.Kowalczyk, J. Adhesion Sci. Technol., 2, 95 (1988).
7. M.Garilli, G.Marletta, O.Puglisi, C.Oliveri, C.Magro and G.Ferla, Mater. Res. Soc. Symp. Proc., 153, 273 (1989).
8. F.Iacona, M.Garilli, G.Marletta, O.Puglisi and S.Pignataro, submitted to J. Mater. Res..
9. see for example Lj.Atanasoska, S.G.Anderson, H.M.Meyer III, Z.Lin and J.H.Weaver, J. Vac. Sci. Technol. A5, 3325 (1987) and references therein.
10. see for example P.S.Ho, P.O.Hahn, J.W.Bartha, G.W.Rubloff, F.K.LeGoues and B.D.Silverman, J. Vac. Sci. Technol. A3, 739 (1985) and references therein.
11. see for example W.C. Stewart, J.Leu and K.F.Jensen, Mater. Res. Soc. Symp. Proc., 153, 285 (1989) and references therein.
12. see for example K.Kelley, Y.Ishino and H.Ishida, Thin Solid Films, 154, 271 (1987) and references therein.

PHOTOELECTRON SPECTROSCOPY MODEL STUDY OF THE INTERFACE BETWEEN POLYIMIDE AND COPPER

P. Bodö, K. Uvdal, and W.R. Salaneck

Department of Physics
Linköping University
S-581 83 Linköping, Sweden

Three different **model molecules**, representative of selected parts of the polyimide monomer unit, phthalimide (PIM), methyl-phthalimide (MPIM), and benzene-phthalimide (BPIM), were deposited on polycrystalline copper in ultra high vacuum. Both thin films and stable adsorbates of these molecules were studied by means of angle-dependent X-ray photoelectron spectroscopy, XPS(θ). The interaction with the copper surface and the orientation of adsorbates appear to be very different for all three of the **model molecules** examined. PIM shows a preferential orientation and a pronounced interaction with the Cu surface via the nitrogen atom. MPIM forms a very stable adsorbate with an estimated thickness of 18 Å. The BPIM adsorbate decomposes on the Cu surface, which results in a large loss of nitrogen.

INTRODUCTION

Polyimide is a polymer which is widely used in the fabrication of microelectronic integrated circuits.[1] For example, polyimide is suitable as a thin film insulating layer in multilayer structures because of its low dielectric constant. An important factor for the quality of semiconductor integrated circuit devices is the chemistry occurring between polyimide and other materials at the interface formed when polyimide is deposited on metals, on ceramics, on polyimide substrates, or *vice versa*. It has, therefore, become very important to carefully examine interactions between polyimide and other materials, especially those of interest to microelectronic devices. However, the interface reaction and bonding of polyimide films on metal surfaces could be significantly different from that of metal deposited on polyimide surfaces. Polyimide films are formed from a solution of polyamic acid (PA), which becomes a solid film upon thermal curing, while a polyimide substrate already is cured when a metal film is deposited. Metal films grew from single atoms or small clusters, which have chemical properties different from those of bulk metal used as substrates. As a consequence of the different conditions when interfaces of "cured PA"/metal and metal/polyimide are formed one can expect different bonding mechanisms, even though the bulk materials are the same in both cases.

Metallized Plastics 2, Edited by K.L. Mittal
Plenum Press, New York, 1991

Adhesion is probably the most frequently studied property, which is directly related to the chemistry of the interface. Thin metal films can be grown on polyimide by evaporation *in situ* in an ultra high vaccum (UHV) environment, a technique that allows metal/polymer interfaces to be examined under well-defined conditions, by means of sophisticated surface analysis techniques, such as ultraviolet and X-ray photoelectron spectroscopies (UPS and XPS).[2] Such XPS and UPS studies have, previously, indicated that reactive metals, for example Cr and Ti, strongly affect polymer carbonyl groups. In some cases, formation of carbide-like species has been reported.[3-7] Adhesion of such reactive metal films to polyimide is relatively good in comparison to adhesion of, e.g., evaporated Cu films, which usually have poor adhesion to polyimide substrates.

The reverse situation, polyimide films grown on metal substrates, is somewhat more complicated to study with XPS in UHV. Polyimide films are normally prepared by spin coating polyamic acid followed by thermal curing. This technique can not be used in UHV. However, it has recently been shown that thin polyimide films can be grown in UHV by co-evaporation of pyromellitic dianhydride (PMDA) and oxydianiline (ODA), which, on the substrate, form polyamic acid, which upon heating imidizes to form polyimide.[8] Using this technique, thin homogeneous polyimide films, suitable for interface studies with XPS, can be prepared under well-defined conditions in UHV. Also in this situation, with polyimide deposited on metal, the interaction of the polymer carbonyl oxygen with metal substrates has been observed.[9-10]

Despite intensive work to clarify the chemical interaction at polyimide/metal and metal/polyimide interfaces, there is still a lack of detailed understanding. The analysis of data obtained from studies of such interfaces is, of course, a complex task. The polyimide repeat unit as well as PMDA and ODA separately are in a spectroscopic sense rather big molecules. We have proposed previously and shown that small **model molecules** chosen as representative of selected parts of the polyimide repeat unit can be used to suggest site specific interactions between polyimide and metals.[11,12] The more detailed information which might be obtained from such studies could give results that can be used to build up a basic knowledge for a better understanding of both metal/polyimide and polyimide/metal interfaces. Thus idealized and simplified model systems, such as those in the present study, have been chosen to be representative of the building blocks of even the precursors to polyimide, namely polyamic acid. By using relatively small **model molecules** it is also possible to do a relevant theoretical modelling to help describe the electronic structure of the interface.[13]

In this work we present a comparison among three **model molecules**, phthalimide (PIM), methyl-phthalimide (MPIM), and benzene-phthalimide (BPIM), deposited *in situ* on copper. The chemical structures of the **model molecules** are shown together with the polyimide repeat unit in Figure 1. XPS was used to identify chemical interactions at the interface, to estimate molecular adsorbate thickness, and to study molecular orientation effects.

EXPERIMENTAL

The copper substrates used were films grown by physical vapor-deposition (PVD) on optically flat Si(100) wafers. Immediately before **model molecule** deposition, the copper surfaces were cleaned by sputtering with 5 keV Ne^{+} ions, until the surface was free from oxygen and carbon according to XPS.

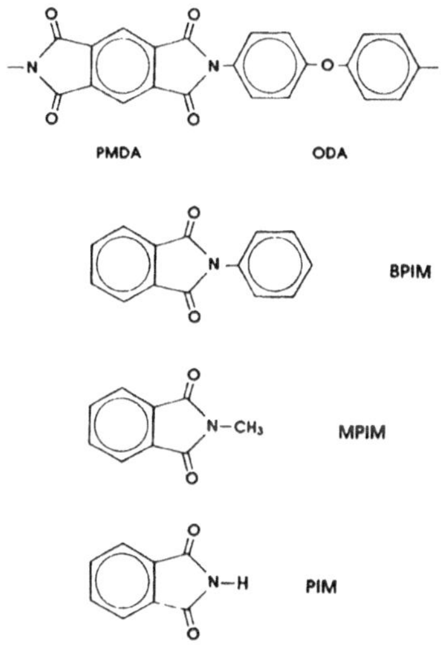

Figure 1. Chemical structure of PMDA/ODA polyimide (PI) monomer unit, benzene-phthalimide (BPIM), methyl-phthalimide (MPIM), and phthalimide (PIM).

The **model molecules**, all of which are solids at room temperature, were evaporated from a simple Knudsen cell source consisted of a borosilicate glass container with a small opening. This source was mounted inside a heatable and coolable copper cylinder, which could be inserted into the vacuum system through a load-lock arrangement. Upon heating, the pressure was allowed to rise from below 10^{-10} Torr up to ≈ 5 x 10^{-9} Torr, upon which rapid cooling of the copper cylinder was started. After the copper cylinder had cooled, but before the cooling reduced the evaporation rate of the molecules, the substrate, held at some specific low temperature, was moved in front of the molecule source. The resulting average film thickness was estimated by observing the attenuation of the XPS Cu 3p intensity, and using the value of Clark *et al.*[14] for the effective elastic mean free path of electrons at 1.17 keV in typical hydrocarbons, 23 ± 2 Å.

XPS spectra were recorded using unmonochromatized Mg K_α radiation (1253.6 eV), and were taken with an analyzer resolution of 0.2 eV, such that the net resolution measured as the full width at half-maximum (FWHM) of Au $4f_{7/2}$ would be 0.9 eV. The spectrometer used in this work is of our own construction and is designed to provide optimum angle-dependent XPS or XPS(θ) for studies of orientational effects.[11,15-17] For high θ-values, the photoelectrons leave the sample surface near the grazing angle and, because of the limited escape depth of the electrons, this is a "surface sensitive" mode. In the "bulk sensitive" mode, for low θ-values, the photoelectrons exit near the surface normal, and hence more information from the "bulk" of the sample is obtained.

All of the spectra of the stable adsorbates contain an overall binding energy shift relative to thin film spectra, which is due to a shift in the work function Φ. This shift, $\Delta\Phi$, can be determined from the cutoff of the secondary electron energy distribution in the UPS valence band spectra. The exact value of the work function shift (typically

$\Delta\Phi \approx 0.4$ eV) is, however, not critical for the discussion below. We focus only on binding energy shifts which are of importance for the interpretation of the data discussed below.

The intensity for each core level was measured as the area under each peak after linear background subtraction. In measuring the relative intensity ratios, the sensitivity factors for the C 1s, O 1s, and N 1s core levels were experimentally determined through measurements on molecular systems of known composition.

RESULTS

Thin films and adsorbates

Each **model molecule** was studied in thin film form to obtain what, in this study, is refered to as bulk information. By "thin film" is meant that the vapor deposited film was thin enough to avoid experimental sample charging, which can cause binding energy calibration problems, yet thick enough that there is negligible contribution from film/Cu interface area in the XPS spectra. This is achieved by having a film thickness somewhat larger than the photoelectron mean free path, which determines the probe depth. In this study we used films having a thickness on the order of 40-100 Å. An estimate of the thickness of the molecular film on the Cu surface after vapor deposition was made by measuring the attenuation of the Cu 3p peak of the Cu substrate. When the Cu 3p level was undetectable or had a very low intensity, we assumed the spectra to be free from any considerable contribution from the interface region. However, the electron mean free path value used for film thickness calculation depends on the density or the packing of molecules and, hence, the obtained thickness value may just serve as an estimate for the real thickness.

Thin films of PIM, MPIM, and BPIM molecules were prepared by exposing the Cu sample to the molecular source for a certain time duration. To be sure to maintain the films at the Cu sample surface, the sample temperature was kept -90°C or lower. The stable adsorbate state, for each type of molecule, was formed by raising the substrate temperature. Stable adsorbates of PIM and BPIM , having an estimated thickness of about 5 Å, were formed when the substrate temperature was somewhere in between -30°C and +30°C. However, MPIM formed a stable

Table I. Compositional Ratios for PIM Thin Film ($\approx$40 Å) and Adsorbate ($\approx$5 Å) in "Bulk and Surface" Sensitive Mode. Calculated Values, Based on the Molecular Structure, are shown to the left.

	40 Å	5 Å	
	θ=0°	θ=0°	θ=80°
C/O 4	3.4	3.9	5.7
C/N 8	7.5	9.6	23
O/N 2	1.9	2.4	4.0
C_H/C_O 3	3.6	5.3	5.6

Table II. Compositional Ratios for MPIM Thin Film (≈40 Å) and Adsorbate (≈18 Å) in "Bulk and Surface" Sensitive Mode at two Different Sample Temperatures, 30 and 100°C. Calculated Values, based on the Molecular Structure, are shown to the left.

	40 Å	18 Å 30°C		18 Å 100°C	
	θ=0°	θ=0°	θ=80°	θ=0°	θ=80°
C/O 4.5	4.6	5.2	5.0	6.8	6.3
C/N 9	8.8	11	9.2	7.9	13
O/N 2	1.9	2.0	1.8	1.2	2.0
C_H/C_O 3.5	5.8	8.0	7.0	6.9	8.3

Table III. Compositional Ratios for BPIM Thin Film (≈65 Å) and Adsorbate (≈5 Å) in "Bulk and Surface" Sensitive Mode. Calculated Values, based on the Molecular Structure, are shown to the left.

	65 Å	5 Å	
	θ=0°	θ=0°	θ=80°
C/O 6	6.8	6.5	5.9
C/N 14	13	60	19
O/N 2	1.9	9.2	3.3
C_H/C_O 6	7.2	16	15

adsorbate with a thickness of about 18 Å, which was intact at a sample temperature as high as +100°C (the highest temperature used in these measurements).

Elemental composition

The elemental composition of the PIM, MPIM, BPIM molecules as thin films and as adsorbates was determined from the C 1s, O 1s, and N 1s peak intensity ratios. Table I,II, and III show the results for PIM, MPIM and BPIM, respectively. As thin films all three molecules have an elemental composition, which corresponds fairly well to expected values (the numbers to the left in each table) for each molecule. As stable adsorbates, at θ=0° the probe depth is relatively large and a preferential orientation of the molecules should not be expected to have any considerable effect on the observed ratios.

The approximately 5 Å thick PIM adsorbate remains rather intact, in comparison with the thin film, except for some decrease in the nitrogen content relative to carbon and oxygen, as shown in Table I. There is also some decrease in the carbonyl carbon intensity, although just a small

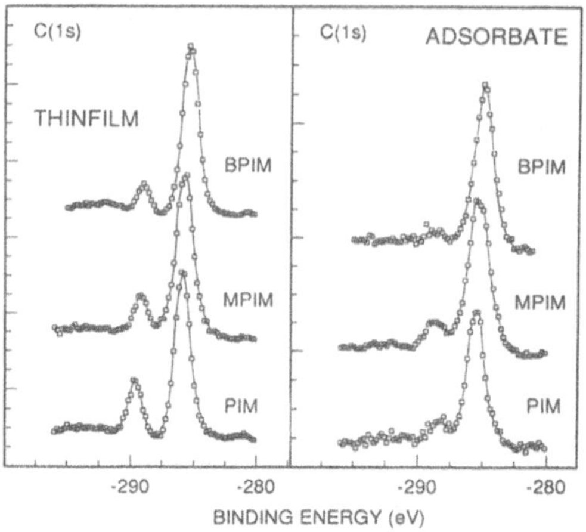

Figure 2. C 1s XPS spectra of PIM, MPIM, and BPIM thin films and adsorbates on copper.

increase in the C/O ratio was observed. MPIM, which differs from PIM only by the methyl group attached to the nitrogen atom, forms a very stable adsorbate with an estimated thickness of about 18 Å. Comparing the thin film with the adsorbate at 30°C both the C/O and C/N ratios as well as the C_H/C_O ratio have increased while the O/N ratio remained about the same. However, the MPIM adsorbate maintained at a substrate temperature of 100°C shows a composition somewhat different from that remaining at 30°C, as shown in Table II. If we still focus on values obtained for θ=0° the main difference is seen as a further decreased oxygen content, since the O/N ratio decreased and the C/O ratio increased when the substrate temperature was increased to 100°C. The nitrgogen content shows a tendency to decrease, relative to carbon and oxygen, for the adsorbate at 30°C, while a slight increase can be observed when going to 100°C. However, one must notice that orientation effects could affect the ratios for an adsorbate as thick as 18 Å. The carbonyl carbon content is lower than the expected value for the thin film as well as for both adsorbates. For the BPIM adsorbate, a dramatic loss of nitrogen was observed, as shown in Table III, but the C/O ratio was almost unaffected. Also for the BPIM adsorbate, the relative intensity of the carbonyl carbon peak was decreased, relative to other C 1s peaks.

Orientation effects

Orientational effects were investigated by comparing the compositional ratios for the adsorbates in the "bulk sensitive" mode (θ=0°) and the "surface sensitive" mode (θ=80°). The most pronounced effect for the PIM adsorbate is the high C/N ratio, C/N=23, in the "surface sensitive" mode, indicating a preferential orientation of nitrogen towards the Cu surface. The C/O ratios also indicate some

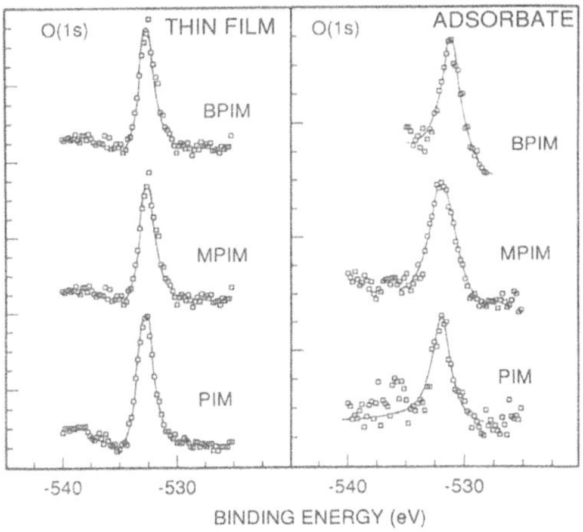

Figure 3. O 1s XPS spectra of PIM, MPIM, and BPIM thin films and adsorbates on copper.

orientation of oxygen towards the Cu surface. For MPIM, no obvious orientation effects are observed for the adsorbate at 30°C. However, the C/N and O/N ratios show a weak preferential orientation of N towards vacuum, but the effect appears reversed for the two adsorbates maintained at 30°C and 100°C, respectively. For the MPIM adsorbate at 100°C, the nitrogen atoms seem to be oriented towards the Cu surface, while the molecule rather has a small tendency to be oriented with the nitrogen away from the Cu surface at 30°C. The BPIM adsorbate, which experiences a considerable loss of nitrogen, seems to have the remaining nitrogen atoms located towards the vacuum, away from the interface.

Chemical shifts

The C 1s spectra of PIM, MPIM, and BPIM thin films and adsorbates are shown in Figure 2. The spectra for thin films consist of two well separated C 1s peaks, one at about 289 eV corresponding to the carbonyl carbons, and another at about 285 eV originating from carbons in the benzene ring, methyl group, and the carbon next to the nitrogen atom. In comparison with the spectra obtained for thin films, the C 1s spectra for all the adsorbates show essentially a different curve shape around the carbonyl carbon peak. For the PIM adsorbate, the carbonyl carbon peak has decreased intensity and is shifted about 1.2 eV towards the main peak. The MPIM adsorbate appears to have considerable amounts of the original carbonyl carbon mixed with a new feature closer to the main peak. The BPIM adsorbate has a carbonyl peak that is unshifted but with decresed intensity.

The O 1s spectra of thin films and adsorbates are shown in Figure 3. All the thin films show one single O 1s feature at about 533 eV with some asymmetry, especially for MPIM, at the low binding energy side of the peak. For the PIM adsorbate, a small shift, less than 1 eV, and a pronounced shoulder at the low binding energy side is observed. The MPIM adsorbate shows a broadening of the O 1s, and for the BPIM adsorbate O 1s is shifted about 1 eV towards lower binding energy.

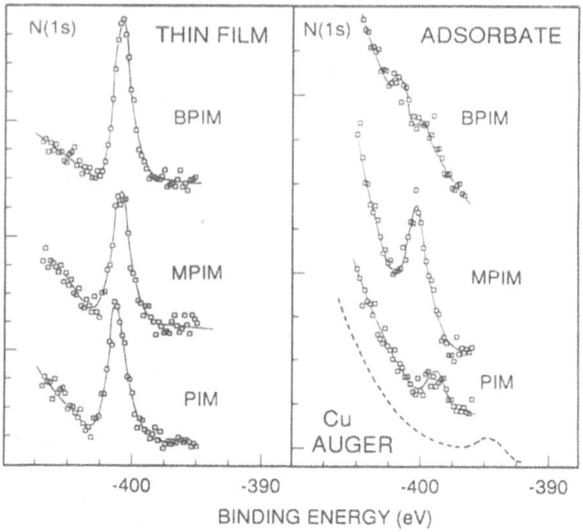

Figure 4. N 1s XPS spectra of PIM, MPIM, and BPIM thin films and adsorbates on copper. The broken line represents the Auger spectrum (background) of clean Cu.

The N 1s spectra, shown in Figure 4, are the most interesting of the three core level spectra. The large sloping background appearing in the adsorbate N 1s spectra comes from a nearby X-ray induced copper LMV Auger spectrum, also shown as a broken line in Figure 4. There are, however, no Cu Auger peaks which interfere with the N 1s XPS peaks. The N 1s of PIM experiences a large shift, ≈ 2.2 eV, when going from a thin film to adsorbate. The MPIM does not show any considerable shift, but only a small shoulder at the low binding energy side of the peak. For the BPIM adsorbate, the N 1s splits into two peaks one of which has shifted about 2.2 eV

DISCUSSION

The angle dependence of the elemental composition ratios C/O, C/N, and O/N show clearly a preferential orientation of the PIM adsorbate. The orientation is such that the PIM molecule turns the nitrogen atom towards the Cu sample surface and the phenyl part towards the vacuum side. However, since the thickness of the adsorbate was estimated to be 5 Å, it is not likely that the PIM molecules are standing close-packed up along the surface normal. Therefore, either the density of molecules remaining on the Cu surface must be low or otherwise there is a layer of molecules oriented at a certain angle relative to the Cu surface, which unfortunately can not be safely determined from available XPS data. The preferential orientation of PIM molecules is further supported by the observed chemical shifts of the core levels. The N 1s exhibits a large shift of about 2.2 eV indicating a relatively strong interaction with the Cu surface. Also, the carbonyl carbon and oxygen core levels are affected by the interaction, but to a smaller degree than the nitrogen. These observed shifts are consistent with previously published data on PIM, where electronic charge transfer from copper to PIM was suggested and supported by numerical estimates.[11]

The MPIM adsorbate shows very interesting behaviour. Unlike PIM and BPIM, the MPIM molecule forms a very stable adsorbate with an estimated thickness of about 18 Å, which is about the size of the molecule itself, and roughly a monolayer of MPIM could be estimated to remain on the Cu surface. But, there are no orientational effects observed for the adsorbate at 30°C, which might indicate an ordered layer of MPIM molecules. The data instead suggest that the adsorbate either consists of several molecules laying flat on top of each other or MPIM molecules randomly standing up and down. However, at 100°C the orientation of the molecules changes so that the nitrogen atoms are preferentially pointing towards the Cu surface. A Cu-N interaction could in this case be assumed to be responsible for the very stable adsorbate, although, the observed core level shifts are not as pronounced as those for PIM and BPIM. We must, however, note that the interface is somewhat hidden by the thick adsorbate. But if a certain site is reactive with the Cu surface at 30°C, then we should at least observe a strong preferential orientation, which is not the case for MPIM at that temperature. We, therefore, suggest that an interaction among the molecules, keeping them together as a film on the surface, also should be considered.

The results of BPIM are somewhat confusing. There is a large loss of N, which implies degradation of the BPIM molecule, but still there is no significant change in the C/O ratio. The remaining nitrogen is preferentially oriented towards the vacuum side, but the N 1s is split into two peaks, one of which is positioned at about the same binding energy as the PIM N 1s, indicating an interaction with the Cu surface. It is, therefore, at this point difficult to suggest a reasonable model for the BPIM adsorption.

The idea of using polyimide **model molecules** to determine interfacial reactions between polyimide and metals is very interesting. But there is still a lot more to be done before we can evaluate the long range relevance of such model systems. The present work is more basic research in its character, and large extrapolation is required to relate present results to "real" systems. Still, it is important to study related new aspects of interface interactions through studies of idealized systems at different levels. For example, a topic often discussed in metal/polymer interface studies is the metal-carbonyl oxygen interaction. The carbonyl carbon C 1s peak in XPS is often reported to be affected when metals are deposited on polymers, and therefore it is frequently concluded that metal/polymer bonding occurs via the carbonyl oxygen.[3-6] However, our results indicate that there is no preferential bonding to the carbonyl oxygen for these **model molecules** on Cu, even though the C 1s peak position and/or intensity is changed in the interface region. A shift and decreased intensity of the carbonyl C 1s does not necessarily have to be due to metal-oxygen bonding. The PIM molecule interacts with the Cu surface via the nitrogen atom and still the carbonyl C 1s is affected, but through delocalization of electronic charge transfer, from the copper substrate to the molecule. However, the N-Cu bonding only occurs for the PIM molecule. It is important to note that the three **model molecules** PIM, MPIM, and BPIM behave differently as adsorbates on Cu. This tendency is a refinement of earlier results.[12] More information is needed to clarify some of the questions raised in this study. It would be especially useful to do complementary studies with methods other than XPS.

CONCLUSIONS

The three **model molecules** phthalimide (PIM), methyl-phthalimide (MPIM), and benzene-phthalimide (BPIM) have been studied in the form of stable adsorbates on copper substrates, as model systems for the polyimide/copper interface. Although these model systems were chosen as a series of successively longer approximations to the imide nitrogen portion of a polyimide, no significant size-dependent trend in the behaviour of the **model molecules** on copper is observed. Instead, the details of the interaction with copper of each individual molecule chosen are unique. However, one particular detail of the features is consistent across the series in that the nitrogen takes an essential part in the interaction in all the model systems. It is important to note that for PIM and BPIM, the primary reaction site with copper substrates is the nitrogen atom, and not the carbonyl group as believed to occur for metals deposited on polyimide. These results indicate the multitude of complexities associated with the adsorption of a full polyimide chain on a clean metal surface.

ACKNOWLEDGEMENT

This work was carried out with support from The Swedish Natural Sciences Research Council (NFR), The Swedish Board for Technical Developement (STU), and Imperial Chemical Industries (I.C.I.), p.l.c., United Kingdom.

REFERENCES

1. See, for example, K.L. Mittal, Editor, "Polyimides: Synthesis, Characterization and Applications," Vols. 1 and 2, Plenum, New York, 1984.
2. K. Siegbahn, C. Nordling, G. Johansson, J. Hedman, P.F. Heden, K. Hamrin, V. Gelius, T. Bergmark, L.O. Verme, R. Manne, and Y. Baer, "ESCA Applied to Free Molecules," North-Holland, Amsterdam, 1969.
3. P.S. Ho, P.O. Hahn, J.W. Bartha, G.W. Rubloff, F.K. LeGoues, and B.D. Silverman, J. Vac. Sci. Technol. A 3, 739 (1985).
4. J.L. Jordan, C.A. Kovac, J.F. Morar, and R.A. Pollak, Phys.Rev. B, 36, 1369 (1987).
5. P. Bodö and J.-E. Sundgren, J. Vac. Sci. Technol. A 6, 2396 (1988).
6. F.S. Ohuchi and S.C. Freilich, J. Vac. Sci. Technol. A 4, 1039 (1986).
7. P. Bodö and J.-E. Sundgren, J. Appl. Phys. 60, 1161 (1986).
8. J.R. Salem, F.O. Sequeda, J, Duran, and W.Y. Lee, and R.M. Yang, J. Vac. Sci. Technol. A 4, 369 (1986).
9. M. Grunze and R.N. Lamb, Chem. Phys. Lett. 133, 283 (1987).
10. S.P. Kowalczyk and J.L. Jordan-Sweet, Chem. Mater. 1, 592 (1989).
11. W.R. Salaneck, S. Stafström, J.L. Brédas, S. Andersson, P. Bodö, S.P. Kowalczyk, and J.J. Ritsko, J. Vac. Sci. Technol. A 6, 3134 (1988).
12. P. Bodö, K. Uvdal, S. Stafström, and W.R. Salaneck, in "Metallization of Polymers," E. Sacher, J.-J. Pireaux, and S.P. Kowalczyk, Editors, ACS Symposium series 440, chapt. 24, 1990.
13. S. Stafström, P. Bodö, W.R. Salaneck, and J.L. Brédas, in "Metallization of Polymers," E. Sacher, J.-J. Pireaux, and S.P. Kowalczyk, Editors, ACS Symposium series 440, chapt. 23, 1990.
14. D.T. Clark, A. Dilks, and D. Shuttleworth, J. Electron Spectrosc. Rel. Phenom. 14, 247 (1978).
15. W.R. Salaneck, R. Bergman, J.-E. Sundgren, A. Rockett, and J.E. Greene, Surface Sci. 198, 461 (1988).
16. E.W. Salaneck, K. Uvdal, H. Elving, A. Askendal, and W.R. Salaneck, J. Colloid Interface Sci. 136, 440 (1990).
17. K. Uvdal, P. Bodö, A. Ihs, B. Liedberg, and W.R. Salaneck, J. Colloid Interface Sci., in press.

ELECTRONIC STRUCTURE OF METAL/POLYMER INTERFACES: ALUMINUM ON CONJUGATED POLYMERS

R. Lazzaroni*, J.L. Brédas

Service de Chimie des Matériaux Nouveaux
Département des Matériaux et Procédés
Université de Mons-Hainaut
21 Avenue Maistriau, B-7000 Mons (Belgium)

M. Chtaïb, P. Snauwaert

Laboratoire Interdépartemental de Spectroscopie Electronique
Facultés Universitaires Notre-Dame de la Paix
61 Rue de Bruxelles, B-5000 Namur (Belgium)

P. Dannetun, M. Lögdlund, K. Uvdal, and W.R. Salaneck

Department of Physics, Linköping University
S-58183 Linköping (Sweden)

Aluminum has been deposited on two types of conjugated polymers, polyalkylthiophene and polyaniline. The polymers were in the undoped, neutral form which possesses semiconducting properties. The chemical structure of the metal/polymer interface has been investigated by X-ray and UV photoelectron spectroscopy. Core level spectra indicate that aluminum reacts with the sulfur atom of the polythiophene chain, strongly perturbing the π-electron system. In the case of polyaniline, the interaction with the metal depends on the oxidation state of the polymer. UPS data, combined with the results of band structure calculations, are interpreted in terms of the Al-induced modifications of the π-electron system.

*: Chargé de Recherches du Fonds National de la Recherche Scientifique (Belgium).

Metallized Plastics 2, Edited by K.L. Mittal
Plenum Press, New York, 1991

1. INTRODUCTION

The interface between metals and polymers has attracted a great deal of interest in the last few years, as polymeric systems have been considered for applications in microelectronics technology.[1] The requirements for such materials are high chemical and thermal stabilities, good processibility, excellent insulating properties, along with the possibility of building metal overlayers showing strong adhesion. In this regard, aromatic polymers, such as polyimide, polyethyleneterephthalate (PET), and polyetheretherketone (PEEK) are most interesting.

Another class of aromatic polymers, the fully π-conjugated polymers, has recently received a great deal of attention for their potential to be included in electronic devices. In these systems, the electrical conductivity (σ) can be tuned over several orders of magnitude, from the insulating to the semiconducting state, and up to the highly conducting metallic regime.[2] In the neutral form, these compounds generally show insulating/semiconducting properties ($\sigma = 10^{-15}$ - 10^{-5} S/cm). Oxidation with electron acceptors (I_2, $FeCl_3$, AsF_5) or reduction with electron donors (Na, Li) dramatically modifies their electrical properties. The highest conductivities are observed for I_2-oxidized polyacetylene ($\sigma > 1.5\text{x}10^5$ S/cm comparable to 6 10^5xS/cm for copper at room temperature);[3] for aromatic compounds such as polyparaphenylene (PPP), polypyrrole (PPy) and polythiophene (PTh), conductivity remains about two orders of magnitude lower. In the pristine state, these compounds represent a new type of semiconductors. Conjugated polymer-based devices, such as Schottky diodes, photovoltaic junctions, field effect transistors, have been proposed recently.[4-8]

A key aspect of these devices is the metal/polymer interface. As in the more conventional polymer systems, where reaction between the organic phase and the metal strongly affects the adhesion properties, chemical reaction between the conjugated polymer and the deposited metal atoms could modify the electronic properties of the junction. In this work, we have investigated the evolution of the electronic structure of two conjugated polymers, polythiophene and polyaniline, as an aluminum layer is gradually deposited onto the surface. Polythiophene is chemically stable both in the doped (oxidized) form and in the neutral form. Here we studied alkyl-substituted polythiophenes (P3ATs, Fig.1a) since these derivatives are soluble in common organic solvents,[9,10] allowing for the preparation of thin polymer films of high chemical purity. It must be noted that the presence of an alkyl group in the 3-position does not affect the basic electronic properties of the polythiophene backbone.[11]

Unlike most conjugated polymers, the electrical conductivity of polyaniline (Fig. 1b) depends both on the oxidation state and on the protonation state of the polymer.[12] Nitrogen atoms along the chain can be in the reduced (-NH-, amine) form or in the oxidized (-N=, imine) form, and the amine/imine ratio can be modified by changing the redox potential. These atoms can also be protonated by acid treatment; the highest-conductive structure is the so-called fully-protonated emeraldine, where the imine/amine ratio is 1 (y=0.5; see Figure 1b) and the protonation level is 50% (all imine nitrogens protonated).[13] In this study, we consider two oxidation states, emeraldine and leucoemeraldine (the all-amine polymer) in the semiconducting non-protonated form.

The electronic structure of the polymer surface exposed to aluminum is investigated with photoelectron spectroscopy (PES). We focus on the very first stages (0-5 Å) of the

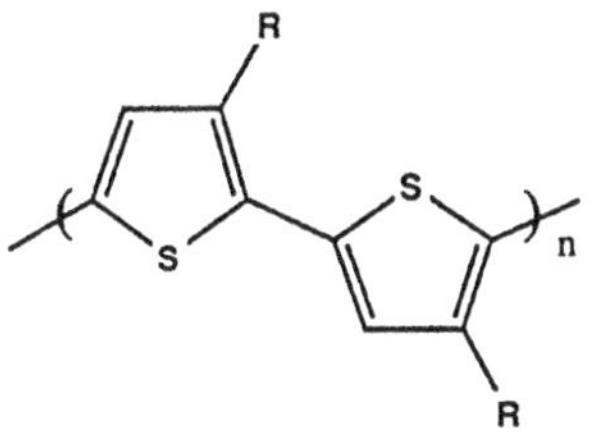

a

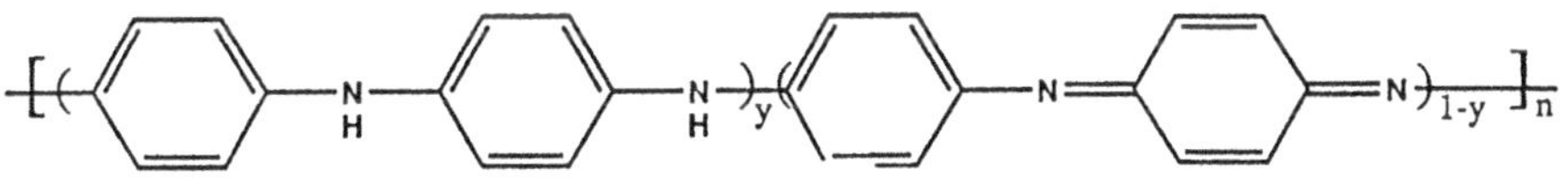

b

Figure 1. Structure of poly-3-alkylthiophene (a; in P3OT, $R=C_8H_{17}$), and polyaniline (b).

metallization process, in order to follow the evolution in the electronic structure of the outer polymer layers. X-ray-induced PES (XPS) is utilized to examine the evolution of the core levels, where changes in the chemical bonding can be detected by shifts in the peak positions and/or by the appearance of new spectral components. XPS secondary structures, such as shake-up satellites, also represent a good probe of the electronic structure of organic compounds.

Photoelectron spectroscopy with a UV source (UPS) is more sensitive to the valence levels. Here, UPS is used to study the changes in the upper π-electronic states. These highly-delocalized levels determine the electronic properties of these polymers, which, in turn, are strongly affected by perturbations of the π system, e.g., those due to charge transfer. The UPS data are interpreted in the framework of our previous studies on the electronic structure of conjugated polymers with a combined theoretical and experimental approach.[14-18] We rely on the results of band structure calculations performed within the Valence Effective Hamiltonian (VEH) methodology. This method has been widely used to obtain good estimates of the electronic properties of polymers.[19-21]

2. EXPERIMENTAL

Aluminum on Poly-3-alkylthiophenes

The P3ATs used in this study were synthesized following the general method described previously.[22] We selected poly-3-octylthiophene (P3OT) as a compromise between good solubility and processibility (which increase with the length of the alkyl chain) and the need to keep the C/S ratio in the polymer as low as possible, in order to prevent the ring contribution to the spectra from being overwhelmed by the contribution from the side chain. Preliminary measurements have shown that the behaviour of the P3ATs towards Al deposition is basically independent of the alkyl chain length.

P3OT was dissolved in spectroscopic-grade chloroform (1mg/ml) and thin polymer films were prepared by spin-coating on optically flat silicon substrates. The samples were introduced into the vacuum chamber and heated up to 180 C for a few minutes in order to remove any residual contamination (solvent, hydrocarbons). The evaporation source consisted of an aluminum-coated tungsten filament which was allowed to degas thoroughly before the experiments. The calibration of the source was performed during a test run on a silicon substrate by monitoring the decrease of the Si2p signal as the metal layer was grown. Assuming the inelastic mean free path of the Si2p electrons to be ≈ 20 Å, one can determine the Al deposition rate as a function of the current applied to the filament. In the case of P3OT, the deposition rate was chosen to be in the 6 - 8 Å/min range. This value does not mean that homogeneous metal films are grown on the surface; it simply represents the average thickness of the deposit, as estimated from the XPS signal attenuation. The pressure in the chamber never exceeded 8×10^{-9} Torr during the evaporation. After the treatment, the samples were moved into the analysis chamber ($P < 10^{-10}$ Torr) where XPS and UPS measurements were carried out.

The spectrometer we used for this part of the work has been described in detail previously.[23] UPS spectra were obtained with monochromatized HeI radiation ($h\upsilon = 21.2$ eV). The XPS source was non-monochromatized MgKα. Its resolution was determined from the width of the Au4f7/2 line to be 0.9 eV. The position of the Au4f7/2 line was also used to calibrate the binding energy scale at 83.8 eV. It is noteworthy that due to a special sample holder geometry, this spectrometer yields high counting rates even for analyses at grazing angles. The standard error in the measurements was ± 0.1 eV.

Aluminum on Polyaniline

Polyaniline films were prepared electrochemically from aqueous acidic solutions, following the procedure reported in a previous paper;[24] polymers obtained at + 0.8 V vs. the saturated calomel electrode were in the emeraldine form. Reduction at - 0.2 V leads to leucoemeraldine. The samples were turned into the base form by a short treatment with 1 M NaOH followed by rinsing in water. They were then introduced into the vacuum chamber and heated briefly at 100 C to remove residual water.

The spectrometer used for these experiments was a Surface Science Instruments ESCA-201 apparatus equipped with a monochromatized AlKα source. XPS spectra were collected on a 600 μm-diameter spot with a resolution of 1.0 eV. When checking the polyaniline films before evaporation, a small oxygen contamination (≈ 1 oxygen atom per 5 rings) was detected. Aluminum was evaporated from a Knudsen cell inside a separate chamber ($P < 5 \times 10^{-10}$ Torr); the deposition rate, which was monitored with a quartz microbalance, was 0.2 Å/min, and the pressure during the evaporation remained below 1 10^{-9} Torr.

3. RESULTS AND DISCUSSION

Aluminum on P3OT

a. Core levels: In the pristine polymer, the C1s and S2p core levels appear at 285.1 and 164.1 eV, respectively. The binding energies (Eb) of C1s and of the S2p major component are not modified by the deposition of the aluminum layer up to 5 Å. This result is in sharp contrast with the data of Ogama and Koezuka on the In/PPy junction.[25] In that case, the 5Å metal overlayer induced a general upward shift of the core levels of 0.4 eV. This has been interpreted as a consequence of band bending at the semiconductor-metal interface. If the same explanation holds for the Al/P3OT system, the absence of core level shift implies that the Fermi level of the semiconductor exactly matches that of the metal.

In order to check this hypothesis, we have determined the position of the Fermi level in pristine P3OT by means of UPS measurements. The total width of the HeI spectrum is the distance between the Fermi level, which is the zero of the energy scale, and the cut-off of the secondary electrons. The difference between this value and the energy of the UV photon (here, 21.2 eV) provides an accurate value for the work function, hence the position of the Fermi level relative to the vacuum level. In the case of P3ATs in general, the Fermi level is located at 3.6 ± 0.2 eV, i.e., 0.7 eV lower than the Fermi level of aluminum (4.3 eV). This is in clear disagreement with the above hypothesis.

It thus seems that band bending is not playing a significant role at this stage. This is likely due to the fact that the amount of Al is too low to display the properties of the bulk metal. Aluminum probably adsorbs as clusters on the polymer surface, as is generally

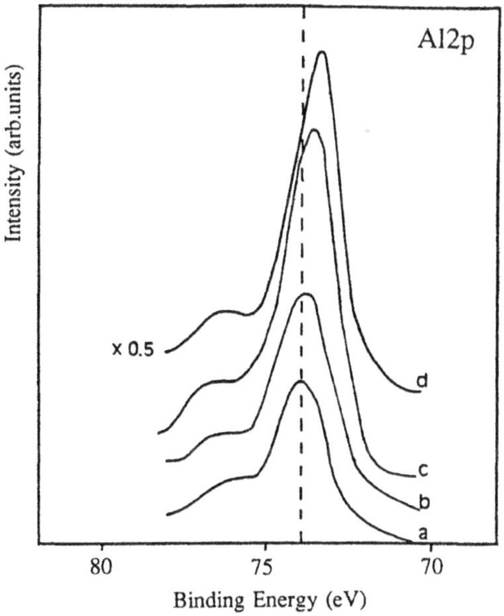

Figure 2. Al2p XPS spectra of the Al/P3OT interface (a: 1Å; b: 2Å; c: 3Å; d: 5Å coverage). The dashed line corresponds to the position of the peak for 1Å-coverage.

the case for common semiconductors, and these islands grow with the deposition time. Even for "5Å-coverage", the situation is still far from that of the true metal.

This is further confirmed by the shift of the Al2p core level to lower binding energies as coverage increases. The peak is located at 74.0 eV after the first evaporation (≈ 1Å) and moves gradually to 73.3 eV for a 5Å-thick layer (Fig. 2), still 0.6 eV away from its position in the bulk metal (72.7 eV).[26] In view of this, we are led to the conclusion that the absence of chemical shift in the core levels simply indicates that no major interaction is taking place neither at the carbon atoms nor on the major part of the sulfur atoms.

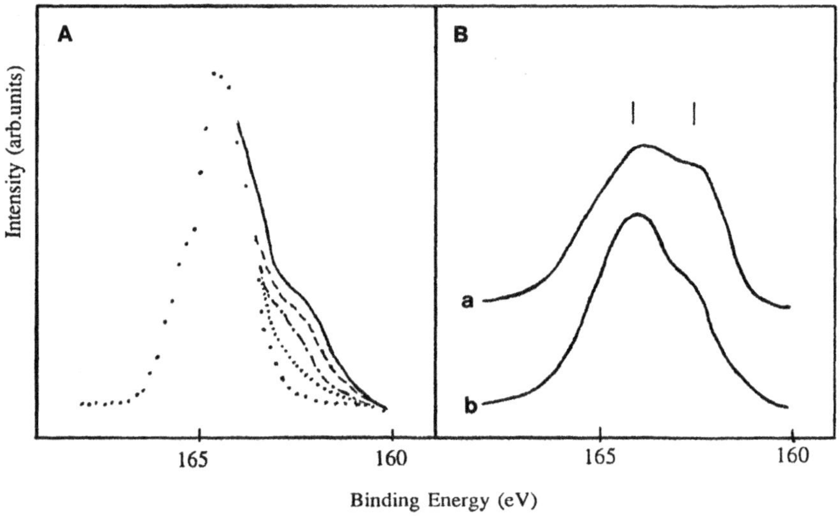

Figure 3. A: S2p XPS spectra of the Al/P3OT interface, collected at normal exit (· · ·: pristine; ·······:1Å; - · - · -: 2Å; - - -: 3Å; ——: 5Å coverage).
B: S2p spectra collected at grazing exit for 3Å (a) and 5Å (b) coverage.

However, some sulfur sites are strongly affected by Al deposition. From Figure 3A, it can be seen clearly that a second component is growing in the S2p spectrum as the amount of aluminum present at the surface increases. This new feature lies approximately at 162.5 ± 0.2 eV, i.e., 1.6 eV lower than the main component, and corresponds to an electron-rich sulfur species. This value is similar to those found in the literature for pure metal sulfides (Na_2S: 162.0 eV;[27] GaS: 162.2 eV;[28] ZnS: 161.7 eV[29]), indicating that the chemical structure of the Al-S complex is close to aluminum sulfide. The intensity of the contribution at 162.5 eV relative to the main peak can not be easily estimated, due to the rather low counting statistics, and to the presence of the spin-orbit coupling (the 2p1/2 component of the 162.5 eV peak largely overlaps with the 2p3/2 component of the peak at 164.1 eV).

At this point, we performed XPS measurements in a configuration where the collected electrons leave the sample with an angle of approximately 80° relative to the normal to the surface. Previous studies of aluminum-containing organic molecules oriented on the surface have shown that, in this geometry, the probing depth is in the 5 Å range.[30] In this configuration, the low binding energy component of the S2p spectrum comes out much more clearly. Figure 3B shows the spectra of tilted P3OT samples at 3Å- and 5Å-coverage. In the latter, the intensity at 162.5 eV is almost as large as that at 164.1 eV. Considering the presence of the spin-orbit satellite, one can estimate that the concentration of the two sulfur species is roughly the same for 5Å coverage. Therefore, we can conclude that, in the upper layer, about half of the sulfur atoms have reacted with aluminum.

The interaction between sulfur and Al, which yields a new electron-rich sulfur species, should also give an electron-poor aluminum species. Although the most important feature in the Al2p spectra (Fig. 2) is the metal-like contribution, a low-intensity component is indeed present at higher binding energy (≈ 76 eV), due to electron-deficient Al atoms. We note that the presence of Al oxide can be dismissed since: (i) a check of the surface before deposition shows that the polymer is free from oxygen, and (ii) the aluminum layer does not oxidize in the vacuum chamber, even after several hours. Moreover, the absolute intensity of the Al2p high-binding energy feature increases as the Al layer grows. This is consistent with the fact that more Al reacts with the polymer and with the increase of the 162.5 eV component in the S2p spectra (Fig. 3).

b. Valence levels: The lowest curve in Figure 4 represents the low-binding energy part of the UPS HeI spectrum of P3OT. We focus on this region since it exclusively contains π-electron levels due to the conjugated chains.[11] The remaining part (5-15 eV) is built mainly from σ states and mostly contains the contribution of the alkyl chain, which possesses 49 σ electrons vs. 17 for the ring. Any effect of Al on the σ states of the ring is thus unlikely to be observed. On the other hand, the side chain does not contribute to the spectrum below 5 eV. The interpretation of the π region in combination with the results of polymer band structure calculations has been extensively reported previously.[16] The most important points are:

- the high-intensity band located at 3.8 eV corresponds to a localized (flat) electronic band (π_2) with large contributions from the sulfur atoms and the β carbon atoms,
- the highest-occupied band (π_3) vanishing around 1 eV is due to π levels delocalized over the whole conjugated carbon backbone without any contribution from the sulfur atoms.

Aluminum deposition strongly modifies the spectrum in the 0-5 eV region, while the inner valence levels are not significantly influenced, except for the appearance of one peak around 7-8 eV (not shown here). The UPS spectrum of metallic Al shows a strong band around 7 eV; then the density of states smoothly decreases towards the Fermi level where a sharp edge is visible.[31] This trend clearly appears in our results, as the contribution of the polymer is gradually covered by the tail of the 7 eV band, leading to an increase in the overall intensity of the signal. The π_2 band seems to shift to lower binding energy upon deposition, then decreases in intensity and vanishes at 5Å-coverage. This shift can be considered to be a genuine consequence of the chemical reaction between Al and sulfur. Since the π_2 band is localized on the sulfur atoms, one would expect it to be influenced by the formation of the Al-S complex. Theoretical calculations are currently underway to explore this point. The decrease in the intensity of the π_2 band at higher coverage can also be interpreted in terms of the perturbation of the π-electron system induced by the reaction with aluminum.

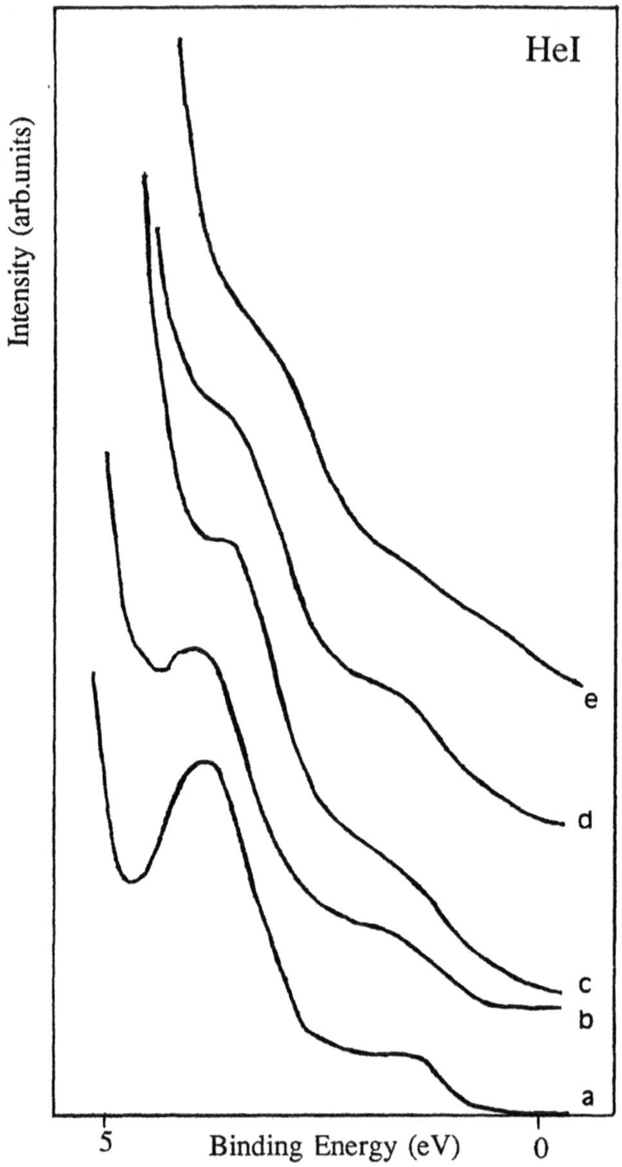

Figure 4: HeI UPS spectra of the Al/P3OT interface
(a: pristine; b: 1Å; c: 2Å; d: 3Å; e: 5Å coverage).

The evolution of the low-intensity π_3 band can hardly be estimated from these spectra. It appears clearly at low coverage (up to 3 Å), but it is no longer observed in the upper curve (Fig. 4e), where a finite density of states appears around the Fermi level. Note the absence of a sharp Fermi edge, again indicating that the electronic structure of the Al layer is still far from that of the "true" metal.

a. Emeraldine: The C1s and N1S core level spectra of polyaniline in the emeraldine form (imine/amine ratio = 1) are shown in Figure 5 (a,c). The C1s spectrum consists of a slightly asymmetric peak containing the contribution of all carbons. The experimental resolution is not sufficient for the C-N component to be observed clearly; hence, it appears as a shoulder on the high-binding energy side of the peak.

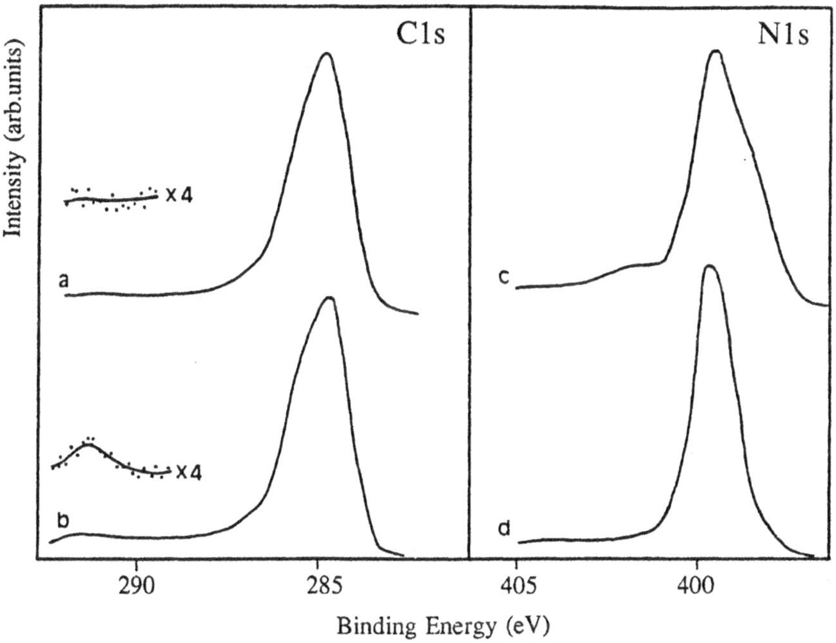

Figure 5. C1s (left) and N1s (right) XPS spectra of pristine emeraldine (a,c) and leucoemeraldine (b,d).

The N1s spectrum exhibits a more complex structure: in addition to the imine and amine components, located at 398.3 and 399.4 eV, respectively, a broad band appears with a low intensity at 402.0 eV. The study of polyanilines in different oxidation states, i.e., with different imine/amine ratios, clearly indicates that this band always lies at the same position and that its intensity is related to the relative amount of imine nitrogens in the polymer.[24] This feature has been assigned to a shake-up satellite related to the imine group. Shake-up satellites are common in aromatic compounds, and correspond to π-π^* electronic transitions in the presence of the core hole. As such, they are very sensitive to the electron density on the molecule. For instance, the energy separation of the shake-up from the main peak and its relative intensity have been shown to be related to the degree of electron delocalization in acenes and polyphenyls.[32,33] The shake-up structure has also been used in the study of the temperature-induced conformational modifications of polyalkylthiophenes.[16,34]

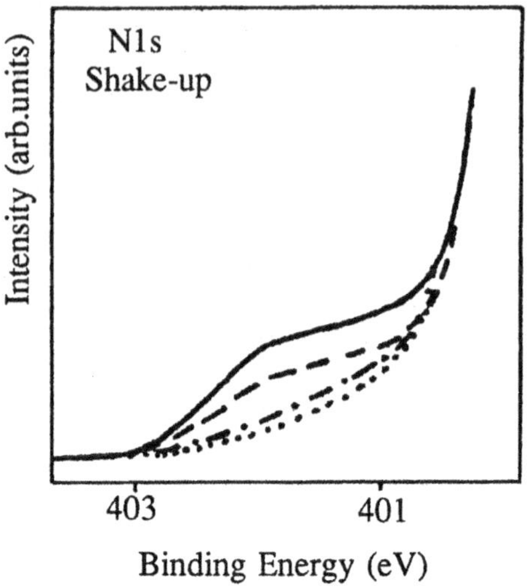

Figure 6. Evolution of the imine N1s shake-up satellite as a function of Al coverage in emeraldine (——: pristine; - - -: 0.5Å; - - - • -; 1Å; ·····: 3Å coverage).

The assignment of the shake-up in the N1s spectrum of emeraldine has been confirmed by the results of recent INDO CI calculations.[35] In the case of an all-imine model molecule for polyaniline, the most intense satellites are calculated to lie in the 3-5 eV region from the main peak. The calculated shake-up pattern of the corresponding all-amine compound, which can be compared to leucoemeraldine (see below), shows only weak lines at higher binding energies.

Upon Al deposition, the imine-related shake-up vanishes gradually (Fig. 6), indicating that the electronic density on the imine nitrogen sites is strongly affected. This suggests that aluminum preferentially interacts with the imine-quinoid-imine part of the chain. This is in agreement with recent theoretical calculations which indicate that imine-quinoid-imine groups along the polyaniline chain are preferentially charged, e.g., upon protonation.[36,37] It is noteworthy that the nitrogen atoms which have reacted with Al do not give rise to a new feature in the N1s spectrum, e.g., as a low binding energy peak. This might be due to the fact that the extra charge density brought in by the aluminum delocalizes over the N=quinoid=N segment. As a consequence, the shift of the carbon and nitrogen atoms would remain too small to be detected. Accordingly, the binding energy of the C1s main peak is constant at 284.7 eV throughout the deposition. However, a low-intensity feature appears at 6.5 eV from this peak upon metallization. This position corresponds to a shake-up transition involving the phenyl ring, which is observed in leucoemeraldine and which is not present in pristine emeraldine.[24] This suggests that the polymer is driven towards a more reduced state upon Al deposition.

The first aluminum layer (0.5-1 Å) appears to be in the oxidized form, with a rather broad Al2p peak centered around 75 eV. This is a clear indication of the charge transfer towards the organic molecules. This peak probably contains a minor contribution from the Al-O species due to the small oxygen contamination. Note, however, that Al deposition does not seem to affect the position and the line shape of the corresponding O1s peak. This behavior is also observed at the Al/polyimide interface;[38] it is explained by assuming that the oxygen atom acts simply as a bridge for the electron transfer from aluminum to the carbon backbone.

At higher coverage (2-4 Å), a sharp band corresponding to unaffected Al atoms starts to grow at 73.2 eV. This value is identical to what is found in the Al/P3OT system; the difference relative to the binding energy in the bulk metal (72.6 eV) reflects the size effect of the Al clusters on the energy of the electronic levels. Note that the absolute intensity of the oxidized component also increases with Al deposition, as the reacting sites on the surface are gradually "filled".

b. Leucoemeraldine: Comparison of the spectra of pristine leucoemeraldine (the all-amine form) with those of emeraldine (Fig. 5b,d) shows that the C1s levels are very similar. However, in leucoemeraldine, a shake-up is located at 6.5 eV from the main line. This band, which is analogous to that found in benzene, aniline, and their oligomers,[33,39] is assigned to an electronic transition within the phenyl ring.

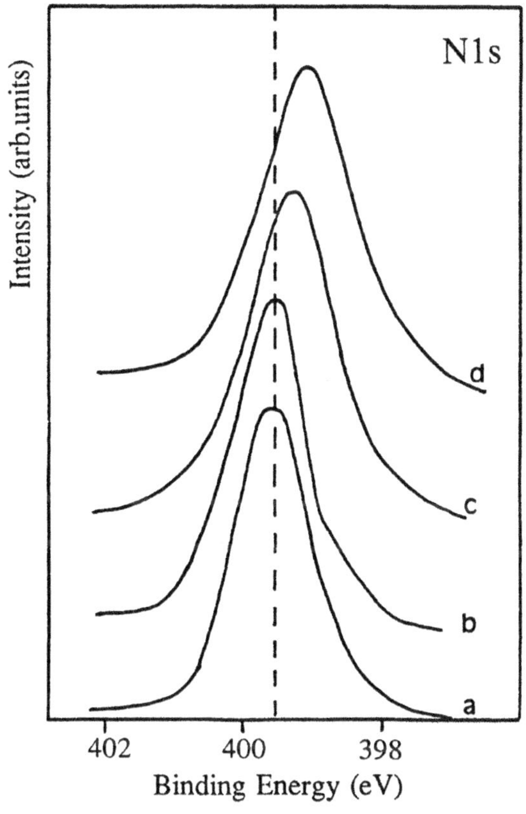

Figure 7. N1s XPS spectra of the Al/leucoemeraldine interface (a: pristine; b: 0.5Å; c: 1Å; d: 3Å coverage). The dashed line corresponds to the position of the peak in the pristine polymer.

The N1s spectra are quite different; in leucoemeraldine, the main peak (399.4 eV) is sharp due to the presence of only one type of nitrogen: the amine species. Accordingly, the shake-up related to the imine group is not present. Again, this evolution quantitatively agrees with the results of theoretical calculations.[35] Deposition of the first Al atoms (< 0.5

Å) has no influence on the N1s spectrum, probably because of the small contribution of the reacted sites relative to the overall intensity. Further deposition leads to a gradual shift in the peak position, from 399.4 eV to 398.6 eV for 3Å-coverage (Fig. 7). This evolution is consistent with a charge transfer from aluminum atoms towards the amine nitrogen. One could expect that, due to the presence of unaffected nitrogen sites and the presence of reacted nitrogen species, the N1s spectrum would display two distinct components, the former being located at the same position as in the pristine polymer and the latter at lower binding energy, as is observed in the S2p spectrum for the Al/P3OT system. In leucoemeraldine, instead, the whole peak is shifted. Along with this shift, the peak width at half maximum increases from 1.3 eV in the pristine system to 1.7 eV for 3Å-coverage. In our opinion, this broadening is a consequence of the existence of a second component too close to the first one to be resolved. As a matter of fact, the upper spectrum (Fig. 7d) can be fitted by the sum of two contributions of approximately the same intensity. The first one corresponds to unaffected nitrogen atoms (399.4 eV), and the second one, which represents the Al-amine complex, is shifted 1 eV downwards. At low coverage, the second component is detected as a shoulder on the low energy side of the peak. In emeraldine, the interaction of Al with the imine group did not induce any significant shift, since the extra electron density can be efficiently delocalized over the neighboring quinoid ring. The fact that a shift is indeed observed in leucoemeraldine indicates that charge delocalization from the amine atom in the Al-N complex to the phenyl ring does not occur to a large extent. This is further confirmed by the fact that the C1s main peak and shake-up satellite are unaffected by the deposition process.

4. SUMMARY

The initial stages of aluminum deposition on two conjugated polymers, polythiophene and polyaniline, have been investigated. In the first case, the metal was gradually evaporated on thin films of poly-3-octylthiophene. Aluminum interacts strongly with the aromatic rings by forming a complex with the sulfur atoms. The reaction also disrupts the π-electron system, as shown by the UPS data, thereby modifying dramatically the electronic properties of the polymer surface.

In the case of polyaniline, the type of interaction with Al depends on the oxidation state of the polymer. In emeraldine, aluminum adsorbs preferentially on the imine sites. The charge transfer to the nitrogen atom then delocalizes over the neighboring quinoid ring, and strongly modifies its electronic structure. When only amine groups are present in the polymer, i.e., in leucoemeraldine, reaction with Al leads to a large increase in the electron density on the nitrogen without any influence on the phenyl groups, as π-delocalization is less efficient in this state.

ACKNOWLEDGEMENTS

We are grateful to Dr. K.L. Mittal for inviting us to present these results at the 2nd Symposium on Metallized Plastics. Research on conducting polymers in Linköping is performed in collaboration with Neste Oy Co. (Finland), with the support of the Swedish Board for Technical Development (STU) and the Swedish Natural Sciences Research Council (NFR).

REFERENCES

1. See, for example, a- K.L. Mittal, Ed., "Polyimides: Synthesis, Characterization, and Applications", Plenum Press, New York, 1984.
   b- K.L. Mittal and J.R. Susko, Eds., "Metallized Plastics 1: Fundamental and Applied Aspects", Plenum Press, New York, 1989.
2. For recent reviews, see: a- T.A. Skotheim, Ed.,"Handbook of Conducting Polymers", Dekker, New York, 1986.
   b- Proceedings of the International Conference on Science and Technology of Synthetic Metals (Santa Fe, June 1988), Synth. Metals 27-29 (1988-89).
3. H. Naarman and N. Theophilou, Synth. Metals 22, 1 (1987).
4. J. Kanicki, in Ref. 2a, pp. 543-660.
5. H. Koezuka and S. Etoh, J. Appl. Phys. 54, 2511 (1983).
6. J.H. Burroughes, C.A. Jones, and R.H. Friend, Nature 335, 137 (1988).
7. G. Gustafson, M. Sundberg, O. Inganäs, and C. Svensson, J. Molec. Electronics, in press.
8. G. Horowitz, D. Fichou, X. Peng, Z. Xu, and F. Garnier, Solid State Commun. 72, 381 (1989).
9. M.A. Sato, S. Tanaka, and K. Kaeriyama, Synth. Metals 14, 289 (1986).
10. S. Hotta, S.D.D.V. Rughooputh, A.J. Heeger, and F. Wudl, Macromolecules 20, 212 (1987).
11. B. Thémans, J.M. André, and J.L. Brédas, Synth. Metals 21, 149 (1987).
12. W.R. Salaneck, I. Lundström, W.S. Huang, and A.G. MacDiarmid, Synth. Metals 13, 291 (1986).
13. J.C. Chiang and A.G. MacDiarmid, Synth. Metals 13, 193 (1986).
14. C.R. Wu, J.O. Nilsson, O. Inganäs, W.R. Salaneck, J.E. Österholm, and J.L. Brédas, Synth. Metals 21, 197 (1987).
15. R. Lazzaroni, J. Riga, J. Verbist, J.L. Brédas, and F. Wudl, J. Chem. Phys. 88, 4257 (1988).
16. W.R. Salaneck, O. Inganäs, B. Thémans, J.O. Nilsson, B. Sjögren, J.E. Österholm, J.L. Brédas, and S. Svensson, J. Chem. Phys. 89, 4613 (1988).
17. M. Lögdlund, R. Lazzaroni, S. Stafström, W.R. Salaneck, and J.L. Brédas, Phys. Rev. Lett. 63, 1841 (1989).
18. R. Lazzaroni, M. Lögdlund, S. Stafström, W.R. Salaneck, and J.L. Brédas, J. Chem. Phys. 93, 4433 (1990).
19. J.L. Brédas, R.R. Chance, R. Silbey, G. Nicolas, and Ph. Durand, J. Chem. Phys. 75, 255 (1981).
20. J.L. Brédas, R.R. Chance, R. Silbey, G. Nicolas, and Ph. Durand, J. Chem. Phys. 77, 371 (1982).
21. J.L. Brédas, B. Thémans, J.G. Fripiat, J.M. André, and R.R. Chance, Phys. Rev. B 29, 6761 (1984).

22. O. Inganäs, W.R. Salaneck, J.E. Österholm, and J. Laakso, Synth. Metals 22, 395 (1988).
23. W.R. Salaneck, S. Stafström, J.L. Brédas, S. Andersson, P. Bodö, S.P. Kowalczyk, and J.J. Ritsko, J. Vac. Sci. Technol A 6, 3134 (1988).
24. P. Snauwaert, R. Lazzaroni, J. Riga, J. Verbist, and D. Gonbeau, J. Chem. Phys. 92, 2187 (1990).
25. T. Ogama and H. Koezuka, J. Appl. Phys. 56, 1036 (1984).
26. C.D. Wagner, W.M. Riggs, L.E. Davis, J.F. Moulder, and G.E. Muilenberg, in "Handbook of X-Ray Photoelectron Spectroscopy", Perkin Elmer Corp., Eden Prairie, Minnesota, 1979.
27. M. Furuyama, K. Kishi, and S. Ikeda, J. Electr. Spectrosc. 13, 59 (1978).
28. J.M. Thomas, I. Adams, R.H. Williams, and M. Barber, J. Chem. Soc. Farad. Trans. II 68, 755 (1972).
29. C.J. Vesely and D.W. Langer, Phys. Rev. B 4, 451 (1971).
30. W.R. Salaneck, R. Bergman, J.E. Sundgren, A. Rockett, T. Motooka, and J.E. Greene, Surface Sci. 198, 461 (1988).
31. S.A. Flodström, L.G. Peterson, and S.B.M. Hagström, Solid State Commun. 19, 257 (1976).
32. J. Riga, J.J. Pireaux, R. Caudano, and J. Verbist, Phys. Scr. 16, 346 (1977).
33. J. Riga, J.J. Pireaux, J.P. Boutique, R. Caudano, J. Verbist, and Y. Gobillon, Synth. Metals 4, 99 (1981).
34. W.R. Salaneck, R. Lazzaroni, N. Sato, M. Lögdlund, B. Sjögren, M.P. Keane, S. Svensson, A. Naves de Brito, and N. Correia, in "Conjugated Polymers: Opportunities in Electronics, Opto-electronics, and Molecular Electronics", J.L. Brédas and R.R. Chance, Eds., NATO-ARW Series E 182, Kluwer, Dordrecht, 1990.
35. B. Sjögren and S. Stafström, to be published.
36. M.C. Dos Santos and J.L. Brédas, Phys. Rev. Lett. 62, 2499 (1989).
37. M.C. Dos Santos and J.L. Brédas, Phys. Rev. B 40, 11997 (1989).
38. P.S. Ho, P.O. Hahn, J.W. Bartha, G.W. Rubloff, F.K. LeGoues, and B.D. Silverman, J. Vac. Sci. Technol. A 3, 739 (1985).
39. P. Snauwaert, R. Lazzaroni, J. Riga, and J. Verbist, to be published.

STUDY OF CHEMICAL INTERACTIONS IN METALLIZED POLYMERS USED FOR MICROELECTRONIC PACKAGING

Hyo-Soo Jeong and Robert C. White

Department of Electrical Engineering and
Microelectronics Sciences Laboratories
Columbia University, New York, N.Y., 10027

Surface modification of PMDA-ODA polyimide (PI) utilized for microelectronic packaging has been performed by the use of low energy (600 eV) Ar ion beam treatment as well as an Ar electron cyclotron resonance (ECR) plasma source (ion energy: 10 eV). This modification was done in a controlled way to understand its effects on enhanced adhesion in metallization. Copper (Cu) was deposited on the Ar ion treated polyimide surface by electron beam evaporation. Interfacial chemical reactions characteristic of the metallized polymer were investigated by x-ray photoelectron spectroscopy (XPS). XPS data revealed that surface modification by Ar ion beam treatment was deeper and more extensive than that by ECR plasma treatment for the same ion dose. The degree of surface modification is found to be a function of ion energy for a given ion dose. This surface modification is observable by XPS even at doses less than 5.6×10^{13} Ar^+/cm^2. Furthermore, enhanced interaction was observed between Cu and the Ar ion treated PI surface, believed to be charge transfer. This indicates enhanced chemical interaction between Cu and ion treated PI as compared to that between Cu and untreated PI.

INTRODUCTION

Polyimide (PI), a high temperature, high performance polymer, has been widely used in microelectronic packaging applications which include inter-layer dielectrics in IC's, inter-metallic insulators in multilevel inter-connections, and thermal-mechanical passivation buffer protection layers.[1,2] As an alternative to ceramics, it has many advantages including: (i) low dielectric constant, (ii) self-planarizing, (iii) no stress-induced cracking, (iv) thermal stability, and (v) ease of processing. Moreover, it can be plasma etched and fully processed at temperatures of up to 400°C.

In spite of these merits, there are some potential problems associated with PI usage, in particular, as inter-metallic dielectrics. Polymers are a totally different class of materials compared to metals and semiconductors. Reliable adhesion between polymer and metal layers is required in packaging applications. Many vapor deposited metals do not adhere well to PI, particularly Cu, the conductor of choice. Surface modification of polymers by ion beam and plasma treatment prior to metallization is one of the ways

Metallized Plastics 2, Edited by K.L. Mittal
Plenum Press, New York, 1991

to obtain strong and stable adhesion in multilayer thin film structures.[3] In general, enhanced adhesion of the metal to the polymer is found in many cases; but little is known regarding changes in the surface chemistry and/or morphology of the polymer effected by this processing. A proposed explanation for enhanced adhesion is, among other things, the increase of surface area due to surface roughening by ion bombardment. Another possible effect is induced chemical change which may enhance adhesion. Therefore, a detailed knowledge of the process is needed. It should be noted that for microelectronics, multilayer thin film reliability is the important parameter rather than strictly adhesion enhancement. Reliability takes into account thermal cycling of the structure, exposure to moisture, and the like, all of which may be dependent on surface modification.

Another interesting process technique for surface modification is the use of ECR plasma discharge. Broad beam plasma sources using ECR discharge have been widely used for reactive ion beam etching[4] and chemical vapor deposition (CVD)[5] in microelectronics fabrication. Table I compares the basic parameters between the Ar ion beam and Ar ECR plasma employed in this study. One of the most prominent ECR plasma characteristics is that it can achieve high ion density at low ion energy which is useful for microelectronic processing.

Table I Basic Parameters of Ar Ion Beam and Ar ECR Plasma.

	Ar ion beam	Ar ECR plasma
beam type	narrow beam	broad beam
electron source	thermionic emission	microwave excitation
operating pressure	7.2×10^{-7} torr	5.0×10^{-4} torr
ion energy	600 eV	10 eV
ion current	0.2 μA	100 μA

The degree of surface modification was compared for the Ar ion beam and Ar ECR plasma exposure at the same ion dose. We employed low energy ions, 600 eV for Ar ion beams and 10 eV for Ar ECR plasma, where nuclear interactions should dominate ion energy loss processes in the polymer and more complex electronic interactions will be reduced. Cu was vapor deposited following the Ar ion beam treatment of the PI surfaces. The reaction of vapor deposited Cu has been studied by many authors[6,7] and it has been found that Cu interacts primarily with the PMDA part of the repeat unit, albeit weakly. This is due to the chemistry of electron rich Cu which does not easily transfer charge to the PI and does not occupy high coordination sites. The reaction characteristics of Cu deposited on PI surfaces were investigated by XPS. In this paper, we explain the XPS observations in terms of Ar ion induced modification of the PI.

EXPERIMENTAL

The PI films under investigation were spin coated on Si wafers from polyamic acid solution PMDA-ODA for a final thickness of 200 Å. The samples were fully cured under N_2 using a standard cure protocol.[8] The samples were annealed in situ for 10 minutes in the preparation chamber [5.0×10^{-8} torr] at 300°C immediately prior to each experimental run to drive off water.[9] After annealing the sample was transferred into the analysis chamber [less than 5.0×10^{-10} torr]. The XPS spectra were taken in ultra-high vacuum. Monochromatized AlK_{α} was used for XPS. Our overall spectrometer resolution is 0.2 eV for Au 4f. All peak positions reported are absolute and referenced to the Au 4f and E_F levels. The thin films of PI are utilized to minimize charging effects.[6,8,10]

For Ar ion beam treatment a 1×1 cm^2 surface area, essentially the entire sample, was rastered with 600 eV Ar ions from an ion gun which is attached to the analysis chamber and differentially pumped. The operating pressure of the ion beam treatment was 7.2×10^{-7} torr in the analysis chamber. The ion current was measured by a Faraday cup and converted into ion doses for a given treatment duration. For Ar ECR plasma treatment, 2.45 GHz microwave was used with microwave power of 120W. The corresponding resonant magnetic field is 875G. Not only positive Ar ions (Ar^+) but electrons and excited Ar neutrals coexist in the Ar ECR plasma. These electrons and ions bombarding the unbiased substrate have average energies on the order of 10~20 eV in the typical ECR plasma employed and could play an important role in the plasma reaction with polymer. The in situ Cu (2Å) deposition by electron beam evaporation followed the Ar ion beam treatment of the PI surfaces. The sample and evaporation source were separated by ~0.4m, and no significant heating of the sample should occur.

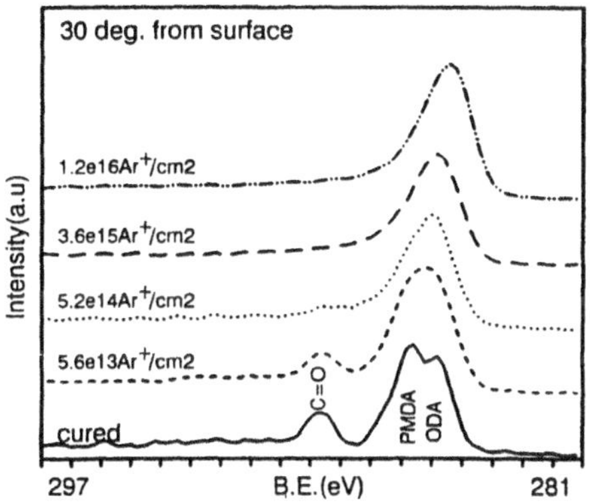

Figure 1. C1s XPS Signal for Ar Ion Beam Treated Polyimide.

RESULTS AND DISCUSSION

i) Surface Modification of PI by Ar Ion Beam and Ar ECR Plasma

The surface modification of PI by low energy (600 eV) Ar ion beam has been completely characterized by XPS of all core levels for ion doses in the range of 5.0×10^{12} to 1.2×10^{16} Ar^+/cm^2.[11,12] We show here the surface sensitive (emission angle is 30° relative to the surface) Cls spectra which display the effects of this modification. These effects are corroborated by Nls and Ols. Figure 1 shows the Cls core level spectra from a PI surface sequentially exposed to Ar ion bombardment of up to 1.2×10^{16} Ar^+/cm^2. The carbonyl (C=O) oxygen or carbonyl containing functionality is removed preferentially with increasing ion dose, leaving material consisting of mostly aromatic carbons with characteristics similar to graphite as evidenced by the loss of intensity of the C=O peak simultaneously with the loss of intensity from the PMDA peak.[11,13] These peaks are identified in the untreated spectrum as C=O (B.E.= 288.7 eV) and PMDA (B.E=285.7 eV). All spectra shift slightly to lower binding energy (B.E) sequentially with increasing ion dose. Two mechanisms lead to this, a slight increase in surface conductivity and removal of strongly electronegative C=O. It should be noted that surface is chemically modified even at very low ion dose of less than 5.6×10^{13} Ar^{+}/cm^{2}. The O1s spectra (not shown) are consistent with this preferential removal of carbonyl oxygen.

Figure 2 shows normal emission C1s spectra from Ar ion and Ar ECR treated PI at high ion dose. The C1s signal from Ar ion treated PI appears "graphitic", while Ar ECR treated C1s emission still is quite intense at the C=O peak. The overall peak shift to lower B.E for Ar ECR

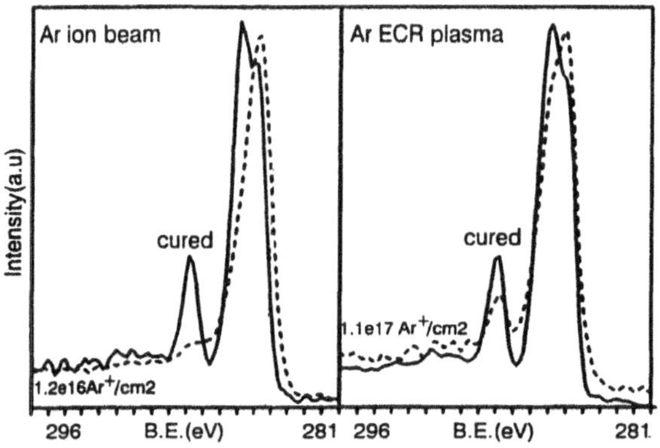

Figure 2. C1s Normal Emission Spectra for Treated Polyimide.

treatment is less than that for Ar ion beam treatment, which is result of the amount of C=O removal as well as the difference in surface conductivity enhancement. We can explain these results in that (i) Ar ECR plasma treatment modifies the surface to a lesser degree than Ar ion beam bombardment for the same ion dose, and therefore (ii) ion energy is the most important parameter affecting surface modification. For the same ion dose, high energy Ar ion effects more surface damage than low energy Ar ion treatment. For the low ion energy regime (less than 100 eV) of ECR plasma, there are still many process unknowns, involving excited neutrals and electrons. Further work is needed to characterize the ECR plasma; however, it appears that ion energy dominates in spite of a large flux of excited neutrals and electrons reaching the PI surfaces.

ii) <u>Chemical Interaction in Cu/PI System</u>

It is known that a weaker chemical interaction occurs at the molecular level between Cu and PI than in the Cr/PI system[6] since Cu is not expected to transfer charge as in the case of Cr. Furthermore, it is believed that Cu initially interacts primarily with the PMDA part of the monomer as does Cr, although more weakly than Cr.

Figure 3 shows the evolution of C1s and O1s core level spectra, respectively, with 2Å Cu deposition after Ar ion treatment [7×10^{13} Ar^+/cm^2] of the PI surface. If the PMDA-ODA were lying more or less flat on the surface, the surface density would be approximately 1.4×10^{14} repeat units/cm^2.[13] Therefore, this dose corresponds approximately to 0.5 Ar^+/repeat unit of the PI. For the C1s, we find a typical shift to lower B.E for the Ar ion treated surface and reduction of the C=O moiety. A slight broadening is observed in the C1s features upon deposition of Cu and a large shift to lower B.E is observed. Statistics and energy resolution do not allow reliable curve fitting analysis for the C1s; however, there is a

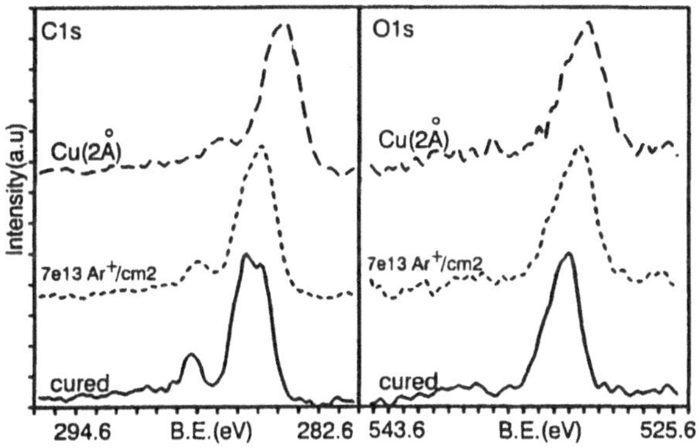

Figure 3. XPS Spectra upon Cu(2Å) Deposition after Ar Ion Treatment of Polyimide(60 deg. from surface).

discernible change in shape for the C=O moiety and the peak-to-valley ratio (valley at 288.1 eV binding energy in the cured C1s spectrum) upon deposition of Cu. The O1s spectra lack the resolution to adequately observe effects on C=O oxygen compared to ether (C-O-C) oxygen upon Cu deposition. However, we find a significant shift to lower B.E of the peak intensity. It is noted, in particular, that both core level peaks shift to lower B.E by larger amount than do those of Cu (2Å) deposited on an untreated PI surface: 1.0 eV vs 0.6 eV for C1s and 0.6 eV vs 0.1 eV for O1s signals [see ref. 6,10]. These stronger peak shifts compared to those of Cu on the untreated PI, and the possible change in peak shapes, indicate enhanced charge transfer from the Cu atoms to the underlying treated PI, implying enhanced chemical interactions. Higher resolution work is necessary if arguments regarding specific chemical interactions are to be made. Such a measurement is difficult at best considering the loss of intensity for N1s and O1s as a function of ion dose. Final state effects in the photoemission experiment due to relaxation or screening could explain enhanced shifts to lower B.E. However, a different molecular geometry and therefore enhanced chemistry would be implied compared to unmodified PI even if final state effects are the explanation. It should be noted that the opposite effect occurs for Au on ion treated PI.[9] That is, core levels of ion treated PI shift to *higher* binding energy upon deposition of Au. Shifts to higher B.E can not be attributed to final state effects. In both cases (Au and Cu) enhanced chemistry is indicated at the interface as a result of prior ion beam treatment.

CONCLUSIONS

We have characterized surface modification of PI by the use of Ar ion beam and Ar ECR plasma/ion sources. Our XPS analysis indicates that (i) Ar ECR plasma treatment effects less surface damage than Ar ion beam bombardment for same ion dose, (ii) Ar ions cause the same type of damage regardless of the source in this energy range, and (iii) the degree of surface damage is a function of ion energy within same ion dose range. In addition, surface modification of PI by Ar ion treatment enhances chemical interaction between Cu and the treated PI surface. Enhanced chemical interaction occurs between Cu and the ion treated PI structure as compared to Cu and the untreated PI. We believe that this observed enhanced chemical interaction contributes to the enhanced adhesion of the metal to the modified polymer.

ACKNOWLEDGEMENT

This work was supported by an IBM Materials and Processing grant and by the Microelectronics Sciences Laboratories of Columbia University.

REFERENCES

1. K.L. Mittal, Ed., "Polyimides: Synthesis, Characterization and Applications," Vols.1 & 2, Plenum Press, New York (1984).
2. D. Wilson and H.D. Stenzenberger, "Polyimides", Blackie & Son, Glasgow (1990).
3. C.A. Chang, J.E.E. Baglin, A.G Schrott and K.C. Lin, Appl. Phys. Lett., 51, 103 (1987).
4. S. Matsuo and Y. Adachi, Jap. J. Appl. Phys., 21, L4 (1982).
5. S. Matsuo and M. Kiuchi, Jap. J. Appl. Phys., 22, L210 (1983).
6. R.C. White, R. Haight, B.D. Silverman and P.S. Ho, Appl. Phys Lett., 51, 481 (1987).
7. R.M. Tromp, F. Legoues and P.S. Ho, J. Vac. Sci. Technol. A3(3), 782 (1985).
8. S.G. Anderson, H.M. Meyer and J.H. Weaver, J. Vac. Sci. Technol. A6(1), 38 (1988).
9. H.S. Jeong and R.C. White, J. Electr. Spectr. Rel. Phenom., 52, 447 (1990).
10. R. Haight, R.C. White, B.D. Silverman and P.S. Ho, J. Vac. Sci. Technol. A6, 2188 (1988).
11. R.C. White and P.S. Ho, in "Ion Beam Deposition, Film Modification and Synthesis", J. Cuomo, S. Rossnagel, and F. Kaufmann, Eds., Noyes Publications (1989).
12. H.S. Jeong and R.C. White, to be published.
13. R.C. White and P.S. Ho, in "Polyimides: Materials Chemistry and Characterization", C. Feger, M. Khojasteh and J. McGrath, Eds., 719, Elsevier, Amsterdam (1989).

XPS STUDY OF THE INTERFACE BETWEEN THERMALLY EVAPORATED ALUMINIUM AND POLYETHYLENETEREPHTHALATE : EVIDENCE FOR OXYCARBIDE SPECIES

M. Bou*, J.M. Martin*, Th. Le Mogne*, L. Vovelle**

*Ecole Centrale de Lyon, Laboratoire de Technologie des Surfaces, URA CNRS 855, BP. 163, 69131 Ecully Cedex, FRANCE
**Rhône Poulenc, C.R.C., BP. 62, 69192 Saint Fons, FRANCE

The interface between *in situ* thermally evaporated aluminium and a biaxially oriented film of polyethyleneterephthalate (PET) has been studied by X-ray Photoelectron Spectroscopy (XPS). The surface of PET has been characterized before and after metallization. The results of metallization show that the bonding of aluminium atoms with PET occurs at carbonyl sites, as evidenced by the rapid attenuation of carbonyl C1s core level photoemission intensity. This chemical reaction leads to the formation of a C-O-Al complex at the interface. With increasing Al coverage, a new C1s species is also observed at a lower binding energy and is supposed to originate from aluminium carbide-like bonding.

INTRODUCTION

Metallized polymer films are useful because of their good barrier properties. They are, for example, frequently used for winding to form capacitors,and in packaging applications. However, the properties of metallized polymers depend on the adhesion of metal to polymer [1]. In particular, the strength of adhesion of metals to polymers has been related to the reactivity and strength of bonding of the first monolayer with reactive surface functional groups (especially oxygen containing groups) [2,3]. Several analytical techniques have been used to investigate the chemistry of the interface, such as ion scattering spectrometry (ISS), Rutherford backscattering spectroscopy (RBS), secondary ion mass spectrometry (SIMS) and X-ray photoelectron spectroscopy (XPS) [2-6]. A number of XPS studies found in the literature for metallization of untreated oxygen-containing polymer surfaces or of oxygen plasma treated polymer surfaces (initially oxygen-free), indicate that the reaction between metal and polymer leads to the formation of metal-oxygen-polymer complexes [2-30]. For example Burkstrand [2] used XPS to characterize chemical interactions at metal/polymer interfaces and correlated his results with adhesion measurements. It was found that the adhesion strength of any metal on an oxygen-containing polymer was greater than on a polymer containing no oxygen. His XPS results suggested that metals reacted with the surface oxygen, resulting in the formation of metal-oxygen-polymer complexes.

Metallized Plastics 2, Edited by K.L. Mittal
Plenum Press, New York, 1991

More recently, DePuydt *et al*[31] investigated adhesion mechanisms between thermally evaporated aluminium on polyethyleneterephtalate using SIMS, ISS and RBS, and their results indicated the formation of Al-O-C at the interface as well. Y. Jugnet *et al*[32] studied differences in interfacial chemical bonding characteristics of evaporated or sputtered Al on silicone rubber film and polyethyleneterephthalate by means of XPS. Their results show that PET is more reactive than silicone and the evaporation or sputtering process did not influence the interface chemistry. The initial adsorption sites are the carbonyl groups and Al-O-C bonds are formed at the interface. The same group also investigated mechanisms of adhesion of aluminized films[33].

In the present work, XPS was used to study chemical reactions at the interface between *in situ* evaporated Al and biaxially drawn PET (figure 1). The results are discussed in terms of the existence of metal-polymer bonding.

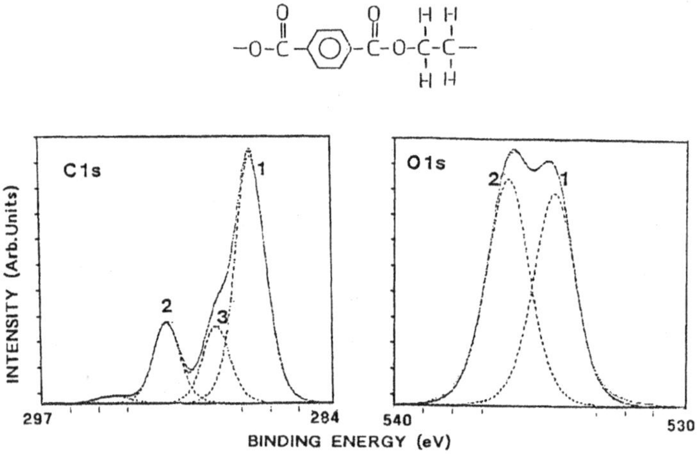

Figure 1. Monomer unit of PET and carbon 1s and oxygen 1s core level spectra of PET film.

EXPERIMENTAL

Instrumentation

The polymer surface and the metal/polymer interface were investigated by using XPS, and the measurements were carried out on a VG Scientific ESCA CLAM 100 spectrometer. The instrument is equipped with two ultra high vacuum (UHV) chambers, one analysis chamber and one film preparation chamber(both chambers have a base pressure better than 1×10^{-8} Pa) and a sample entry load-lock allowing rapid sample transfer from atmosphere into the UHV system .

In this study, an unmonochromatized MgK_α X-ray source (1253.6 eV) was used. The spectrometer was operated in the fixed analyzer transmission mode with a pass energy of 50 eV. The probed area of the sample was about 1 cm^2. The photoelectron spectra were collected at an electron take-off angle of 15° relative to the surface normal.

Transmission electron microscopy study was carried out using a Philips EM 420 microscope. The samples were prepared by ultramicrotomy.

Sample Preparation

The samples were prepared by using a 12 µm thick, biaxially oriented and thermoset at 235°C with 0 % relaxation PET films (Terphane R°, Rhône-Poulenc). The polymer did not contain any fillers or additives. No surface treatment or cleaning were done prior to metallization.

Metallization Conditions

The aluminium was evaporated from a resistively heated helical tungsten filament at a vacuum of 10^{-7} Pa. The resulting Al film thickness was calculated from the attenuation of the C1s intensity of PET , and by using a value of 34 Å for the effective elastic mean free path of electron [34]. Using the calculated film thickness, the average deposition rate was estimated to be about 1 Å/s. The distance between the source and the sample was about 2 cm. In order to minimize substrate heating at such a short source-to-sample distance,the aluminium source was positioned inside a cold finger,which was cooled by liquid nitrogen, Five subsequent deposits (three seconds each) were carried out, and the sample was analyzed after each deposit in order to follow the evolution of the interface chemistry as a function of Al coverage .

Peak Resolution

We will emphasize the coverage-dependent emission from the C1s, O1s and Al2p core levels. Analysis of these spectra includes background subtraction and interactive lineshape decomposition based on summed Gaussian-Lorentzian functions (55 % - 45 %), which provide the best fit.

The best fit is achieved for using the full width at half-maximun (FWHM) of 1.7 eV for each curve fitted C1s lineshape (with standard deviation (s.d) 0.2 eV). A FWHM of 1.8 eV (s.d = 0.05 eV) is found to yield the best fit of the oxygen involved in O=C and O-C. During the evolving interface analysis, we try to keep the same widths for C1s and O1s which characterize PET. The FWHM of the oxygen involved in the Al-O-C complex is about 2.2 (s.d = 0.2 eV). The FWHM of each Al2p component is 1.3 (s.d = 0.1 eV) for metallic Al and 2.3 (s.d = 0.1 eV) for oxidized state of Al.

The π-π^* transition peak observed in C1s spectra at about 291.2 eV was included in the curve-fitting routine, but not in the calculation of relative concentrations.

The binding energy of the C1s peak originating from C in the C-H environment was set at 284.8 eV [35] and used as a reference for calibrating the other peak positions. We want, however to point out that this method is not entirely rigorous because the carbon peak of the benzene rings and the one due to the C-H contamination are at the same position. Hence we also used energy difference between two peak positions : ΔE.

The evaluation of the C/O ratio of the PET substrate was based on the measured photoelectron peak areas [35].

RESULTS AND DISCUSSION

XPS Analysis of the As-received PET surface

The C1s and O1s core level photoemission spectra of PET are shown in figure 1. The C1s spectrum shows three distinctly resolved peaks (experimental peaks are as dots, simulated peaks are as dashes, and fitted spectra are as solid lines). Based on published data on PET, peak 1 corresponds to the carbon involved in C-H and in the benzene rings (284.8 eV), peak 2 is assigned to C atoms involved in carbonyl bonds C=O (288.3 eV), and peak 3 is attributed to the carbon of C-O (286.6 eV). The O1s photoelectron signal contains two resolved peaks, peak 1 corresponds to O=C (531.7 eV) and peak 2 is assigned to O-C (533.4 eV) [36,37].

The relative concentration of each type of functional group was determined from the area under the corresponding components in the total C1s and O1s spectra shown in Figure 1.The results, presented in Table I, show that the concentration of C=O and C-O, and the relative ratio of carbon and oxygen derived from the experimental XPS data are in good agreement with the theoretical stoichiometry (ten carbon and four oxygen atoms are taken into account for the ideal unit stoichiometry). However,

Table I. Atomic composition of PET film.

	C1s			O1s		C/O
	C=O	C-O	C-H	O=C	O-C	
theory	14.3	14.3	42.8	14.3	14.3	2.5
experiment	13	12	44.6	13.6	15.4	2.45

the contribution of each carbon component to the C1s spectrum shows a slight C-H enrichment (1.8 % relative), compared with the PET molecular structure and can be attributed to surface contamination from environment. Actually, as the maximum thickness probed by the XPS technique is approximately 50 Å (which is roughly three times the mean free path of a C1s photoelectron in a polymer [37]), the contribution of a monolayer contamination film in the carbon photoelectron peak is expected to be relatively low.

Accordingly, a slight deficiency of C atoms involved in carbonyl groups and in C-O is observed. Extrinsic factors such as the curve fitting, photoionization cross section values and sensitivity factors may also be sources of small discrepancies.

Concerning the comparative study of C and O respectively involved in carbon-oxygen bonding (Table I), experimental and theoretical data are also in good agreement.

In addition,X-ray induced degradation of the PET sample surface was also investigated, and no changes were observed after 7 hours of analysis. Since our experiments last for less than 6 hours, we assume that there are no degradation effects to be considered.

Al/PET Interface

Figure 2 shows extended scan spectra of the Al2s,Al2p, C1s and O1s core levels as a function of metallization time. The binding energies of all components, with charging effect correction are listed in Table II. As expected, C1s and O1s intensities decrease while the emission signals of both Al2s and Al2p increase with increasing Al coverage. It is interesting to notice that characteristic Al plasmon loss structures at about 90 eV and 106 eV begin to appear at an Al thickness of about 10 Å. These plasmon losses indicate the metallic nature of the Al. At very low coverage, the plasmon loss structure is absent. This observation could be explained by the formation of Al clusters.

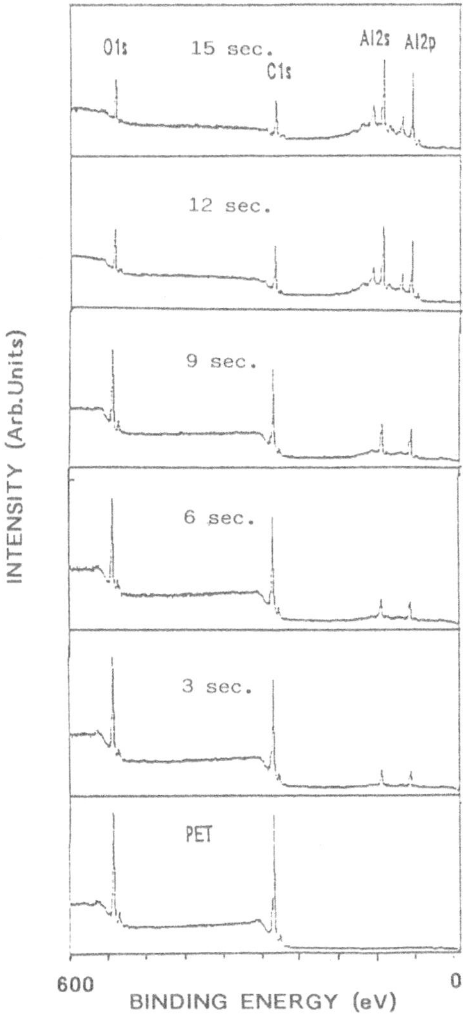

Figure 2. Extended scan showing the Al2s, 2p, C1s and O1s core levels as a function of metallization time.

Table II. Binding energies (B.E.) and assigment of the Al2p, C1s and O1s components obtained from the resolution of the spectra. B.E. are in eV ± 0.2 eV. Time represents the metallization time.

	Al2p		C1s						O1s	
Time (sec.)	II Al-O	I Al	I CH	II C-O	III C=O	IV Al-C-O	$\pi-\pi^*$	V New. Comp.	I O = C + Al-O-C	II O-C
0	-	-	284.8	286.4	288.7	-	291.3	-	531.7	533.3
3	74.42	72.25	284.8	286.7	289	283.3	291.5	-	531.8	533.6
6	74.44	72.18	284.8	286.7	289	283.3	291.5	-	531.8	533.6
9	74.38	72.18	284.8	286.7	289.1	283.3	-	-	531.7	533.5
12	73.06	72	284.8	286.7	289.3	283.7	-	281.8	531.5	533.7
15	74.3	72.4	284.8	286.5	-	283.8	-	281.9	531.8	533.7

The atomic percentage of each detected element and its atomic ratio as a function of Al coverage before and after metallization are presented in Table III.

The amounts of C1s and O1s decrease whereas that of Al2p increases. We also note that the C/O ratio remains fairly constant with Al coverage. We can conclude that only pure aluminium is deposited and that it does not react with carbon or oxygen atoms present in preparation chamber during evaporation. Further information about the interfacial layer are obtained from the variation of the atomic percentage of each resolved component in the high resolution spectra for the C1s, O1s and Al2p core levels, shown in Figure 3.

Table III. Atomic compositionof Al2p, C1s and O1s with increasing time of metallization.

time (sec)	atomic %			atomic ratio		
	Al2p	C1s	O1s	C/O	Al/O	Al/C
0	0	71.5	28.6	2.5	0	0
3	11.7	62.3	26	2.4	0.4	0.2
6	16.3	58.7	24.2	2.4	0.7	0.3
9	23.7	53.7	22.6	2.4	1.1	0.4
12	46.2	38.3	15.5	2.5	2.9	1.2
15	54.7	32	13.2	2.4	4.1	1.7

Evolution of Al2p: As can be seen from Figure 3, the Al2p photoelectron peak consists of two resolved components. The first one, located at 72.2 eV, is attributed to metallic Al. This value is somewhat different from the binding energy, 72.9 eV, obtained for a clean aluminium reference sample, and also from the value, 72.9 eV, reported by DeKoven and Hagans [30]. However, it is close to the value (72.5 eV) published by Atanasoska *et al* [12]. This discrepancy is probably due to charging effects and energy relaxation phenomena. The second component is found at 74.4 eV, which is a chemical shift of 2.2 eV compared with the metallic Al peak. This value is higher than the one we found for our alumina reference (1.5 eV). Therefore, this component is not alumina and can be attributed to a Al-O-C chemical bonding [3,11,22,30].

It is interesting to note that even just after the first deposit (3Å estimated thickness), we already have both metallic and oxidized Al, whereas the interface is not completely formed. This result may be explained by the existence of Al clusters at the surface of PET. A TEM micrograph of a 3 Å thick Al layer deposited on PET (see Figure 4) confirms the existence of Al clusters with a size of roughly 1 nm.

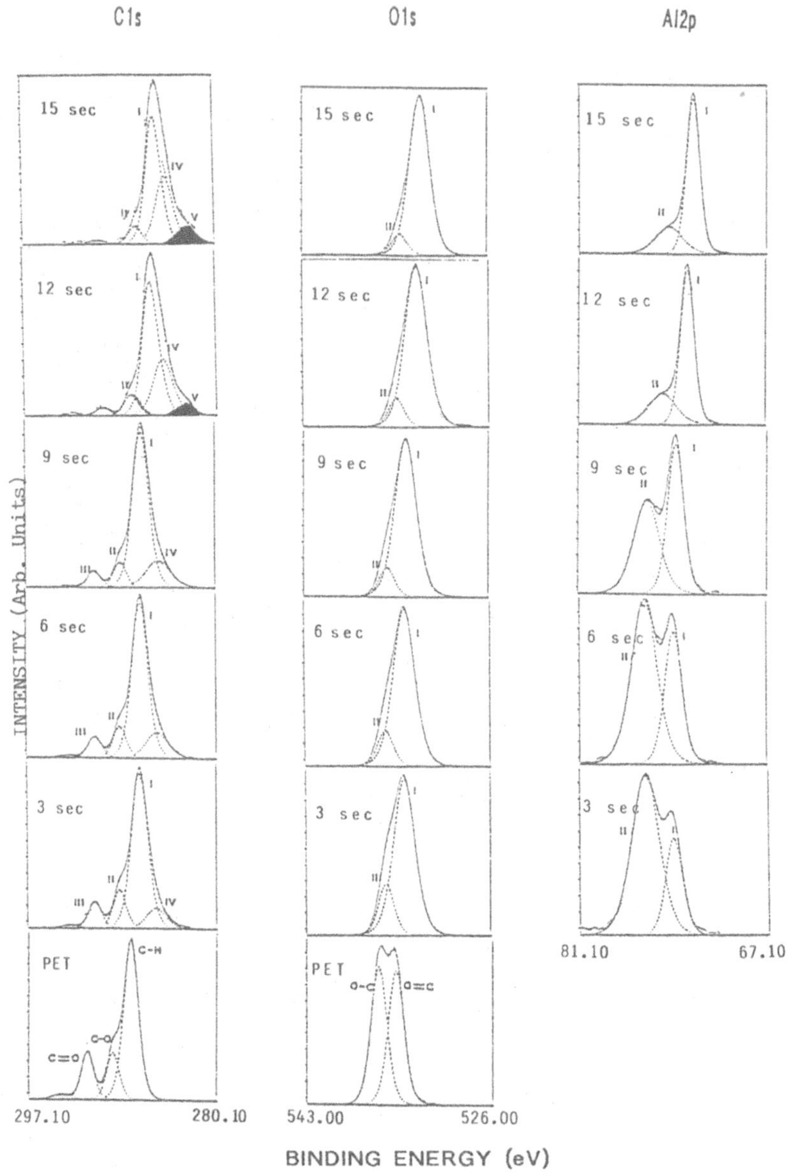

Figure 3. Al2p, C1s and O1s core level spectra recorded as a function of metallization time (binding energies are given without charge effect correction : see Table III).
Expérimental spectra.
Sum of curve fitted peaks.
Components obtained from the curve fitting.

Evolution of O1s. In Figure 3, we show O1s spectra at different stages of metallization. The O1s spectrum of unmetallized PET consists of two resolved peaks corresponding to O=C and O-C separated by about 1.6 eV. After metallization the O1s spectrum still contains two peaks. The one located at higher binding energy corresponds to O-C after Al deposition, and the second peak is assumed to be not only due to O=C, since the FWHM now is higher than that of O=C clean PET. We can consider this second peak as a combination of two contributions. Because of the small chemical shifts observed in the oxygen peaks, it is difficult to

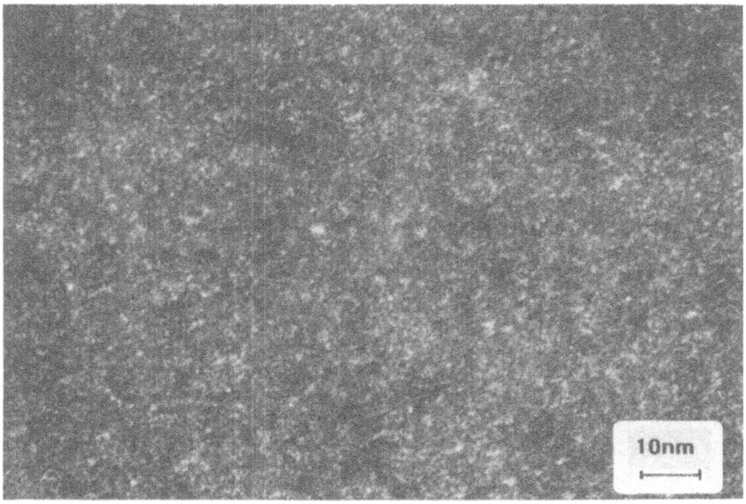

Figure 4. Dark field transmission electron micrograph of metallized PET (deposition time is 3 seconds).

determine the position of a new peak by curve fitting. Nevertheless, we are able to detect the decrease in the O-C peak intensity after the first Al deposit and note that it is almost unchanged after further deposition of Al. However, for the subsequent decay of substrate oxygen (carbonyl groups) we fit three oxygen components to experimental spectra. In the new fitting we vary the relative intensities of the O1s features, while relative peak positions are held fixed, and introduce a reaction-induced O state. Figure 5 illustrates this tentative fitting, and shows the decay of O-C and O=C. The peak position of the new component is in good agreement with the literature [30]. This way, the decrease in the intensity of both substrate oxygens can be followed. But, with increasing Al coverage the decreased O1s intensity makes it difficult to obtain a good fitting.

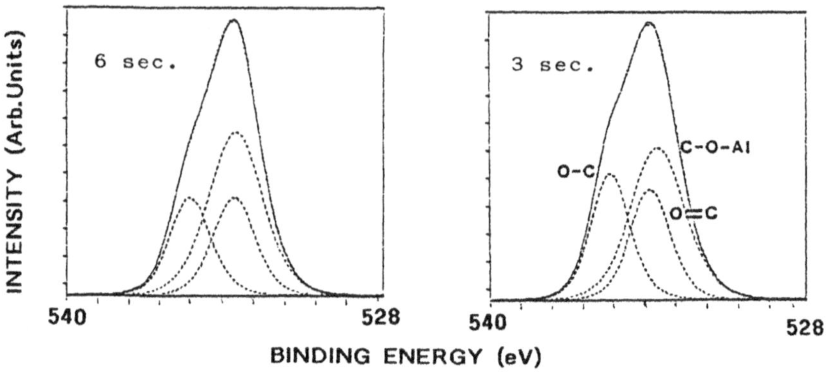

Figure 5. O1s core level spectra, including an Al-O-C feature.

Evolution of C1s. Additional information about the chemical reaction sites following Al/PET interface evolution can be determined from the attenuation of each of the observed C spectral features. To examine the attenuation of each C component, we decomposed the C1s spectra for clean PET into its principal components. In Figure 3 the C1s core level for representative Al coverages are shown. It is found possible to fit peaks to the experimental spectra for all coverages by varying only the relative intensities of the substrate induced features from clean PET (relative positions of peak I, II and III were held fixed), and by introducing two reaction-induced C states. The first one appears at 283.5 eV (IV), attributed to Al-O-C, and persists for all subsequent coverages [32]. The second reaction-induced feature is relatively weak and appears at 281.9 eV (V), and is even more pronounced with the increased Al coverage. As can be seen from Figure 3, the C1s core level gradually changes in shape as the Al coverage increases and there is a preferential decay of the carbonyl carbon emission at 288.7 eV (III), which completely vanishes at high coverage.

In Figure 6, we show the variation of C1s atomic ratios of C-O/CH, C=O/CH, and C-O-Al/CH as a function of Al coverage. This figure shows the rapid attenuation of C=O, and we also note the significant decrease in C-O emission and the increase of C-O-Al. From these data we can conclude that the reaction of Al with PET primarily occur at the carbonyl sites, but the decrease of C-O bond is not yet explained. The new component appearing at 281.2 eV is supposed to be an aluminium carbide-like chemical bond (Al-C) [30,39], similar to that reported for other metal/polymer systems [21,30,39,40]. After three seconds of deposition, the Al-C component is more visible in the Al2p spectrum than in the corresponding C1s spectrum. On the other hand, at higher Al coverage, the Al-C component is stretched out in the C1s spectrum and is less visible in Al2p spectrum (Figure 7). Indeed, at the earlier metallization stage, the C1s photoelectrons come from the PET and the interface layer, so the Al-C signal is masked by the other signals in C1s spectrum. As the Al coverage increases, the interface region will be more dominant in the C1s spectrum. Unfortunately, the Al-C species formation is not understood yet, and further work is in progress to examine this type of interaction.

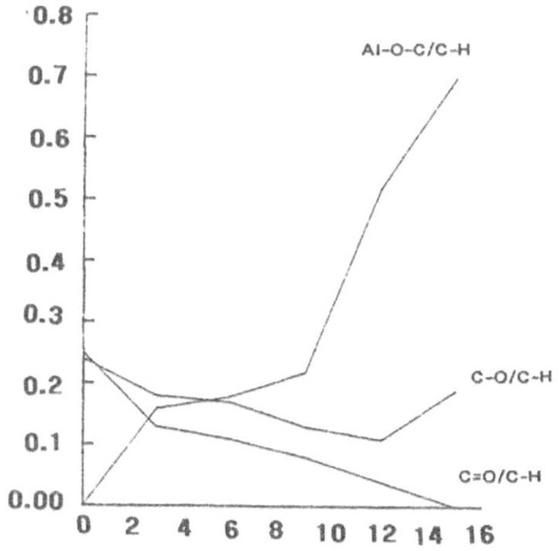

Figure 6. Atomic ratios of C-O/CH, C=O/CH and C-O-Al/CH as a function of metallization time.

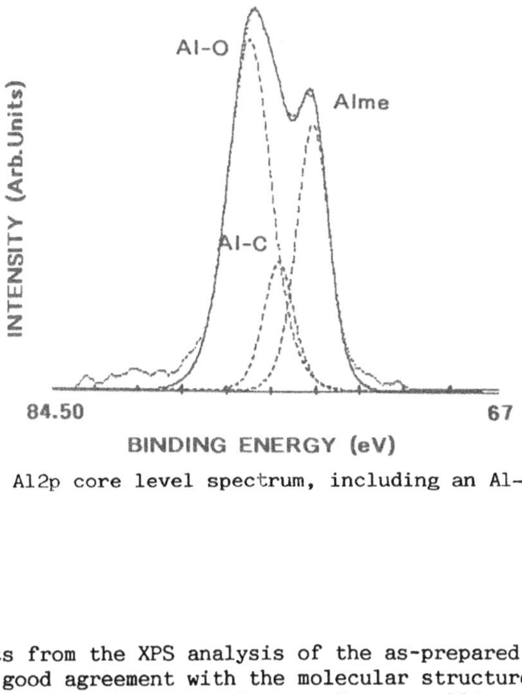

Figure 7. Al2p core level spectrum, including an Al-C feature.

CONCLUSION

The results from the XPS analysis of the as-prepared PET film surface are in good agreement with the molecular structure, except for a small amount of hydrocarbon contaminants. Upon metallization, our results show the Al metal atoms are very reactive with the PET surface. The interaction was observed to occur mainly at the C=O linkage of PET, evidenced by its rapid attenuation, and Al oxide species were observed as the final product. The decrease in the ether linkage peak indicates that this group might participate in the interaction as well. Formation of Al carbide-like species is also observed as the Al coverage increases. For comparison we suggest further examination of other Al/polymer interfaces using complementary analytical techniques.

REFERENCES

1. K.L. Mittal, J. Vac. Sci. Technol., 13, 19 (1976).
2. J.M. Burkstrand, J. Appl. Phys., 52, 4795 (1981).
3. P.S. Ho, P.O. Hahn, J.W. Bartha, G.W. Rubloff, F.K. Legoues and B.D. Silvermann, J. Vac. Sci. Technol A3, 739 (1985).
4. J.M. Burkstrand, Appl. Phys. Lett., 33, 387 (1978).
5. J.M. Burkstrand, J. Vac. Sci. Technol., 16, 1072 (1979)
6. J.M. Burkstrand, J. Vac. Sci. Technol., 15, 223 (1978).
7. J.M. Burkstrand, J. Vac. Sci. Technol., 16, 363 (1979).
8. J.M. Burkstrand, J. Vac. Sci. Technol., 15, 223 (1978).
9. J.M. Burkstrand, J. Vac. Sci. Technol., 21, 70 (1982).
10. F. Soeda, K. Hayashi and A. Ishitani, J. Electron Spectros. Related Phenomena, 27, 205 (1982).
11. J.W. Bartha, P.O. Hahn, F. Legoues and P.S. Ho, J. Vac. Sci. Technol., A3, 1390 (1985).
12. L.J. Atanasoska, S.G. Anderson, H.M. Meyer III, Z. Lin and J.H. Weaver, J. Vac. Sci. Technol., A5, 3325 (1987).
13. H.J. Leary and D.S. Campbell, Surface Interface Anal., 1, 75 (1979).
14. B.D. Silverman, P.N. Sanda, P.S. Ho and A.R. Rossi, J. Polym. Sci., Polym. Chem. Ed., 23, 2057 (1985).
15. B.D. Silverman, J.W. Bartha, J.G. Clabes, P.S. Ho and A.R. Rossi, J. Polym. Sci., Polym. Chem. Ed., 24, 3325 (1986).
16. S.G. Anderson, H.M. Meyer III, Lj. Atanasoska and J.W. Weaver, J. Vac. Sci. Technol. A6, 38 (1988).
17. J.J. Pireaux, C. Gregoire, P.A. Thiry, R. Caudano and J.C. Clarke, J. Vac. Sci. Technol., A5, 598 (1987).
18. N.J. DiNardo, J.E. Demuth and T.C. Clarke, J. Vac. Sci. Technol.,A4, 1050 (1986).
19. N.J. DiNardo, J.E. Demuth and T.C. Clarke, Chem. Phys. Lett., 121, 239, (1985).
20. C.A. Kovac, J.L. Jordan and R.A. Pollack, Mater. Res. Soc. Symp. Proc., 72, (1986).
21. F.S. Ohuchi and S.C. Freilich, J. Vac. Sci. Technol. A4 1039 (1986).
22. Lj. Atanasoska, H.M. Meyer III, G. Anderson, and J.H. Weaver, J. Vac. Sci. Technol., A6, 2175 (1988).
23. N.J. Chou and C.H. Tang, J. Vac. Sci. Technol. A2, 751 (1984).
24. P.O. Hahn, G.W. Rubloff, J.W. Bartha, F. Legoues, R. Tramp and P.S. Ho, Mater. Res. Soc. Symp. Proc., 40, 251 (1985).
25. J.L. Jordan, C.A. Kovac, J.F. Morac and R.A. Pollack, Phys. Rev., B, 36, 1369 (1987).
26. P.N. Sanda, J.W. Bartha, B.D. Silverman, P.S. Ho and A.R. Rossi, Mater. Res. Soc. Symp. Proc., 40, 283 (1985).
27. P.N. Sanda, J.W. Bartha, J.G. Clabes, J.L. Jordan, C. Feger, B.D. Silverman and P.S. Ho, J. Vac. Sci. Technol., A4, 1035 (1986).
28. H.M. Meyer III, S.G. Anderson, Lj. Atanasoska and J.H. Weaver, J. Vac. Sci. Technol., A6, 30 (1988).
29. S.G. Anderson, H.M. Meyer, and J.H. Weaver, J. Vac. Sci. Technol., A6 (4), 2205 (1988).
30. B.M. DeKoven and P. Hagans, Appl Surface Sci, 27, 199 (1986).
31. Y. DePuydt, P. Bertrand and P. Lutgen, Surface Interface Anal, 12, 86 (1988).
32. Y. Jugnet, J.L. Droulas, Tran Minh Duc, and A. Pouchelon, Postdeadline poster, European Conference on Applications of Surface and Interface Analysis, (1989).

33. Y. Jugnet, Tran Minh Duc, and A. Pouchelon, International Conference on Metallization of Polymers, Montreal, Canada (1989).
34. D.T. Clark, in, "Polymer Surfaces", D.T.Clark and W.J.Feast, editors p 309, Wiley, New York, (1978).
35. D. Briggs and M.P. Seah, "Practical Surface Analysis by Auger and X-ray Photoelectron Spectroscopy", (1983).
36. D.W. Dwight, J.E. McGrath and J.P. Wightman, J Appl Polym Sci : Appl Polym Symp, 34, 35, (1978).
37. D.T. Clark, CRC Critical Reviews in Solid State and Materials Sciences, 1, (1978).
38. D.T. Clark, J.W. Feast, W.R.S. Musgrave and I. Ritchie, J. Polymer Sci : Polym Chem. Ed. 13, 857 (1975).
39. L. Porte, J. Appl. Phys. 60, 635 (1986).
40. P. Bodo and J.E. Sundgren, J. Vac. Sci. Technol. A2, 1498 (1984).

XPS-STUDY OF METAL-POLYMER INTERFACES AFTER POLYMER SURFACE TREATMENT BY ION AND PLASMA TECHNIQUES

S. Nowak, R. Mauron, G. Dietler and
L. Schlapbach

Physics Department
University of Fribourg
CH -1700 Fribourg
Switzerland

A study of the effects of surface treatment of polypropylene on the deposition of thin metal films is presented. The polypropylene samples were treated in a low-pressure air plasma in one case and sputtered by argon ions in the other case. XPS was used to detect compositional surface modifications. Plasma treatment leads to surface oxidation, whereas the sputtering results in oxygen-free surfaces for in-situ analysis. After surface treatment, thin films of Mg, Ca and Ce were deposited. Depending on the surface treatment, we find indications for different bonding mechanisms and very different sticking probabilities. For the case of oxidized surfaces, the formation of a metal-oxygen-polymer complex is anticipated, whereas for oxygen-free surfaces a metal-carbon bond is likely to be formed.

INTRODUCTION

Technological applications of polymers frequently involve metallization of the polymer substrate in order to achieve certain surface properties. Examples of such applications are encountered in electronic devices, packaging as well as for decorative coatings.[1] The adhesion of such metal films onto polymers is often too low. Therefore, a number of pretreatments to enhance the low intrinsic adhesion property of polymers have been developed. Among these pretreatments, surface modification by ion and plasma techniques has received increased interest.[2-5] The effect of the surface treatment includes cleaning by ablation of low molecular weight material, activation of the surface, change in surface polarity and wetting characteristics, crosslinking in a surface layer, and structural modification. While it is accepted that these mechanical, physical and chemi-

Metallized Plastics 2, Edited by K.L. Mittal
Plenum Press, New York, 1991

cal modifications affect the ultimate adhesion strength of an interface, there remains a lack of understanding of the detailed processes that take place.

In a plasma, the surface is exposed to a broad spectrum of ions, electrons, neutrals and electromagnetic radiation, where the energy of these species depends on the operating parameters such as gas nature and pressure, power density and potential distributions. The individual contributions of the different effects to the surface modification can, therefore, become rather complicated to study. For the case of an ion treatment this aspect can be simplified since the experimental parameters are reduced to the nature and energy of the ions used, and the ion current impinging on the surface. However, the differences in the nature of the two kinds of surface treatments have to be borne in mind when studying their surface modification effects.

A number of studies have been carried out to investigate the nature of the interface formed between a polymer and a metal deposit. Burkstrand[6] studied the interaction of different metals with several kinds of polymers, in particular with respect to the oxygen content of the polymer. A particularly well documented system is the case of interfacial reaction during metallization of polyimide.[7-9] The effect of plasma treatment on the adhesion of metal films and the nature of the interface formed has been studied for the case of polyethylene by Gerenser.[10] Bodö and Sundgren[11] have studied similar aspects for the case of an ion treated polyethylene surface. They found indications for dehydrogenation effects, surface crosslinking and graphitization of the polymer surface.

In this contribution, we have studied the effects of ion and plasma treatments of polypropylene on the deposition of thin metal films. X-ray photoelectron spectroscopy (XPS) was used to investigate the surface composition before and after surface treatment and metal deposition. The metal concentration on the polymer surface as well as the interfacial reaction were investigated as a function of the surface treatment.

EXPERIMENTAL

Materials

The substrate material used for the present investigation was isotactic polypropylene (PP) [Propaten HF 26 (ICI)]. This base material was mixed and extruded with 0.2% ®Irganox 1010 (Ciba-Geigy) and 0.1 % Ca-stearate at 260 °C. After cooling, the PP pellets were pressed to plates in a heat mold at 230 °C. A degree of crystallinity of typically 50% is deduced from Differential Scanning Calorimetry.

Surface treatment

Plasma treatment of the polypropylene samples was carried out in a rf plasma cleaner (PDC-23G from Harrick Scientific). The reactor consists of a 10 cm diameter glass barrel pumped by a mechanical forepump to a base pressure lower than 0.5 Pa. The plasma is obtained by inductive coupling at 13.56 MHz. Plasma treatment was carried out in air at different pressures and power levels. The following table gives the main operating parameters which were used in this study. Between the plasma treatment and the surface analysis, the samples were in contact with the laboratory ambient.

Plasma Treatment Conditions

RF	:	13.56 MHz
Power	:	≤ 100 W
p	:	0.8 - 40 Pa
flow of air	:	0.3 - 30 sccm
time	:	15 seconds

The ion treatment was performed with a sputter gun installed in the preparation chamber of an ESCA spectrometer. Thereby, surface analysis and metal deposition could be performed in-situ without ambient contact of the sample. The energy of the Ar^+ ions was varied between 20 eV and 3 keV for different treatment times.

Thin films of different metals, Mg, Ca and Ce, were deposited by thermal evaporation in the preparation chamber of the spectrometer. The amount of metal deposit was monitored by a quartz crystal balance.

Surface Analysis

X-ray photoelectron spectroscopy (XPS) was performed in a VG ESCALAB 5 spectrometer at a base pressure lower than 10^{-8} Pa. Experiments were carried out with non-monochromatized Mg Kα (hυ = 1253.6 eV) and Si Kα (hυ = 1740 eV) radiation. With the PP samples (10 x 10 x 0.8 mm) at an electrical power input of 200 W (10 kV, 20 mA), the pressure rose up to 10^{-6} Pa. This pressure rise is associated with polymer degradation effects by the x-ray source.[12] The charging of the polypropylene samples was corrected by setting the energy of the main component of the C 1s peak at 285.0 eV.

RESULTS

In Fig. 1 a comparison of the surface composition of untreated, plasma and ion treated polypropylene is presented. A small amount of oxygen (≈ 5 at.%) on the untreated surface is observed. Upon plasma treatment, the O 1s intensity increases and the C 1s peak shows a shoulder to the higher binding energy side which is characteristic of CO functionalities. This surface oxidation feature depends on the pressure used for the plasma treatment, and it increases as the pressure is reduced. Fig. 2 reflects this behaviour. In an earlier contribution, it was found that the pressure dependence of the observed surface oxidation was correlated with the plasma energy density.[12] As the plasma treated sample was in contact with the ambient, it is difficult to distinguish an oxidation during the plasma treatment from a reaction of an activated surface with ambient oxygen. In contrast, the ion treatment, shown in Fig. 1, leads to a surface which is virtually oxygen-free for in-situ analysis. However, the C1s peak of this sample has a larger line width than for the untreated surface.

Fig. 3 shows survey spectra of plasma treated surfaces after receiving similar deposits of calcium (typi-cally 10 monolayers). Only a trace of Ca can be observed on the un-treated surface. As the surface oxidation increases with the plasma treatment, an increasing Ca content at the surface is found. We can, therefore, relate the surface concentration of Ca with the amount of surface oxidation.

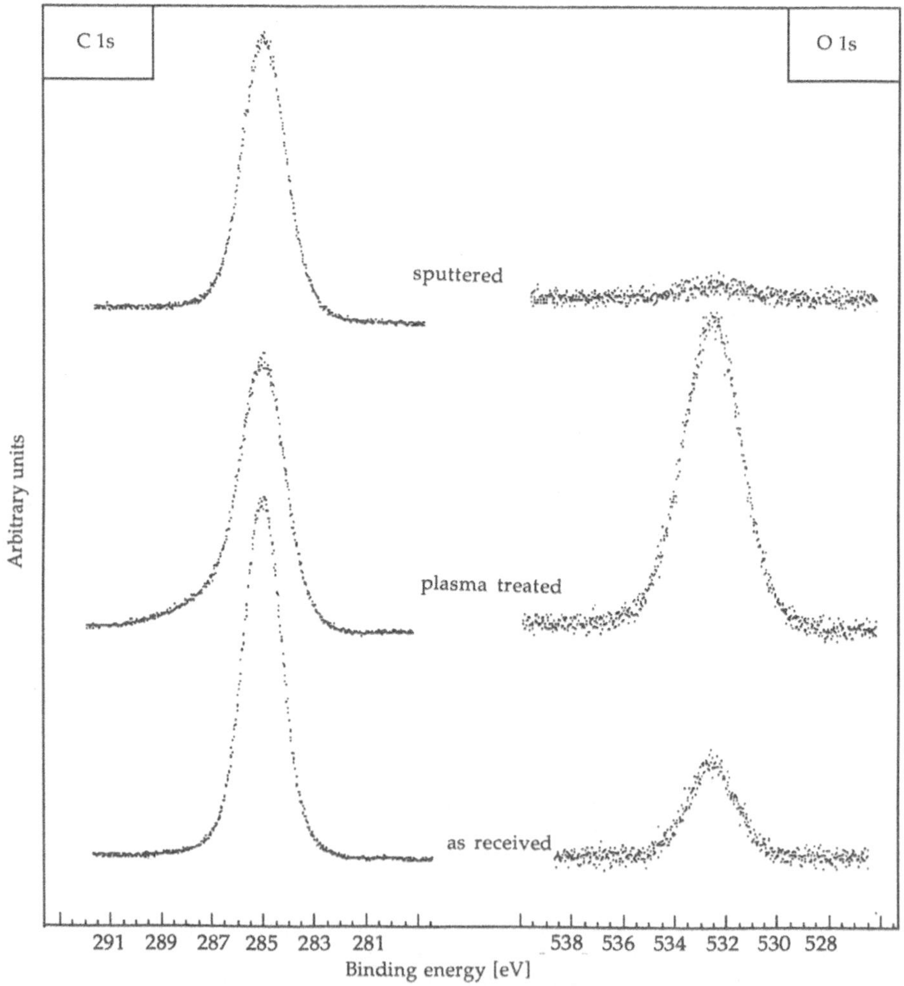

Fig.1 XPS spectra of C1s and O1s before and after polypropylene surface treatment; bottom: surface as received, center: plasma treated surface, and top: sputtered surface.

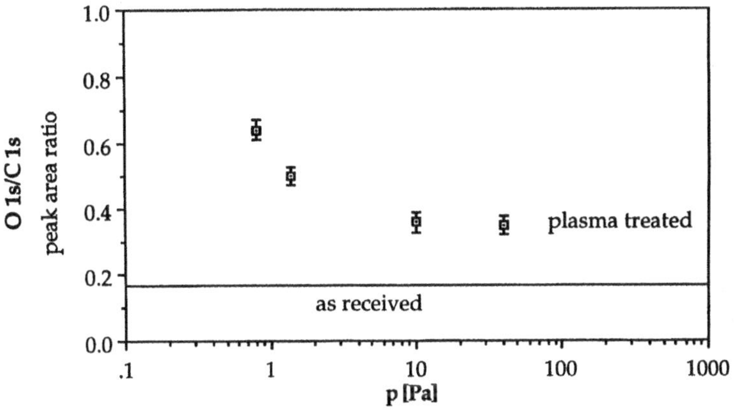

Fig.2 Peak area ratio O1s/C1s as a function of pressure during the plasma treatment of polypropylene.

A similar behaviour is observed for a thin film of cerium, shown in detail in Fig. 4. In addition to the plasma treatment, the effect of an ion treatment is also shown, again for similar deposits of Ce of typically 10 monolayers.

As for the case of Ca we observe an increasing intensity of the Ce $3d_{5/2}$ peak as the surface oxidation due to the plasma treatment increases. In contrast, the oxygen-free surface after ion treatment shows an even larger metal content. Depending upon the kind of surface treatment, the Ce is found in different chemical states. For the case of the oxidized surface, the doublet feature of the Ce $3d_{5/2}$ peak can be associated with final-state effects in the photoelectric process, characte-ristic of the formation of an oxide of Ce.[13,14] The double feature of the O 1s peak supports this interpretation and can be explained by the presence of carbon-oxygen bonds as well as a metallic oxide. From these observations we deduce the for-mation of a metal-oxygen-polymer complex similar to that proposed by Burkstrand[6] or Gerenser[10]. In contrast, the oxygen-free surface leads to a completely different picture which may be characterized by a metallic nature of the Ce $3d_{5/2}$ peak. The carbon 1s peak for this situation shows a slight shoulder toward the lower binding energy side which can possibly be assigned to the formation of cerium carbide. Based on these results we may state that the two kinds of surfaces obtained after different surface treatments lead to different bonding mechanisms at a metal-polymer interface.

Similar results were also obtained for thin films of Mg. In this case, the plasma treated surfaces showed only very little metal concentration and substantial amounts of Mg could only be found for the ion treated surfaces. The results of the three metals used during this study on the polypropylene surfaces were compared to similar films on a reference aluminium surface. For each of the three metals, similar amounts were

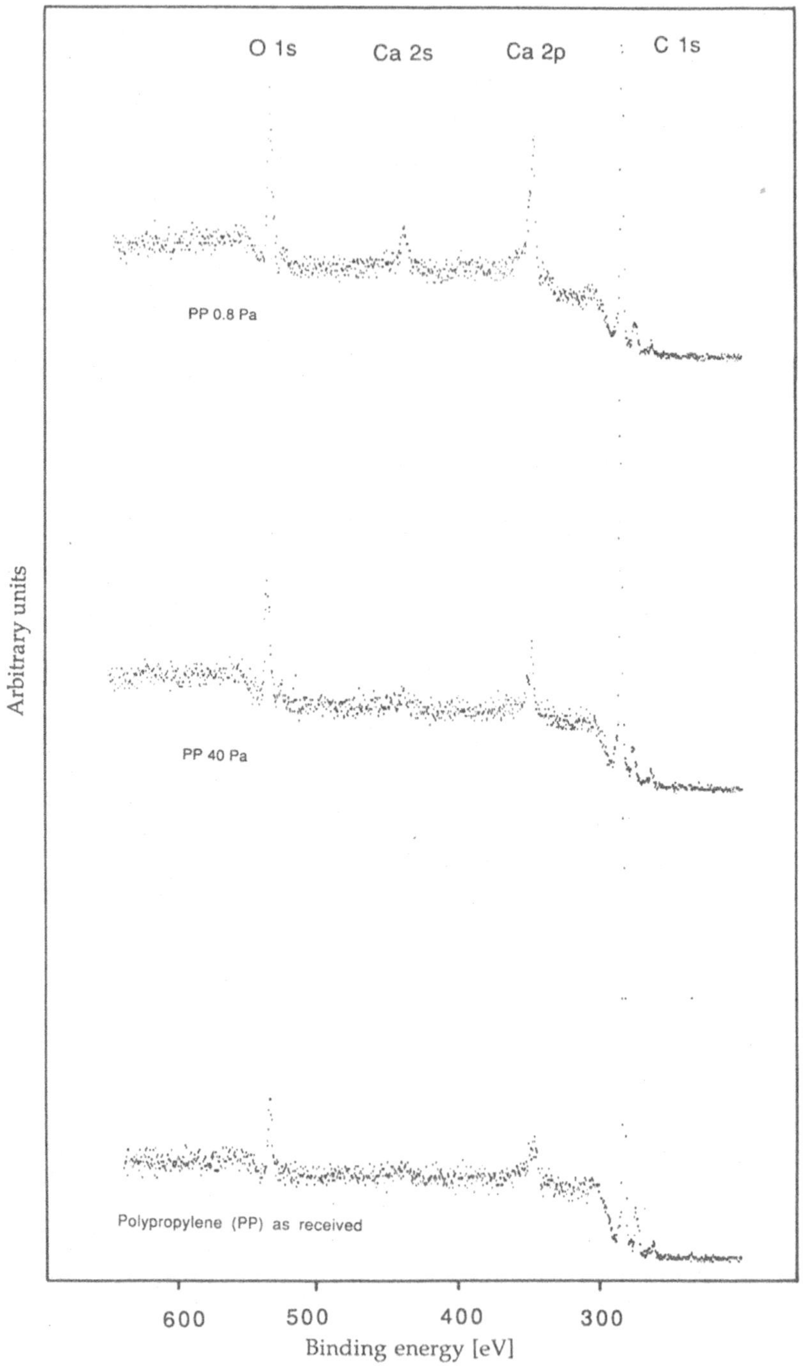

Fig.3 Survey spectra of polypropylene surfaces after evaporation of Ca; bottom: evaporation on surface as received, center: evaporation on plasma treated surface at 40 Pa, and top: evaporation on plasma treated surface at 0.8 Pa.

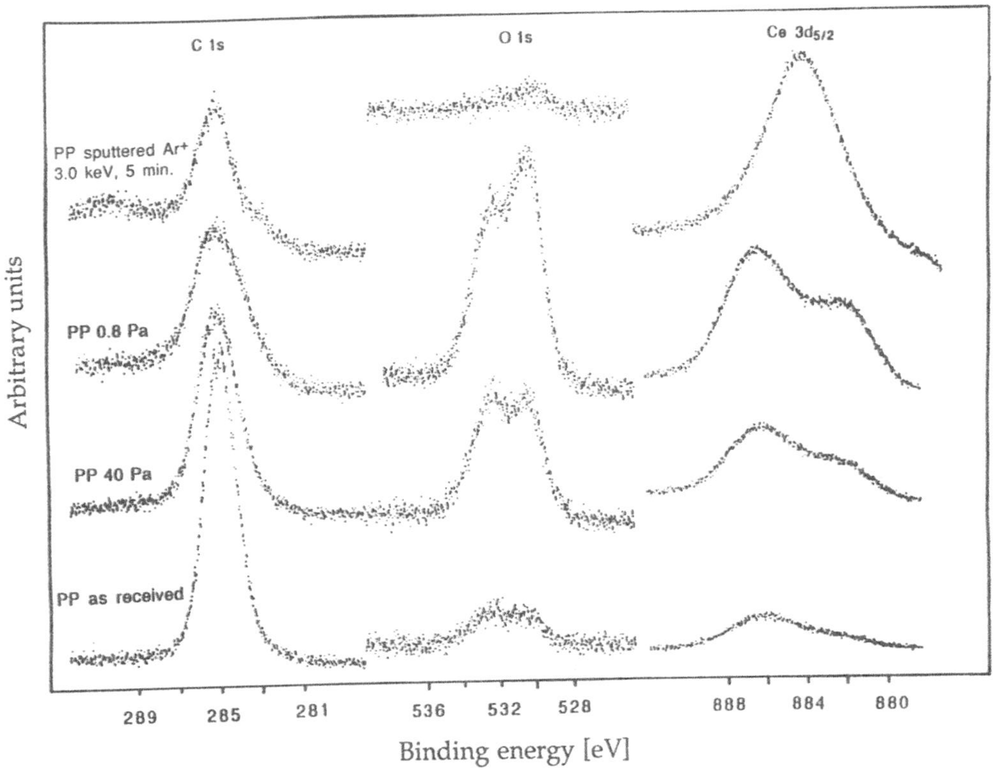

Fig.4 XPS spectra of C1s, O1s and Ce $3d_{5/2}$ after evaporation of Ce; bottom: evaporation on a polypropylene surface as received, second: evaporation on plasma treated surface at 40 Pa, third: evaporation on plasma treated surface at 0.8 Pa, and top: evaporation on sputtered surface with 3.0 keV Ar^+ ions.

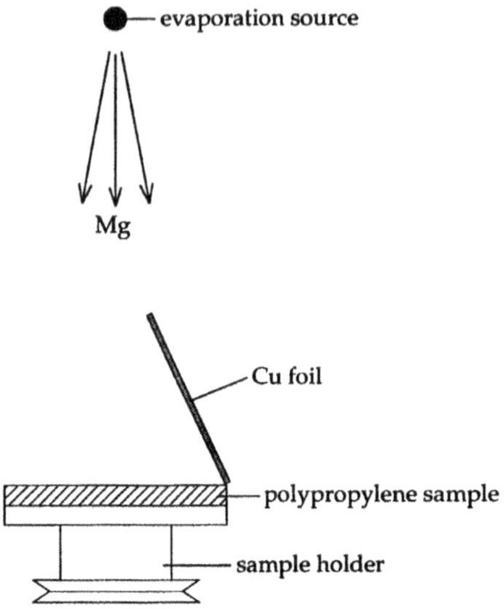

Fig.5 Schematic set-up of the qualitative experiment to determine the sticking probability of a metal on a polymer surface.

evaporated onto the polypropylene surfaces as well as onto the aluminium surface. We compare the metal concentration after evaporation on the polypropylene surface to that which is found on the reference aluminium surface at the same amount of deposited material. A progressive increase in the amount of metal on the polypropylene samples following the surface treatments as compared to the reference surface is found in successive order for Mg, Ca and finally Ce.In other words, if a relative concentration c(metal) is defined as c(metal) = metal concentration on polypropylene / metal concentration on reference, then $c(Mg) < c(Ca) < c(Ce)$. This result has to be interpreted in terms of the reactivity and the sticking coefficient of the considered metal.

We have also investigated to what extent the sticking probability of a metal on a polypropylene surface is modified by the applied surface treatment. For this purpose, we have carried out a qualitative experiment which is shown schematically in Fig. 5. After ion treatment of the polypropylene sample, a film of Mg is deposited on this surface. A copper foil is fixed next to the polypropylene sample in such a way that one side faces the polypropylene surface without being exposed to the Mg source. In this way, any Mg which can be detected on the copper foil must have been reflected from some surface which is exposed to the evaporation source. In Fig. 6 a typical result of such an experiment is presented. For an untreated polypropylene surface no Mg can be detected on the polypropylene, but some Mg is found on the copper surface. A similar polypropylene sample which had been ion treated shows a clear feature of a Mg film. The amount of Mg which can be detected on the copper surface is, however, reduced by ≈ 50%. This result

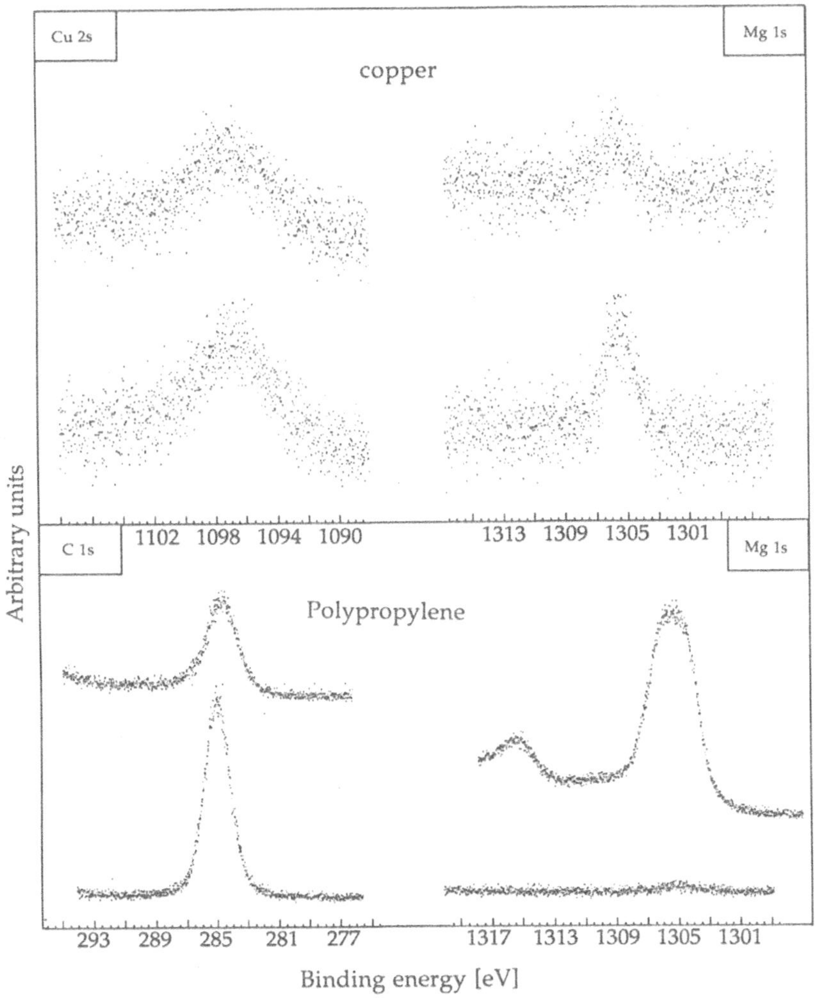

Fig.6 XPS spectra of Cu2s, Mg1s of the copper foil and C1s and Mg 1s of the polypropylene surface; bottom: evaporation on surface before treatment, and top: evaporation on surface after ion treatment.

indicates that the sticking probability of the ion treated polypropylene surface is indeed strongly modified.

First, ion treatments of surfaces were carried out with ion energies up to 3 keV. Later, we observed that much lower energies were capable of producing the same effects. We have measured substantial Mg concentration down to ion energies of less than 20 eV. No energy threshold value was observed down to these energies. However, the time necessary to achieve the desired effect depends on the ion dose which impinges upon the polypropylene surface, as shown in Fig. 7. The ion dose is determined approximately by measuring the sputter current on the polypropylene surface. There apparently exists a clear minimum dose where the sticking coefficient is modified in a substantial way.

DISCUSSION

The results of this study indicate that depending on the nature of the surface treatment and on the resulting surface state, different types of bonding mechanisms between metal and polymers take place. These differences are not only qualitative, but also of a quantitative nature. Increased surface oxidation by plasma treatment leads to larger metal concentrations, but a surface free of oxygen after ion treatment may accept even more metal. This means that the sticking probability of a metal on a polymer surface for these kinds of surface treatments can be modified as a consequence of either formation of a metal-oxygen-polymer complex, or a metal-polymer bond. The question which arises is to what extent the variation in the surface metal concentration is representative of the adhesion at this interface. The metal-polymer interaction resulting from the surface treatment has to be enhanced on the one hand, whereas a degradation of the polymer must be avoided as far as possible, on the other hand. It is known that ion treatments can lead to graphitized surfaces.[11] In this context, it is interesting to note that relatively modest ion energies are capable of producing substantial modifications to the sticking probability of the metal to the polymer. A plasma treatment will generally be of a less energetic nature than the ion treatment. Experiments therefore have to be carried out to enable direct comparison of the different surface treatments in-situ.

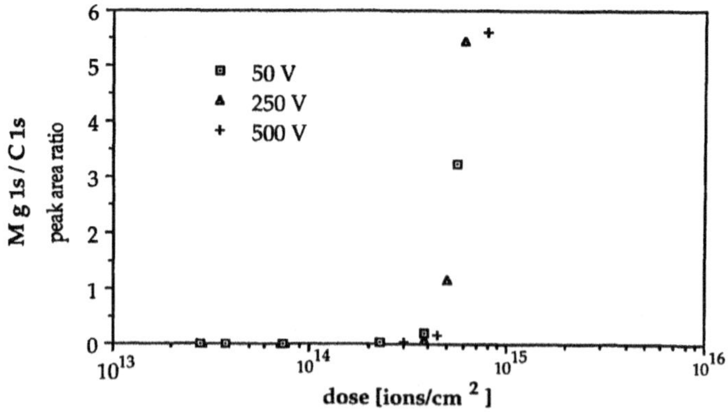

Fig.7 Peak area ratio Mg1s/C1s on a polypropylene surface as a function of the ion dose at different ion energies.

Besides modified sticking probabilities, other effects such as diffusion of the metal into the polymer bulk, or changes in the surface topography as a consequence of the applied surface treatment are possible. These effects may, in turn, have their influence upon the observed interfacial phenomena. Further experiments have to be carried out to clarify the role of these effects. No attempt was made so far in this study to investigate the growth process of a metal film on a polymer surface. The observed results have to be correlated with measurements of the actual adhesion of a metal film on a polymer surface from a macroscopic point of view, as well as with regard to the nature of the adhesion mechanism.

CONCLUSIONS

Surface treatments resulting in different states of a polymer surface, for example rich in oxygen or free of it, lead to different bonding mechanisms at a metal-polymer interface. Oxidized surfaces tend to form a metal-oxygen-polymer complex, whereas oxygen-free surfaces are likely to form a direct metal-polymer bond. We have found evidence for a mechanism by which the sticking probability of a metal on a polymer surface is modified substantially by the applied surface treatment. This effect shows no energy threshold value down to ion energies of 20 eV. A clear minimum dose is, however, necessary for the observed effect.

ACKNOWLEDGEMENTS

We would like to thank Dr. T. Greber and P. Gröning for fruitful discussions. Sample preparation and plasma treatment were carried out by Dr. H.P. Haerri and I. Rime (Ciba-Geigy, 1700 Fribourg, Switzerland). This work was supported by the Swiss National Science Foundation, NFP 24 and by Ciba-Geigy.

REFERENCES

1. K.L. Mittal and J.R. Susko, Editors, "Metallized Plastics 1: Fundamental and Applied Aspects", Plenum Press, New York, 1989
2. D.M. Brewis and D. Briggs, Polymer 22, 7 (1981)
3. L.J. Gerenser, J. Adhesion Sci. Tech. 1, 303 (1987)
4. B.J. Bachman and M.J. Vasile, J. Vac. Sci. Technol. A7, 2709 (1989)
5. E.M. Liston, J. Adhesion 30, 199 (1989)
6. J.M. Burkstrand, J. Appl. Phys. 52, 4795 (1981)
7. N.J. Chou and C.H. Tang, J. Vac. Sci.Technol A2, 751 (1984)
8. J.L. Jordan, C.A. Kovac, J.F. Morar and R.A. Pollak, Phys. Rev. B 36, 1369 (1987)
9. J.G. Clabes, J. Vac. Sci.Technol A6, 2887 (1988)
10. L.J. Gerenser, J. Vac. Sci.Technol A6, 2897 (1988)
11. P. Bodö and J.-E. Sundgren, J. Appl. Phys. 60, 1161 (1986)
12. S. Nowak, H.-P. Haerri, L. Schlapbach and J. Vogt, Surface Interface Anal. 16, 418 (1990)

13. G. Praline, B.E. Koel, R.L. Hance, H.-I. Lee and J.M. White, J. Electron Spectr. Rel. Phen. 21, 17 (1980)
14. O. Gunnarsson and K. Schoenhammer, Phys. Rev. B 28, 4315 (1983)

BEHAVIOR OF METAL - POLYIMIDE INTERFACES UNDER DEGRADATIVE AMBIENTS

K. Seshan and R.H. Lacombe

IBM E. Fishkill Facility

Hopewell Junction, NY 12533

During wire bond reliability testing, samples which contained multilayer thin film metal bonding pads on polyimide were subjected to thermal cycling in deleterious ambients. Subsequently the bonds were pull tested. Wire bond pull test results showed that the region between the metal and the polyimide had degraded. This degradation is evidenced by low pull strengths and apparent interfacial failures.

In this study Cr and Ti were compared in their roles as adhesion layers to the polyimide. Wire bond pull test showed that both metals were prone to degradation and that oxidizing ambients accelerated failure. The degradation of metal to polyimide adhesion has been shown to occur in a transition layer between the metal and the polyimide. Interactions between metal, polyimide and the ambient evidently occur within this layer causing the metals to oxidize and leaving the oxide particles dispersed in a transition layer.

In this study the influence on the mechanical properties of a uniformly dispersed second phase in the transition layer was examined. Finite element stress modelling was used to simulate the mechanical consequences of a uniform dispersion of second phase material within this transition layer. The modelling shows that uniform dispersions do not give rise to changes in the stress levels which would account for the observed changes in wire bond pull strength. This study is confined to linear elastic models. It is concluded that more complex processes must be operative, such as nonuniform distributions, local stress concentrations, or changes in the intrinsic fracture toughness properties of the transition layer.

INTRODUCTION

Polymeric materials are being increasingly used in microelectronics. Most popular of the polymeric materials are members of the polyimide family. Polyimides have low dielectric constant (2.5-4). They can withstand elevated temperatures, up to 400° C, and are resistant to solvents when fully cured. These are attractive properties from the

Metallized Plastics 2, Edited by K.L. Mittal
Plenum Press, New York, 1991

viewpoint of microelectronic applications.[1-3] In this study of wire bond reliability, a multilayer metal bonding pad was deposited onto fully cured polyimide. Wires were then bonded to this pad and pull tested. A cross sectional view of the samples tested is shown in Fig.1 below. The samples were cycled in oxidizing ambients up to 380° C. When tested after thermal cycling the adhesion between the metal and the polyimide was observed to degrade, as measured by the wire bond pull test. This paper deals with the degradation of the interface between the metal and the polyimide.

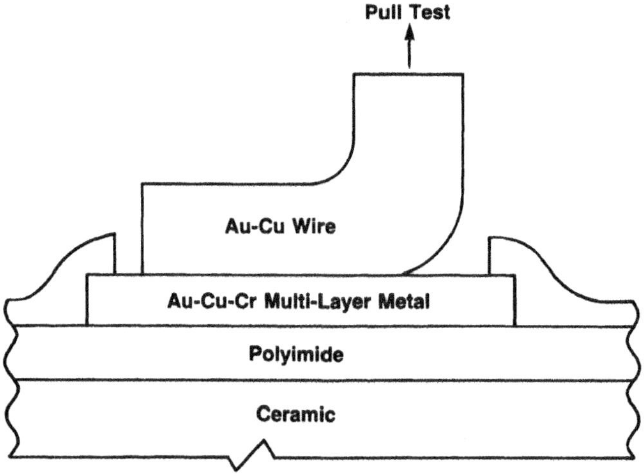

Figure 1. Cross section of a typical multilayer thin film bonding pad, 100μm square, defined by photolithography and subsequent etching. The multilayer sequence is Au/Cu/Cr/PI with Cr as the adhesion layer to the polyimide. Both PMDA-ODA and PMDA-BTDA polyimides were studied. Wires, 35μm in diameter, were ultrasonically bonded to the thin film pad. The bonds were tested by pulling the wires at 90° and the pull strengths were recorded.

There is an extensive literature on the subject of polyimides.[4] The basic chemistry of bonding between a metal and the polyimide is discussed in the paper by Kovac et al.[5] Ho and co-workers[6-8] have discussed the atomistic and quantum mechanical details of bonding. The mechanical stress and adhesion modelling studies of Kim et al.[9] deal with a detailed analysis of the pull and peel tests. Papers by Mittal,[10] and Buchwalter[11] show the decisive effect which surface condition and preparation can have on "practical adhesion". Experimental work on adhesion, and surface modification reported by Horwath, Ruoff and co-workers[12-14] supports the view that surface modification is an important issue. Furman et al.[15] have used the SIMS technique to study the effects of ambient exposure. This particular investigation is of special interest to the work at hand.

A review of all the literature on this subject is beyond the scope of the present work. Rather, we infer on the basis of the papers mentioned above that there is a transition layer between the adhesion metal and the polyimide. In particular, we follow up on the hypothesis of Kovac et al.[5] who claim that there could be discrete oxide particles within this transition layer even at the very early stages of thin film growth. The surface modification work of Paik and Ruoff, and others[12-14] is further evidence for the role of the transition layer in determining the adhesion. Lastly Furman et al.[15] showed that the effects of moisture can be explained on the basis of interactions within this transition layer. We also assume that the interaction of ambients also occurs by the same mechanism. We further assume that the wire bond pull test measures "practical adhesion"[10,11] which is defined as the force or the work required to remove the multilayer coating from the substrate, and which depends critically on the strength of the transition layer.

In the previous work the results of wire bond pulling after the pad was subjected to thermal cycling were reported.[16] This involved pull testing of bonded wires. That study showed that ambient exposure degrades the interface. A thermochemical explanation was offered at that time. In this paper we have used finite element techniques to answer the question: Can the early degradation of the pull strength be explained by changes in the mechanical properties of the transition layer caused by a uniform distribution of a second phase at the metal-polyimide interface? Some of the results of that earlier paper are briefly quoted here as they provide a background for the present work.

EXPERIMENTAL

Ambient Sensitivity; Wire Bond Results

The sensitivity of the metal/polyimide interface to ambients is illustrated in Figures 2 and 3. These results have been explained in detail in previously published wire bond results.[16] There it was shown that the Ti-PI and the Cr-PI interfaces exhibited predominantly cohesive type failure at time zero when the metal was first deposited. Such failures occur near the strength of the bonding wire (about 40g) and always tear up some of the polyimide which is found adhered to the underside of the wire bonding pad. The polyimide on the underside is always thin and requires careful surface analysis to be detected.[15]

After thermal cycling in an oxidizing ambient, with temperature excursions up to 380° C, the failure occurs at low strengths and with a different failure mechanism. The failures become predominantly interfacial. This means that separation was found to occur near the interface or in the transition layer between the copper and the polyimide. There is now a clean separation between metal and polyimide. SIMS analysis shows that there is very little polyimide still left on the underside of the separated metal pads. This failure mode occurs at a low strength of 20g as opposed to the 40g for the cohesive fails. This is also consistent with the observations of Furman et al.[15] Using water made with the isotope of oxygen,[18] they showed that moisture percolated through the polyimide to the metal-polyimide interface. From this fact they inferred that oxygen from the water could be involved in oxidation reactions near the interface.

These results are illustrated in Fig. 2 and 3. It is to be noticed that in Fig. 2, only the fraction of interfacial failures is reported. Notice how the proportion of interfacial failures increases with thermal cycling. In fig. 3, the strength of those bonds showing interfacial failure is plotted with respect to thermal cycles. Notice once more that interfacial adhesion is decreased by thermal cycling.

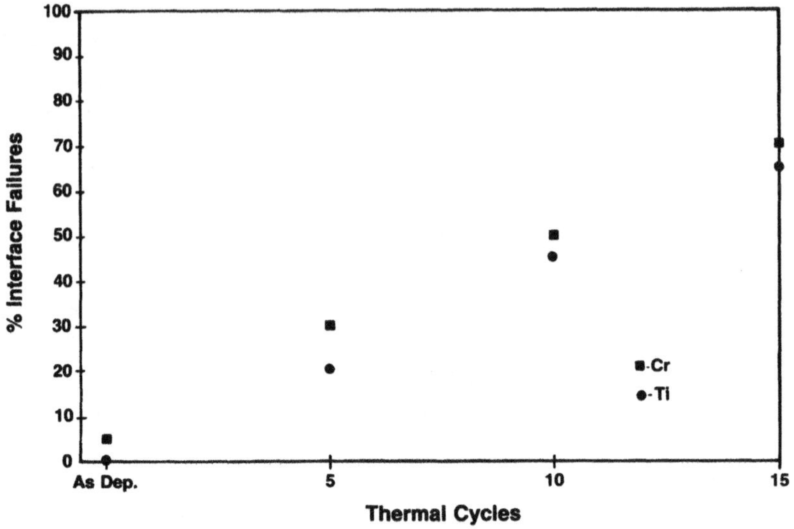

Figure 2. Degradation of the metal/polyimide interface with continued thermal cycling. At each point 30 bonds were pulled and the percentage of failures and the load at failure were recorded. This figure shows that for both Ti and Cr the percentage of interfacial failures increases with exposure to degradative ambient.

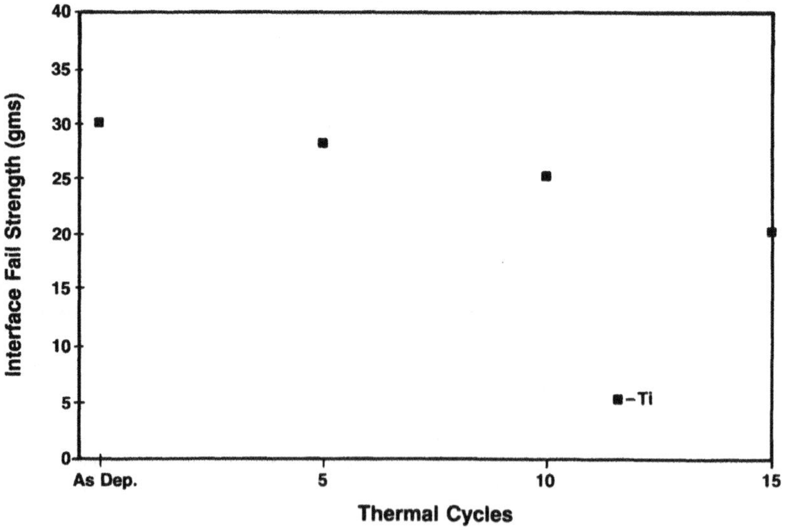

Figure 3. Strength distribution corresponding to Fig.(2). Only the data for Ti are shown. Cr follows a similar trend. At each stage of thermal cycling there is a decrease in the wire bond pull strength which is symptomatic of a degrading interface.

The conclusions from the ambient exposure experiments together with the SIMS work of Furman et al. are that with continued thermal cycling, the oxidizing gases percolate to the "transition" layer between the metal layer and the polyimide. Here a certain amount of the adhesion metal, Cr or Ti, is oxidized. The result is that the adhesion of the metal multilayer to the polyimide decreases. Presumably the oxidation products are left dispersed in the transition layer. This is the starting point for the stress modelling, which is described below.

Stress Modelling

At this juncture we had to resolve what role the interfacial inclusions were playing in the observed adhesion degradation. There are at least two potential mechanisms whereby the transition layer inclusions could lead to drastically modified mechanical strength near the interface:

1. Particles or voids can have a dramatic effect on the elastic properties of the transition layer which could result in significantly higher stress levels which would, in turn, promote premature mechanical failure.

2. The particles or voids could lead to a significant degradation in the intrinsic strength of the transition layer region thereby making it more prone to failure.

The second of the two hypotheses is a very difficult question beyond the current capabilities of mathematical modeling and thus is better settled by careful experiments. The first hypothesis can be framed as a precise boundary value problem and can thus be addressed by mathematical techniques. In this work we used the ANSYS[17] finite element code as a tool to investigate the stress/strain response of a simple model of a wire bond pad subjected to thermal and tensile loads. The basic model is depicted in Fig. 4. In this figure we have a copper pad on top of a layer of polyimide which is on top of a massive ceramic substrate. In between the copper pad and the polyimide is a thin transition layer whose elastic properties will be changed because of the presence of voids or hard particles.

The stress loads are applied in two steps to simulate the actual load sequence which pads on real substrates see. The first load step is a thermal load brought about by cooling the structure down from an elevated temperature (400° C). This step imposes a significant stress upon the structure due to thermal expansion mismatches between the copper, the polyimide and the ceramic. Such thermal expansion mismatch stresses are generally present in all structures of this kind. In addition to this thermal load an entirely mechanical tensile load of 100g is applied normal to the top surface of the copper pad. This load basically simulates the effect of pulling off the wire bond. No special significance need be attached to the precise value of the applied load since the main thrust of this work is to compare what happens when the elastic properties of the transition layer material are changed. The applied load is kept constant throughout and only has to be reasonably representative of the load state experienced under actual wire bond pull testing conditions.

Fig. 5 shows the displacement response of the pad after first thermal loading followed by thermal plus mechanical loading. At this stage the transition layer does not exist in that its mechanical properties are taken to be identical to those of the underlying polyimide. The state of stress and deformation represented by Fig. 5 serves as a fiducial state for the remainder of the calculations. From this point on we proceed as follows:

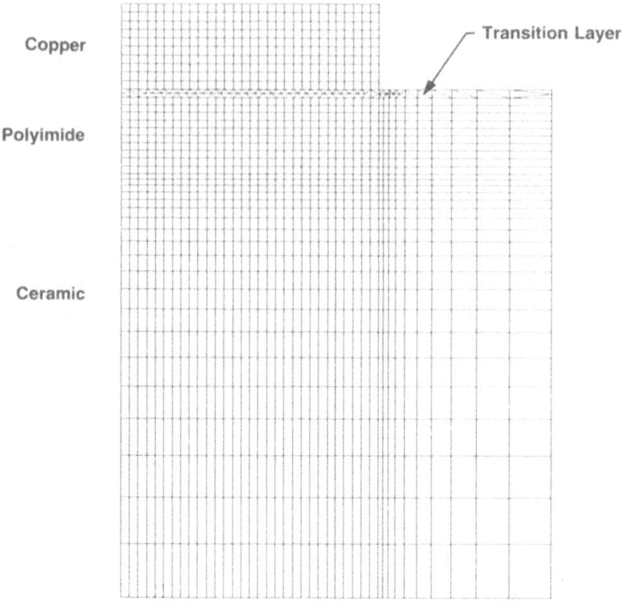

Figure 4. Finite element idealization of copper pad on polyimide/ceramic substrate. Model is axisymmetric with the left hand boundary as the symmetry axis. Note the 1um thick transition layer between the polyimide and the copper. This layer is meshed with a thickness of three 6 noded triangular elements in order to accurately estimate the deformations and stresses in this region.

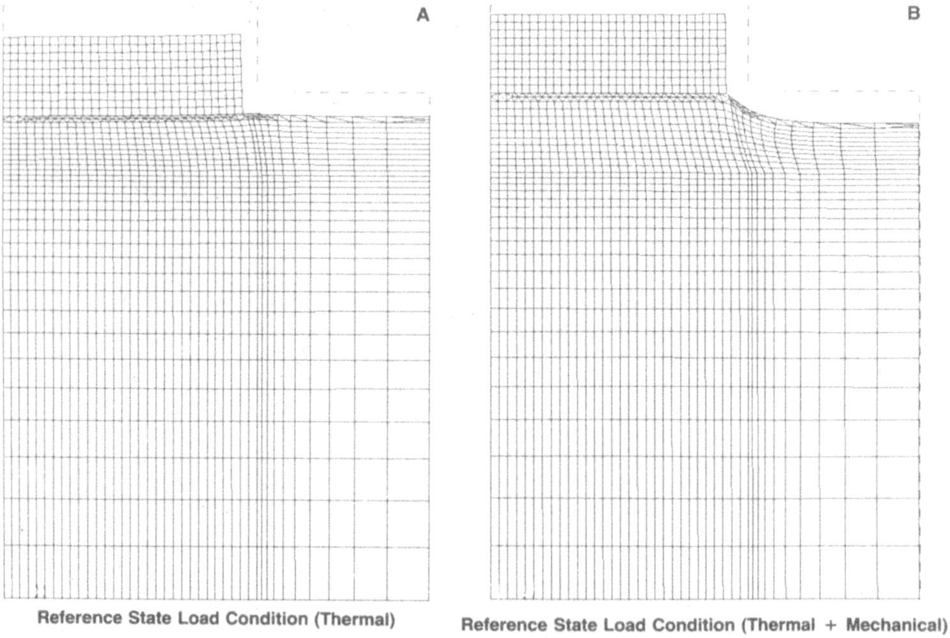

Figure 5. Deformation state of pad after thermal and mechanical loading (not to scale). a) Deformation after thermal loading alone. b) Deformation after thermal and mechanical loads are applied.

1. The transition layer depicted in Fig. 4 is created by altering the elastic properties of this zone by filling either with voids or hard particles.

2. Thermal and mechanical loads are applied to the pad structure in the identical fashion as was done for the unaltered system depicted in Fig. 5.

3. Stresses and deformations are recomputed for the altered system which is identical to the configuration in Fig. 5 except for the changes in the transition zone.

4. The stresses and deformations of the unaltered state are subtracted out leaving only the residual stresses and deformations which arise from the altered conditions in the transition zone.

5. All stresses and deformations are plotted out along the midline of the transition zone where the effect of the altered elastic properties of this region are expected to be largest.

By proceeding in the manner outlined above we had hoped to be able to detect those changes in the stress and deformation fields which are due to the presence of the added inclusions, be they particles or voids. It was hoped that by using this subtractive procedure our conclusions would be independent of the particulars of the reference load state we used and therefore reflect only effects arising from the presence of the inclusions.

A special procedure is required to model the presence of inclusions in the transition zone. In order to accomplish this, we rely on the theory of Farber and Farris[18] who have worked out a scheme for calculating the elastic properties of polymer materials which have been uniformly loaded with some additive. These workers have presented what is essentially an incremental technique for computing the change in the elastic constants of a material as an inclusion of known mechanical properties is added in a stepwise fashion. An important feature of this approach is that only the properties of the pure filler and binder materials need to be known in order to compute the elastic constants of the mixture. In this work we garnered our data on the hard particle inclusion from the handbook of Samsonov.[19] Handbook values were used for the properties of copper, polyimide and ceramic materials.

RESULTS

A number of loading conditions were investigated to see what effects would occur in the transition layer region. In what follows we will report on two cases where the transition region was loaded in one case with 10% voids and in another case 10% titania particles. We had initially expected that loading with voids would soften the transition region and thus cause the stresses to decrease. On the other hand adding hard titania particles was expected to stiffen the transition zone thus causing the stresses to increase. Somewhat to our surprise we found that neither of the above occurred and that in fact the stress distribution in the transition zone was sensibly unchanged irrespective of whether particles or voids were added. Figures 6 and 7 show the resultant stress and deformations in the transition zone for the voids and titania cases, respectively. Each Figure plots 3 representative field quantities along the midline of the transition layer. The first quantity is the residual maximum principal stress represented by SIG1. This is the largest stress component which can exist at any given point. The next quantity is the

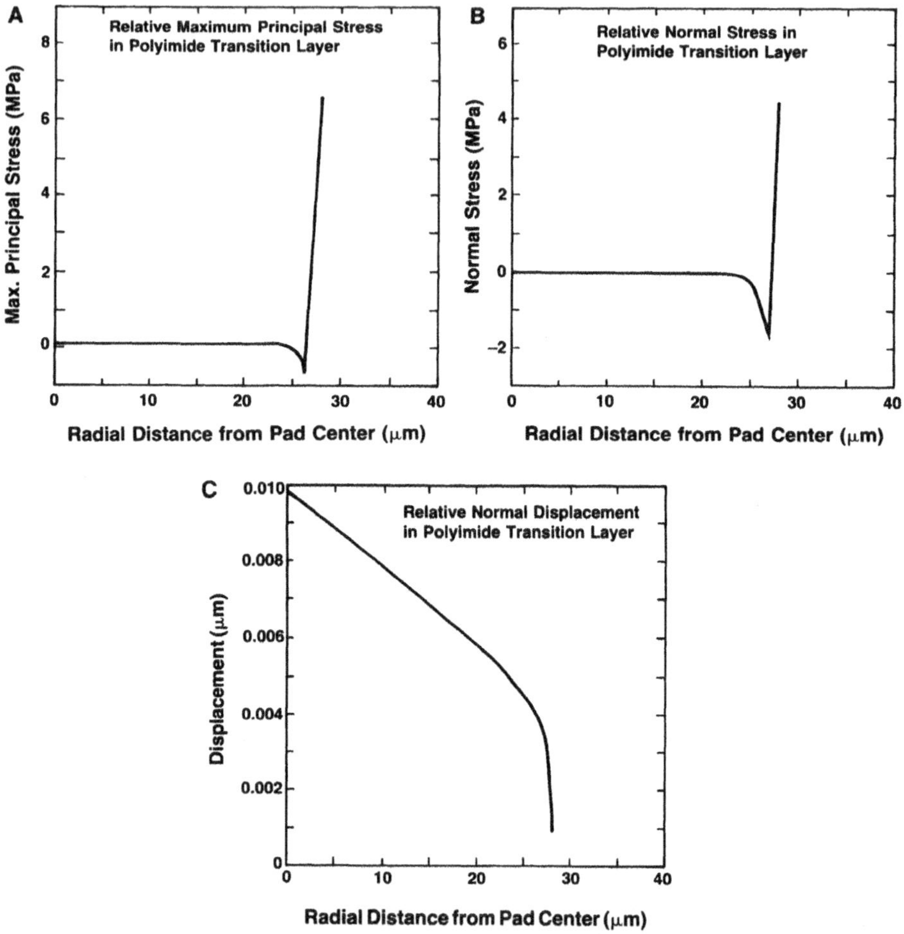

Figure 6. Residual stress and deformation fields along the midline of the transition zone which has a 10% void content. A) SIG1 the maximum principal stress. B) SY the normal stress perpendicular to the interface. C) UY the normal displacement. Note that UY shows a positive residual, whereas SIG1 and SY show virtually no residual.

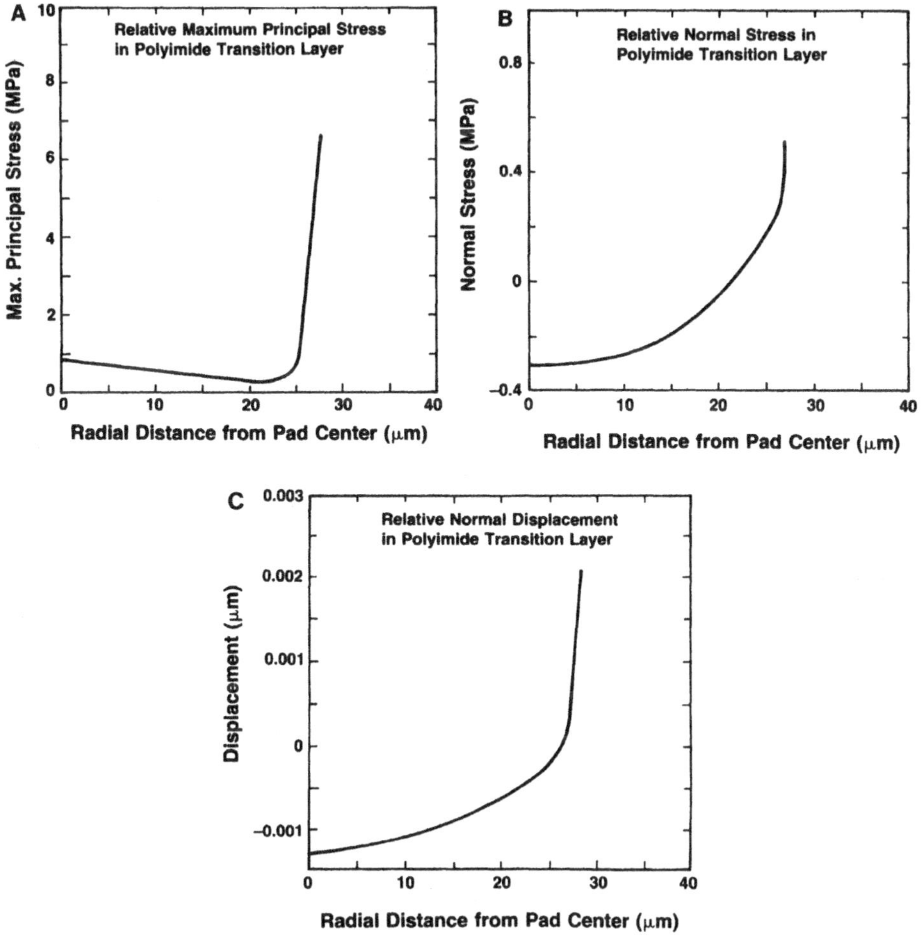

Figure 7. Same as Figure 6 except that the transition layer is filled with titania particles. Note that the situation is opposite to that of Figure 6 in that there is now a negative displacement field; however, the stress residuals are still very small.

residual normal stress represented by SY. This stress component would be expected to play a role in delamination due to pulling loads. The final quantity is the residual normal displacement. This quantity naturally reflects the amount of deformation in the vertical direction which the transition layer exhibits on being exposed to pull forces.

We discuss the case of void loading first. Figure 6C shows the residual normal displacement which occurs for 10% void loading. We note a significant positive residual displacement which accords with our intuition that voids will soften the transition zone and thus give rise to larger deflection on loading. However, Figures 6A and 6B for the stress residuals show essentially no change over the reference condition. This would seem to contradict our intuition about a softer phase giving rise to lower stresses.

Figure 7 exhibits analogous results for the case of titania loading. Referring to Figure 7C the normal residual deformation is again in accord with common experience by showing a negative residual deformation for the harder phase. Thus, the loaded transition layer deformed less than the unloaded case. However, the stress components represented by Figures 7A and 7B show virtually no change over the reference state and give nearly zero for the residual principal and normal stresses.

On reflecting over these results we concluded that the transition layer was in effect compensating out the stress field by adjusting only the deformations. Thus for the voids case the transition layer was made softer which means the elastic modulus is lowered. However, the system responds by allowing a larger deformation and thus by Hooke's law the resulting stress remains essentially unchanged. The opposite occurs when the transition layer is loaded with titania. The layer is now stiffer which gives rise to smaller deformations which again by Hooke's law leaves the stresses unchanged.

CONCLUSIONS

1. Copper wire bond pads suffer a dramatic decrease in adhesion strength to underlying polyimide when subjected to thermal cycling in an oxidative environment.

2. Thermal cycling can give rise to a transition layer between the copper and the polyimide which can contain inclusions either in the form of voids or oxidized metal particles.

3. The presence of inclusions in the transition layer alters the elastic properties of this region but the system compensates so as to keep the local stress distribution constant.

4. In view of the above statements, we conclude that the main degradative effect of the inclusions is to lower the intrinsic mechanical strength of the transition layer.

REFERENCES

1. M. Terasawa, S. Minami and J. Rubin, Intl. J. Hybrid Microelectronics, 6, 607 (1983).
2. T. Watari and H. Murano, IEEE Trans. CHMT-8, No. 4, 462-467 (1985).
3. L. B. Rothman, J. Electrochemical Soc., 130(5), 1131 (1983).

4. K. L. Mittal, Editor, "Polyimides: Synthesis, Characterization and Applications," Vols. 1 and 2,Plenum Press, New York (1984).
5. C. A. Kovac, J. L. Jordan-Sweet, M. J. Goldberg, J. G. Clabes, A. Viehbeck and R. A. Pollak, IBM J. Res. Dev. 32(5), 603 (1988).
6. F. Faupel, C. H. Yang, S. T. Chen and P. S. Ho, J. Appl. Phys., 65, 1911 (1989).
7. P. S. Ho, P. D. Hahn, J. W. Bartha, G. W. Rubloff, F. K. LeGoues, B. D. Silverman, J. Vac. Sci. Technol. A3(3), 739 (1985).
8. P. S. Ho, B. D. Silverman, R. A. Haight, R. C. White, P. N. Sanda, A. R. Rossi,IBM J. Res. Dev., 32, 658 (1988).
9. Y. H. Kim, G. F. Walker, J. Kim and J. Park, J. Adhesion Sci. Technol., 1, 331 (1987).
10. K. L. Mittal, "Adhesion Measurement of Thin Films, Thick Films and Bulk Coatings" K. L. Mittal, editor,ASTM STP 640,American Society For Testing and Materials, Philadelphia, 1978.
11. L.P. Buchwalter, J. Adhesion Sci. Technol., 1, 341 (1987).
12. R. S. Horwath, R. L. Headrick, and J. W. Wilson, Proc. Intl. Ion Engineering Congress, 3, 1743-6 (1983).
13. K. W. Paik, and A. L. Ruoff, J. Adhesion Sci. Technol., 2, 245 (1988).
14. A. L. Ruoff, E. J. Kramer, and C-Y Li, IBM J. Res. Dev. 32(5), 626 1988).
15. B.K. Furman, S. Purushothaman, E. Castellani, S. Renick, and D. Neugroschl, in " Metallization of Polymers," E. Sacher, J.J. Pireaux and S. P. Kowalczyk, editors, ACS Symposium Series No. 440, p. 297(1990).
16. K. Seshan, S.N.S Reddy, S. K. Ray Proc. 37th IEEE Electronics Components Conference, 604 (1987).
17. ANSYS is the trademark of Swanson Analysis Systems, Huston, PA.
18. J. N. Farber and R. Farris, J. Appl. Polym. Sci. 34, 2093 (1987).
19. G. Samsonov, Editor," The Oxide Handbook, " 2nd Edition, Plenum Press, New York (1982).

STUDY OF TITANIUM/POLYIMIDE INTERFACE IN A REDUCING ENVIRONMENT

Kabul S. Sengupta and Howard K. Birnbaum

Department of Materials Science and Engineering and Materials Research Laboratory, University of Illinois at Urbana-Champaign, Urbana, Illinois 61801

The system investigated was Au/Cu/Ti/Polyimide. Heat treatment of these films in reducing ambient introduced a host of problems resulting in film failure. SIMS depth profiles were determined for samples annealed in hydrogen (an essential component of forming gas), at different concentrations and temperatures. It was observed that the hydrogen segregates at the diffusion barrier (Ti), and causes delamination under certain conditions. SIMS, Auger, XPS and X-ray diffraction techniques were used to characterize the interface and to locate the delamination plane in the metal/polymer system.

INTRODUCTION

In recent years, the study of interface formation has become an important area of investigation. In the semiconductor industry, interfaces between metals and/or insulators are of extreme importance as they play a central role in efforts to achieve further miniaturization of semiconductor devices. During the fast switching of devices, high temperatures and high current densities are imposed.[1] A host of additional problems arise during the processing of the packages in inert or reducing ambient; such as thermomigration, hillock growth,[2] reactions between the metal and the substrate, reactions between the ambient and the system, and delamination due to thermal mismatch or interfacial stresses caused by the reactions.

In the Au/Cu/Ti/Polyimide composite, copper is the conductive layer which is usually deposited on the polyimide dielectric. Previous studies have shown copper to be nonreactive with the polyimide and to diffuse into the polyimide at elevated temperatures[3-5] resulting in poor electrical characteristics. However, this problem was circumvented by introducing a thin film of a highly reactive metal like titanium (or chromium) between the copper and the polyimide. Titanium (and chromium) have low mutual solubilities with copper and thus would be able to serve both as a diffusion barrier as well as an 'adhesive' layer between the copper and the polyimide.[6-10] In the present study surface analysis techniques such as Secondary Ion Mass Spectrometry (SIMS), Auger Electron Spectroscopy (AES), X-ray Photoelectron Spectroscopy (XPS) together with X-ray Diffraction (XRD) have been used to investigate the system-ambient interactions at elevated temperatures, especially in hydrogen and vacuum.

Metallized Plastics 2, Edited by K.L. Mittal
Plenum Press, New York, 1991

EXPERIMENTAL PROCEDURES

The polyimide used was DuPont RC5878, with a PMDA-ODA type molecular structure. The RC5878 coated silicon wafer first received a 90°C bake in an IR oven for 30 mins. It then received a soft cure at 150°C for 30 mins followed by 230°C for 30 mins in a nitrogen ambient. A final cure at 300°C for 30 mins followed by 400°C for 40 mins was performed on the sample to complete imidization. This final cure was performed in a forming gas ambient (90% nitrogen, 10% hydrogen). The thickness of the polyimide was about 12-13μm. Before metallization, the polyimide was sputter cleaned with an argon ion beam at 1 kV and 150 mA for 20 mins which corresponds to a dose of about 5×10^{14} ions/cm^2. A 300 nm titanium layer was first deposited followed by a 5μm copper film which was capped off by a 0.5μm gold layer. The gold layer was mainly to prevent the oxidation of copper during storage. The order in which the films were deposited are shown schematically in Fig.1. The metallizations were carried out at a substrate temperature of 448 K using electron beam evaporation technique at a pressure of 1×10^{-7} Torr. All the layers were deposited without breaking the vacuum. Since the effect of hydrogen on the interfacial stability was to be studied, the metal/polymer composites were annealed in purified deuterium (the isotope chosen for analysis in the Secondary Ion Mass Spectrometer) at pressures of 760 Torr D_2 at 623 K for 24 hours, and 648 and 688 K for 2 hours. Specimens were also annealed in an ultrahigh vacuum of 5×10^{-9} Torr for 48 hours for comparison. To study the effect of sputter cleaning and substrate temperature during deposition, a 100 nm titanium layer was vapor deposited onto the unsputtered polyimide at room temperature.

Secondary Ion Mass Spectrometry

The Secondary Ion Mass Spectrometry analysis was carried out in a Cameca IMS-3f Ion Microprobe using Cs^+ ions at an accelerating voltage of +12.6 kV as the primary beam and using negative secondary ion detection for higher secondary deuterium ion yield. The sample was maintained at a constant bias of -4.5 kV; thus the relative accelerating voltage for the Cs^+ ion was the difference between these voltages, i.e. 17.1 kV. The incoming primary beam of 50μm diameter was rastered over a 250μm x 250μm area. Secondary ions were produced over the entire rastered area but a mechanical aperture was used to monitor only a 10μm diameter circular area in the center of the crater, thus eliminating secondary ion counts from the crater

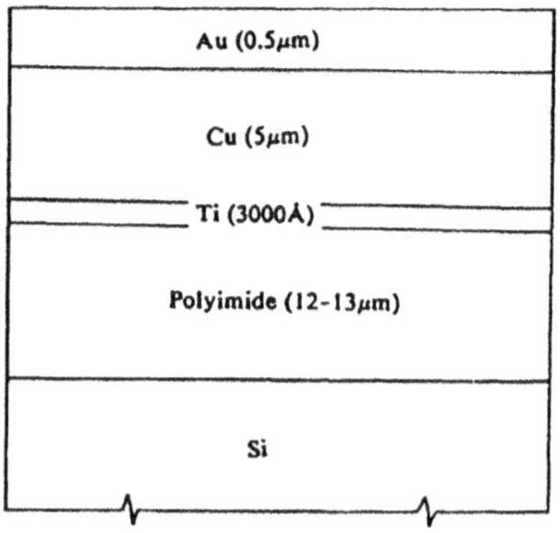

Figure 1. Schematic diagram of the metal/polymer system (not drawn to scale).

walls which would adversely affect measurement of the depth profile. The primary beam current was set at about 240 nA.
The specimen was held at 140 K using a liquid nitrogen cooled stage to prevent deuterium redistribution during the profile analysis.[11]

Auger Electron Spectroscopy

A Physical Electronics 660 spectrometer operated at a primary electron beam energy of 3 to 5 keV, depending on the specimen charging, was used for the Auger analysis of the delaminated specimens. The sputtering was done with a 3 kV argon ion beam so as to minimize the ion beam induced, atomic mixing effect prevalent in the SIMS technique.

X-ray Photoelectron Spectroscopy

A Physical Electronics 5400 XPS system was used to characterize the Ti-O-C ternary that was detected on the metal side of the delaminated surface. All the spectra were taken immediately after peeling. Unmonochromatized MgK_α radiation ($h\nu$=1253.6 eV) was used as the excitation source, and the power was set at 400 watts (15 kV and 26.6 mA). High resolution spectra were taken using a 17.9 eV pass energy, step size = 0.1 eV, with an analysis area of 1.1 mm diameter (FWHM on Au $4f_{7/2}$ transition ~1·0 eV). The photoemission detector angle was held at 45 degrees. The system vacuum never exceeded beyond 1.5×10^{-9} Torr during spectra acquisition. The x-ray source to sample distance was kept at 1.5 cm. Deconvolution was carried out by the commercial computer software developed by Perkin-Elmer. Calibration of the binding energies was done by the internal standard technique using a C_{1s} signal from carbon atoms bound to other carbon and hydrogen atoms for which the C_{1s} energy is 284.6 eV.

RESULTS AND DISCUSSION

Secondary Ion Mass Spectrometry Analysis

Figure 2 shows a depth profile of the metal/polymer composite which was annealed in 760 Torr D_2 for 2 hours at 648 K. The profiles clearly show segregation of deuterium at the titanium layer, consistent with higher deuterium solubility in titanium than in copper or gold.[12] Also evident is the interdiffusion of gold, copper, and titanium during the 648 K anneal. This interdiffusion occurred throughout the titanium layer. The apparent diffusion of copper and the gold into the polyimide is an ion beam induced, atomic mixing effect. A similar effect was seen in the profile of the unsputtered sample for which the metallization was carried out at room temperature and which did not experience any elevated temperature anneal. The depth resolution of the SIMS technique degrades with depth as could be seen by viewing the sputtered pit with Scanning Electron Microscopy. At an incident primary beam voltage of 17.1 kV it is expected that the metal atoms are embedded into the polymer resulting in the broad profile.

X-ray analysis showed that the titanium layer had converted into the fct deuteride, $TiD_{1.9}$. Titanium is known to undergo a volume expansion of 20-23% on conversion to deuteride.[13] This created large interfacial stresses resulting in delamination during exposure to the deuterium atmosphere. Despite the exothermic heat of deuterium solution in titanium, which results in a lower deuterium concentration in titanium at higher temperatures in equilibrium with a given gaseous deuterium pressure[14], delamination occurred more readily at higher temperatures. This may be due to the barrier effectiveness of the gold and copper overlayers to deuterium entry which decreases as the temperature increases.

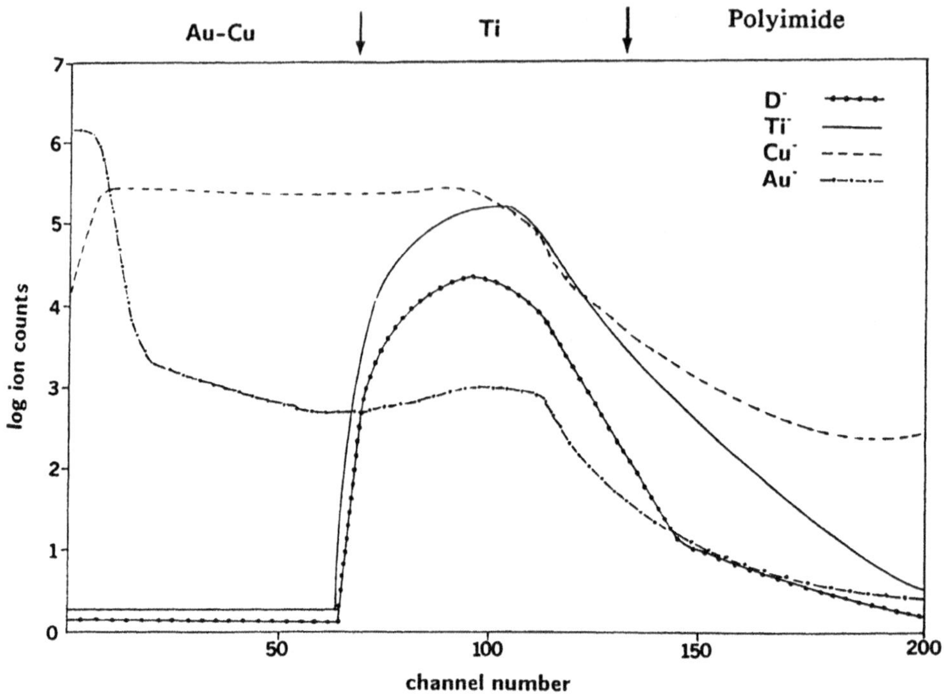

Figure 2. SIMS depth profiles of Au, Cu, Ti, and D in Au/Cu/Ti/Polyimide specimen annealed in 760 Torr D_2, for 2 hours at 648 K showing D segregation at the titanium layer. Approximate location of the interfaces are indicated by the arrows.

Auger Electron Spectroscopy Analysis

Three samples 1, 2 and 3, annealed at 623, 648 and 688 K respectively, in pure deuterium gas at 760 Torr were studied to determine the location of the delamination plane. Examination of the delaminated specimens showed C, N and O on the polyimide side and C, O and Ti on the metal side. The spectra of the specimen which delaminated after annealing at 648 K are shown in Figs.3a and 3b. The other two samples showed similar AES spectra.The origin of the carbon and the oxygen on the metal film is of interest as fracture through the polyimide would be indicated, if they were from the polyimide; whereas if they were due to air contamination, the failure plane would be at the metal/polyimide interface. Since all specimens were given equivalent treatments after the anneal, delamination at the titanium/polyimide interface would result in an equal thickness of carbon and oxygen if these resulted from air contamination. Depth profiling of sample 1 (annealed at 623 K) indicated the presence of a thin carbonaceous layer of about 16 nm thickness between the delamination plane and the titanium film as shown in Fig.4a. At the 648 K annealing temperature the carbonaceous layer was thicker, about 30 nm (Fig.4b); whereas at the highest annealing temperature (688 K), the carbonaceous layer was 70 nm thick (Fig.4c). The AES data (Fig.4) indicated a greater extent of interdiffusion of the titanium and the polyimide as the annealing temperature was increased. The role of deuterium in this interdiffusion is not known and the increased extent with increased temperature may be solely a thermal effect. The data in Fig.4 also show that significant copper was prevented from diffusing into the polyimide by the

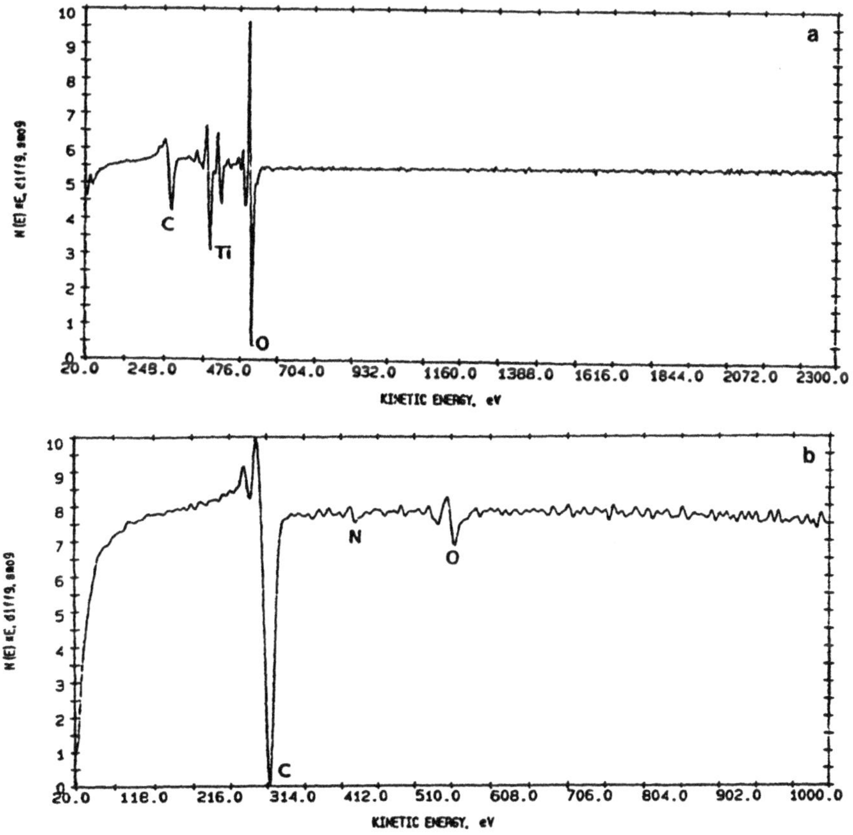

Figure 3. AES spectra from the delaminated surface on the a) metallization side, and b) polyimide side of Au/Cu/Ti/Polyimide specimen annealed in 760 Torr D_2, for 2 hours at 648 K.

titanium film. The level of the copper and gold at the titanium/polyimide interface shown in Fig.2 is too small to be detected with the AES sensitivity.

Thus, in all three cases the delamination occurred in the interdiffusion zone/polyimide interface as shown schematically in Fig.5.

X-ray Photoelectron Spectroscopy Analysis

The O_{1s} spectra from the delaminated metal film side indicate mainly Ti-O and Ti-OH bonds as shown in Figs.6a for specimen delaminated at 688 K.[15] No Ti-C type bond was found as shown by the Ti_{2p} spectrum which indicates titanium dioxide (Fig.6b). The Ti-O and Ti-OH bonds reflect oxide formation and titanium-water reaction, respectively. Since the depth profile indicates fracture occurred in the interdiffusion zone/polyimide interface, this oxide probably occurred during the bonding of the titanium to the polyimide and/or during the annealing. The presence of C=O and C-O-C bonds in the O_{1s} spectra and the nature of the C_{1s} spectra (not shown) indicated the presence of polyimide on the metal side although its composition was not stoichiometric.

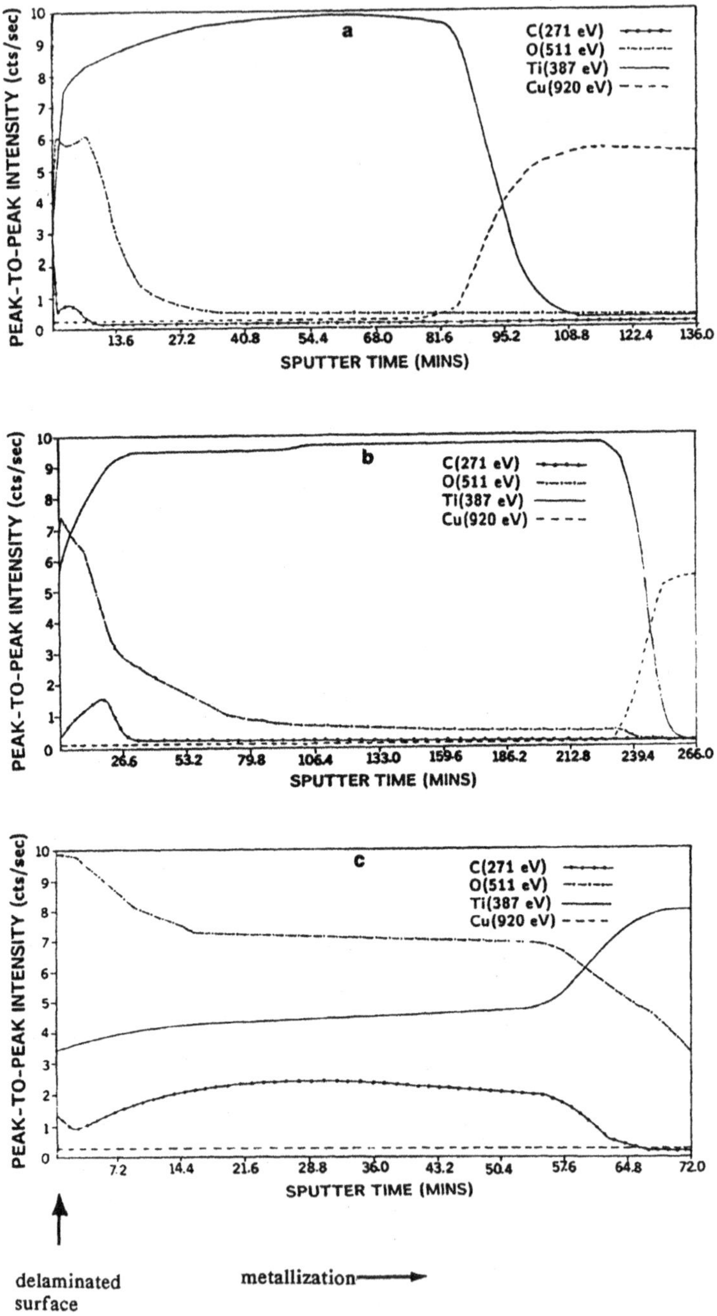

Figure 4. AES depth profiles of principal elements Ti, Cu, C and O from the delaminated surface on the metallization side for Au/Cu/Ti/Polyimide specimen annealed in a) 760 Torr D_2, for 24 hours at 623 K, b) 760 Torr D_2, for 2 hours at 648 K, and c) 760 Torr D_2, for 2 hours at 688 K. Note differences in the sputter time scales.

The Ti-OH bonds (Fig.6a) are formed during the anneal in deuterium gas as a result of diffusion of H_2O contained in the polyimide[16] to the metal/polyimide interface. The diffusion of H_2O is shown by the data in Fig.7 for a titanium/polyimide specimen grown at room temperature without a prior ion-sputtering of the polyimide. The titanium was sputtered up to the titanium/polyimide interface in the AES vacuum system. The specimen was then heated in-situ and the Ti, O, and C AES signals were monitored. Figure 7 shows the rapid rise in the oxygen signal at about 548 K due to the diffusion of H_2O to the interface. A residual gas analyzer also showed the rise in the mass 18 signal at 548 K, consistent with the release of H_2O at this temperature.

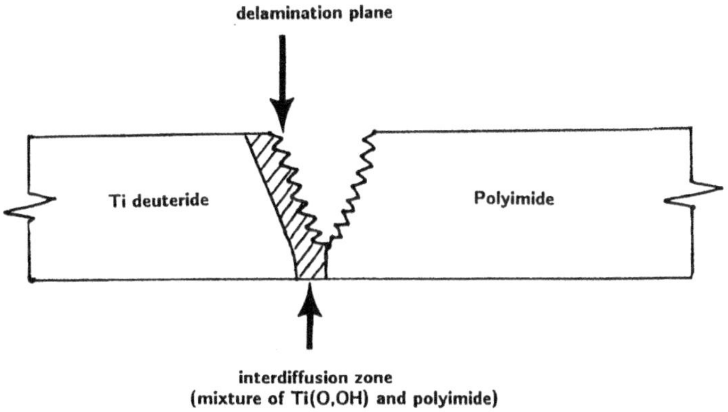

Figure 5. Schematic of the failure plane of the metal/polyimide composite.

The conclusions drawn from the XPS data are complicated by the fact that the specimens were exposed to air during transfer from the hydrogen annealing furnace to the XPS. However, the AES data suggest that the Ti-O and the Ti-OH bonds formed prior to delamination rather than on exposure to the ambient during transfer to the XPS.

None of the samples from the set in which the titanium was deposited at room temperature without first sputtering the polyimide delaminated on annealing in 760 Torr of deuterium up to temperatures of 688 K. Since the titanium formed the deuteride in these specimens, as in the previous series the increased metal/polyimide bonding on annealing in deuterium gas reflects the influence of sputtering. Sputtering of the polyimide reduces the carbonyl content through the breakage of the carbonyl (C=O) bonds and the evolution of oxygen. This phenomenon was observed by others[17,18] as well as in our experiments. The importance of this stems from the fact that it is this functional group that interacts with the titanium atoms to form strong Ti-O type bonds.

In samples on which the metal was deposited at room temperature without sputtering of the substrate, the carbonyl functional groups remained intact thus contributing to an extensive Ti-O-C bond formation. Transformation of the titanium to the deuteride state could not delaminate this strong interfacial bond.

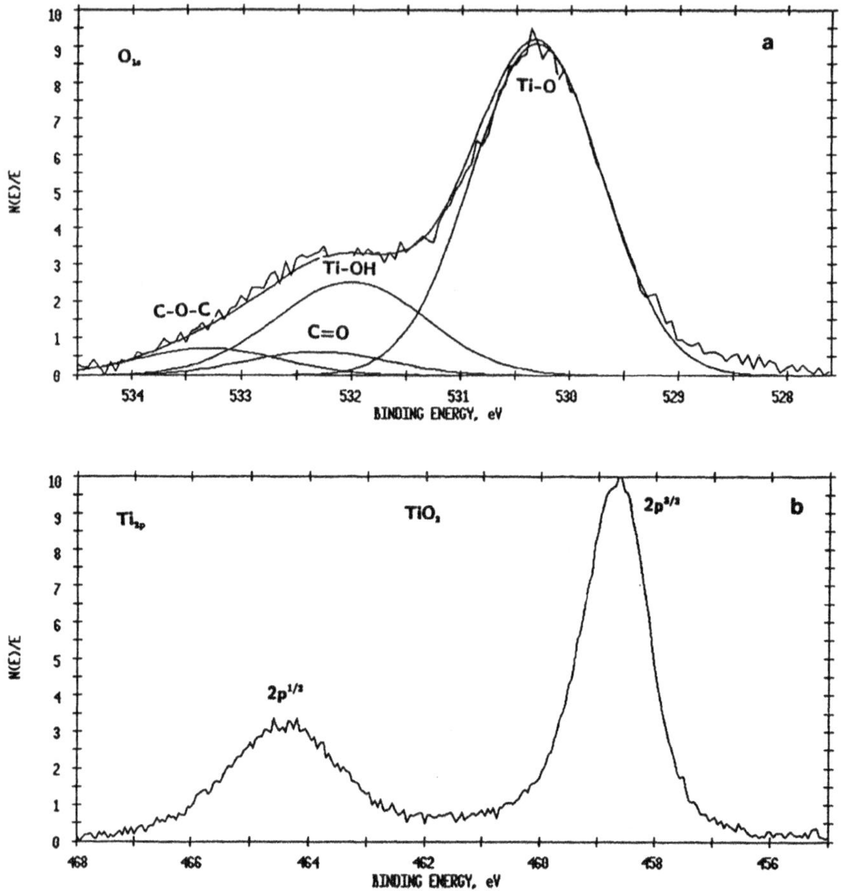

Figure 6. a) XPS O_{1s} spectra from the delaminated surface on the metallization side of the Au/Cu/Ti/PI specimen annealed in 760 Torr D_2, for 2 hours at 688 K, and b) XPS Ti_{2p} spectra on the same surface as in a). The spectra were deconvoluted using the commercial computer software developed by Perkin-Elmer.

CONCLUSIONS

Sputter cleaning of the polyimide substrate has a large effect on the stability of the titanium/polyimide interface during its heat treatment in hydrogen. The sputtering process depletes the oxygen at the surface of the polyimide, leading to fewer Ti-O-C bonds during the formation of the interface and hence to a weaker composite. When such a composite is annealed in a hydrogen ambient, transformation of the titanium to the hydride state creates enough interfacial stress due to volume expansion near the interface, resulting in failure.

Titanium deposition on an unsputtered polyimide substrate is expected to result in a strong interface due to extensive Ti-O-C bond formation. Neither the transformation of the titanium to the hydride state, nor the outdiffusion of water resulted in the composite failure.

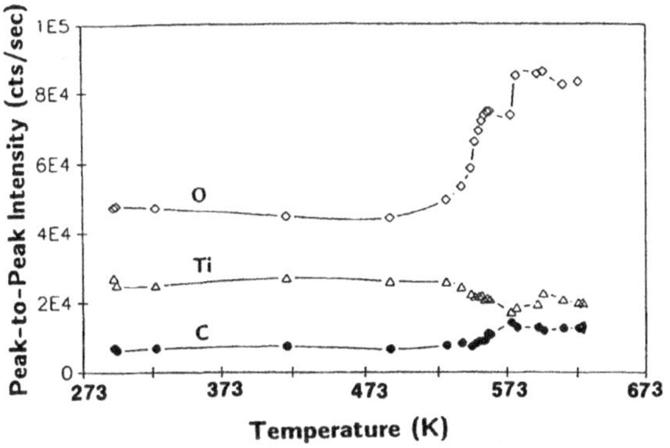

Figure 7. AES peak-to-peak intensity of Ti, O, C at varying temperatures at the titanium/polyimide interface formed at 300 K after sputtering through the titanium to the interface.

ACKNOWLEDGEMENTS

The authors would like to acknowledge the support of Department of Energy through contract DOE AC02-76ER01198. The use of the facilities of the MRL Center for Microanalysis of Materials which is partially supported by Dept. of Energy, Office of Basic Energy Sciences, is also gratefully acknowledged. Also they express their appreciation to Dr. K. Seshan of IBM Corp., East Fishkill Facility for providing some of the samples.

REFERENCES

1) R. J. Jensen, J. P. Cummings, and H. Vora, IEEE Trans., CHMT-7, 384 (1984).
2) D. Gupta, Mater. Res. Soc. Symp. Proc., 47, 11 (1985).
3) R. M. Tromp, F. K. LeGoues, and P. S. Ho, J. Vac. Sci. Tech., A3(3), 782 (1985).
4) Y. H. Kim, G. F. Walker, J. Kim, and J. Park, J. Adhesion Sci. Technol., 1, 331 (1987).
5) R. C. White, R. Haight, B.D. Silvermann, and P. S. Ho, Appl. Phys. Lett., 51(7), 481 (1987).
6) F. S. Ohuchi and S. Freilich, J. Vac. Sci. Tech., A6(3), 1004 (1988).
7) P. N. Sanda, J. W. Bartha, J. G. Clabes, J. L. Jordan, C. Feger, B. D. Silvermann, and P. S. Ho, J. Vac. Sci. Tech., A4(3), 1035 (1986).
8) N. J. Chou, D. W. Dong, J. Kim, and A. C. Liu, J. Electrochem. Soc., 131, 2335 (1984).
9) F. S. Ohuchi and S. C. Freilich, J. Vac. Sci. Tech., A4(3), 1039 (1986).
10) M. J. Goldberg, J. G. Clabes, and C. A. Kovac, J. Vac. Sci. Tech., A6(3), 991 (1988).
11) B. Ladna, C. M. Loxton, and H. K. Birnbaum, Acta Metallurgica, 34, 899 (1986).
12) A. McQuillan, Proc. Roy. Soc. London, Ser. A, 204, 309 (1950).
13) W. Pardee and N. Paton, Metall. Trans., 11A, 1391 (1980).
14) Z. M. Turovtseva and L. L. Kunin, "Analysis of Gases in Metals", Consultants Bureau, New York, 1971.
15) C. N. Panagopoulous, J. Less Common Metals, 134, 237 (1984).

16) A. Wilson, in "Polyimides: Synthesis, Characterization and Applications", K.L. Mittal, editor, vol. 2, p.715, Plenum Press, New York, 1984.
17) P. Bodo and J. E. Sundgren, J. Vac. Sci. Tech., A6(4), 2396 (1988).
18) R. G. Nuzzo, Y. H. Wong, and G. P. Schwartz, Langmuir, 3, 1136 (1987).

PART III. PLASTIC SURFACE MODIFICATIONS

ION BEAM INDUCED CHEMICAL REACTIONS AT POLYMER SURFACES

S. Pignataro and G. Marletta

Dipartimento di Scienze Chimiche
Università di Catania
Viale A.Doria 6, 95125 Catania (Italy)

The interaction of ion beams with polymer surfaces induces physical and chemical modifications which are more and more understood in terms of solid state chemical reactions. Several types of relatively well defined (non-chaotic !) reactions have been shown to occur. These can be classified in a classical way as: dehydrogenation, decarbonilation, decarboxylation, ejection of other stable molecules, reduction of functional groups, loss of aromaticity via ring opening, loss of heteroatoms, backbone rearrangements (chain scission, cross-linking), etc.

The type and the products of the reactions which occur on a given polymeric substrate may be tuned by changing the type, the energy and the fluence of the ions. In comparison to ordinary thermal reactions, ion beam reactions can be confined in a well defined spatial region of the bombarded polymer. The size and the shape of these chemically and, therefore, physically modified regions can be tailored by taking advantage of the focusing capability of the ion beams. This may lead to the patterning of chemical synthesis on polymer surfaces with the consequent patterning of physical characteristics such as conductivity, solubility,optical, and dielectric properties. Relevant examples of such concepts are reported for several polymers, including polyethersulfone, polyimide and polystyrene, bombarded with rare gas ions at energies ranging from 50 to 300 keV.

INTRODUCTION

In the last few years, the study of the effects induced by keV or MeV heavy ions on polymers has attracted growing attention[1,2], both because of the fundamental interest in nonconventional ion beam effects (in compari-

Metallized Plastics 2, Edited by K.L. Mittal
Plenum Press, New York, 1991

son to the "conventional" photon, electron, or neutron effects) and for potential technological applications.

As to the fundamental studies, we may mention that the physics and chemistry of ion-polymer interactions is a very suitable case for revealing eventual non-linear effects due to the energy deposition in the high density excitation limit[3,4]. In particular, in the case of heavy ions the Linear Energy Transfer (LET) is expected to be two or three orders of magnitude higher than for conventional radiation[5], the relative chemical yield (G-value) is expected to be considerably higher but not linearly dependent upon the LET[5,6] and a non-negligible rôle is expected to be played by different energy deposition mechanisms[6-8]. Thus, the fundamental interest in ion-polymer interaction resides in the possibility to use the polymer targets as a very sensitive detector for the energy deposition processes.

As to the technological applications, it has been shown that almost all the relevant properties of polymers can be dramatically modified by ion beams: rheological properties, i.e. solubility and molecular weight distribution[9], electrical conductivity[10,11], optical properties[12], mechanical properties and biocompatibility[13].

A particularly interesting application of ion beam treatments concerns the use of inert or reactive ion beams to promote or improve the adhesion of polymer surfaces to metal films[14-18]. As an example of the typical ion beam induced effect on adhesion, it has been recently shown that 800 eV Ar or N_2 ion milling of polyimide (PMDA-ODA) surfaces increases the peel strength by a factor of 5-6 with respect to the as cured polymer surface[19], the effect being assigned to the increase of the polarity of the ion bombarded surfaces. However, in general the mechanism responsible for the increased metal-polymer adhesion after ion bombardment is not yet established, the various proposals including: the formation of a crosslinked or graphitic layer which may improve the mechanical strength of the surface[19,20], the increase in the surface area (i.e., the roughening of the polymer surface)[20], the formation of chemical sites which bond to the metal or the increase of surface free energy (i.e., the modification of the chemical structure of the polymer surface)[19,21,22] have been proposed.

It is noteworthy that, in spite of such relevant fundamental and technological interest in ion-polymer interactions, the complexity of the chemical effects in these very reactive targets has prevented the understanding of basic mechanisms which have been understood for "conventional" ionizing radiation. In particular, there does not exist a simple, well established pattern of chemical reactions accounting for the large variety of ion-induced chemical effects. For instance, the modification of the chemical state of ion bombarded polymers has been generally discussed in terms of carbonization and/or graphitization of the organic targets[2], paying little attention to the chemical pathways for such processes and to the chemical characterization of such "char" phases.

In the present paper a more detailed discussion and explanation of previous data from this laboratory will be provided. In particular, several types of reactions will be discussed and a correlation of many of these with a critical ion dose regime will be evidenced. Thus, it will be shown that few and relatively well defined reactions (including dehydrogenation, loss of CO_2, ejection of stable hydrocarbon molecules as well as reduction of functional groups, loss of heteroatoms, backbone rearrangements (chain scission and crosslinking), etc.) are produced by ion bombardment in each ion dose regime, the reactions being "ordered" as a function of the total ion dose or, equivalently, as a function of the total deposited energy. Our results strongly suggest that chemical reactions are not simultaneously occurring in a kind of "plasma-like" region along the ion track, but that discrete reactions are successively induced, progressively building-up the new chemical structure of the ion-bombarded polymers. In this sense, as opposed to the "plasma-like" processes, we may define the ion-induced chemical reactions as "non-chaotic" phenomena. In this paper we also propose possible relationships among the various chemical processes and the related property modifications of technological importance. Finally, an example of an ideal maskless patterning of optically and electrically active layers onto the same polymer films is proposed.

EXPERIMENTAL

BTDA-ODA polyimide (PI) obtained as Pyralin 2555 from Du Pont de Nemours, polyethersulphone (PES) and polystyrene (PS) obtained from Aldrich and BDH, respectively, were used in this study. The PES and PS polymers were pure for laboratory purposes. All the polymers were used in the form of films of about 1 μm thickness spin coated on 5-inch Si wafers respectively, from the original solution of the precursor BTDA-ODA polyamic acid in N-methylpyrrolidone and of PES and PS in chloroform. The concentration was adjusted in order to obtain the desired film thickness. A baking of 15 minutes at 90 °C was used to remove the solvent from the films of polyamic acid. The films were then cured to obtain the full imidization as follows: 20' at 140 °C, 90' at 300 °C and 15' at 400 °C in N_2 atmosphere. The complete imidization was monitored by conventional IR technique for the bulk of the films and by XPS for the film surfaces. Only a mild baking of 5 minutes at 40 °C was used for PES and PS films. In each case no traces of Cl were detected on the surface of these films.

The samples were irradiated with rare gas ions (He, Ar,Xe) at 50-300 keV energies in a Varian DF-300 ion implanter at various doses from $5x10^{12}$ up to $1x10^{16}$ ions$\cdot$cm^{-2}. The ion beams were rastered over the whole wafer. The typical beam current densities were between 0.5 and 5.0 μA$\cdot$cm^{-2}.

The ion bombarded samples were characterized by XPS and REELS techniques, using a Kratos ES 300 electron spectrometer, equipped with 1) a twin Mg/Al anode, 2) a Kratos electron gun, operating 50-5000 eV electron beams at variable currents (between 100 nA and 10 μA). Some in situ treatments were performed in the XPS apparatus, using a PHI ion gun (model 04161) operating between 0.5-2.0 keV with typical beam current densities of 5-6 μA$\cdot$cm^{-2} or a 11NF Fast Atom Bombardment (FAB) source of Ion Tech Ltd operating between 1.0-10.0 keV with beam current densities ranging from 6 to 15 μA$\cdot$cm^{-2}. The XPS spectra were referenced to the hydrocarbon component of C 1s peak of polymers, assumed to be at a Binding Energy (B.E.) of 284.8 eV. After the irradiation treatments, in which the B.E. of the most reduced hydrocarbon-like component is expected to be shifted towards lower values, the referencing was performed using 3d peaks of Ag in electrical contact with the irradiated surfaces. Data acquisition and data processing were performed using routine programs on a HP-1000 computer. The Shirley-type non-linear background subtraction was used. Overlapping peaks were resolved by the peak synthesis method based on gaussian components[11].

RESULTS AND DISCUSSION

In previous papers we showed that chemical reactions may be selectively produced on a polymer surface by ion bombardment depending on the fluence range and the energy deposition mechanisms[11,23-27]. These results lead to the concept of "non-chaotic" modification of ion bombarded surfaces. In particular, the experimental results we reported showed that the chemical modification of a polymer target occurs through few basic, successive and well defined chemical modifications mainly involving the two levels of the polymer structure, i.e., a) the chemical structure of the monomer units, and b) the spatial arrangement of the basic polymer units within the individual polymer molecules (the so-called "microstructure").

Figure 1 summarizes the relationship between ion fluence, type of predominant chemical modifications and related physical properties. A relatively ordered succession of the modification events is apparent,

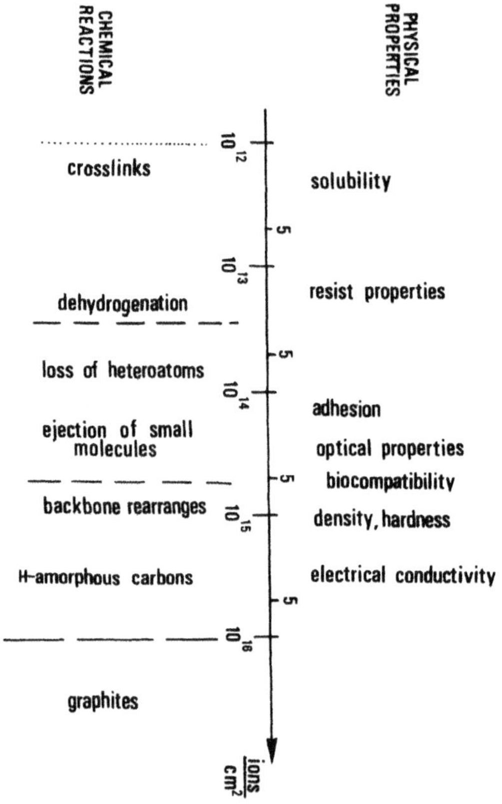

Figure 1 .Qualitative trends of ion fluence vs. typical chemical reactions and related physical properties.

going from very low up to high ion fluences. The point to emphasize based on Figure 1 is that the chemical modification of a polymer target under ion bombardment is no longer related to the average density of energy deposited per unit volume, as in "classical" radiation chemistry with γ or energetic electrons[28], but it is related to the local density of deposited energy in a characteristic volume (ion track) around the physical trajectory of the energetic ion. This is due to the fact that the trajectory of an ion is relatively well defined in comparison to conventional projectiles such as energetic photons or electrons. In fact, in the case of ions the energy absorption process is characterized by a very high density of events and, therefore, a close spatial correlation of such events (Å or tenths of Å) with respect to the large distance between two photon- or electron-induced adsorption sites (hundreds or thousand of Å)[28].

In ion tracks the typical energy released for keV ions on a polymer target may range between 6-7 eV/Å for light ions (H or He) at 10 keV and a few hundreds eV/Å for heavy ions (e.g., about 300 eV/Å for Bi 600 keV on polyethylene) at a few hundreds of keV[29]. Accordingly, the depths of modification range from few tenths up to thousands of Å, depending on the ion energy and mass. In each case the energy deposited per molecule is larger than that required to break several bonds. On the other hand, the typical radius of such ion tracks (which depends on the ion mass and energy) may range between few Å and few tens of Å, thereby defining two crucial fluence regimes: the single track regime, in which there is no overlap between the various modified regions (typically between 10^{11} and 5×10^{13} ions/cm^2) and the track overlap regime, in which each point on the surface has been struck at least once[30] (fluence above 5×10^{13} ions/cm^2).

In Figure 1 the first chemical events observed around 10^{12}-10^{13} ions/cm^2 are the changes in the solubility of bombarded polymers, due to the simultaneous occurrence of dehydrogenation, chain scissioning and crosslinking.

Recent results from our laboratory discuss the relative importance of the different processes of chain scissioning and cross-linking for some polymers[9]. Figure 2 reports the case of monodisperse polystyrene (PS) irradiated with He^+ 100 keV at doses ranging between 10^{12} and 5×10^{13}

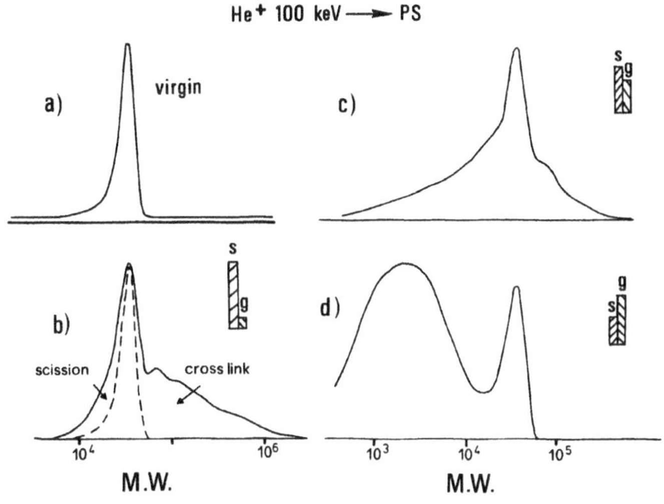

Figure 2 . Gel Permeation Chromatograms of the soluble fraction of polystyrene irradiated with He^+ 100 keV: a) virgin; b) 5×10^{12}; c) 1×10^{13}; and d) 5×10^{13} ions/cm^2. "s" and "g" in figure indicate the soluble fraction and the gel fraction, respectively (reproduced from ref.9 with permission of the authors).

ions/cm². The irradiated polymer was dissolved in chloroform, and the solution was analysed with a Gel Permeation Chromatographic technique (GPC), while the gel fraction was simultaneously estimated. The very sharp shape of the Molecular Weight Distribution for such monodisperse polystyrene allowed the unequivocal identification of the two opposing processes of chain scission and intermolecular crosslinking at various doses. The crosslinking process is shown to dominate, depleting the solution of the high M.W. fractions and yielding finally the insoluble gel fraction; while the low M.W. fraction due to the chain scission process is seemingly increasing in the residual soluble fraction. It is noteworthy that also when most of the irradiated polymer layer is made insoluble, an appreciable fraction of the original M.W. fraction is still present in the solution. This fact, together with the observation that the GPC trace shows a predominance of the dimers and trimers of the original M.W., is an indication of the importance of the intramolecular cross-linking processes, which do not affect the M.W. distribution (persistence of original M.W.) or affect it very little (dimers or trimers).

The phenomenon described above typically occurs and is completed in the fluence range previously characterized as the single track regime. In terms of the polymer structure, we may state that the ion bombardment in this fluence range mainly affects the microstructure, i.e. the spatial arrangement of the polymer units and the macrostructure, i.e. the sol-gel fraction.

The transition from the single track to the track overlap regime for projectile of keV energy arises between the fluence of 10^{13} and 10^{14} ions/cm². In this fluence range the chemical structure of the monomer units is heavily modified. Few basic modification patterns are observed[23-27]: modification of the functional groups either involving the modification of the chemical state or the loss of heteroatoms present such as

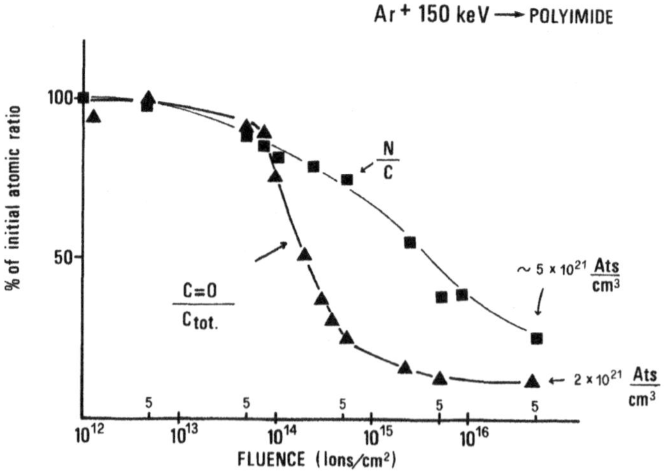

Figure 3. Percent of residual C=O groups and N atoms in BTDA-ODA polyimide bombarded with Ar^+ 150 keV as a function of ion fluence. The data are obtained from the analysis of XPS O 1s band and N 1s bands for the bombarded surfaces. The data are normalized to the initial ratios.

oxygen, nitrogen, and sulphur; the destruction of the aromaticity via the opening of the present aromatic rings, or their partial hydrogenation; extensive processes of backbone rearrangement and crosslinking, leading to a densification of the irradiated polymer layer; and the formation of volatile molecular species, due to the fragmentation of the monomeric structure.

Figure 3 presents an example of typical depletion reactions, induced in polyimide (BTDA-ODA) with Ar^+ 150 keV, and their effects on the carbonyl groups and the imidic nitrogen. The ion-beam induced carbonyl depletion from the very stable imide moiety has been recently explained in terms of the imide ring isomerization to an isoimidic form and a subsequent easy elimination of CO_2[11]. This explanation was based on XPS data, showing that the depletion of the carbonyl is related to the simultaneous growth of a nitrogen site having a component at 399.3 eV in the XPS N 1s band, in agreement with the formation of a C=N- bond expected for the isoimidic structure[31] as well as for the decarboxylated species[11,23]. It is to emphasize that, while the depletion of heteroatoms and the related processes of backbone rearrangement are usually defined as "carbonization", i.e., a violent process, leading essentially to a graphite-like phase losing any memory of its initial chemical structure[2], the data in figure 3 and the suggested chemical mechanism for depletion of C=O in no way suggest that such a drastic process is occurring during ion bombardment. In fact, we can observe as much as 65% of the initial N and about 20% of the initial C=O groups still present at the polymer surface after a bombardment fluence of $1x10^{15}$ ions/cm^2, when the polymer starts to show an ohmic conductivity. It is also noteworthy that at very high fluence ($5x10^{16}$ ions/cm^2), when each point on the surface has been struck on average of 100 times, concentrations of $5x10^{21}$ (N atoms) and $2x10^{21}$ (C=O groups) per cm^3 are detectable on the bombarded surface.

Figure 3 also contains other important informations, concerning the different behaviours observed respectively for the depletion of the C=O groups and N atoms. In fact, the depletion trend of C=O group shows a sharp threshold, and it is substantially completed (close to the saturation value) in a relatively narrow fluence range ($5x10^{13}$-$5x10^{14}$ ions/cm^2). On the contrary, the depletion trend of the N atoms is a very slow process (on the fluence scale) and no saturation point is observed. The two different trends may be explained in terms of processes activated by two different energy deposition mechanisms, i.e., the abrupt depletion of the carbonyl can be attributed mainly to the electronic part of ion energy deposition, while the N depletion can be explained as due to the collisional part of the energy deposition. The problem may be related to the effect of the preferred channel for dissipation of the deposited energy. In particular, according to the Lindhard-Scharff-Schiott approach[30], an incoming ion may deposit energy mainly through two mechanisms, i.e., by interaction with the electronic system of the target (electronic term) or by direct successive collisions with the target atoms (collisional term). In turn the following fundamental events may be produced:

-- electronic interaction of atoms or molecules, ranging from the production of an electronically excited state to the production of ionized-excited states.
-- collision-induced breaking of molecular bonds in a random way inside the solid, with the production of free radicals, ions, etc.

Each ion-solid interaction involves both types of energy deposition mechanisms and, therefore, both the classes of reactions are expected along the total ion path inside the polymer, with a different relative importance due to the fact that at higher energies (i.e. at the beginning of the ion track) the electronic interactions dominate, while at lower energies (i.e., at the end of the ion track) the collisional interactions are dominant.

Let us discuss briefly these statements for the present case. From data in figure 3 we can evaluate the effective cross sections respectively for the C=O and for the N depletion processes. Taking into account the fluence at which 50% of C=O and N are depleted, we find for the C=O depletion a value around $2x10^{14}$ ions/cm^2 corresponding to an effective track radius which is at least fourfold (5.1 Å) of that obtained for N depletion (around $3x10^{15}$ ions/cm^2) corresponding to an effective track radius of 1.2 Å. Therefore, the C=O depletion is suggested to be due to an effective transfer of energy away from the core of the ion track, typically related to the secondary electrons or to the migration along the monomer units of electronically excited states, i.e. to the electronic terms of energy deposition.

On the contrary, the process of N depletion may occur only in the case of an effective pathway of production of N or N_2 species, to be seen by XPS as modification of the N/C atomic ratio. Such kind of processes have to involve the formation of -N=N- systems at the expense of the imide moiety, probably through an interchain reaction, followed (at fluences of the order of 10^{15} ions·cm^{-2}) by the N_2 ejection and formation of an amorphous hydrogenated network. In such a process the rupture of the imidic ring is needed. This rupture is energetically unfavourable via electronic excitations, as demonstrated for instance by the very high resistance of polyimides to electron irradiation[32], but it may occur easily through a collisional event either between the primary ion and a carbon atom or between recoiling carbon atoms in the collision cascade (so-called "nuclear" term of energy deposition). Therefore, the high fluence needed for N depletion and the high stability of imide rings strongly suggest that the N-loss occurs within the area affected by primary collisional events.

Summarizing, chemical reactions may occur at different rates, in different fluence ranges, due to the different effects of the ion energy deposition mechanisms. Accordingly, analogous phenomena of dependence of ion-induced chemical reactions upon the energy deposition mechanisms have also been observed for reactions in PES ,poly-2-vinylpyridine (PVPY), polyvinylcarbazole (PVK) and PS[23-27], as well as in frozen benzene and cyclohexane[6-8].

It should be pointed out that the fluence dependence of ion-induced reactions is governed by the total deposited ion energy, depending on the ion mass and energy. Large amount of total deposited energy may accelerate the reaction rate as well as temperature effects induced by the ion beams. Figure 4 shows as an example the rate of S depletion in PES irradiated with Xe^+ 100 keV (total energy deposited of the order of 230 eV/Å) and He^+ 200 keV (total energy deposited of the order of 24 eV/Å). The relative depletion traces are shifted in the fluence range of less than one order of magnitude, indicating that Xe bombardment is more effective than He bombardment. However, the shift in fluence is not proportional to the total deposited energy difference, rather, as it has been demonstrated in a related paper[27], the reaction rate is governed almost exclusively by the electronic term.

The ion bombardment at fluence higher than 10^{15} ions/cm^2 strongly modifies the primary chemical structure of the polymer units. This fluence range is characterized as "reimplantation" regime[27], because each point of the surface is repeatedly struck by successive ions. In this fluence regime highly reactive species such as aromatic radicals, trapped fragments, electronically excited species, etc. may react with each other, producing a very complex situation. The physical properties which are most affected in this fluence range are the electrical conductivity, the optical response of the materials and the modification of the surface morphology (i.e., texturization).

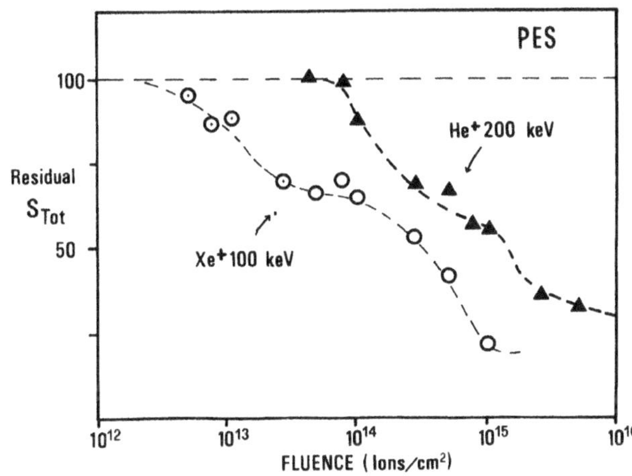

Figure 4. Percent of residual sulfur in PES bombarded using He^+ 200 keV and Xe^+ 100 keV as a function of the fluence.

One of the most interesting application of the ion beam modification of polymers may be related to the modification of the above described physical properties in spatially defined regions. Let us describe an ideal experiment of maskless patterning. Figure 5 shows the idealized scheme of a device coupling optical and electrical functions. Electrically conducting and optically active strips of micrometer dimensions are patterned onto an insulating polymer film. We would like to show preliminary results demonstrating that this kind of selective modifications may be obtained for the same polymer just by changing the relative ion fluence.

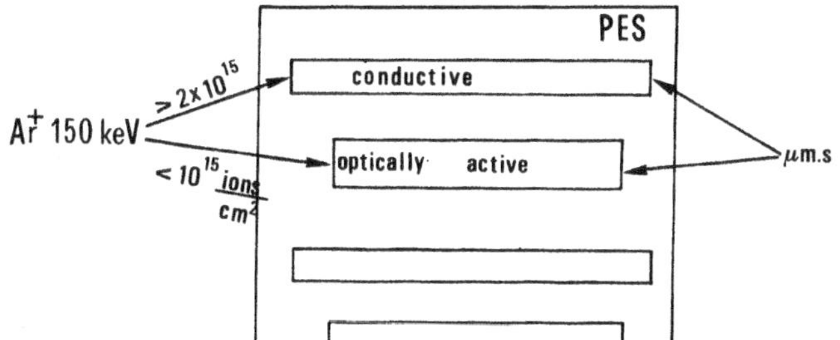

Figure 5. Scheme of the "ideal" maskless patterning of optically and electrically active regions (of micrometers dimensions), produced by different fluences of ion bombardment.

In particular, the chemical structure and the optical properties of a polymer can be modified in a substantial way by ion bombardment. Figure 6 shows the Energy Loss function ($-\mathrm{Im}\ 1/\varepsilon$) derived with standard procedure of background subtraction and elastic peak removal[33] from the Reflection Electron Energy Loss Spectra (REELS) obtained for PES before and after bombardment with Ar^+ 150 keV at a fluence of 1×10^{15} ions/cm^2. The Energy Loss function is dominated by the excitation structure at about 6 eV, due to the $\pi \rightarrow \pi^*$ excitations involving the phenyl rings, and by the large structure due to $\sigma + \pi$ electron excitation (bulk plasmon) at about 21.8 eV before the bombardment and at 23.6 eV at 1×10^{15} ions/cm^2. The shift of the bulk plasmon indicates that the density of valence electrons (and, in turn, the macroscopic density of the irradiated layers) of the bombarded polymer is strongly increased. The intensity decrease of the structure at 6 eV is indicative of the partial ion bombardment induced destruction of the original phenyl rings. Also reported in Figure 6 are the real and imaginary parts of dielectric constant obtained from the Energy Loss Function for the two samples. Such dielectric functions, in turn, determine the refractive index and the reflectance (or the absorbance) of the sampled layers[34].

We may note that there is a dramatic modification of the complex dielectric constants with ion bombardment fluence, i.e., a dramatic change in the expected refractive index and in the related optical properties. The important point is that the bombarded layer still keeps a good degree of transparency,a mandatory condition for optical use[35].

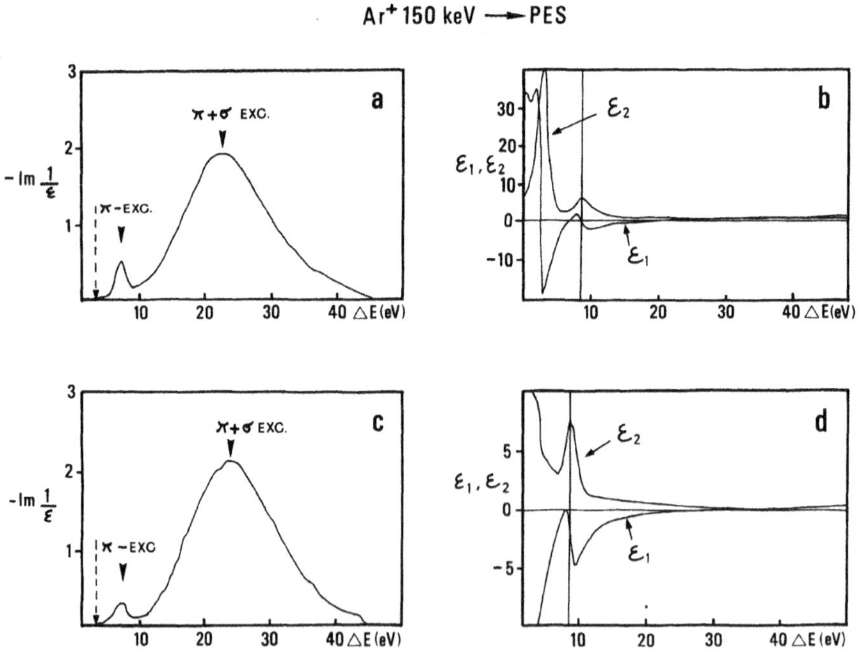

Figure 6. a) Energy loss function ($-\mathrm{Im}\ 1/\varepsilon$) and b) complex dielectric constants obtained from REELS spectra of PES before bombardment and c) and d) after bombardment with Ar^+ 150 keV at a fluence of 1×10^{15} ions/cm^2. The energy of the primary electron beam (E_p) was 1200 eV, with a FWHM of the elastic peak of 0.9 eV.

The bombarded samples do not exhibit any appreciable conductivity (note the high band gap).

Figure 7 shows the low-excitation energy region of the raw REELS spectra for PES bombarded with Ar^+ 150 keV at fluences of $1x10^{15}$, $2.5x10^{15}$ and $5x10^{15}$ ions/cm^2. The samples bombarded at $2.5x10^{15}$ and $5x10^{15}$ ions/cm^2 behave as ohmic conductors (10^{-1}-10^{-2} $\Omega\cdot$cm). The effects reported in Figure 7 concern the filling of the conduction band. In particular, we can observe the transition from a situation in which the low-excitation energy tail is determined by the interband transitions ($\pi \longrightarrow \pi^*$) or by transitions from defect states in the band gap to electronic states in the conduction band ($D,D^* \longrightarrow \pi^*$) to the situation in which the excitation tail is dominated by intraband transitions within the partially filled conduction band. These transitions are responsible for the characteristic exponentially decaying tail (Friedel excitonic tail)[35].

Therefore, in a very narrow fluence range, on the same originally insulating polymer film, optically modified polymeric strips can be produced together with ohmically conducting polymer strips just using focused ion beams and controlling the bombarding ion doses.

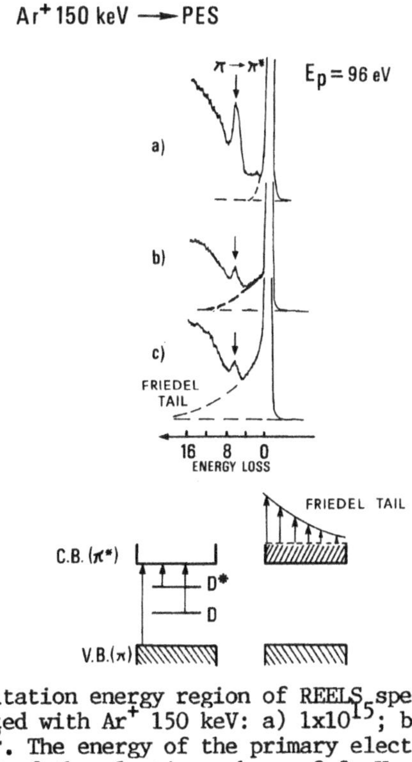

Figure 7. Low-excitation energy region of REELS spectra obtained for PES irradiated with Ar^+ 150 keV: a) $1x10^{15}$; b) $2.5x10^{15}$; c) $5x10^{15}$ ions/cm^2. The energy of the primary electron beam was 96 eV. The FWHM of the elastic peak was 0.3 eV. At the bottom are reported details of the possible interband (left side) and intraband (right side) transitions.

CONCLUSIONS

Ion bombardment induces well defined reactions in polymers involving the gradual modification of the original chemical structure. The polymer structure is gradually modified both at the level of primary chemical structure of monomer units and at the level of the chain structure within well defined fluence (or energy) ranges, characteristic of each type of chemical reaction. The rates of the different kinds of reactions and the fluence and energy ranges in which these are completed mainly depend on the predominant energy deposition mechanisms, as well as on the nature of the intermediates involved.

The ion bombarded polymers have been shown to undergo a continuous evolution of the primary chemical structure to a "final" carbonaceous material still keeping "memory" of its primary stoichiometry and chemical structure. Some examples have been reported of optically and/or electrically modified layers, which can be obtained in a relatively narrow fluence range on the same polymer film by just tuning the ion fluence. This paper presented some evidence of the occurrence of a "mild", "non-chaotic" chemistry and a contribution to the rationalization of the large and still ill-defined ion-induced phenomenology in polymers.

ACKNOWLEDGMENTS

The authors thank Dr.F.Iacona (IMETEM-CNR, Catania, Italy) for help with experimental work and Dr.C.Oliveri (SGS-Thomson, Catania, Italy) for assistance with ion implantation facilities. MPI (Rome) and CNR (Rome) are gratefully acknowledged for financial support.

REFERENCES

1. W.L.Brown, Nucl.Instrum. Methods, B 37/38,270(1989).
2. T.Venkatesan,L.Calcagno,B.S.Elman and G.Foti, in "Ion Beam Modification of Insulators", P.Mazzoldi and G.Arnold, Editors, p.301, Elsevier,Amsterdam,1987.
3. S.Tagawa, Proceedings of the 12th International Symposium on Ion-Assisted Technology (ISIAT), T.Takagi, Editor, p.475, Kyoto University Press, Tokyo, June 5-7 1989.
4. W.L.Brown, Radiation Effects, 90, 115 (1986).
5. S.Tagawa,N.Kouchi,Y.Aochi and H.Shibata; "Adv. in Resist Technology and Processing III", SPIE Symp., 22 , 631 (1986).
6. O.Puglisi,G.Marletta,G.Foti and S.Pignataro, Chem.Phys., 75, 417 (1983).
7. O.Puglisi,G.Marletta,S.Pignataro,G.Foti,A.Trovato and E.Rimini, Chem.Phys.Letters,78, 207 (1981).
8. O.Puglisi,G.Marletta,A.Torrisi,L.Torrisi and G.Foti, Radiation Effects, 65, 11 (1982).
9. O.Puglisi, Mater.Sci.Eng., B 2, 167 (1989); and references therein.
10. a)T.Venkatesan, Nucl.Instrum.Methods, B 7/8, 461 (1985). b) T.Hioki, S.Noda, M.Sugiura,M.Kakeno,K.Yamada and J.Kawamoto, Appl.Phys. Letters, 43, 30 (1983); c) S.Noda and T.Hioki, Carbon, 22, 359 (1984); d) J.Davenas,X.L.Xu,G.Boiteux and D.Sage, Nucl.Instrum. Methods B 39, 754 (1989).
11. G.Marletta,C.Oliveri,G.Ferla and S.Pignataro, Surface Interface Anal., 12, 447 (1988).
12. a)J.Davenas,G.Boiteux,X.L.Xu and E.Adem,Nucl.Instrum.Methods, B 32, 136 (1988); b)D.Fink,M.Muller,L.T.Chadderton,P.H.Cannington, R.G.Elliman and D.C.McDonald, Nucl.Instrum.Methods,B 32, 125 (1988).
13. a) J.D.Carlson,J.E.Bares,A.M.Guzman and P.P.Pronko, Nucl. Instrum.

Methods,B 7/8, 507 (1985); b)Y.Suzuki,M.Kusakabe,M.Iwaki and M.Suzuki,Nucl.Instrum.Methods, B 32, 120 (1988).
14. P.Bodo and J.E.Sundgren, J.Vac.Sci.Technol., A 2, 1498 (1984).
15. T.A.Tombrello, Mater.Res.Soc. Symp. Proc., 25, 173 (1984).
16. J.E.E.Baglin, A.G.Schrott, R.D.Thompson, K.N.Tu and A.Segmuller, Nucl.Instrum. Methods, B 19/20, 782 (1987).
17. I.H.Loh and J.K.Hirvonen, Mater.Res.Soc. Symp. Proc., 108, 241 (1988).
18. P.A.Ingemarsson, Nucl.Instrum.Methods, B 44, 437 (1990).
19. R.Flitsch and D.Y.Shih, J.Vac.Sci.Technol., A 8, 2376 (1990).
20. A.J.Kinloch, Adhesion and Adhesives, p.129, Chapman and Hall, New York, 1987.
21. R.G.Nuzzo, Y.H.Wong and G.P.Schwartz, Langmuir, 3, 1136 (1987).
22. B.J.Bachman and M.J.Vasile, J.Vac.Sci.Technol., A 7, 2709 (1989).
23. G.Marletta,C.Oliveri and S.Pignataro, Nucl.Instrum.Methods,B 39, 792 (1989).
24. G.Marletta,C.Oliveri and S.Pignataro, Nucl.Instrum.Methods,B 39, 773 (1989).
25. G.Marletta,A.Licciardello,L.Calcagno and G.Foti, Nucl. Instrum. Methods, B 37/38, 712 (1989).
26. G.Compagnini,R.Reitano,L.Calcagno,G.Marletta and G.Foti, Appl.Surface Sci., 43, 228 (1989).
27. G.Marletta, Nucl.Instrum.Methods, B 46, 295 (1990).
28. a)W.Schnabel, in "Aspects of Degradation and Stabilization of Polymers", H.H.G.Jellinek, Editor, p.149, Amsterdam, Elsevier,1978; b)M.Hagiwara and T.Kagiya, in "Degradation and Stabilization of Polymers", H.H.G.Jellinek, Editor, Vol.1,p.358 Amsterdam, Elsevier, 1983.
29. J.P.Biersack in "Ion Beam Modification of Insulators", P.Mazzoldi and G.Arnold, Editors, p.301,Elsevierer,Amsterdam,1987.
30) J.A.Davies,in "Surface Modification and Alloying by Laser, Ions and Electron Beams",J.M.Poate,G.Foti and D.C. Jacobson, Editors, p.189, Plenum Press, New York, 1983.
31. P.L.Buchwalter and A.I.Baise, in "Polyimides:Synthesis, Characterization and Applications", K.L.Mittal, Editor,Vol.1, p.537, Plenum Press, New York,1984.
32. J.E.Ferl and E.R.Long,Jr., IEEE Trans. NS-28, 4119 (1981).
33. G.Mondio, F.Neri, A.Arena, G.Marletta and F.Iacona, Surface Sci., in press.
34. H.Raether,"Excitation of Plasmon and Interband Transitions by Electrons", Springer Tracts in Modern Physics, vol.88, Springer, New York, 1979.
35. G.Mondio, G.Marletta and G.Foti, Appl.Phys.Letters, submitted.

SURFACE PRE-TREATMENT AND METAL COATING OF COMMERCIAL POLYIMIDE STUDIED BY SURFACE ANALYTICAL TECHNIQUES

K. Horn, A. Ortega, K. Doblhofer, A.M. Bradshaw, R. Schulz*, and H. Meyer*

Fritz-Haber-Institut der Max-Planck-Gesellschaft
D-1000 Berlin 33, Germany
*Schering AG, Dept. GT-FO, D-1000 Berlin 65, Germany

The influence of the chemical pre-treatment of commercial polyimide foil on the adhesion of copper deposited by electroless and electrochemical processes has been studied using SEM, XPS, EDX, and IR spectroscopy. The results show that a thin surface layer, chemically different from the pure material, is important for metal adhesion. This layer, which can be influenced by chemical etching, exhibits distinctly different features in the x-ray photoelectron spectrum.

INTRODUCTION

Polyimides (PI) are high performance polymers with excellent thermal and chemical stability, as well as outstanding electrical properties, and are of growing importance as insulating materials in the electronics industry. For such applications, it is important to characterize the chemistry of the metal-polymer interface and to understand the processes which influence the adhesion between metal and polymer[1]. In the present paper we report on surface and interface studies of commercial PI foils under different conditions of pretreatment and after coating with copper using different deposition processes. We have used scanning electron microscopy (SEM), x-ray photoelectron spectroscopy (XPS), electron microprobe (EDX), and infrared (IR) spectroscopy. We compare our results from as-received commercial samples with previous studies on samples prepared under better defined conditions.

EXPERIMENTAL

Several instruments were used for the experiments. SEM pictures and EDX spectra were recorded on an ISI-60 electron microscope (ISI Corp. Japan) at electron beam energies of 10 and 30 keV, using a Si(Li) detector. In separate runs, SEM pictures were recorded with a Hitachi 4000 X scanning electron microscope yielding superior resolution at about 200 Å, and capable of operation at beam energies down to 500 eV. Core level photoemission was carried out in a VG photoelectron spectrometer equipped with an Al Kα/Mg Kα twin x-ray anode, a hemispherical electron energy analyser and an electron gun for charge compensation. Spectra were recorded at a combined energy resolution of about 0.8 eV. The samples were introduced into the vacuum chamber via a fast load lock. The elapsed time between specimen treatment (including chemical treatment) was less than two hours in each case. Operating pressure during measurements was about 7×10^{-9} mbar. The infrared transmission and reflection-absorption spectra were recorded using a spectrometer consisting of a glow-bar as infrared source, a circular variable interference filter (OCLI Corp. USA) as monochromator, and a HgCdTe infrared detector (Santa Barbara Research Center, USA); this instrument is described in detail elsewhere[2]. In the majority of the experiments reported here polyimide foils (PMDA-ODA, KAPTON [9]) of 50 μm thickness were used.

RESULTS AND DISCUSSION

XPS studies of polymer surfaces and interfaces can provide a detailed understanding of the chemical processes which influence adhesion from an analysis of chemical shifts on the relevant core level lines. They also enable the detection of small quantities of material in the surface or interface region because of the small sampling depth of this technique. There have already been a number of ESCA studies of the interaction of metals with polyimide surface[3-8]. Some of these were carried out on polyimide samples which were prepared by spinning a solution of polyamic acid onto a suitable substrate (e.g. a silicon wafer), with subsequent imidization in high vacuum by heating to 360°C, or by vapor deposition of thin films onto metallic substrates. Thin layers of metals were then evaporated onto these surfaces under high or ultrahigh vacuum conditions, with the aim of detecting changes in the C 1s, O 1s, and N 1s core level lines. While such model studies are very important for a basic understanding of the chemical interactions which may take place in these systems, the reaction of metals deposited through chemical and electrochemical deposition techniques onto commercial polyimide foils may not be directly comparable with the samples used in such studies. In the applications of polyimide in electronics, coating with copper is usually performed by electrochemical plating; for this purpose, the pre-deposition of small amounts of palladium has proven most successful. After an initial cleaning step, the commercial samples are exposed

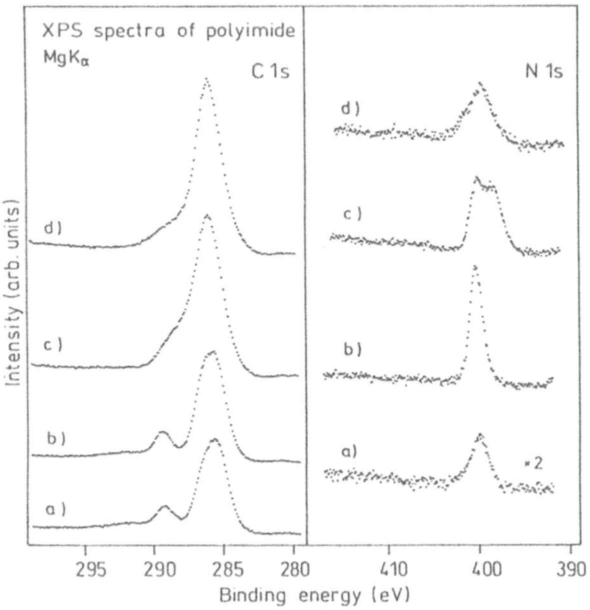

Figure 1. XPS spectra of the carbon 1s and nitrogen 1s levels, from a commercial polyimide (KAPTON) foil (a), a peeled-off copper surface with a polyimide residue (b), a PI foil after chemical etching by $KMnO_4$ (c) and after treatment with an agent containing palladium complexes (d).

to an alkaline activator containing palladium complexes. This step leads to a deposition of small amounts of Pd on the surface, which cause strong adhesion of the electroless deposited Cu layer deposited in the next step. Finally, a thick Cu layer is electrodeposited onto the substrate. Obviously, the surface of a commercial foil subjected to this treatment may deviate significantly from polymer surfaces prepared by spin-coating.

Consider the XPS spectra of the carbon 1s and nitrogen 1s regions of a polyimide foil in Figure 1. The bottom spectra (a) are from an as-received commercial foil[9]; the familiar spectrum of PI can be recognized, with a broad peak around 285 eV binding energy, to which the aromatic rings and other hydrocarbon constituents of PI contribute, and the split-off peak due to the carbon atoms in the carboxyl groups at around 287.5 eV. The N 1s region shows only one peak, which is plausible since there is only one nitrogen atom. Spectra (b) are from the "inside" surface of a copper-polyimide composite; here, the polyimide was peeled off from the thick (about 30 µm) copper layer. We have previously shown[10] that cohesion failure inside the polyimide layer rather than adhesion failure at the copper-PI interface occurs, and that thick (several thousand Å) PI layers remain on the copper surface after the peeling process. The spectra (b) were taken from such a residue, which is likely to reflect the composition of the PI foil in the bulk. The spectrum from this sample has the additional advantage that charging does not present a problem, so that a common energy reference can be used. In the C 1s region, the spectral features are quite similar in (a) and (b), except that the main peak shows a much more pronounced asymmetry indicative of the two separate lines, which are not quite resolved under our experimental conditions (ref. 4 shows a spectrum in which, by using synchrotron radiation, the peaks are actually resolved). The relative intensities of the main peak and the C=O peak in (a) and (b) are very similar, however. This is not true for the N 1s region, where the peak from the peeled-off PI surface is four times more intense. This result points to a loss of nitrogen in the topmost surface region, in line with the view that the surface composition of this commercial PI foil is indeed different from that of the bulk. No other contamination of the untreated foil (a) was evident from a wide-scan overview spectra recorded before each separate run. Spectra (c) were taken after a 5 min. chemical etch of the PI foil in an alkaline agent containing potassium permanganate, and show obvious changes. The C 1s region is now much better defined, with one broad peak in the 285 eV region and a badly defined shoulder where the carboxyl carbon signal was so clearly split-off in spectra (a) and (b). However, the biggest difference occurs in the N 1s region: here, an extra peak shifted by about 2 eV towards lower binding energy is now seen. This result indicates that the nitrogen atoms are most strongly affected by the etching agent. One possibility might be that the bond between the nitrogen atom and the carbon atom in the carboxyl group of the imide ring is opened due to hydrolysis, giving rise to a carboxylic acid group and to a N-H bond at the nitrogen site. Such hydrolysis reactions are known to occur in basic environments. The other possibility might be a severing of the polymer chains at the nitrogen sites. From the sign of the binding energy shift the nitrogen atom is now bonded to a more electropositive species.

Figure 2. Scanning electron micrograph of a polyimide foil having been exposed to the activating agent; electron beam energy 800 eV, magnification as indicated. The spot-like structures, not seen in the micrographs of the clean PI foil, are due to Pd clusters deposited on the surface by the activation process.

The final spectrum represents a PI surface etched with potassium permanganate and other proprietary etching agents, and subsequently exposed to an alkaline activator containing palladium complexes (NEOGANTH™ [12]). There is little influence of this process on the carbon 1s region of the spectrum, but the extra peak on the N 1s has now disappeared, and a broad structure indicative of a multitude of chemical environments exists for the nitrogen atoms; the relative intensity of the N 1s signal has also decreased considerably. The difference in the N 1s XPS line shape indicates that it is again the nitrogen atom which is involved in the bonding with the Pd deposit. Further studies of the chemical pre-treatment steps are planned and it is hoped to clarify the influence of each step separately.

In a recent study of uv laser-induced polyimide surface degradation by XPS, Chtaib et al.[11] have found similar changes in the carbon and nitrogen 1s XPS lines for high laser fluences for which photodecomposition is expected. In particular, a new line at lower binding energy on the N 1s peak was observed. Chtaib et al. present a model of laser-induced surface reaction with the ambient atmosphere which is quite similar to the explanation given above in terms of imide ring breaking by hydrolysis.

Since the palladium layer deposited on top of the PI foil is the principal agent for subsequent chemical copper deposition, it is of interest to examine its structure using surface sensitive techniques. Figure 2 shows a scanning electron micrograph of a polyimide foil exposed to the activating agent[12]. The micrograph, taken at a beam energy of 800 eV, reveals metallic clusters with an average size of about 200 to 1500 Å . This picture was recorded without any deposition of gold or carbon for the elimination of charging, such as is commonly carried out in scanning electron microscopy. The low beam energy enabled us to determine the morphology of the Pd deposit. Recently, the structure of Pd particles on highly oriented pyrolytic graphite (HOPG) deposited by the same activation process[12] has been studied in scanning tunneling microscopy (STM) by Besenhard et al.[13]. The authors observe clusters of Pd with a size distribution between 10 and 50 Å, smaller than those found here. Besenhard et al. were able to distinguish between the clusters and the substrate on account of the superior resolution of the STM, enabling single atoms of the substrate to be resolved, whereas the structures seen in the SEM picture of Figure 2 may to some extent also be due to the PI substrate; different sizes of Pd particles on HOPG and PI are also possible, on the other hand. The presence of a metallic species on the surface can be directly inferred from the picture. XPS spectra of the Pd 3d level (not shown here) show a characteristic asymmetry, also indicating that the majority of the Pd is present in metallic form on the surface, probably in a layer about 15 to 25 Å thick. This does not exclude the possibility of local clustering, however. Some of the Pd atoms form bonds with the substrate, but they will not be detected in the Pd signal because of the dominance of the metallic emission.

Spectrum (b) of Figure 1 has demonstrated the existence of a polyimide residue on the copper surface exposed by peeling. This residue is also apparent from infrared (IR) transmission and reflection-absorption experiments. IR spectroscopy is a standard method for the investigation of the chemical bonds in materials, and a large body of reference data for polymers exists[14]. However, this technique has not been widely used for metal-polymer interface studies, although a different kind of vibrational spectroscopy, high resolution electron

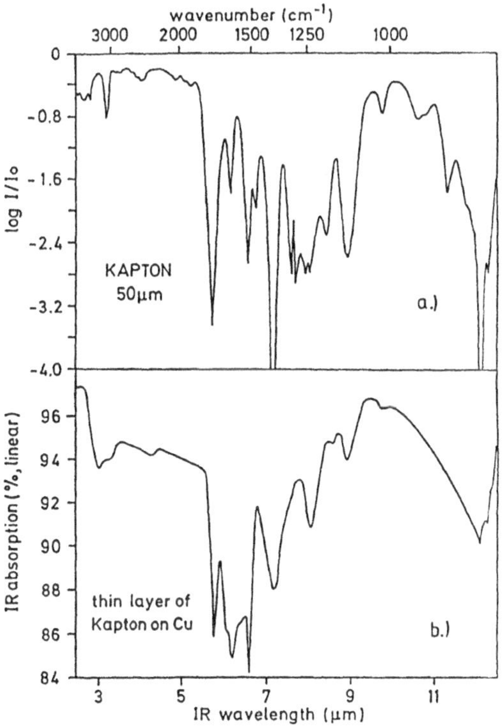

Figure 3: a) IR transmission spectrum from a 50 µm polyimide foil, in the region from about 800 to 3500 cm^{-1}. b) IR reflection-absorption spectrum of thin polyimide residue on a copper coating, produced by peeling off the polyimide foil. Note the linear % absorption scale.

energy loss spectroscopy, has been used for study of surfaces of spin-coated polyimide[15]. We have investigated the transmission as well as the reflection-absorption spectra of the bulk foil and the polyimide residues on top of the peeled-off copper surfaces. A transmission spectrum of a 50 µm thick polyimide foil is shown in the upper part of Figure 3. Note the logarithmic intensity scale, which allowed us to determine the precise wavelength of all but the most intense absorption bands. The strongest bands in the spectra are assigned to the carbon-oxygen vibration at 1724 cm^{-1}, the carbon-carbon stretching vibration in the aromatic rings (1504 cm^{-1}), and the carbon-hydrogen bending vibration (825 cm^{-1}). In the region around 3000 to 3400 cm^{-1} bands due to water and N-H groups can be observed, although some structure here may be due to the aromatic C-H stretching vibrations.

The lower spectrum of Figure 3 shows an IR reflection-absorption spectrum from a peeled-off copper surface. Note the linear transmittance scale in this part of the Figure. All major bands from the polyimide are also present here, proving that the peeling process does indeed leave a polyimide residue on the surface. The relative intensities of several bands are, however, different from those in the upper spectrum. This indicates that the chemical composition of this surface layer differs from the bulk, in agreement with the XPS spectra shown in Figure 1. As described above, such compositional changes are expected to result from the cleaning, pretreatment/con-ditioning and activating processes to which the polymer surface is subjected before copper deposition. It should also be taken into account that the

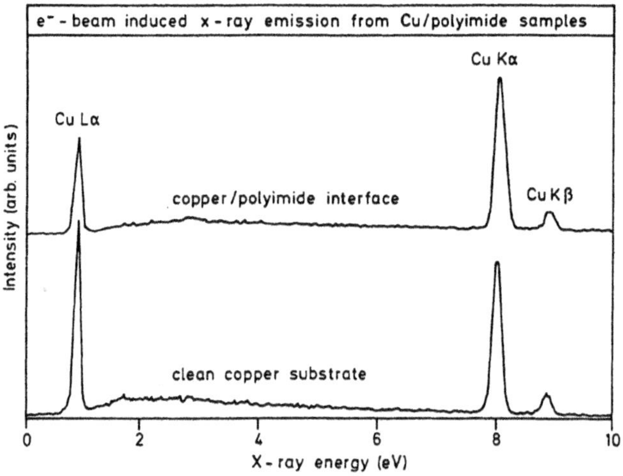

Figure 4. EDX spectra from a clean copper surface, and from a copper surface prepared by peeling off from polyimide foil. Note the difference in the relative intensity of the Cu Lα and Kα emission lines.

polyimide foil could exhibit a certain degree of order in the surface region, for example, with the PMDA-ODA moieties lying parallel or in some other geometry near the interface. Since the reflection IR spectra were recorded by means of polarisation modulation[2], such a preferential ordering could influence the data. Thus a quantitative evaluation of the relative band strengths on the basis of the spectra in Figure 3 is difficult without reference to overlayers of different thicknesses.

The IR and XPS results indicate that, for the copper-PI system studied here, and under the conditions of pre-treatment to which the PI was subjected, disruption of the metal-polymer bond occurs through cohesive failure rather than interfacial failure. Thus the interaction of pure PI with a metal is less important for the general question of adhesion strength than the bond between copper and the (presently ill-defined) surface layer in these samples. The thickness of the PI layer remaining on the copper after peeling is also of interest. In the photoelectron spectrum of Figure 1 (b), no signal from the underlying copper was found. This indicates that the residual overlayer is thicker than about 60 Å, taking into account the escape depths for the Cu 2p photoelectrons when excited with Mg Kα radiation. The IR data shown at the bottom of Figure 3 cannot be readily interpreted in terms of layer thickness since in order to obtain this information for reflection-absorption data, the angle of incidence needs to be known with rather high accuracy. In this situation, electron probe microanalysis (EDX) spectra, recorded in combination with scanning electron micrographs (SEM), have provided some important information. The method is based on EDX spectra such as shown in Figure 4. The bottom curve shows the x-ray emission spectrum from a clean copper substrate with its Cu Kα, Kβ, and Lα emission peaks. The top curve shows the same spectrum taken from a copper surface produced by peeling, in which a PI deposit remains on the copper. Here, the intensity ratio between the Cu Kα and Lα peaks is found to be quite different from that found for the clean substrate. For samples showing a strong contrast in SEM the relative strength of these two lines varied from sample to sample and also across the surface of the same sample. The Kα/Lα ratio, which is determined by the ionization cross section of these levels for electrons of a certain kinetic energy, as well as the detector efficiency for different x-ray wavelengths (among other factors), is influenced by the wavelength-dependent absorption of x-rays in any overlayer material which may be present on the surface. If the overlayer is homogeneous, and of known elemental composition, its x-ray absorption coefficient can be obtained from the literature . The change in the Kα/Lα ratio can be visualized as follows: for two x-ray lines with such different wavelengths as the Cu Kα and Lα lines, the x-ray absorption coefficient will in general be different. For the present system, the Lα radiation will be much more strongly attenuated than the Kα line because of the proximity of the former to the K absorption edges of the principal constituents of polyimide. All other factors which affect the ratio of Kα to Lα radiation can thus be used to determine the attenuating overlayer thickness. For a quantitative evaluation of our spectra we can simplify an otherwise rather complicated process by considering only relative intensities. This procedure has been outlined in more detail [10]. For the present system thicknesses of the polyimide residue of the order of several hundred nanometers are obtained; this has to be seen against a total thickness of the foil of 50 µm.

In summary, we have studied the surfaces and peeled-off interfaces of PI using surface sensitive methods. We find that chemical pre-treatment of the surface gives rise to characteristic changes in its photoelectron spectrum, and that peeling of the copper layer is due to cohesive failure in the material rather than adhesion failure at the metal/polyimide interface.

REFERENCES

1. K.L.Mittal, editor, "Polyimides: Synthesis, Characterization, and Applications", Plenum Press New York 1984.
2. O.Hofmann, K.Doblhofer, and H.Gerischer, J.Electroanal.Chem. 161, 337(1984).
3. P.O.Hahn, G.W.Rubloff, J.W.Bartha, F.LeGoues, R.Tromp, and P.S.Ho, Mater. Res. Soc. Proc. 40, 247(1984).
4. J.L.Jordan, C.A.Kovac, J.F.Morar, and R.A.Pollak, Phys.Rev. B36, 1369 (1987).
5. N.G.Chou, D.W.Dong, J.Kim, and A.C.Liu, J.Electrochem.Soc. 131, 2335 (1984).
6. Y.-H.Kim, G.F.Walker, J.Kim, and J.Park, J. Adhesion. Sci. Technol. 1, 331 (1987), and references therein.
7. R.C.White, R.Haight, B.D.Silverman, and P.S.Ho, Appl. Phys. Lett. 51, 481 (1987).
8. S.Chambers, V.A.Loebs, and K.K.Chakravorty J. Vac. Sci. Technol. 8, 875 (1990).
9. KAPTON Technical Bulletin, Du Pont de Nemours International, Geneva, Switzerland.
10. K.Horn, A.M.Bradshaw, K.Doblhofer, S.Krause, G.Weinberg, H.M.Seidenspinner, and R.Schulz, Fresenius Z. Anal. Chemie 333, 590 (1989).
11. M.Chtaib, E.M.Roberfroid, Y.Novis, J.J.Pireaux, and R.Caudano, ACS Symposium Series, to be published.
12. German Federal patent No. 21-16398.
13. J.O.Besenhard, U.Krebber, J.K.H.Horber, N.Kanani and H.Meyer, J. Electrochem. Soc. 136, 3608 (1989).
14. H.W.Siesler and K.Holland-Morowitz, "Infrared and Raman Spectra of Polymers", Marcel Dekker, New York 1980.
15. J.D.DiNardo, J.E.Demuth, and T.C.Clarke, J. Chem. Phys. 85, 6739 (1986).

CHARACTERIZATION OF A SURFACE CHEMICALLY-MODIFIED POLYIMIDE

Richard R. Thomas, Stephen L. Buchwalter
and L. Paivikki Buchwalter
IBM Research Division, Thomas J. Watson Research Center
P. O. Box 218, Yorktown Heights, New York 10598

The polyimide, PMDA-ODA, has been surface (sub-surface) modified by base hydrolysis followed by protonation to leave a thin film of polyamic acid over a fully cured polyimide. The modified polyimide is characterized, in this study, by a variety of techniques including X-ray photoelectron spectroscopy, capacity (titratable amic acid groups/volume of modified polyimide) and contact angle measurements as a function of pH. XPS data clearly showed the formation of the Pd^{2+}-carboxylic acid complex on the modified polyimide. Contact angle data, for various pH water, gathered on the modified polyimide, indicated the presence of carboxylic acid groups which ionize and are more hydrophilic when the pH of the probe liquid was greater than 7.

INTRODUCTION

Polyimides are a class of polymers that meet many of the requirements for polymers used in the electronics industry. [1,2] Polyimides are generally characterized by excellent thermal, chemical and mechanical stability. However, due to the somewhat inert surface of polyimides, it is difficult to form adhesive bonds to these materials. If the desired bulk properties of polyimides are to be retained then an alternative to this problem is surface modification. Surface modification schemes include the use of ion beams, [3-6] photografting, [7] plasmas, [8-12] particle beams, [13] sputtering, [14] $^{60}Co(\gamma)$ -irradiation [15] and wet chemical methods. [16-19] The surface (sub-surface) of polyimides has been modified by base hydrolysis followed by protonolysis to leave a thin film of polyamic acid over a fully cured polyimide.

Modified polymers are often poorly understood with regards to thickness and extent of modification, chemical composition, extent of cross-linking, morphology and orientation of functional groups at the surface. While X-ray photoelectron spectroscopy has provided a

Metallized Plastics 2, Edited by K.L. Mittal
Plenum Press, New York, 1991

wealth of information regarding the chemical composition of the modified surface (or near-surface),[20,21] a great deal of scientific understanding is still lacking. Recent studies using other surface science techniques, such as secondary ion mass spectroscopy,[22] as well as contact angle measurements on modified polymer surfaces[23,24] have appeared in the literature. Many of these techniques, though, probe either just the surface (≃ 1 monolayer) or provide a spatially-averaged image of the modified polymer depending on sampling depth. Surface science techniques which require sputtering for analysis do not reach steady state measurements until well after a gross perturbation of a thin modified layer.

EXPERIMENTAL

Materials

Polyimide substrates were prepared on 2.12 inch diameter x 0.12 inch thick polished glass disks (Swift Glass Co., Elmira, NY) to provide mechanical stability. The disks were cleaned in basic ethanol/water, rinsed with water and dried before use. Sodium hydroxide (J. T. Baker Chemical Co., ' Baker Analyzed ' ™ Reagent) was used as received. Palladium nitrate hydrate was A. C. S. Reagent grade from Aldrich Chemical Co. and was used as received. Glacial acetic acid (Ashland Chemical Co., reagent grade) was diluted to the appropriate concentration. Deionized water used was further purified by treatment with a Millipore Milli-Q ™ Water System.

Preparation of Polyimide Films

Polymer films were prepared from Du Pont Pyralin ™ PI 2545 (an amic acid polymer prepared from 1,2,4,5-benzenetetracarboxylic acid dianhydride and 4,4′-oxydianiline). After pretreatment of the glass substrates with a silane coupling agent (to prevent stress delamination of the films), the polyamic acid films were prepared by spin coating (3000 RPM, 30 seconds using a 16 wt % solution of the polyamic acid in 1-methyl-2-pyrrolidinone). The films were dried by baking in a convection oven at 80 ° C for 30 minutes. The polyamic acid to polyimide cyclodehydration reaction was performed by heating to 150 °C (30 minutes), 200 °C (30 minutes), 300 ° C (30 minutes) and 400 ° C (60 minutes) followed by slow cooling in a furnace under flowing nitrogen. The thickness of the fully cured polyimide films was found to be 5 μm.

Polyimide Modification and Palladium Exchange

Polyimide modification was performed by hydrolysis in an aqueous 0.25 M NaOH solution for the appropriate amount of time (2-5 hours). At this stage, a portion of the polyimide had been converted to the sodium salt of polyamic acid. The salt was converted to the amic acid form by treatment with an aqueous 0.1 M acetic acid solution for an equal period of time. The palladium salt of polyamic acid was prepared by immersion in an aqueous 1 mg/mL solution of palladium nitrate hydrate. The substrates were vertically rack-mounted in a Teflon carrier (Fluoroware, Inc.) during processing and rinsed thoroughly with water between each step. Due to the nature of the modified polyimide, the samples were handled in such a fashion as to avoid contact of the surface with any contaminants and to avoid scratching. The samples were stored under an atmosphere of flowing nitrogen and there were no effects of storing or aging observed on XPS or contact angle data. All chemical modifications were done at room temperature and the films were air dried when completed.

X-ray Photoelectron Spectroscopy

XPS data were obtained with a SSX 100-05 spectrometer from Surface Science Laboratories using monochromatized Al-K $_{\alpha}$ x-ray radiation. Polyimide films were prepared and modified on single crystal silicon wafers. A flood gun was used to compensate for sample charging.

Contact Angle Measurements

The contact angles of various pH buffer solutions (Hydrion Buffer, Aldrich Chemical Co.)on the polyimide surfaces were measured with a Ramé-Hart Model100-00contact angle goniometer.The measured surface tensions for the buffer solutions varied between60and73 dynes/cm. A drop volume of 20 μL was used for advancing angles and 10 μL for receding.

Determination of Modified Film Capacity and Reaction Stoichiometry

A set of 12 substrates was treated (see Polyimide Modification and Palladium Exchange) by two hour hydrolysis and two hour acidification. After extensive rinsing (to remove any residual acetic acid) and air drying, the substrates were immersed into 500 mL of deionized water. The water was purged with nitrogen and the pH of the solution was monitored after the addition of known amounts of an aqueous NaOH solution. The pH was measured with an Orion Research 811 microprocessor-controlled pH meter using a 91-02 pH electrode with automatic temperature compensation. The meter was calibrated with a pH 7.00 buffer solution before each titration. The NaOH titrating solution was standardized by titration with potassium hydrogen phthalate. [26] The inflection point of the titration curve was determined from the derivative of the second degree interpolation polynomial passing through three successive points. The experimental data gave inflection points that were consistently in the pH range 7.1-7.5. The inflection point corresponded to a modified film capacity of 6.8 $\pm$ 1.8 X 10 $^{-6}$ moles of amic acid functional groups/volume for two independent determinations (for 273.24 cm^2 of film). Under identical conditions, using a five hour hydrolysis and five hour acidification, the film capacity was measured to be 8.8 $\pm$ 3.7 X 10 $^{-6}$ moles/volume.

The reaction stoichiometry was determined in the following manner. A set of polyimide films was hydrolyzed, as above, for two hours and acidified for two hours. The modified polyimide was allowed to ion exchange in a 1 mg/L solution of palladium nitrate hydrate. An ion exchange experiment was conducted in which the palladium was quantitatively removed by immersing the film in an alkaline solution. [27] The total palladium recovered after complete exchange was measured, by inductively coupled plasma-atomic emission spectroscopy, and found to be 7.01 X 10 $^{-6}$ moles.

RESULTS AND DISCUSSION

Modification of Polyimide

The polyimide studied was a copolymer of 1,2,4,5-benzenetetracarboxylic acid anhydride and 4,4′-oxydianiline prepared by thermal cyclodehydration of the polyamic acid (Scheme 1). This polyimide is commonly referred to as PMDA-ODA. The fully cured polyimide is inert to most solvents but has been shown to be etched by base hydrolysis. [28] Hydrolysis by a dilute aqueous base, such as NaOH, does not readily etch the polymer but, rather, leaves behind a layer of the sodium salt of polyamic acid (Scheme 2, Step I). The

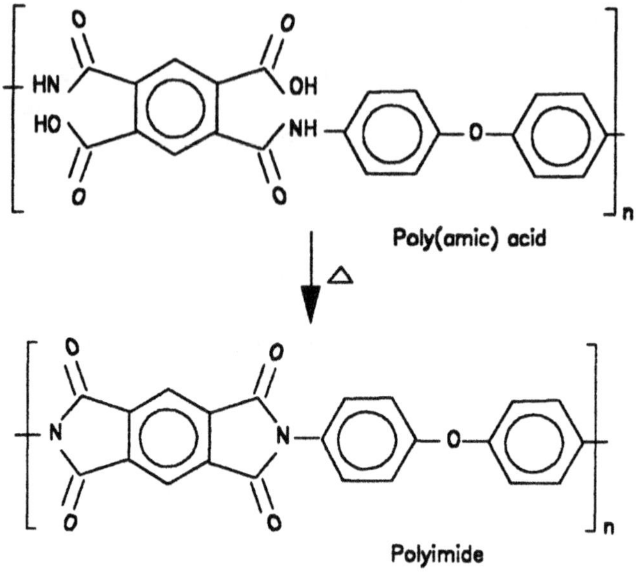

Scheme 1. Thermal cyclodehydration reaction of polyamic acid to polyimide.

sodium salt is converted to the free carboxylic acid by proton exchange with an acid such as acetic (Scheme 2, Step II). The proton is readily exchangeable by a number of metallic salts including Pd $^{2+}$ (from $Pd(NO_3)_2$) (Scheme 3, Step I). The carboxylic acid groups of the

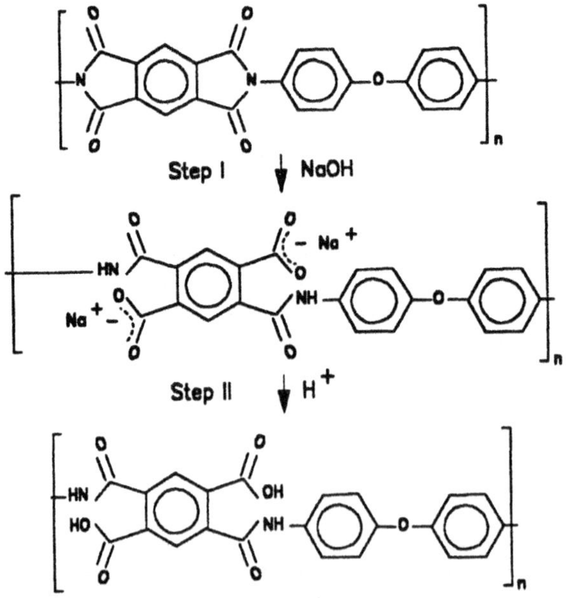

Scheme 2. NaOH hydrolysis of polyimide (Step I) to form sodium salt of polyamic acid followed by acidification (Step II) to form polyamic acid.

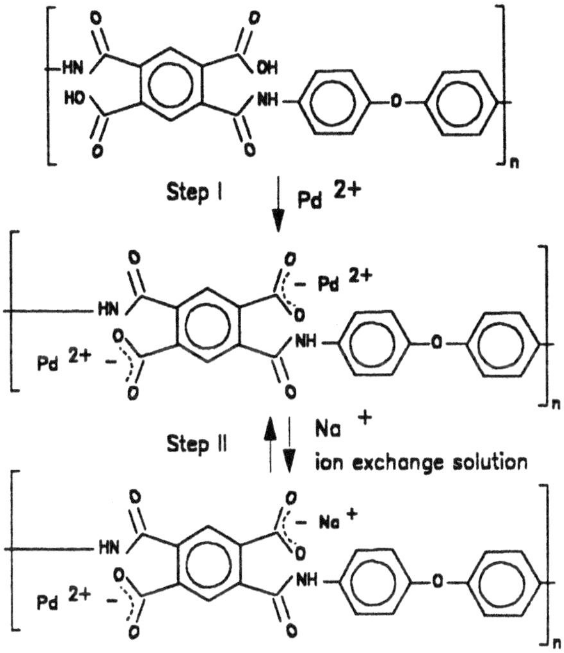

Scheme 3. Ion exchange with Pd^{2+} to form palladium salt of polyamic acid (Step I). Sodium ion exchange (Step II) when contacted with ion exchange solution.

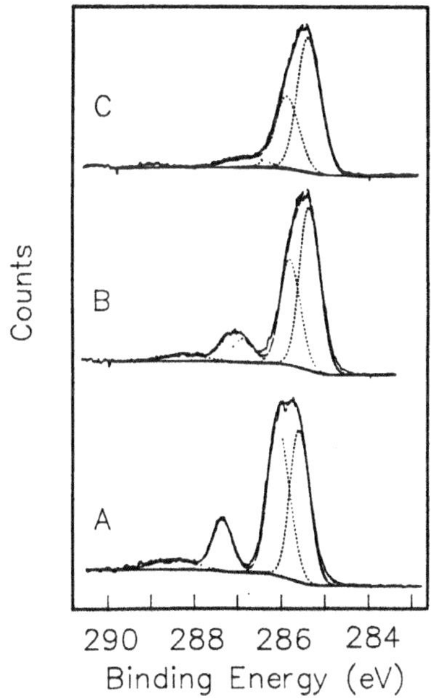

Figure 1. C 1s XPS spectra of virgin PMDA-ODA polyimide (A), after hydrolysis and acidification (B), and after exchange with Pd^{2+} (C).

modified polyimide act as complexing ligands for the metal. [29] This material is referred to as the palladium complex of polyamic acid.

The hydrolysis and palladium exchange on PMDA-ODA polyimide was examined by X-ray photoelectron spectroscopy (XPS). Shown in Figure 1 are the C 1s signals for the virgin polyimide (A), after two hour hydrolysis and acidification (B), and after formation of the palladium complex of polyamic acid (C). The hydrolysis is indicated by the significant broadening of the C=O signal in the C 1s data, which is primarily due to the introduction of acid and amide C=O groups to the surface. This interpretation is further supported by the O 1s spectra shown in Figure 2. It is clear from the figure that the -O- signal, at higher binding energy, has increased relative to the C=O oxygen intensity at lower binding energy. It is true, however, that this relative increase in the -O- peak intensity can also be due to ether-type O 1s species. As will be discussed later, the hydrolyzed surface has a contact angle that is strongly dependent on pH.

After exposure to a $Pd(NO_3)_2$ solution, the C 1s spectrum is drastically altered by a significant reduction in the C=O intensity. It is suggested that this is due to a RCO_2Pd^+-type species, where R represents the polyimide backbone. Pd^{2+} is capable, *vide infra* , of complexing with two carboxylic acid groups in a square planar arrangement. Due to geometrical considerations for bonding, each carboxylic acid group would have to come from a different polyimide chain. Palladium complexation would effectively result in a cross-link between hydrolyzed chains. Therefore, palladium complexed at the polyimide/air interface might be expected to be ligand (carboxylic acid)-deficient. Electroneutrality would then require that NO_3^- (from $Pd(NO_3)_2$ used in the ion exchange) be present with Pd^{2+}. This is supported by the N 1s data of the polyimide surface after Pd^{2+} exchange (Figure 3). As is evident from the data, a multitude of nitrogen-containing species are present at the Pd^{2+}-exchanged surface. The lowest binding energy peak at ≃ 400.4 eV may be due to the amide species, while a detailed explanation of the additional peaks in the N 1s spectrum requires further study. It is accepted, however, that the peak at ≃406.7 eV can be assigned to $-NO_3^-$ [30], which should not be an unexpected species on the Pd^{2+}-exchanged modified polyimide surface. The presence of Pd^{2+} is further supported by the Pd $3d_{5/2}$ binding energy peak at ≃ 338.3 eV, as shown in Figure 4. The existence of $RCO_2\,Pd^+$ in the hydrolyzed polyimide as the major palladium species present has also been shown with capacity measurements on the modified film as well as by ion exchange studies. A second Pd^{2+} species, in low amounts, is suggested by the XPS data. The identity of this species is not clear at this time. The thermal treatment of this polyamic acid in the presence of several metal salts has been discussed previously. [31] By varying the hydrolysis time, the depth of PMDA-ODA modification can be controlled. This has been explored with film capacity measurements.

The exchange stoichiometry, along with modified film capacity, were determined by potentiometric titration. The capacity being defined by the number of carboxylic acid groups/modified volume of polyimide. A 273.24 cm^2 (12 substrates) film of fully cured PMDA-ODA was treated with an aqueous 0.25 M NaOH solution for two hours followed by a two hour treatment with 0.1 M acetic acid. This film was immersed in water and titrated with an aqueous NaOH solution. The titration curve obtained is shown in Figure 5. It is clear from this figure that the modified film is of the weak acid variety ($pK_A \simeq 7.5$) as inferred from the shape of its curve. From the inflection point (maximum $d(pH)/d(NaOH)$,) it was calculated that 8.90×10^{-6} moles of amic acid functional groups/unit volume were present. The titration of two films which were subjected to identical hydrolysis treatments gave an average $6.80 \pm 1.85 \times 10^{-6}$ moles of amic acid functional groups. An identical area of film

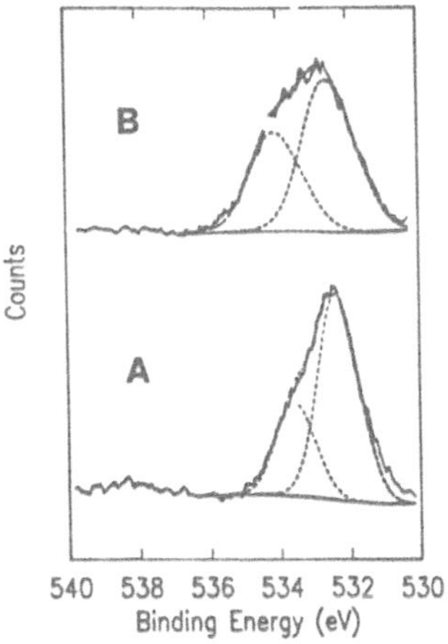

Figure 2. O 1s XPS spectra of PMDA-ODA polyimide (A), and after hydrolysis and acidification (B).

which was hydrolyzed under the same conditions for five hours had a similar pH titration curve but was calculated to have $8.83 \pm 3.70 \times 10^{-6}$ moles of amic acid functional groups for two independent experiments. This indicates that the extent of hydrolysis is not linear with time but represents a factor of 1.30 ± 0.65 increase in depth of hydrolysis compared to the two hour treatment. Using a two hour hydrolysis time, followed by acidification and palladium exchange, under experimental conditions which resulted in complete exchange of sodium for palladium gave 7.01×10^{-6} moles of palladium.[27] Since the ratio of number of palladium ions to amic acid groups much more closely approaches 1 than 0.5, (ratio = 1.0 ± 0.30) the stoichiometry of the exchange can be approximated by [32]

$$RCO_2Pd^+ + Na^+ \rightarrow RCO_2Na + Pd^{2+} \quad (1)$$

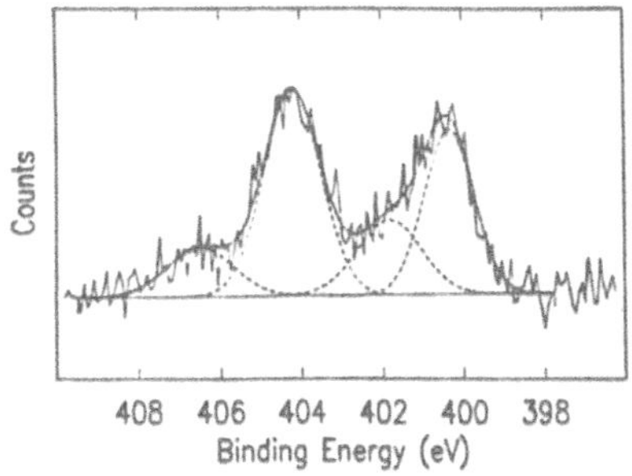

Figure 3. N 1s XPS spectrum of PMDA-ODA polyimide after hydrolysis and acidification and exchange with Pd^{2+}.

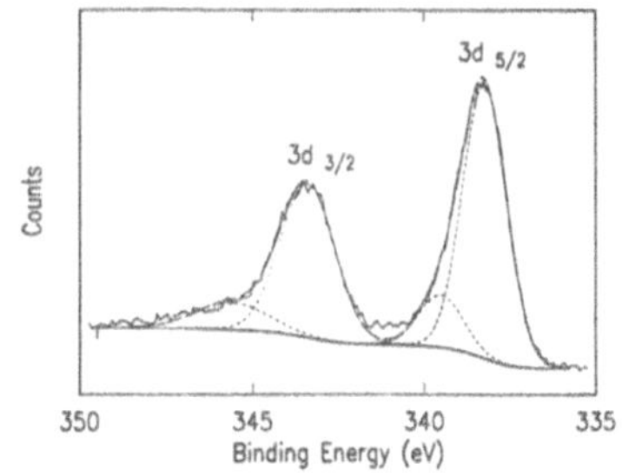

Figure 4. Pd 3d XPS spectrum of hydrolyzed PMDA-ODA polyimide after exchange with Pd^{2+}.

This stoichiometry is important for three reasons. The carboxylic acid group present on the hydrolyzed polyimide would be expected to be a better ligand for palladium than any monodentate ligands present. The chelate effect [33] associated with the amic acid groups would suggest that, if possible and in the presence of only monodentate ligands, the thermodynamically most stable configuration for palladium in this case would be $(RCO_2)_2Pd$. Several metal salts having valences > 1 have been observed to cause the gellation of ionogenic polymers, [34] presumably through a cross-linking reaction. However, these cases have been observed in solution where the polymer is expected to have a good deal of mobility. The palladium ions are unable to be satisfied through coordination to two carboxylic acid groups but must only be bound through one with another counter anion present.

Contact Angle Measurements

The contact angles of liquids on polyimide and modified polyimide surfaces can be used to provide a great deal of information regarding the types of functional groups that are present on these surfaces. [23,24] In particular, the results of a study on the acid-base behavior of oxidized polyethylene using contact angles are useful. [23] The wettability of oxidized (carboxylic acid-rich) polyethylene was found to be critically dependent on the pH of the wetting solution. The ionization of the surface carboxylic acid groups could be observed by measuring contact angles on this surface as a function of pH. The onset of ionization was observed as an increase in wettability when the pH of the probe liquid was > 7. The surface of hydrolyzed polyimide is expected to be rich in carboxylic acid groups. This was observed experimentally in a study of the pH dependence of the contact angles for various solutions on the modified polyimide. Shown in Figure 6 are the observed advancing contact angles of the modified polyimide compared to the native polymer. For comparison, the receding angles are shown in Figure 7. This film was hydrolyzed for two hours in 0.25 M NaOH; however, films which were hydrolyzed for shorter periods of time show a similar trend. The data indicate that the effective acid dissociation constant, pK_a^{eff}, lies in the range 6-8 and is comparable to that of other organic acids. [35] The contact angle data can be compared to the potentiometric titration data on the modified films which show the pK_a^{eff} for the amic acid groups to be in the 7-8 range.

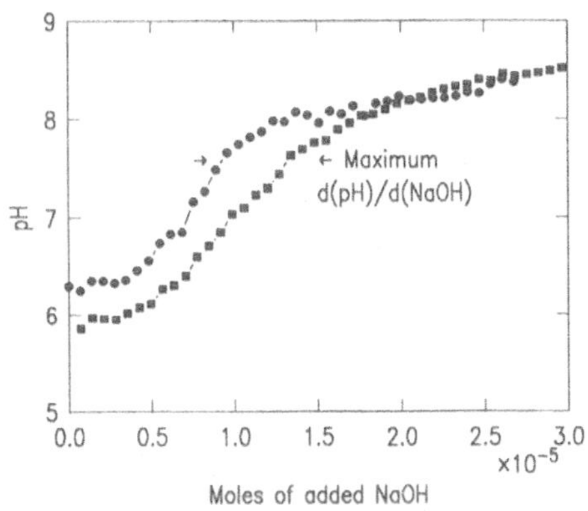

Figure 5. Typical potentiometric titration curves for modified polyimide capacity measurements. Curve obtained after two hour hydrolysis and acidification (●), and after five hour hydrolysis and acidification (■).

CONCLUSIONS

While many of the questions, posed in the Introduction, regarding the nature of modified polyimides remain to be answered, several conclusions can be drawn from the current study. The polyimide PMDA-ODA can be hydrolyzed and then acidified to give a modified polymer that is rich in amic acid groups. The modified polyimide is not entirely polyamic acid since polyamic acid is etched readily in alkaline solutions. The nature of these amic acid groups can be examined by monitoring contact angles as a function of pH. The data indicate that a substantial fraction of the amic acid groups is oriented towards the water/polyimide interface. The trend in contact angle data suggests that there are no other ionizable functional groups present at the surface unless they are coincidentally obscured by the ionization of carboxylic acid groups. The quantity of these groups can be determined by using

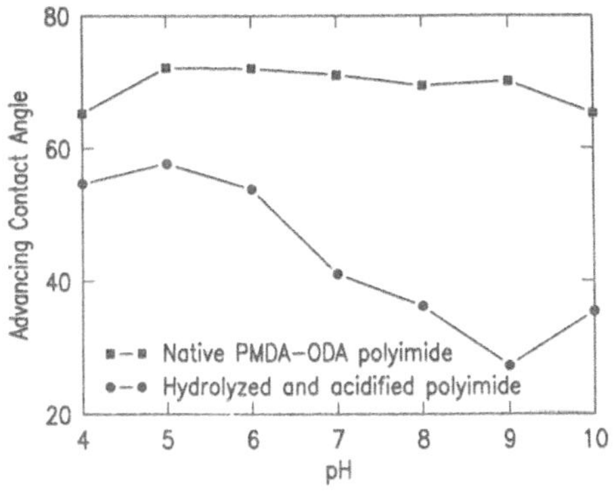

Figure 6. Advancing contact angles, of buffered water, as a function of pH for native PMDA-ODA polyimide (■), and hydrolyzed and acidified polyimide (●).

potentiometric titrations to measure modified film capacity. The modified polyimide is an ion exchange resin and is capable of coordinating a variety of ions.

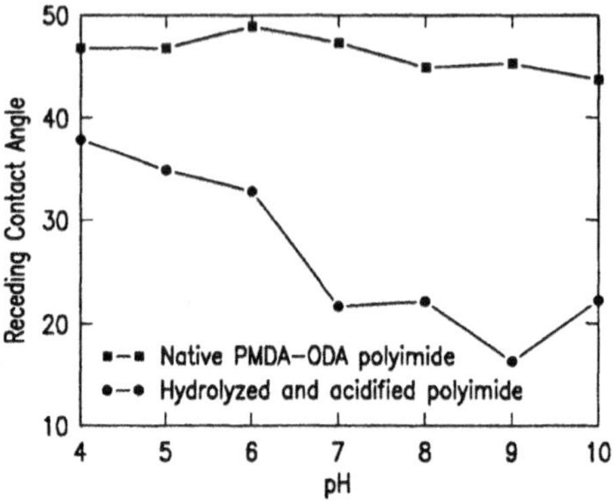

Figure 7. Receding contact angles, of buffered water, as a function of pH for native PMDA-ODA polyimide (■), and hydrolyzed and acidified polyimide (●).

REFERENCES

1. K. L. Mittal, Editor, "Polyimides: Synthesis, Characterization and Applications," Vols. 1 and 2, Plenum Press, New York, 1984.
2. W. D. Weber and M. R. Gupta, Editors, "Recent Advances in Polyimide Science and Technology," The Society of Plastics Engineers, 1987.
3. B. T. Werner, T. Vreeland, Jr., M. H. Marshall, Y.Qui and T. A. Tombrello, Thin Solid Films, 104 , 163 (1983).
4. G. M. Sessler, J. E. West, F. W. Ryan and H. Schonhorn, J. Appl. Polym. Sci., 17 , 3199 (1973).
5. J. A. Taylor, G. M. Lancaster and J. W. Rablais, Appl. Surface Sci., 7 , 503 (1978).
6. W. K. Fisher and J. C. Corelli, J. Polym. Sci.: Polym. Chem. Ed., 19 , 2465 (1981).
7. R. D. Goldblatt, J. M. Park, R. C. White, L. J. Matienzo, S. J. Huang and J. F. Johnson, J. Appl. Polym. Sci., 37 , 335 (1989).
8. H. Schonhorn and F. W. Ryan, J. Adhesion, 1 , 43 (1969).
9. A. Celerier and J. Machet, Thin Solid Films, 148 , 323 (1987).
10. T. Hjertberg, B.-Å. Sultan and E. M. Sörvik, J. Appl. Polym. Sci., 37 , 1183 (1989).
11. A. M. Mance, R. A. Waldo and A. A. Dow, J. Electrochem. Soc., 136 , 1667 (1989).
12. E. Occhiello and F. Garbassi, Polymer, 29 , 2277 (1988).
13. R. Micheal and D. Stulik, J. Vac. Sci. Technol., A4 , 1861 (1986).
14. J. L. Grant, D. S. Dunn and D. J. McClure, J. Vac. Sci. Technol., A6 , 2213 (1988).
15. H. Vanni and J. F. Rabolt, J. Polym. Sci.: Polym. Phys. Ed., 18 , 587 (1980).
16. T. J. McCarthy and K.-W. Lee, Macromolecules, 21 , 3353 (1988).
17. C. A. Costello and T. J. McCarthy, Macromolecules, 17 , 2940 (1984).
18. A. J. Dias and T. J. McCarthy, Macromolecules, 18 , 1826 (1985).
19. A. J. Dias and T. J. McCarthy, Macromolecules, 20 , 2068 (1987).
20. H. Yasuda, H. C. Marsh, S. Brandt and C. N. Reilley, J. Polym. Sci.: Polym. Chem. Ed., 15 , 991 (1977).

21. D. T. Clark and W. J. West, J. Macromol. Sci.-Revs. Macromol. Chem., C12 , 191 (1975).
22. R. D. Goldblatt, G. J. Scilla, J. M. Park, J. F. Johnson and S. J. Huang, J. Appl. Polym. Sci., 35 , 2075 (1988).
23. S. R. Holmes-Farley, R. H. Reamey, T. J. McCarthy, J. Deutch and G. M. Whitesides, Langmuir, 1 , 725 (1985).
24. G. M. Whitesides and G. S. Ferguson, Chemtracts-Org. Chem., 1 , 171 (1988).
25. The term capacity as used in the study of ion exchange resins is defined as the number of titratable functional groups/resin volume. In the present case, the volume of the modified polymer is unknown and, at best, can only be estimated.
26. W. J. Blaedel and V. W. Meloche, "Elementary Quantitative Analysis: Theory and Practice," 2nd ed., Chapter 18, Harper & Row, New York, 1957.
27. R. R. Thomas and M. M. Plechaty, Diffusion of ions in a surface chemically-modified polyimide, this proceedings volume.
28. H. J. Fink, U. S. Patent, 3,395,057, July 30, 1968.
29. For example, see: R. C. Mehrotra and R. Bohra, "Metal Carboxylates," Academic Press, London, UK, 1983.
30. G. E. Muilenberg, Ed., "Handbook of X-ray and Ultraviolet Photoelectron Spectroscopy," Perkin-Elmer Corp., Eden Prairie, MN, 1979.
31. H. G. Linde and R. T. Gleason, J. Polym. Sci. Polym. Phys. Ed., 26 , 1485 (1989).
32. The data cannot eliminate the possibility that $[(RCO_2)_2Pd]$exists in the modified polymer. The stoichiometry would then be given by

$$X\,(RCO_2)_2Pd \;+\; (1-X)\;RCO_2Pd^+ \;+\; (1+X)\;Na^+ \;\rightarrow\; (1+X)\;RCO_2Na \;+\; Pd^{2+}$$

where $0 < X < 1$.
33. F. A. Cotton and G. Wilkinson, "Advanced Inorganic Chemistry," 5th ed., Chapter 2, Wiley & Sons, New York, 1988.
34. R. R. Thomas, unpublished results (1987).
35. J. T. Davies and E. K. Rideal, "Interfacial Phenomena," pp. 237-238, Academic Press, New York, 1963.

DIFFUSION OF IONS IN A SURFACE CHEMICALLY-MODIFIED POLYIMIDE

Richard R. Thomas and Miro M. Plechaty

IBM Research Division, Thomas J. Watson Research Center
P. O. Box 218, Yorktown Heights, New York 10598

The polyimide, PMDA-ODA, can be surface (sub-surface) modified by base hydrolysis followed by protonation to leave a thin film of polyamic acid over a fully cured polyimide. The hydrolysis/acidification scheme leaves the modified polyimide with a layer rich in carboxylic acid (and amide) groups. The carboxylic acid groups are effective in complexing ions from aqueous solution. In particular, the complexation of sodium and palladium ions has been examined. When in contact with an aqueous electrolyte solution, the palladium-complexed modified polyimide is subject to ion exchange. The kinetics of ion exchange between palladium and a sodium ion-containing electrolyte solution have been investigated. The exchange process is slow and appears to be diffusion-controlled.

INTRODUCTION

The phenomenon of ion exchange occurs whenever a polyelectrolyte or a polymer containing ionogenic groups is placed into contact with an electrolyte solution. Ion exchange phenomena have been studied extensively on a variety of resins. [1] These studies have focused primarily on network polymers that are cross-linked to various degrees and contain a variety of functional groups. Kinetic and thermodynamic studies on ion exchange have provided a great deal of information on diffusion, equilibria and structure of ion exchange resins. These types of studies have not been performed, to any extent, on surface (sub-surface) modified polymers. Ion exchange kinetic experiments on typical ion exchange resins reflect bulk macroscopic properties that are dependent upon the degree of cross-linking and chemical groups, and these are determined by chemical groups of the monomer and how the resin was prepared. As a consequence, ion exchange occurs throughout the entire region. Surface modified polymers can have functional groups that are not present on the native polymer and the regions of modification can be exceedingly thin.

EXPERIMENTAL

The experimental details of the polyimide film preparation have been described elsewhere. [2] Films of PMDA-ODA polyimide were prepared from the polyamic acid precursor

Metallized Plastics 2, Edited by K.L. Mittal
Plenum Press, New York, 1991

by spin coating on glass disks. The films were cured to a temperature of 400 °C under flowing nitrogen. Hydrolysis was performed using aqueous NaOH for specified periods of time followed by treatment with acetic acid. Palladium complexation was accomplished by ion exchange with an aqueous solution of palladium nitrate. The ion exchange solution consisted of EDTA (40 g/L), sodium sulfate (20 g/L), sodium formate (20 g/L) and sodium hydroxide to pH 11.70. The exchange solution was heated to the specified temperature (±1°C) with a jacketed and thermostated vessel. The substrates were immersed into the exchange solution and the concentration of evolved palladium was monitored as a function of time by inductively coupled plasma- atomic emission spectroscopy. A fresh ion exchange solution and new substrates were used for each kinetic experiment. Experimental conditions used for the film hydrolysis, sodium ion and palladium ion exchange are shown in Table I. Experimental conditions for the kinetic measurements are given in Table II.

The conditions used for the standard hydrolysis and ion exchange experiment form the basis by which other experiments are compared and were chosen simply for convenience. The $\Delta[Na]$ experiment refers to the ion exchange reaction in which $[Na]$ has been increased from 0.45 to 0.675 M. The Δl experiment contains a longer hydrolysis and acidification time to increase the depth of polyimide modification. The ΔT experiments are based on standard hydrolysis and acidification times but with the ion exchange solution maintained at the cited temperatures.

RESULTS AND DISCUSSION

PMDA-ODA polyimide can be hydrolyzed with aqueous NaOH and then acidified with acetic acid to form a thin layer of polyamic acid over the fully cured polyimide. [2] The polyamic acid layer complexes palladium according to Equation (1) (Scheme 1, Step I).

$$RCO_2H + Pd(NO_3)_2 \rightleftarrows RCO_2Pd(NO_3) + HNO_3 \qquad (1)$$

where R represents the polymer backbone. When this palladium-complexed film is placed into contact with a sodium ion- containing electrolyte solution, an ion exchange reaction occurs (Scheme 1, Step II).

$$RCO_2Pd^+ + Na^+ \underset{k_{-1}}{\overset{k_1}{\rightleftarrows}} RCO_2Na + Pd^{2+} \qquad (2)$$

According to the nature of the experiment, the ion exchange of palladium for sodium will occur through three processes. The overall process dictates that sodium ions will first

Table I. Experimental Conditions for Hydrolysis and Ion Exchange.

Experiment[1]	Film hydrolysis[2] time (min)	Film protonation[3] time (min)	Palladium ion exchange into film[4] time (min)
Standard	120	120	30
$\Delta[Na]$	120	120	30
Δl	300	300	75
ΔT_1	120	120	30
ΔT_2	120	120	30

[1]All experiments were performed at room temperature. [2]$[NaOH] = 0.25$ M. [3]$[CH_3CO_2H] = 0.1$ M. [4]$[Pd(NO_3)_2] = 1$ mg/mL.

Table II. Experimental Conditions for Kinetic Measurements of Ion Exchange.[1]

Experiment	[Na] in ion exchange solution (M)	T (°C)
Standard	0.45	72
$\Delta[Na]$	0.675	72
Δl	0.45	72
ΔT_1	0.45	60.5
ΔT_2	0.45	81

[1]After treatments summarized in Table I.

have to diffuse into and through the modified polyimide. Secondly, the actual exchange of palladium by sodium has to occur. Lastly, the palladium ions will have to diffuse out of the modified film into solution. If the rate for all of these processes is of the same order of magnitude then the kinetic problem becomes quite difficult to solve. If one of these steps is rate determining then a kinetic analysis is somewhat simplified. Our approach is based upon the assumption that one of these steps is rate determining. The diffusion of sodium ions into or palladium ions out of the modified polyimide, as being rate determining, are referred to as diffusion-controlled cases. Kinetics that is governed by the chemical exchange of palladium for sodium is referred to as the reaction-controlled case.

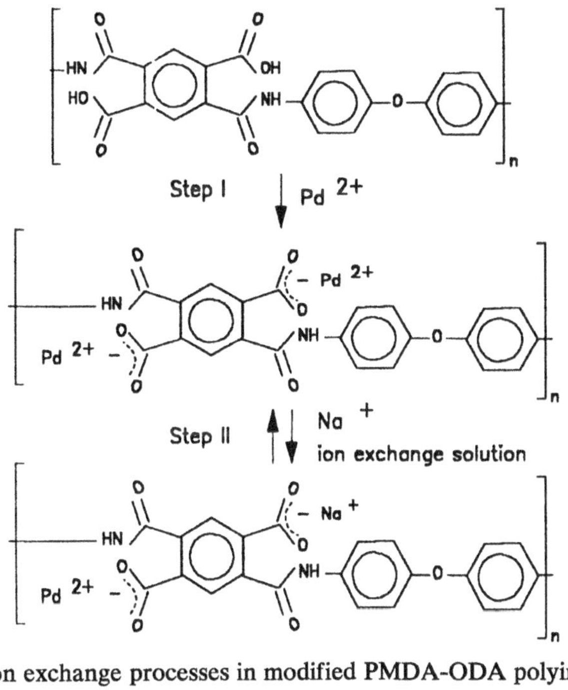

Scheme 1. Ion exchange processes in modified PMDA-ODA polyimide. Ion exchange with Pd^{2+} to form palladium salt of polyamic acid (Step I). Sodium ion exchange (Step II) when contacted with ion exchange solution.

The case for reaction-controlled ion exchange was examined first. Experimentally, the conditions for a pseudo-first-order kinetic expression [3] were established by keeping $[Na]$, in the ion exchange solution, large and invariant with time. At early stages in the exchange process, the forward reaction would dominate and lead to the kinetic expression

$$\frac{d[RCO_2Pd^+]}{dt} = -k'_1 [RCO_2Pd^+]t \tag{3}$$

with the solution

$$\ln [RCO_2Pd^+] = -k'_1 t$$

or

$$\ln [Pd^{2+}] = k'_1 t \tag{4}$$

since

$$\frac{-d[RCO_2Pd^+]}{dt} = \frac{d[Pd^{2+}]}{dt}$$

where k'_1 is the pseudo-first-order rate constant. The data for exchange, plotted according to Equation (4) and shown in Figure 1, were not found to give the necessary linear relationship. Furthermore, increasing $[Na]$ from 0.45 M to 0.675 M, in the exchange solution, had no effect on the rate of exchange. This is seen in a comparison of the experimentally determined k'_1 values for the two experiments at early stages in the exchange reaction. [4]

The case of diffusion-controlled kinetics was examined next. For an infinite slab of modified polyimide, Fick's first and second laws of diffusion lead to the differential equation

$$\frac{\partial C}{\partial t} = D \frac{\partial^2 C}{\partial x^2} \tag{5}$$

where C is the concentration of the diffusing material (in this case Pd^{2+}), t is time, D is the diffusion coefficient and x is the spatial coordinate normal to the surface. Under the experimentally appropriate boundary and initial value conditions, (the region $0 < x < l$ is initially at a uniform concentration, $[Pd]_0$, and the surface of the modified polyimide, l, is kept at a constant concentration $[Pd]_1$), the solution to Equation (5) is given by [5]

$$F = \frac{Q_t}{Q_\infty} = 1 - \frac{8}{\pi^2} \sum_{n=0}^{\infty} \frac{1}{(2n+1)^2} \exp[-(2n+1)^2 Bt] \tag{6}$$

where F is the fractional attainment of equilibrium, Q_t is the amount of palladium released from the modified polyimide at time t, and Q_∞ is the amount after attainment of equilibrium. B is the reduced diffusion coefficient and is given by

$$B = \frac{D\pi^2}{4l^2} \tag{6a}$$

where D is the diffusion coefficient of the diffusing species and l is the depth or diffusion layer thickness of the modified polyimide.

Shown in Figure 2 are the evolved palladium versus time data obtained for the various ion exchange experiments listed in Table II. Shown in Figure 3 are the fractional attainment of equilibrium data for the same experiments. However, the diffusion coefficient was found to vary with time and is due to the coupled diffusion of palladium and sodium through the

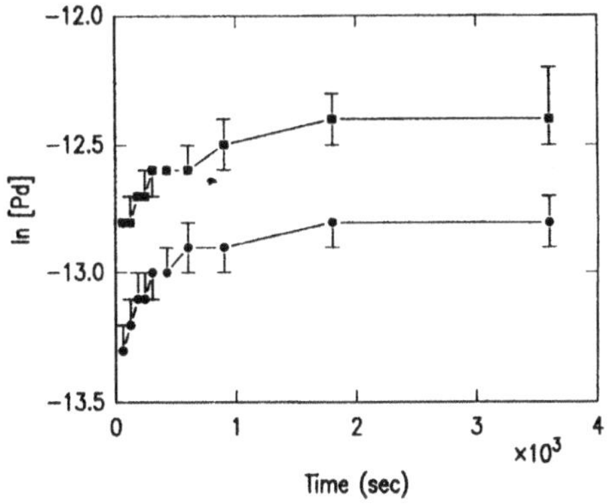

Figure 1. Pseudo-first-order kinetic plot for reaction-controlled case of ion exchange for standard ([Na] = 0.45 M) (●), and Δ [Na] ([Na]= 0.675 M) (■) experiments.

modified polymide. The coupled diffusion of the two species is a consequence of electroneutrality. The variable diffusion coefficient is predicted by the Nernst-Planck equations for diffusion. However, the Nernst-Planck equations are nonlinear and have not been solved analytically. [6] Shown in Figure 4 are the reduced diffusion coefficient versus time data for the various ion exchange experiments. Equation 5 can then be written as [7]

$$\frac{\partial C}{\partial t} = D(t)\,\frac{\partial^2 C}{\partial x^2} \tag{7}$$

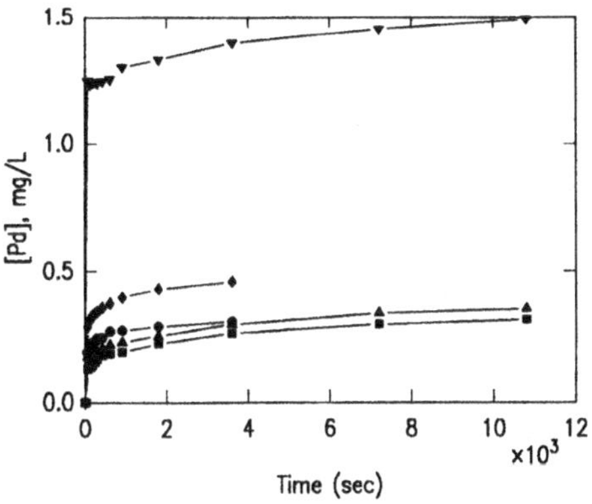

Figure 2. Evolved palladium with time for the ion exchange experiments. Standard (2 hour hydrolysis, [Na]= 0.45 M, 72 °C) (●); [Na]= 0.675 M (Δ[Na]) (◆); 5 hour hydrolysis (Δl) (▲); 60.5 °C (ΔT_1) (■) and 81 °C (ΔT_2) (▼) ion exchange experiments.

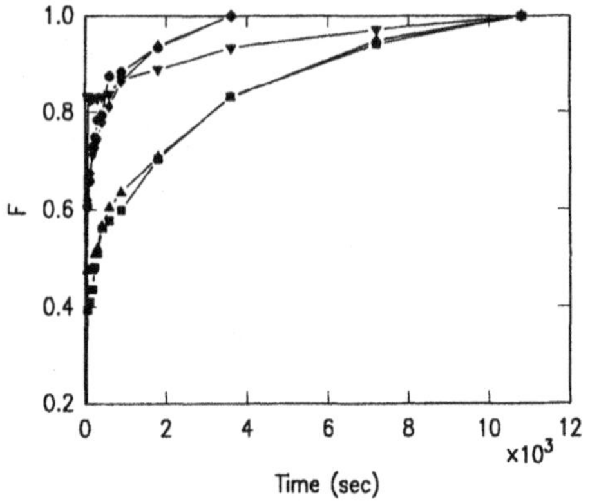

Figure 3. Fractional attainment of equilibrium with time data for standard (2 hour hydrolysis, $[Na] = 0.45$ M, 72 °C)(●); $[Na] = 0.675$ M ($\Delta[Na]$) (◆); 5 hour hydrolysis (Δl)(▲); 60.5 °C) (ΔT_1) (■); and 81 °C (ΔT_2) (▼) ion exchange experiments.

where D is now expressed as a function of time. With a change of variables, T is defined as

$$dT = D(t)\,dt \tag{8a}$$

or

$$T = \int_0^t D(t')\,dt' \tag{8b}$$

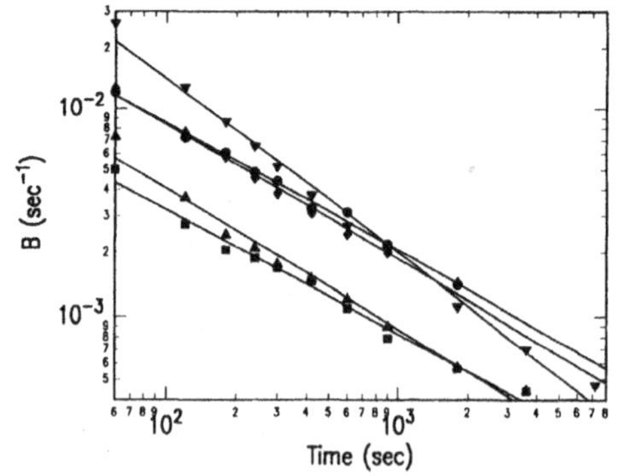

Figure 4. Reduced diffusion coefficient versus time for the various ion exchange experiments. Standard (2 hour hydrolysis, $[Na] = 0.45$ M, 72 °C)(●); $[Na] = 0.675$ M ($\Delta\ [Na]$) (◆); 5 hour hydrolysis (Δl)(▲); 60.5 °C (ΔT_1)(■); and 81 °C (ΔT_2) (▼) experiments.

Equation (7) then reduces to

$$\frac{\partial C}{\partial T} = \frac{\partial^2 C}{\partial x^2}$$

with D now referring to an interdiffusion coefficient for both diffusing species. The variable interdiffusion coefficient can be treated mathematically, by using Equation 8a or 8b, and substituted into Equation (6). Experimentally a linear relationship was established from a log-log plot of B *versus* t (see Figure 4) and provided solutions for B in the form of power law equations.

$$B_t = B_0 \, t^a$$

where B_t is the reduced interdiffusion coefficient at time t, B_0 is the reduced interdiffusion coefficient at time 0 and a is a coefficient from the least squares fit of the experimental data. These data are presented in Table III. Substituting B_t for B into Equation 6 gave satisfactory results for diffusion-controlled kinetics of ion exchange based on an interdiffusion coefficient. A typical plot of Equation 6, with B_t substituted for B, fitted with experimentally obtained data is shown in Figure 5.

The solutions for diffusion-controlled kinetics do not predict a rate dependence on the concentrations of any exchanging species. This invariance was observed experimentally. The solutions do predict a depth dependence on the rate of diffusion-controlled ion exchange. The kinetics of exchange was examined for films which were hydrolyzed for different times (to different depths). Based on comparison to other measurements, the kinetic expression for diffusion-controlled ion exchange accurately predicts the observed depth dependence. Equation 6a predicts that if a given hydrolysis time results in a modification of l_1 thickness then

$$l_2 = \sqrt{\frac{B_{t_1}}{B_{t_2}}}\; l_1 \qquad (9)$$

where l_2 is the thickness of the modified polyimide for a hydrolysis time t_2; B_{t_1} and B_{t_2} are the reduced interdiffusion coefficients for hydrolysis times t_1 and t_2, respectively. The kinetic data obtained for the films which were hydrolyzed for two (standard) and five (Δl) hours (see Table III) under identical conditions are compared. Solving Equation 9 with the power law equations for B_t for the two different hydrolysis times gives $l_2 = 1.50\, l_1$. For comparison, the capacity measurements, [2] on identically modified films, indicate that five hour hydrolysis results in a depth of modification that is 1.30 $\pm$0.65 times as great as that of the films which were hydrolyzed for two hours. [2] The application of Equation 5 is contingent on the assumption that diffusion occurs through a homogeneous medium or that each dx section of this medium is the same as the preceding one. The rough correlation of kinetic and film capacity data suggests that there is a homogeneous distribution of amic acid functional groups in the modified polyimide.

The measured interdiffusion coefficients for Pd^{2+}/Na^{+} exchange were in the range $10^{-10} - 10^{-15}\ cm^2 \cdot \mathrm{sec}^{-1}$. A substantial tortuosity of the diffusion path compared to values obtained in the absence of a matrix is implied. Compared to the measured diffusion coefficients for Na^+ and di- and trivalent metal ions of comparable size, the interdiffusion coefficients observed in this study are small. Extensive or complete hydrolysis would be expected to cause extensive dissolution of the polyimide. This was not experimentally observed. In

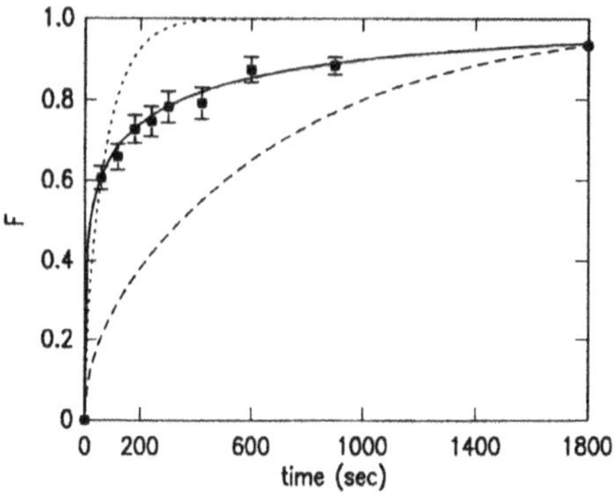

Figure 5. Fractional attainment of equilibrium versus time fits for standard ion exchange experiment. $B = 1.2 \times 10^{-2}$ sec^{-1} (— — —); $B = 1.4 \times 10^{-2}$ sec^{-1} (- - - - -); and $B_t = 0.1469$ $t^{-0.6182}$ sec^{-1} (——).

addition, the concurrent release of palladium-coordinated modified polyimide chains into solution would likely lead to intractable reaction kinetics. Partial hydrolysis may account for the low magnitude of interdiffusion coefficients. This would result in an increase in distance between diffusion sites (amic acid groups) as the mobile ions traverse the film to the solution interface and cause a relative increase in magnitude of interdiffusion coefficients. However, the effect of swelling upon conversion to the amic acid form has to be considered. Extensive swelling of the polyimide upon modification would allow the polymer to imbibe large portions of water from the exchange solutions. This would provide easily accessible hydrophilic diffusion channels for the mobile ions and would be expected to cause a relative increase in the interdiffusion coefficients.

It has been reported that the density of a polyimide is directly proportional to the final imidization temperature. [8] Furthermore, in a study on the swelling behavior of PMDA-ODA polyimide in a variety of solvents, it was noted that the imidization temperature had a marked influence on solvent uptake. [9] The samples which had seen the higher imidization temperatures exhibited slower swelling kinetics. This solvent diffusion process reflects the compo-

Table III. Time-dependent Equations for B_t.

Experiment	$B_t = B_0 t^a$		
	B_0	a	δ^1
Standard	0.1469	-0.6182	$2.6 X 10^{-4}$
Standard [2]	0.1303	-0.5936	$2.0 X 10^{-4}$
$\Delta[Na]$	0.1688	-0.6522	$3.9 X 10^{-4}$
Δl	0.06246	-0.6064	$7.0 X 10^{-4}$
Δl [2]	0.05521	-0.5988	$5.1 X 10^{-4}$
ΔT_1	0.03773	-0.5448	$3.3 X 10^{-4}$
ΔT_2	0.6908	-0.8454	$1.6 X 10^{-3}$

[1] Standard deviation from least squares fit. [2] Repeat of experiment under identical conditions.

sition of the cured polymer. The data presented in this study suggest a model for the modified polyimide in which hydrolysis proceeds by opening only a fraction of the imide rings to give the carboxylic acid derivative (in a step which is the reverse of imidization) yet which does seem to reflect the original density and/or order of the native cured polyimide.

CONCLUSIONS

Base hydrolysis and acidification of PMDA-ODA polyimide has been shown to create a region which is rich in carboxylic acid groups (polyamic acid). The depth of polyimide modification can be controlled by changes in hydrolysis conditions. The carboxylic acid groups are capable of complexing metal salts including palladium and sodium. When the palladium-complexed modified polymer is placed into contact with a sodium ion-containing electrolyte solution, ion exchange occurs. A kinetic study of this ion exchange process shows that the exchange of palladium for sodium occurs through diffusion-controlled kinetics. The kinetics were found to obey the solution to Fick's laws of diffusion after accounting for a variable diffusion coefficient. The variable diffusion (interdiffusion) coefficient arises from the coupled diffusion of two different species (Pd^{2+} and Na^{+}). The measured interdiffusion coefficients for the coupled diffusion of Pd^{2+} and Na^{+} are in the range 10^{-10} - 10^{-15} $cm^2 \cdot sec^{-1}$. The magnitude of these values is small compared to self-diffusion coefficients for these species in the absence of a matrix.

REFERENCES

1. F. Helfferich, "Ion Exchange," McGraw-Hill, New York, 1962.
2. R. R. Thomas, S. L. Buchwalter and L. P. Buchwalter, Characterization of a surface chemically-modified polyimide, this proceedings volume.
3. J. I. Steinfeld, J. S. Francisco and W. L. Hase, "Chemical Kinetics and Dynamics," p. 41, Prentice Hall, New Jersey, 1989.
4. Statistically linear data were obtained for times up to 900 seconds in both exchange reactions; however, the slopes, and hence pseudo-first-order rate constants, were nearly equal.
5. J. Crank, "The Mathematics of Diffusion," Chapter IV, Oxford University Press, London, UK, 1956.
6. F. Helfferich and M. S. Plesset, J. Chem. Phys., 28 , 418 (1958).
7. Reference 5, Chapter IX.
8. T. P. Russell, J. Polym. Sci. Polym. Phys. Ed., 22 , 1105 (1984).
9. E. Gattiglia and T. P. Russell, J. Polym. Sci., Polym. Phys. Ed., 27 , 2131 (1989).

MODIFICATION OF POLYMER SURFACES BY DUAL FREQUENCY PLASMA

J.E. Klemberg-Sapieha, L. Martinu, O.M. Küttel,
and M.R. Wertheimer

"Groupe des Couches Minces" and
Department of Engineering Physics, Ecole Polytechnique
Box 6079, Station "A", Montréal, Qc H3C 3A7, Canada

Several commercial polymers (polyethylene, polyimide, polytetrafluoroethylene, polyvinylchloride and polycarbonate) have been treated by low temperature glow discharge plasmas in various gases, namely N_2, NH_3, O_2, Ar and CF_4. These surface modifications were performed in "pure" microwave (MW, 2.45 GHz) or in combined microwave/radiofrequency (MW/RF, 2.45 GHz/13.56 MHz, "dual-frequency") plasma. Plasma diagnostics using electrostatic probes has demonstrated intense surface bombardment by energetic ions in the MW/RF case. Important systematic changes in the surface composition (determined by XPS), water wettability (by contact angle goniometry), and adhesion of evaporated aluminum films were observed for different RF-induced substrate bias values, and for the different plasma gases. The modified surface-chemical structure is correlated with contact angle hysteresis of water drops; this helps to identify which surface characteristics are connected with the wettability and with adhesion properties, and how they are influenced by plasma-surface interactions. The contribution to the latter by VUV radiation from the plasma is also examined, as is "ageing" of the plasma treated surfaces.

INTRODUCTION

Over the years, several methods have been developed to modify polymer surfaces for improved adhesion, wettability, printability, dye-uptake, etc. These include wet-chemical treatments, and exposure to flames, corona discharges, and glow discharge plasmas. Among these, low pressure, low temperature nonequilibrium plasma activation has proven increasingly attractive for the treatment of polymer surfaces[1,2]. Due to inelastic collisions of electrons (typical energies 1 - 10 eV) in low pressure (0.01 - 1 Torr) discharges, the plasma contains ions (in the same concentration as electrons, to preserve quasineutrality), and neutral molecules, namely the original ground state parent gas molecules and also chemically reactive species such as metastable excited molecules and molecular fragments (free radicals). These particles, and photons emitted from excited species, can interact with the surfaces exposed to

Metallized Plastics 2, Edited by K.L. Mittal
Plenum Press, New York, 1991

the plasma. Depending mainly upon the type of feed gas, this can result in deposition, etching (chemical and/or ion-assisted), or modification of the surface structure and surface-chemical composition ("grafting"). All these can drastically alter the macroscopic surface properties such as wettability, and adhesion to other materials.

Since Liston[2] has recently published an excellent review of plasma treatment for improved bonding, there is no need to repeat this here. Instead, we wish to describe two sets of experiments from our laboratory pertaining to different aspects of plasma-induced bonding enhancement; the first is directly related to the theme of this symposium: metallization of polymers. The second topic, dye uptake by plasma-treated polymer surfaces would appear, at first glance, to deviate rather strongly from that theme; we will show, however, that they have an important feature in common, chemical bonding at the interface. Among the four mechanisms of adhesion[3] [mechanical/interlocking, weak boundary layer, electrostatic, chemical reaction], there is an ever-growing body of evidence in favor of the latter, and it is indeed the preferred mechanism, since it should theoretically lead to the strongest bonds. Regarding metallized polymers, Burkstrand[4,5] was among the first to show that evaporated metals can react with polymer surfaces, in particular with those containing oxygen in their structure, or at least having oxidized surface sites. In the intervening years, numerous authors have employed a variety of surface analytical techniques to confirm that strong metal (M)-to-polymer bonds occur on account of the formation, for example, of M-O-C types of chemical linkages: Pireaux and coworkers[6] have shown this using High-Resolution Electron Energy Loss Spectroscopy (HREELS) for the case of aluminized polyimide, while Gerenser[7] has used XPS to demonstrate the existence of Ag-O-C and Ag-N-C complexes on plasma-treated polyethylene. These are but two examples in a rather voluminous literature on the subject[8,9].

Returning to the topic of plasma processes, the effectiveness of a given plasma-assisted surface treatment depends primarily on the nature of the feed gas, and on a number of externally controllable parameters: pressure, power, gas flow rate, frequency of the electrical energy used to excite the discharge, reactor geometry, etc. These "external" variables, in turn, affect the "internal" plasma parameters which control the overall processes, namely the electron density n_e, the average electron energy $\langle E \rangle$, the electron energy distribution function f(E), and the plasma potential V_p.

Glow discharges may also be distinguished on the basis of the applied field frequency: it has recently been shown that microwave (MW) plasma produces high concentrations of chemical precursors in the gas phase, on account of a higher population of electrons in the energetic "tail" of f(E)[10,11]. This leads, for example, to substantially higher deposition rates than at lower excitation frequencies[12], and to more efficient surface treatment, both in direct plasma exposure[13], and in "remote" reactors, where the surface is exposed to the plasma effluent downstream[14]. On the other hand, processing at lower excitation frequencies (for example, radio-frequency RF) implicates enhanced interactions of charged species with surfaces exposed to the plasma[15]. This results from ion bombardment caused by a larger difference between V_p and the surface potential V_s than is encountered in the case of MW plasma. In typical arrangements using parallel plate electrodes, the powered electrode (which is capacitively coupled to the RF power supply) develops a high negative DC self bias voltage V_s, owing to the much higher mobility of electrons in the plasma, compared with ions. The value of V_s can readily exceed several hundred volts, and it can be exploited for sputtering, reactive ion etching, and controlled ion-assisted deposition[16], to name but a few categories of technologically important processes.

Recently, the method termed "dual-frequency plasma deposition" has been used to demonstrate the growth of thin films with excellent properties, the deposition occurring at high rates near ambient temperature[17]. In this method, MW energy provides a high concentration of precursors in the plasma, while RF power applied to the substrate creates a DC bias V_s, hence controlled bombardment by ions. Investigation of the beneficial effects of this dual-frequency approach was extended for surface-modifying a series of commercial polymers in different feed gases[18]. The effect of ion energy and ion flux on the concentration of functional groups on polyethylene and polyimide surfaces treated in N_2 or NH_3 plasmas is discussed in more detail in ref. 19.

As mentioned above, the remainder of this paper will mainly deal with two sets of investigations based on surface modification of polymers with dual frequency plasma, the objectives being strong adhesion of metal and of dye molecules.

EXPERIMENTAL METHODOLOGY

Plasma Treatments

Surface treatments were carried out in a "dual-frequency" microwave (MW, 2.45 GHz)/radiofrequency (RF, 13.56 MHz) apparatus described elsewhere[18]. This reactor contained a 15 cm diameter substrate holder/electrode facing a fused silica window through which MW power was applied from a slow wave applicator. The substrate holder was capacitively coupled to an RF generator, and the self-developed, negative DC substrate bias voltage V_s[15] was measured (with reference to ground potential) using an RF choke. The reactor was turbomolecularly pumped to its base pressure prior to plasma treatments; these were performed in flows of N_2, NH_3, Ar, O_2 or CF_4 (10-50 sccm) at a constant pressure in the range 80-200 mTorr. The MW power was kept constant at 150 W in all experiments, while RF power (0-250 W), resulting in V_s values ranging from 0 to -500 V, was adjusted separately for each polymer/gas combination. Plasma exposure times were 15 or 30 seconds.

Plasma diagnostic experiments have been carried out using a Langmuir probe to measure the concentration of electrons and ions in the bulk of the plasma. A double-grid electrostatic analyser[20] at the RF powered electrode was used to measure the flux and energy of ions impinging upon the electrode (the substrate holder) as a function of V_s. Some of these measurements will be reported here, but for more details the reader may consult reference 20.

It has been pointed out by several authors, and recently summarized by Liston[2], that vacuum ultraviolet (VUV) photons may play an important role in plasma-surface interactions, on account of photochemically induced bond-breakage and subsequent reactions. In order to test this in our apparatus, we have fabricated a metal cell which can be placed on the substrate holder during a plasma run. A polymer sample in the cell is not directly exposed to the plasma, but to its radiation via 5 cm diameter interchangeable filters with various cutoff wavelengths λ_c, (or photon energies); we have used fused silica (λ_c = 160 nm), CaF_2 (120 nm) and LiF (105 nm). Gas exchange between the cell and the plasma reactor occurred through fine (1 mm ϕ) holes in the cell wall.

Characterization of Modified Surfaces

The following commercial polymer substrates have been investigated: low density polyethylene (PE, Dow Chemical Canada Inc.), polyimide (PI, DuPont Kapton H), polytetrafluoroethylene (PTFE, DuPont Teflon), polycarbonate (PC, Mobay Corp.) and surface-lubricated (with glycerol ester) polyvinylchloride (PVC, Canadian Occidental Petroleum Ltd). After plasma treatment, the samples were exposed to ambient atmosphere for approximately 5-10 minutes while being transferred to the following analytical instrument for surface characterization: the surface chemical composition was determined by X-ray photoelectron spectroscopy (XPS) in a VG-ESCALAB 3 MkII system using Mg-K_α radiation at 90° angle of incidence. The surface wettability was characterized by measuring the advancing (θ_a) and receding (θ_r) contact angles for water drops, using a micrometric syringe and a telescopic goniometer (Ramé-Hart, model NRL-C.A.). The measurement was repeated several times at different locations on each sample surface, and the average value was determined. The adhesion of 250 nm thin evaporated aluminum films was evaluated using the adhesive-tape peel-test[21]. The type 810 Scotch^R-brand tape was applied to the metallized polymer at constant pressure, then removed by peeling at a 90° angle. This was repeated at several different locations over the sample surface, and the observed adhesion was classified by measuring the percentage area of metal removed: 100% thus designates poor adhesion, all metal removed; while 0% signifies excellent adhesion, no metal removed. It is recognized that this test is not a rigorous one, and that it serves at best as a semi-quantitative evaluation of adhesion. An important merit, however, is its convenience of use.

RESULTS AND DISCUSSION

Wettability and Composition - Effect of V_s

The effect of V_s on θ_a, θ_r, and on the atomic concentration ratios (ACR) at the surface has been investigated for all polymer/gas combinations. Since the results were found to differ principally on the basis of the various plasma treatments rather than based on the initial polymer structure, we present here for the sake of brevity mainly data pertaining to PE, PI and PTFE. Results are shown in Figs. 1 and 2, for O_2 and CF_4 plasma, respectively. These graphs indicate systematic variations in surface composition, and in θ_a and θ_r values, as a function of V_s, starting from V_s = 0 ("MW", or "pure" microwave plasma, with no applied RF power). The data points labelled "N" near the ordinate axis correspond to the clean, untreated, virgin polymer surface.

In another paper on this subject[18], we have divided the range of increasing V_s values into three different regimes, designated "A", "B" and "C", as follows: "A": $V_s = 0$; "B": $0 < |V_s| \lesssim |-50V|$; "C": $|V_s| \gtrsim |-50V|$.

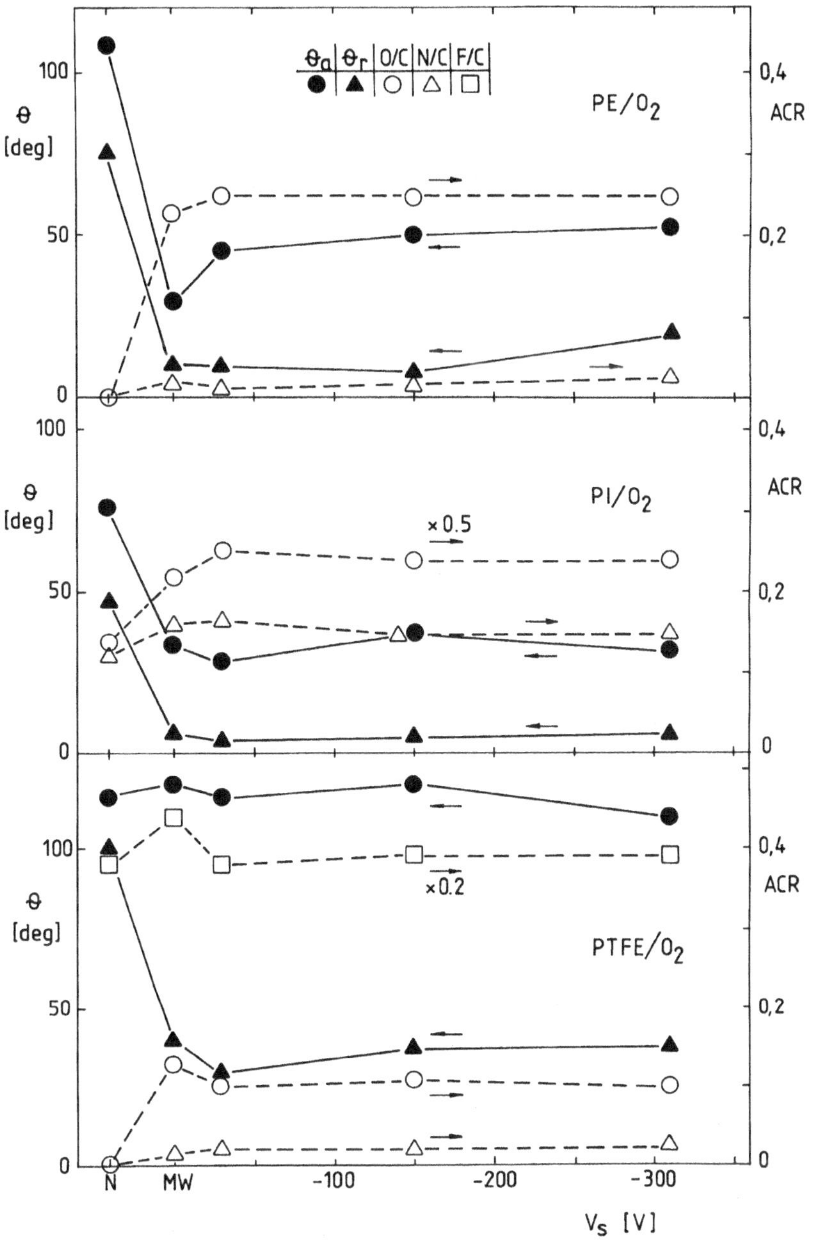

Fig. 1. Effect of substrate bias V_s on the surface characteristics of PE, PI and PTFE treated by MW/RF plasma in O_2 at 80 mTorr: advancing (●) and receding (▲) water contact angles; atomic concentration ratios (ACR): N/C (△), O/C (○), F/C (□). "N" denotes untreated polymer surface.

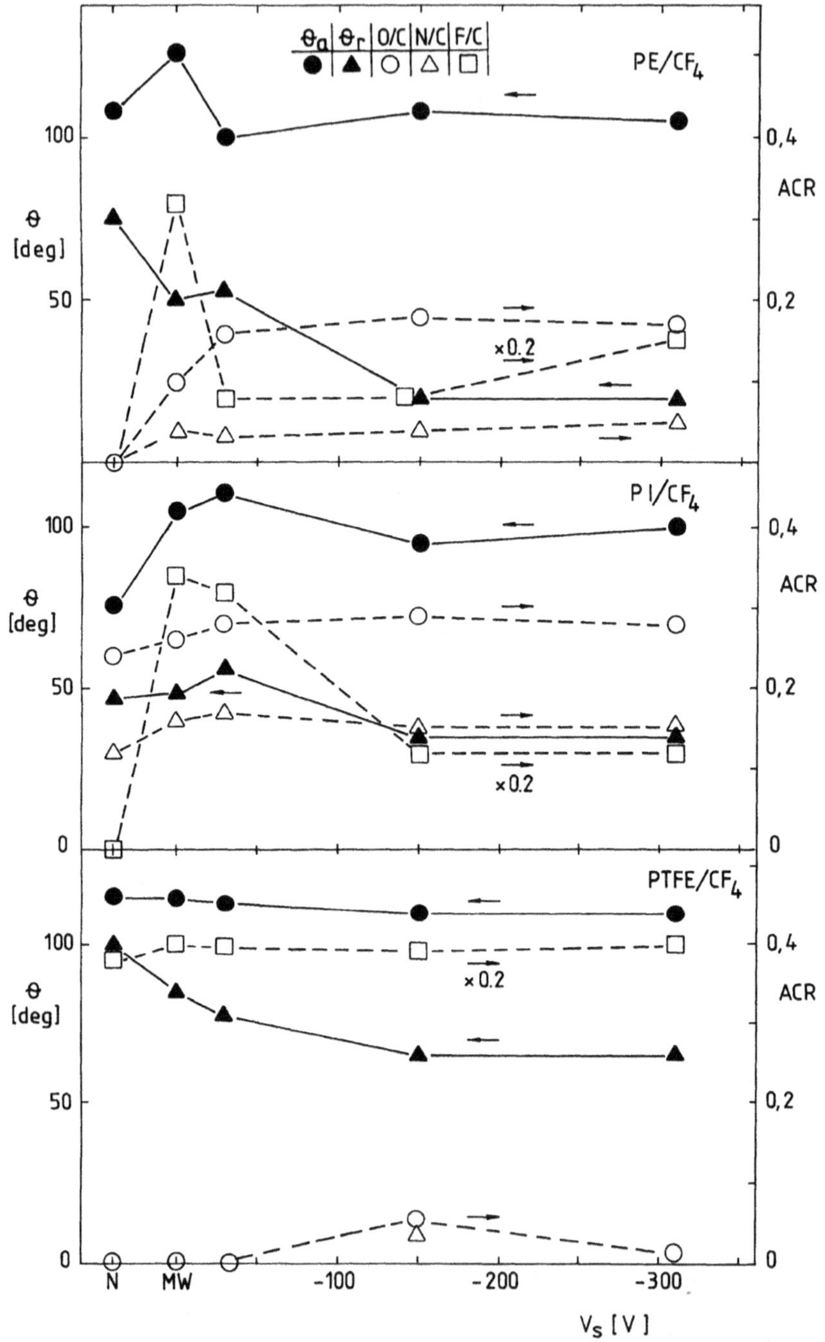

Fig. 2. Effect of substrate bias V_s on the surface characteristics of PE, PI and PTFE treated by MW/RF plasma in CF_4 at 80 mTorr. Symbols are the same as in Fig. 1.

In "A", the "pure" microwave plasma provides an abundant source of reactive species, which diffuse to the polymer surface. Since this surface is floating (a few volts positive with respect to ground), little ion bombardment occurs due to the small difference $V_p - V_s \sim 20$ V, and chemical reaction occurs mostly by attachment of radical species from the plasma. In regimes "B" and "C", on the other hand, the polymer surface is at a more negative potential with respect to the plasma, leading to a more pronounced bombardment by positive ions, as these accelerate through the plasma sheath[15]. Depending upon the mean ion energy, the effects of this bombardment can include "sputtering" of loosely (or strongly) bound surface species, and bond breakage, giving rise to increasing concentrations of chemically reactive receptor sites at the surface. The flux of ions ϕ_i arriving at the surface, and its variation with V_s, depends very strongly upon the type of gas as we may note from Fig. 3, measured at the same conditions as in Figs. 1 and 2. Considering the case of O_2 plasma, even though ϕ_i increases sharply with V_s, the O/C ratio at the surface of all three polymers shown in Fig. 1 rapidly reaches a constant level. This indicates that the principal oxidizing species are neutrals, most probably atomic oxygen. XPS shows that the O is chemically bound, mostly in the form of carbon singly bonded to oxygen.

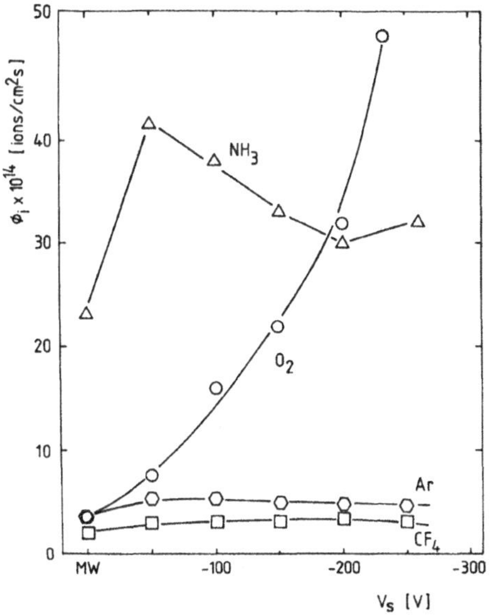

Fig. 3. Ion fluxes in MW/RF plasmas in NH_3 (△), O_2 (○), Ar (⬡), CF_4 (□) at 80 mTorr as a function of substrate bias V_s.

In the case of CF_4 plasma, ϕ_i is seen to vary little with V_s; on the basis of Fig. 2, however, we may surmise that energetic ions striking the polymer surface have a very important effect in this type of plasma: For both PE and PI, MW plasma gives rise to substantial surface fluorination, but much of this fluorine is again lost on account of ion bombardment as V_s is increased. Another very important feature may be noticed in connection with Fig. 2, namely the occurrence of surface oxidation (see PE and PTFE, in particular). This may arise from residual gases in the vacuum system and gas released from microcavities in the polymer, but it is believed to be mainly due to reaction of plasma-induced radical sites on the polymer surface with atmospheric oxygen, when the samples are removed from the vacuum chamber. The resulting polar oxygen-containing groups have an extremely strong influence on water wettability, which can also be observed in Fig. 1: Upon exposure to O_2 plasma, the originally hydrophobic polymer surfaces (see θ_a and θ_r for "N") become hydrophilic, as manifested by the sharp declines in θ_a and θ_r values. The only exception is PTFE, for which θ_a remains essentially unchanged. From Fig. 2 it is noted that CF_4 "MW" plasma renders the PE and PI surfaces even more hydrophobic on account of their becoming fluorinated, while oxygen uptake (due to the above-explained "post-plasma oxidation") leads to a reduction of θ_r.

The hysteresis in θ_a and θ_r values, both for untreated polymer surfaces and following any given type of plasma treatment, can be readily explained with the help of the so-called heterogeneous surface structure model, proposed by Johnson and Dettre[22]: Figure 4 shows values of θ_a (upper curves) and θ_r (lower curves) plotted versus N/C surface ratios (determined by XPS) for untreated and NH_3 plasma treated PE and PI, respectively. The data points on the extreme left for each polymer correspond to the untreated materials; while PE of course originally contains no nitrogen in its structure, this is not the case for PI. When changing V_s during the polymer exposure to NH_3 plasma, the surface becomes enriched with chemically bonded nitrogen, mostly in the form of amine and amide groups[19]. These, like carbonyl, are polar and hence hydrophilic. In agreement with the above- named model for heterogeneous surfaces, θ_r is seen to decrease rapidly when the hydrophilic C-N linkages are created, while θ_a remains high (on account of the hydrophobic surface component, namely C-H, C-F or C-Cℓ bonds, depending upon the particular polymer).

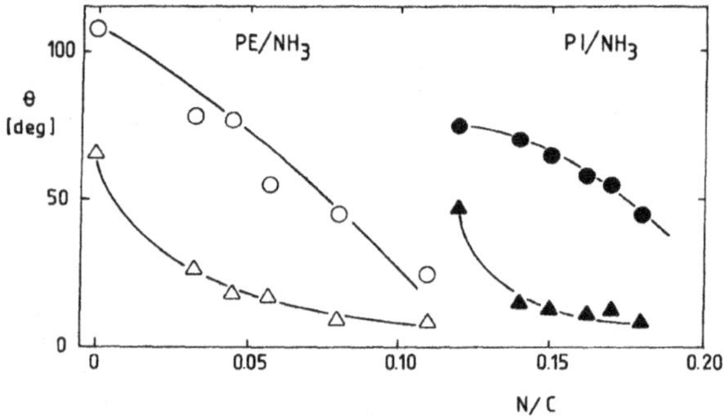

Fig .4. Effect of XPS atomic concentration ratio N/C on the advancing (○,●) and receding (△,▲) water contact angles for PE (○,△) and PI (●,▲) treated in NH_3 MW/RF plasma.

<u>Adhesion Improvement due to Plasma Treatment</u>

Table I shows the preliminary results of peel tests conducted on aluminized samples of the five untreated materials (second column), and following plasma treatments in four different gases, both in "pure" MW and in "dual frequency" plasmas. Since only the first two materials, PC

Table I. Adhesion of 250 nm evaporated $A\ell$ Films on Plasma-treated and Untreated Polymer Surfaces*

		NH_3				Ar				O_2				CF_4			
			V_s [V]				V_s [V]				V_s [V]				V_s [V]		
Polymer	Untreated	MW	-30	-150	-225	MW	-25	-150	-350	MW	-30	-150	-350	MW	-30	-150	-300
PC	90	25	45	65	2	30	50	40	10	1	0	0	0	3	0	1	0
PVC	95	70	40	65	90	0	85	70	75	1	0	0	0	85	90	80	5
PTFE	0	5	50	--	0	0	0	0	60	0	0	0	0	7	0	10	0
PI	0	--	0	0	0	0	5	30	60	0	1	2	0	10	15	10	10
PE	5	8	0	10	0	--	2	--	2	0	0	0	0	0	0	0	0

* Numerical values indicate the percentage area of evaporated $A\ell$ removed by the Scotch-tape peel-test: "100" - no adhesion, "0" - excellent adhesion.

and PVC, displayed poor metal adhesion in their untreated state, we will focus most of our comments on plasma induced changes upon these two substrates. Of course, the adhesion is governed by a complex interrelation of surface properties including wettability, chemical interactions and morphology. The latter effect is presumed to be of minor importance since no substantial changes were observed by SEM on the polymer surfaces treated for 15 or 30 seconds. Nevertheless, several important observations can be made on the basis of the tabulated results, even if we keep in mind the semiquantitative nature of the data, pointed out above:

(i) For a given plasma gas, adhesion may improve with increasing ion bombardment (see PC/NH_3 and PVC/CF_4), or the reverse may occur (e.g. PVC/Ar); no further experiments have so far been performed in order to help explain these observations.

(ii) For both PC and PVC, all O_2 plasma treatments resulted in excellent adhesion, presumably on account of the formation of C-O-Al types of chemical bonds[6,8,9].

(iii) For the case of PC (but not for PVC), CF_4 plasma treatment (and to a lesser degree, NH_3 plasma) led to good adhesion.

Observation (iii) above has some particularly interesting implications. We have shown elsewhere[18] that PC and PVC behave very similarly to PE, PI and PTFE in regard to the response of θ_a and θ_r to the various plasma treatments: NH_3, N_2 and O_2 plasma give rise to hydrophilic surfaces, while CF_4 plasma results in the expected opposite. On the basis of (iii), then, one is drawn to conclude that there exists no direct correlation between water wettability and enhanced adhesion, in the present cases. This conflicts rather sharply with one's intuitive expectations: indeed, for the case of lap shear bond strengths of polyphenylene sulfide samples (θ = 80°) with epoxy adhesive, Liston[2] (see his Fig. 3) found that bond strength increased monotonically with decreasing contact angle (increasing wettability), resulting from plasma treatments in seven different gases or gas mixtures. Nevertheless, he observed (but offered no explanation) that treatment in CF_4 + 4% O_2 resulted in a very hydrophobic surface ($\theta \simeq$ 98°) displaying the second highest bond strength! Therefore, it appears that wetting is a necessary but not sufficient condition for high adhesive strength, and that other contributing effects (e.g. weak boundary layers) should also be considered (for a review, see ref. 23).

While we only discuss wettability here, modification of surface morphology is also an important consideration. Having discussed metal adhesion to polymer surfaces, let us now briefly describe a rather different, but technologically important, aspect of plasma-induced bonding enhancement: dyeability. Non-polar polymers such as polyolefins (PE or polypropylene, PP) cannot be treated with cheap, water-based acid dyes, since the dye molecules do not adhere to the surface. Now, there is a growing incentive to use these, rather than "conventional" procedures, for both economic and environmental reasons. This can be accomplished by appropriate plasma-induced modification of the polymer surface. If "basic" (Lewis-base) moieties can be grafted to the surface, then acid-base interactions[24] between these and the (Lewis acid) dye molecules can give rise to strong ("chemical") dye adhesion to the polymer surface. Surface amination using plasma of NH_3, N_2, or volatile amines has been shown[25] to give the desired results, hexamethylenediamine plasma being a particularly effective treatment.

VUV Photochemistry and Ageing Effects

In this last section we describe, first, a series of experiments designed to distinguish VUV photon-induced surface-chemical changes from those resulting when the polymer specimen is directly immersed in the plasma. In the latter case, the situation which applied throughout the preceding sections, the observed chemical changes, of course, are the result of multiple plasma effects listed in the Introduction, which may act cumulatively or probably even synergistically. In Table II we show the amount of N uptake in the surface of PE samples exposed directly to NH_3 and to N_2 plasmas ("pure" MW, 150 W, p = 200 mTorr, t = 30 s), and with the various filters inserted between the plasma and the polymer surface. Several interesting observations can be made:

(i) Clearly, direct exposure to the plasma is the most effective; as reported elsewhere[19], higher N uptake results from N_2 than from NH_3 plasma exposure, and the same applies to O "contamination".

(ii) On the other hand, the contribution to N uptake resulting from VUV is higher in NH_3 plasma. Probable reasons are that large amounts of atomic hydrogen in NH_3 plasma give rise to more intense Lyman α emission at 121.5 nm. Dissociation of NH_3 molecules (bond energies 3.4 - 4.8 eV) accompanied by formation of hydrogen is efficient compared with that of the strongly-bound N_2 molecule (9.8 eV).

(iii) Both for N_2 and NH_3 plasmas, the N (and O) uptake increases systematically with decreasing cutoff wavelength of the filter (increasing photon energy). This, of course, is expected, since the lower cutoff materials (CaF_2 and LiF) transmit more VUV radiation, and "hotter" photons. Nevertheless, even in the LiF case, the maximum N uptake is only about 20% of that resulting from direct plasma exposure.

Table II. Nitrogen uptake on PE by direct NH_3 and N_2 plasma exposure, and by plasma UV radiation; elemental concentrations shown are in atomic %.

	NH_3			N_2		
	C	N	O	C	N	O
Direct plasma	74	21	5	66.5	26	7.5
SiO_2 (λ_c=160nm)	97	2.8	~0.2	98.6	1.3	~0.1
CaF_2 (λ_c=120nm)	96.3	3.2	~0.5	98.2	1.7	~0.1
LiF (λ_c=105nm)	95.2	4.3	~0.5	96.7	2.3	~1.0

Finally, it is interesting to address the question of "ageing" of plasma-modified surfaces, as it is known[2] that the beneficial effects of plasma treatment decrease with storage time, especially when treated specimens are exposed to hot, humid conditions. Once again, considering NH_3 plasma treated PE and PI, Figs. 5 and 6 show how θ_a and θ_r values evolve with exposure time to the indicated storage conditions. At normal relative humidity (RH), θ_a and θ_r rise to steady-state values after a few days. Further rises occur when RH is increased to 90%, but even then θ_a

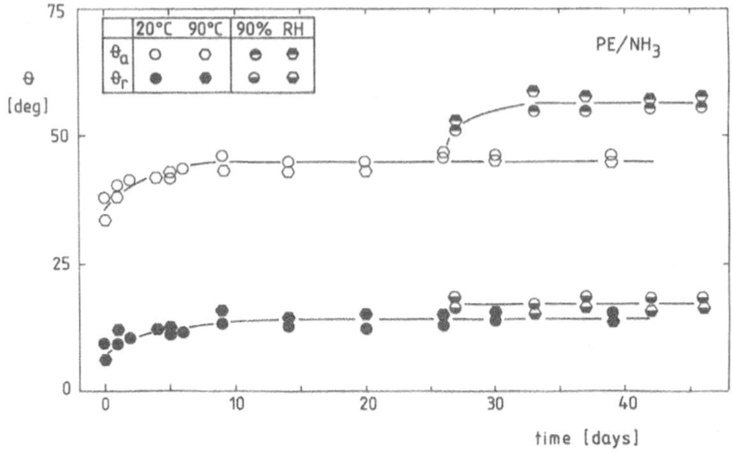

Fig. 5. Advancing (θ_a) and receding (θ_r) water contact angles for PE treated at substrate temperatures of 20°C and 90°C in MW plasma of NH_3, plotted versus storage time in ambient atmosphere and at an elevated RH.

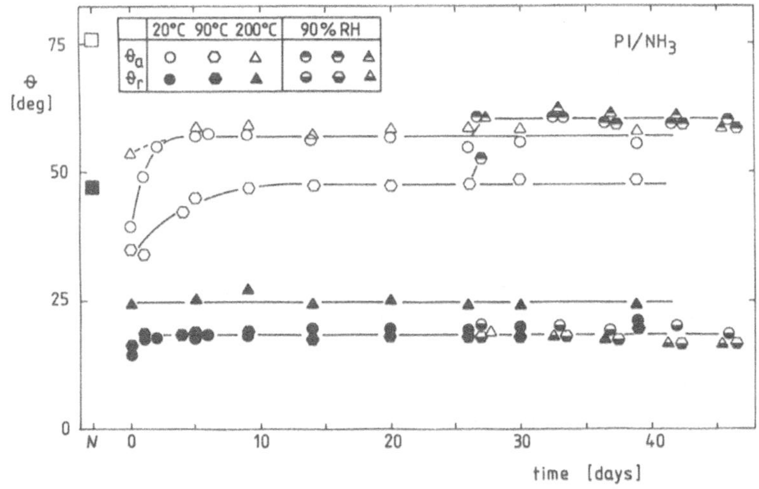

Fig. 6. Advancing (θ_a) and receding (θ_r) water contact angles for PI treated at substrate temperatures of 20°C, 90°C and 200°C in MW plasma of NH_3, plotted as a function of storage time in ambient atmosphere, and at an elevated RH.

and θ_r remain well below the θ values for the untreated polymers. It is logical to expect that these wettability changes can be associated with observable variations in the corresponding surface chemical structures of the polymers. This is indeed the case, as illustrated in Fig. 7; it shows the C(1s) XPS spectra of N_2 plasma treated PE after 1 hr, 14 days and 60 days. In Table III the time evolutions of surface composition are given in detail. Clearly, "ageing" appears to involve a loss of N, accompanied by an approximately equivalent increase in O concentration. The C(1s) spectra in Fig. 7 were deconvoluted into four components[19] that represent the C-C and C-H bonds (C1), amine C-N (C2) groups, imine C=N (C3) and amide N-C=O (C4) groups. During the aging process the imines presumably transform into amides (decrease in C3), and more C-O and C=O species are formed, which contribute to peaks C3 and C4, respectively. These changes, and their exact chemical interpretation, are currently the object of further studies in this laboratory.

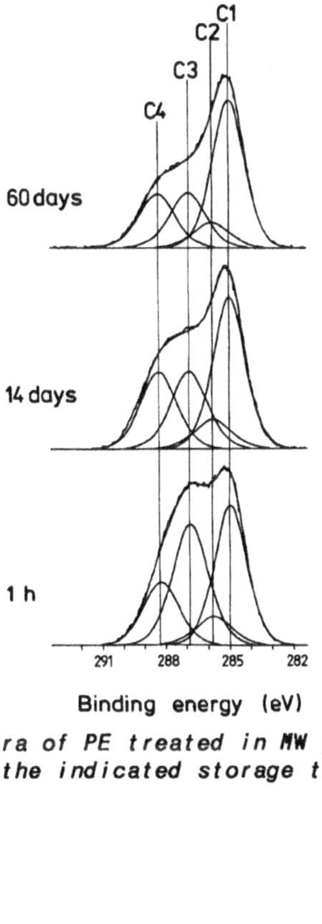

Fig. 7. C(1s) XPS spectra of PE treated in MW plasma in N_2. Spectra were recorded after the indicated storage times in ambient atmosphere.

Recently, the ageing process on polymer surfaces treated in O_2 plasma has been discussed in terms of the tendency to reach thermodynamic equilibrium[26]. It was shown that ageing proceeds differently at different depths, probed by various complementary techniques: contact angle measurements and SSIMS (~5 Å), and XPS (~50-100 Å). Ageing depends on the thermodynamic conditions, determined, for example, by composition and molecular weight of the original polymer surface region, and on temperature.

Table III. Ageing of PE treated by MW plasma in N_2 gas.

Time		1h	14 days	60 days
Concentration [at. %]	C	54	47	51
	N	38	34	30
	O	8	19	19

CONCLUSIONS

We have shown that MW and dual-frequency (MW/RF) plasma treatments constitute efficient approaches for surface modification of polymers. An important parameter for controlling the surface composition is the RF-induced substrate bias voltage V_s. Surface-compositional changes, in turn, can account for transitions from a dominantly hydrophobic to a dominantly hydrophilic surface state of the treated polymer, or vice versa. This has been clearly demonstrated by the correlation between surface composition determined by XPS, and receding and advancing water contact angles, θ_r and θ_a. Conversely, measured values of θ_r and θ_a can be used to provide information regarding surface composition. On the other hand, we have found little or no correlation between water wettability and adhesion strength of evaporated aluminum films. Like other workers, however, we have observed enhanced adhesion of $A\ell$ to most polymers following O_2 plasma treatment, probably on account of C-O-$A\ell$ chemical linkages. In the case of aminated surfaces (resulting from N_2, NH_3, or organic amine plasma treatments), enhanced adhesion of acidic dye molecules may also be explained in terms of chemical bonds.

The use of various VUV filters has allowed us to assess the relative contribution of photochemistry to surface nitriding: In the case of polyethylene, the maximum N uptake attributable to VUV irradiation (~20%) was observed for NH_3 plasma. In agreement with other workers' observations, we have also found that plasma-treated surfaces undergo "ageing", especially at elevated temperature and humidity. For the case of nitrided polyethylene, evolution with time of θ_a and θ_r values have been correlated with XPS-determined changes in chemically bonded N and O. Finally, the described dual-frequency plasma technique is very versatile, and it can be applied for deposition, etching, or surface modification merely by changing the operating parameters of the plasma reactor. This technology can readily be scaled up to large-area industrial requirements.

ACKNOWLEDGEMENTS

The authors are grateful to Dr. E. Sacher for many useful discussions and to Mr. G. Jalbert for expert technical assistance. This work was supported in part by the Natural Sciences and Engineering Research Council of Canada, and by the fonds Formation des Chercheurs et Aide à la Recherche of Québec.

REFERENCES

1. D.T. Clark, A. Dilks and D. Shuttleworth, in "Polymer Surfaces", D.T. Clark and W.J. Feast, editors, p. 185, Wiley, New York, 1978.
2. E.M. Liston, J. Adhesion 30, 199 (1989).
3. K.L. Mittal, J. Vac. Sci. Technol. 13, 19 (1976).

4. J.M. Burkstrand, Appl. Phys. Lett. 50, 1152 (1978).
5. J.M. Burkstrand, J. Vac. Sci. Technol. 15, 223 (1978).
6. J.J. Pireaux, C. Gregoire, M. Vermeersch, P.A. Thiry and R. Caudano, Surface Sci. 189/190, 903 (1987).
7. L.J. Gerenser, J. Vac. Sci. Technol. A6, 2897 (1988).
8. E. Sacher, J.J. Pireaux and S. Kowalczyk, editors, "Metallization in Polymers", A.C.S. Symp. Series 440, 1990 (in press).
9. K.L. Mittal and J.R. Susko, editors., "Metallized Plastics 1: Fundamental and Applied Aspects", Plenum Press, New York, 1989.
10. M.R. Wertheimer and M. Moisan, J. Vac. Sci. Technol. A3, 2643 (1985).
11. M. Moisan, C. Barbeau, R. Claude, C.M. Ferreira, J. Margot-Chaker, J. Paraszczak, A.B. Sa, G. Sauvé, and M.R. Wertheimer, J. Vac. Sci. Technol., accepted for publication.
12. M.R. Wertheimer, M. Moisan, J.E. Klemberg-Sapieha, and R. Claude, Pure Appl. Chem. 60, 815 (1988).
13. M. Neusch and J. Kieser, Vacuum 34, 959 (1984).
14. R. Foerch, N.S. McIntyre, R.N.S. Sodhi and D.H. Hunter, J. Appl. Polym. Sci. 40, 1903 (1990).
15. B. Chapman, "Glow Discharge Processes", Wiley, New York, 1980.
16. H. Biederman and L. Martinu, in "Plasma Deposition, Treatment and Etching of Polymers", R. d'Agostino, editor, Academic Press, New York, 1990.
17. L. Martinu, J.E. Klemberg-Sapieha, and M.R. Wertheimer, Appl. Phys. Lett. 54, 2645 (1989).
18. M.H. Bernier, J.E. Klemberg-Sapieha, L. Martinu and M.R. Wertheimer, in ref. 8.
19. J.E. Klemberg-Sapieha, L. Martinu, O.M. Küttel and M.R. Wertheimer, J. Vac. Sci. Technol., submitted.
20. O.M. Küttel, J.E. Klemberg-Sapieha, L. Martinu and M.R. Wertheimer, Thin Solid Films 193, (1990) in press.
21. J. Strong, Rev. Sci. Instrum. 6, 97 (1945).
22. R.E. Johnson and R.H. Dettre, J. Phys. Chem. 68, 1744 (1964).
23. K.L. Mittal, in "Adhesion Science and Technology", L.-H. Lee, editor, Part A, pp. 129-168, Plenum Press, New York, 1975.
24. F.M. Fowkes, J. Adhesion Sci. Technol. 1, 7 (1987).
25. L. Cop, J. Jordaan, H.P. Schreiber and M.R. Wertheimer, US Pat. 4,744,860 (1988), U.K. Pat. 2,188,947 (1987).
26. F. Garbassi, M. Morra, and E. Occhiello, these proceedings.

ELECTROCHEMICAL REDUCTION OF PMDA-ODA AND ITS EFFECT ON METAL/POLYIMIDE INTERFACIAL RELIABILITY

M.J. Schadt

IBM Systems Technology Division
1701 North Street
Endicott, New York 13760

A. Viehbeck

IBM Thomas J. Watson Research Center
P.O. Box 218
Yorktown Heights, New York 10598

Numerous polymers, including many polyimides, are known to exhibit reversible electrochemical redox behavior. This phenomenon has been observed for thermally imidized films in aqueous electrolytes, and for chemically and thermally cured polyimides in nonaqueous electrolytes. Here we report that even relatively thick chemically imidized polyimides exhibit reversible redox behavior in simple aqueous electrolytes. In addition, the effect of this electroactivity on metal to polyimide adhesion was investigated by forming metal electrodes on Kapton®' polyimide film and applying a cathodic bias in aqueous electrolyte solutions. UV-visible electronic absorption spectroscopy, Attenuated Total Reflectance-Fourier Transform Infrared Spectroscopy (ATR-FTIR) and Cyclic Voltammetry (CV) demonstrate that the species formed at the solvent saturated polyimide/metal interface is the singly reduced or "radical anion" form of Kapton®' paired with a charge-balancing counterion. The rate of formation of the radical anion has been found to be consistent with calculated hydrated-cation radial size and measured equivalent conductance data, relative cation acidity, and solution pH. When alkali metal counterions are used, reversible redox behavior of PMDA-ODA is preserved with polyimide mediated electron transfer into the bulk polyimide, or to oxygen (in O_2 containing solutions), possible. Electron transfer to oxygen is believed to result in a locally high concentration of hydroxide anions at the metal/polyimide interface. Temperature and humidity exposure accelerates the rate of polyimide hydrolysis by this nucleophile, resulting in substantially increased rates of metal to polyimide adhesion degradation. When counterion cations are used, which are more acidic in nature, the stability of the radical anion and cation complex is increased, effectively slowing the rate of polyimide mediated electron transfer. At pH below 4.5 a highly stable covalent radical-alcohol complex is suggested to result from direct protonation of the radical anion.

Metallized Plastics 2, Edited by K.L. Mittal
Plenum Press, New York, 1991

INTRODUCTION

The use of polyimides as both dielectric materials and as passivating films in microelectronic applications has become increasingly ubiquitous[2-3]. In many applications, metal circuit lines are placed in direct, intimate contact with polyimide dielectrics. While the dry film dielectric performance of these polyimides offers significant advantages, polyimides, when saturated with solvent and subjected to electrical bias conditions in the presence of mobile charged ions, can undergo electrochemical reaction.

The electrochemical activity of polyimides was first reported by Haushalter and Krause[4]. They showed that Kapton® polyimide film derived from 1,2,4,5-pyromellitic dianhydride (PMDA) and 4,4'-oxydianiline (ODA) undergoes reversible reduction/oxidation (redox) reactions in electrolyte solutions to yield radical anion and dianion forms. Mazur et al.[5] did a detailed study of the electrochemical properties of chemically imidized aromatic PMDA-derived polyimides and model compounds in nonaqueous solutions recording cyclic voltammograms, and identified the UV-visible absorbance bands of the radical anion and dianion forms. However, reversible redox behavior was not observed for chemically imidized films in these aqueous electrolytes and was presumed to be the result of decreased solvent swelling. It was recently shown that thin, thermally cured PMDA-ODA films on electrodes are electroactive in aqueous electrolytes.[6] However, no work using chemically cured films in aqueous electrolyte has been reported until now, probably due to the results obtained by Mazur et al.[5] CV data of thermally cured PMDA-ODA obtained in nonaqueous solutions were found to yield irreproducible results.[6] While the implication is made that the nonaqueous system should be more easily reduced than similar aqueous systems given increased polymer swelling (and, in turn, increased ion diffusion), it could not be shown that thermally imidized polyimides undergo reversible redox chemistry in nonaqueous solutions. Viehbeck et al.[7-8] further investigated the redox properties and effects of charge delocalization for different polyimides and imide model compounds in aprotic solutions and found that by using an electrochemically generated reducing agent to mediate electron transfer from the cathode to the polyimide surface, many polyimides could be reduced from the polymer/electrolyte interface into the bulk polymer at diffusion controlled rates.

Because polyimide materials are used in a variety of microelectronic applications, these dielectrics are subjected to electric field bias conditions during terminal product usage. Often the conditions of use require that the product function over a wide range of environmental conditions. Such conditions include elevated temperature, high humidity, and the presence of ionic contaminants. Additional exposure to these types of conditions often occurs as a result of routine fabrication processes. While the net effect of such exposures on long term product reliability is often known, in many cases the specific chemical or mechanical mechanism by which reliability is compromised is not fully understood. The work contained herein focuses on chemically imidized and thermally treated PMDA-ODA supplied commercially as Du Pont Kapton® polyimide film. As the ability to reduce polyimide films at an electrode surface or by contact with an appropriate reducing agent in an electrolyte solution has been demonstrated,[4-8] so the previously assumed inert nature of this dielectric is questioned. Here we endeavor to record and explain the interfacial chemistry of the chemically imidized (and thermally treated) polyimide PMDA-ODA supplied as Du Pont Kapton® when placed in contact with a cathodically biased metal electrode in the presence of aqueous solvent and simple ionic contaminants, and to explain the effect this chemistry has upon the integrity of the metal/polyimide interfacial bond.

EXPERIMENTAL

Starting Materials

Kapton® 50.8 μm thick H-type films were purchased from E.I. Du Pont de Nemours Inc. Starting materials and solvents were purchased from Aldrich Chemical Co.; acetonitrile (ACN) and N,N-dimethylformamide (DMF), were obtained anhydrously in Sure/Seal™[1] bottles and used as received. The polyamic acid of PMDA-ODA was supplied by DuPont as 2545 Pyralin®[1]. The chloride and bromide salts of potassium, sodium, lithium, and ammonium were obtained from Fisher Scientific Company. Tetramethylammonium tetrafluoroborate was obtained from Aldrich. All salts were a minimum of reagent quality and used as received.

Polymer Films

Polyimide films were made by casting or spin-coating the precursor polyamic acid solution onto conductive substrates. Imide cyclization was achieved by either heating the films to 400 °C for 30 min. when thermally imidized films were desired, or through a chemical dehydration process involving immersion in a 1:1 mixture of acetic anhydride and pyridine for the production of chemically imidized films. The PMDA-NBA (NBA= n-butylamine) model compound was synthesized.[9]

Fabrication of Test Structures on Kapton®

Electrodes were formed *in-situ* on the polyimide surface by photopatterning a vacuum-deposited metal seed layer of Cr/Cu on Kapton®. The metallized starting material is available from several commercial sources. Upon this metal, a photomask was patterned and the electrodes formed by acid Cu electroplating. Typical Cu electrode thickness measured 38 μm. Photoresist was then removed and the thin vacuum-deposited layer sub-etched[10] to isolate individual electrodes. Typical electrode line width measured ≃ 50.8 μm.

Electrochemical and Spectroscopic Methods

Cyclic voltammetry (CV) was carried out using a conventional three-compartment cell and saturated calomel reference electrode (SCE). Anode and cathode compartments were separated by a fine porosity fritted glass disk. All experiments were done under a nitrogen blanket at 23 ± 2 °C. Nonaqueous solutions were made from anhydrous solvents and 0.1 M tetrabutylammonium tetrafluoroborate (TBAFB) as supporting electrolyte unless otherwise stated. Aqueous salt solutions were made using 18 MΩ deionized water and 0.2 M salts. UV-visible measurements were made using an HP 8452A (Hewlett-Packard) diode array spectrophotometer. Surface ATR-FTIR spectra were obtained on an IBM Instruments IR44 Fourier Transform spectrometer using a KRS-5 (45°) crystal. Spectroscopic and CV analysis of thin films was accomplished by spin coating the precursor polyamic acid onto conductive tin oxide coated quartz wafers, or, in the case of CV analysis only, by dip coating on the appropriate electrode. Samples could then be studied by cathodic bias in the appropriate electrolyte solution.

Samples on tin oxide/quartz were placed directly into the spectrophotometer sampling beam. To obtain spectra at the metal/polyimide interface, *in-situ* fine-line metal electrodes were peeled from the polyimide surface following cathodic bias at -1.2 V for the stated time and electrolyte concentration.

Peel Force Measurement

Peel data were obtained on an Instron model A1026 90° peel tester using a 50-gram load cell. Linewidths were optically measured and peel force values corrected for variation from a nominal 203 μm line width. The average peel values reported were obtained from 15-20 individual peels.

RESULTS AND DISCUSSION

Thin film polyimides (< 2000 Å) can readily undergo reversible reduction/oxidation reactions at an electrode surface in an electrolyte solution.[2-6] Work using comparatively thick film polyimides has also been reported,[7-8] however this work has until now been limited to nonaqueous electrolytes. These papers describe the CV response for 12 μm thick chemically imidized PMDA-ODA,[8] and the UV-visible electronic absorption spectra of 12.7 μm thick Kapton® following reduction with an electrochemically generated reducing agent.[7] The UV-visible spectra and CV response for 8 μm thick thermally imidized film on conductive tin oxide electrodes has also been reported.[7] The electrochemistry of thick polyimide films in aqueous electrolytes has been much more difficult to study however, due to the poor swelling of these films in aqueous media, and the barrier these thick films present to diffusing counterion cations. The results presented below offer evidence supporting reversible redox electrochemisty in thick PMDA-ODA films in aqueous electrolytes. This work further attempts to gain insight into cation effects on the redox mechanism, and the mechanism at work which causes decreased cathode to polymer bonding once reduction of polyimide at the interface has occurred. Cation incorporation in, and redox effects on polyimides imply possible long term reliability concerns.

Aqueous Cyclic Voltammetry of Thick Film PMDA-ODA

In order for reversible redox behavior to occur, cations must be readily available at the reducing interface for radical anion/counterion ion-pair formation. Figure 1 shows schematically a typical electrochemical cell used for measuring the electrochemical properties of polyimides. CV results obtained on relatively *thick* (12 μm) films demonstrate that even these thick films are capable of passing electrolyte and solvent to the polyimide/metal-cathode interface. Figure 2 shows the CV response obtained for aqueous reduction of chemically cured PMDA-ODA polyimide on a stainless steel electrode in N_2 purged KCl electrolyte. Although the CV contains a large uncompensated resistance and some degree of irreversibility due to the kinetic limitations toward cation diffusion posed by the thick polymer, two anodic and two cathodic responses are clearly evident.

Due to the poor CV resolution there is a high degree of uncertainty involved in attempting to calculate half-wave potentials. Considering this uncertainty, we feel that the estimated half-wave potentials obtained from this CV (-0.58 V for radical anion formation and -0.92 V for dianion formation) are in reasonable agreement with results obtained earlier from thin thermally cured films (-0.70 and -0.96 V respectively)[6] and for thin chemically cured polyimide in nonaqueous electrolytes (-0.73 and -1.13 V respectively)[8]. This result offers clear evidence for cation and solvent

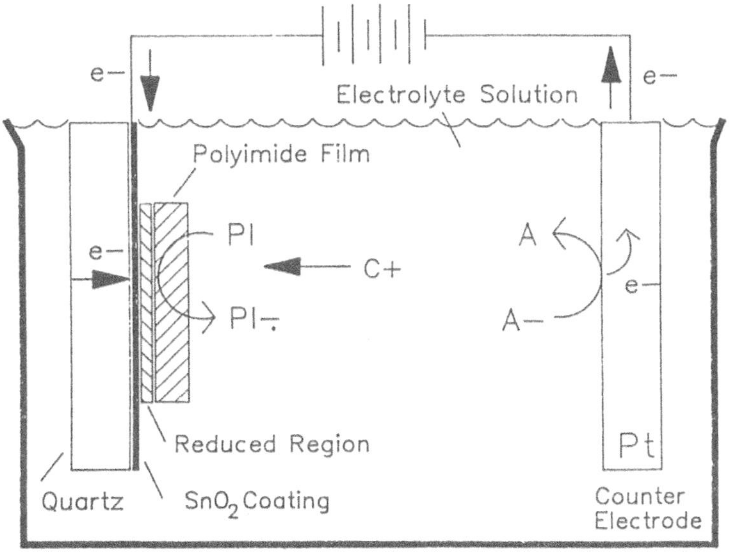

Figure 1. Schematic of electrochemical cell used for measuring the electrochemical properties of polyimides.

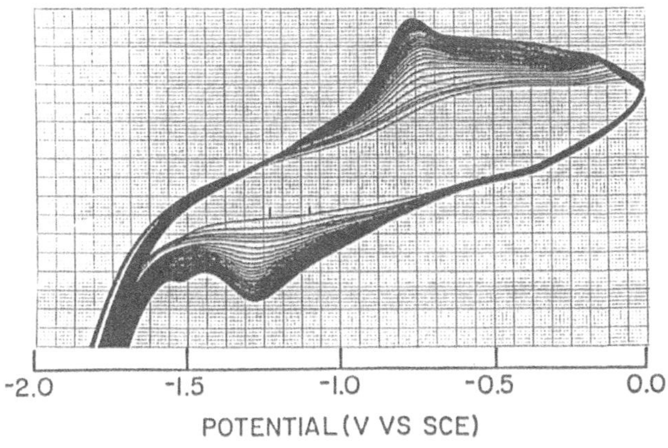

Figure 2. Cyclic voltammogram obtained from reduction of a 12 μm chemically imidized film of PMDA-ODA polyimide coated onto a stainless steel electrode. Trace was obtained in aqueous 1M KCl solution with a scan rate of 2 mV/second.

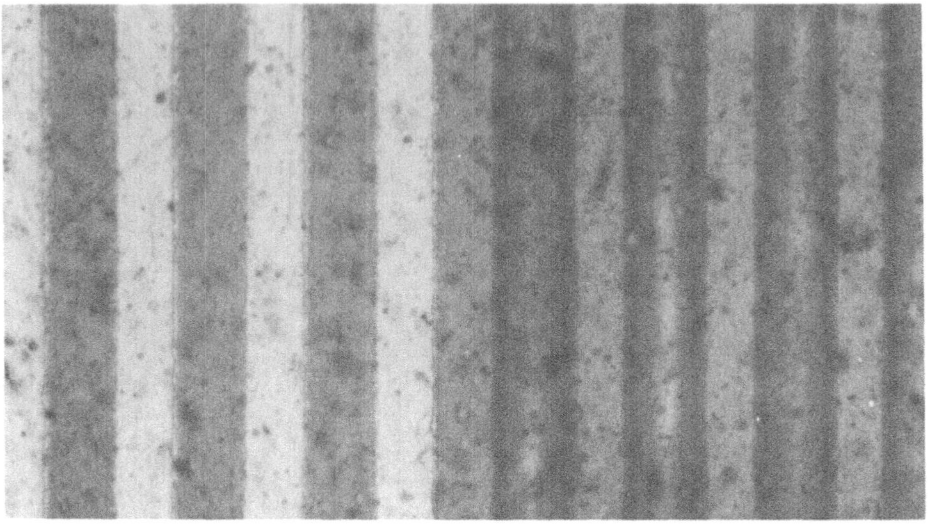

Figure 3. Optical photograph of 63.5 μm wide Cu electrodes cathodically biased in 0.2 M KCl solution. Dark lines were biased and show evidence of the radical anion form extending from the metal/polyimide interface into the bulk polymer. Lighter lines at left were unbiased for comparison.

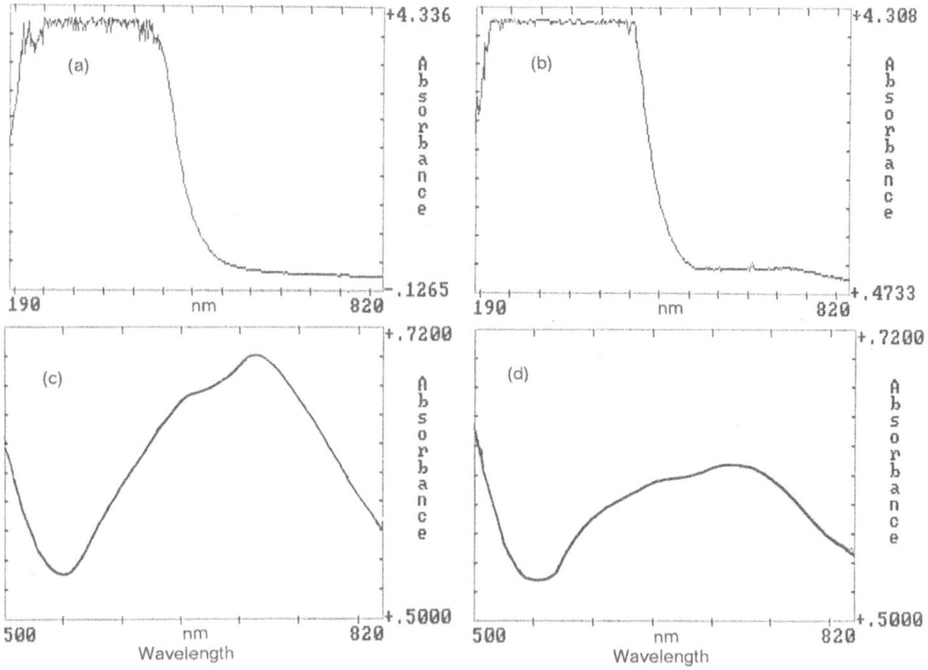

Figure 4. UV-vis spectra of 50.8 μm Kapton following circuitry bias and peel in 0.2 M aqueous KCl; (a) before bias, (b) immediately following bias and peel, (c) subtraction spectrum of (b)-(a), and (d) subtraction spectrum obtained from subtraction of (a) from data obtained 40 minutes following electrode peel.

diffusion to the polyimide/cathode interface, and reversible polyimide redox behavior even in thick-polyimide/aqueous-electrolyte systems.

Electrochemical Reduction of Thick Film Kapton®

The UV-visible and FTIR spectral characteristics and CV response of electrochemically reduced PMDA-ODA polyimides have been reported previously.[4-8] The results presented here will be discussed in relation to this earlier work.

Figure 3 shows an optical photograph of 63.5 μm wide electrodes on 50.8 μm thick H-type Kapton® cathodically biased in 2M KCl solution for 10 minutes at -1.2 V. These lines are arranged on a 127 μm pitch so that sufficient polymer could be affected for spectroscopic analysis. In addition, the narrow electrodes could be easily removed due to their relatively small interfacial contact area. In this photograph, an obvious darkening of the polyimide is observed beneath the biased Cu electrodes. Here the sample is viewed from the non-metallized side of the polymer and the metal/polyimide interface is visible. The electrodes which do not show this surrounding darkness were not electrically biased and are for comparison only. Visual inspection shows this darkening to be green in color. Under bias conditions the green color first appears at the interface formed by the intersection of metal, polyimide, and electrolyte solution. With continued bias, the green discoloration is observed to increase in both area (spreads into the bulk polymer) and optical density (increasing polyimide/unit area is affected). Evidence that this green color is the result of radical anion formation comes from the following.

Following potentiostatic cathodic bias (-1.2 V) for two hours in aqueous 0.2 M KCl salt solution, the sample electrodes were peeled away and spectroscopic measurements made. Figures 4a-4d show the UV-visible spectra obtained from the above experiment at a time a) immediately preceding reduction, b) immediately following reduction, c) b minus a subtraction spectrum immediately following reduction, and d) subtraction spectrum obtained from data 40 min. following reduction. Absorbance maxima at 666 and 720 nm are in excellent agreement with those reported earlier as characteristic of the radical anion.[6] The time-dependent drop in absorbance indicates that disappearance of the radical anion is occurring. That this disappearance is atmosphere-dependent is supported by data obtained from samples having comparable reduction times but which were not peeled until one to two hours following reduction. During this waiting period the samples were stored in aqueous unpurged electrolyte solution. These samples showed higher remaining concentrations of radical anion polyimide indicating that the radical anion, when isolated to some degree by the electrode from air, is longer lived. Further evidence in support of air sensitivity is discussed later.

Figure 5 shows the ATR-FTIR spectra of a virgin 50.8 μm Kapton® film and one following electrolysis and electrode removal. The absorbance maxima at 1772 and 1725 cm^{-1} are attributed to imide and ring carbonyl absorptions and are present in both spectra since only a small portion of the bulk polyimide has been reduced. Growth in absorbance in the 1656 and 1647 wave number regions is consistent with the formation of the radical anion species and in agreement with earlier work.[8] Spectra obtained in HCl gave no detectable changes in transmittance relative to a virgin film. Under acidic conditions (pH $\simeq$ 4), UV-visible spectra characteristic of radical anion/cation pair formation were also unobtainable. These results indicate that PMDA-ODA reduction was either blocked, or the radical anion reacted preferentially with protons.

The electronic absorption results presented above demonstrate that the green coloration appearing at the metal electrode/polyimide interface is due to the formation of radical anion polyimide paired with a charge balancing counterion. The observed mode of color growth parallels the potassium ion distribution in this layer as measured by Auger electron spectroscopy[11]. Here, the regions of highest optical density correspond to

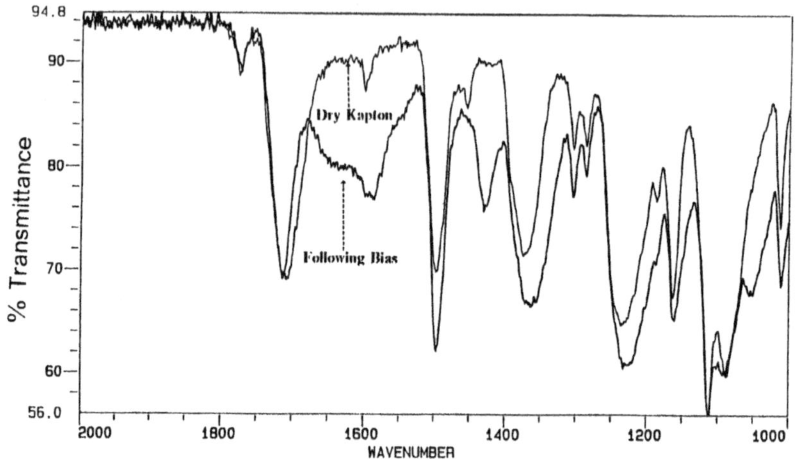

Figure 5. ATR-FTIR of virgin 50.8 μm Dry Kapton®, and Kapton® following -1.2 V electrode bias for 20 minutes in aqueous 0.2 M KCl. Spectra were obtained following removal of the in-situ electrodes to expose the metal/polyimide interface to the sampling beam.

regions having the greatest potassium ion concentration. The reduction/ion diffusion profile in the polymer and beneath electrodes is consistent with an ion-pair (formation) driven diffusion process. As potassium cations diffuse to the reduction interface they allow for the reduction of PMDA-ODA and concurrently form an electrically neutral ion-pair with the reducing polymer. This charge neutralization gradient creates a net ion gradient which promotes subsequent cation diffusion. Profiles are consistent with anticipated solvent penetration into the Kapton® and are localized radially from the metal, polyimide, and solution interface, and are skewed through the polyimide toward the bulk solution (since solution swelling of Kapton® is prerequisite for ion movement). Concurrently, the reducing interface moves into the bulk polymer as electrons can now electrochemically exchange across this solvent-saturated/ion-pair rich polyimide salt region.

To further investigate the effects of cations and pH in this electrochemical system, the rate of reduction of 50.8μm thick Kapton® film was monitored by UV-visible spectrophotometry as a function of counterion species and pH. These results are discussed in detail below.

Ionic Effects on Kapton® Reduction

Dry polyimides have excellent dielectric properties. Electron conduction across a dry, 50.8 μm thick Kapton® film requires approximately 212.5 V/μm.[12] However, electrochemical reduction of polyimides occurs at much lower applied potentials, as evidenced from the CV data obtained for several polyimides derived from differing precursor diimines.[8] The much lower magnitude of the applied potential required for electrochemical reduction illustrates the role of solvated cations in this process.

Cations allow a much slower polymer reduction/electron exchange mechanism to reduce the polyimide film in a layer upon layer fashion via ion pair formation. Conductivity measurements indicate an increase of 11-13 orders of magnitude and support an *interchain electron hopping*[6] mechanism. According to this mechanism, cations supplied by the electrolyte enter the film at the polyimide/solution interface. Thus, the rate of reduction of the polymer can be limited by the diffusion rate of the cations through the polymer film to the reducing interface. Because less polymer swelling occurs in aqueous solvent systems, polymer reduction rates in these systems may exhibit a much greater cation-size rate dependence. In addition, these effects are enhanced as the thickness of the polyimide film is increased. Here the electrode/polymer/electrolyte system is analogous to a circuitized PMDA-ODA film which is immersed in an electrolyte solution under voltage bias conditions. Thus we see that the presence of solvated cations allows an electrochemical pathway to form which may eventually support current passage through the polyimide film. The rate of reduction may either be limited by cation diffusion into the film, or by the electron hopping (self-exchange) between neighboring diimide species. Although we have not determined here which is the limiting step, we do report the effects of solvated cation size, cation acidity, and pH on the reduction phenomenon and discuss them in more detail below.

Cation size dependence was studied using 0.2M H, Li, K, Na, NH_4, and $N(CH_3)_4$, each from its respective chloride salt. The ordering below, based on the rate of formation of radical anion as determined from UV-visible absorbance data and Beer's law considerations is $K^+ > Na^+ >> Li^+ > NH_4^+ > N(CH_3)_4^+ >> H^+$, consistent with results reported elsewhere.[6] For metal cations the rate of diffusion through the polyimide film should decrease with increasing hydrated cation radial size, and this is indeed observed spectrophotometrically. Hydrated radii have been calculated based on their limiting mobility from Stokes' law as Li^+, 2.37; Na^+, 1.83; and K^+, 1.25 Å.[13] This observation is also in good agreement with equivalent conductance data[14] for 0.1 gram-equivalent salts per cubic liter of solvent with LiCl, 95.86; NaCl, 106.74; and KCl, 128.96. The case of NH_4^+ is special and we believe closely related to the results obtained at lower pH. NH_4^+ has a hydrated radial size of 1.25 Å and an equivalent conductance of 128.75, virtually identical to the respective values for K^+. Despite these similarities, there is very rapid radical anion formation using K^+ counterions and relatively little radical anion formation when NH_4^+ cations are substituted. Likewise, radical anion formation is not observed when a high concentration of protons is present. We believe there are plausible explanations for these observations.

Radical anion formation in PMDA-ODA occurs with primary involvement of the carbonyl functionality resulting in a net negative charge at carbonyl oxygen. Proton interaction at this site should result in a high degree of covalent interaction between hydrogen and oxygen and give rise to a more stable radical anion/proton complex, one which is not as easily dissociated to allow for electron exchange to nearby polyimide-- a process requisite for electron exchange and for further reduction into the bulk polyimide. One might consider this to be the formation of a radical-alcohol species rather than a radical anion polyimide paired with a counterion cation. This mechanism would be dependent upon the kinetics of the system as well as the thermodynamic stability of the complex formed; however, results are consistent with the pH dependence (H^+ concentration) observed. At neutral and higher pH's where a large excess concentration of alternate cations (i.e. K^+) exists, proton effects are negligible due to their relatively low concentration. At lower pH, concentration effects bring more protons into the film and result in the formation of much more of the relatively stable polyimide radical-alcohol species. The slow rate of polyimide reduction into the bulk polyimide film from the cathode surface when NH_4^+ is present may be the result of the formation of a similar complex between hydrogen (bonded to nitrogen in NH_4^+) and the carbonyl oxygen of radical anion polyimide. In this case, hydrogen to oxygen bonding interactions would be less covalent in nature than in the case of lone protons; however, we would expect a significant contribution to bonding from covalent interaction compared with that occurring for alkali metal cations. To explain this complex formation consider the propensity of three species to undergo dissociation liberating protons. Comparing the pKa's[14] of HCl, NH_4Cl, and H_2O we see that a dissociation giving rise to a free proton by each species occurs most readily for HCl, less readily for NH_4^+, and least readily for H_2O. Adding HCl to our system results in the formation of hydronium cations which are highly acidic and most probably result in proton transfer to radical anion polyimide and in the formation of the polyimide radical-alcohol species. Formation of this layer is postulated to limit or block further electron exchange from such sites. Water, however, which forms the solvation sphere of hydronium ions, ammonium ions, and alkali metal cations is, only very weakly acidic, and is most probably not sufficiently acidic to facilitate proton transfer to radical anion polyimide. Evidence for this comes from the reversible redox behavior of the radical anion (CV data) when paired with a hydrated alkali metal cation. Thus, for hydrated alkali metal cation species, the alkali metal cation/polyimide radical anion interaction most probably remains largely ionic in nature and capable of supporting further polyimide reduction by an electron exchange mechanism. Because ammonium ions occupy the interim acidity range between HCl and H_2O, and in fact are somewhat acidic, it is not unreasonable to postulate that ammonium ions may either transfer protons directly to radical anion polyimide, or may remain as ammonium ions but interact differently with radical anion polyimide. This interaction will not be largely ionic as is the case for the alkali metal cations, nor will it be largely covalent as in the case of radical-alcohol formation, but will exist as an ion pair complex with partial covalent character. This interaction being directly attributable to the more acidic nature of hydrated ammonium cations compared to that of hydrated alkali metal ions. This ammonium/radical anion complex would therefore be expected to be more stable than a corresponding radical anion/alkali metal cation pair, and somewhat less stable than a radical-alcohol species formed from protonation of the radical anion directly. The stability ordering necessarily implies that the reverse reaction (oxidation of the radical anion polyimide) required for electron exchange to a neighboring polyimide and reduction into the bulk should occur most readily when alkali metal cation-containing ion-pairs are formed, and most slowly when radical-alcohol species are formed. This discussion is indeed consistent with our experimental observations with regard to the rate of PMDA-ODA radical anion formation as a function of hydrated cation radial size, cation equivalent conductance, cation acidity, and solution pH.

An important realization which is derived from the above discussion is that the reduction of PMDA-ODA to its radical anion can be limited by either choosing metal cations of larger size which are less mobile within the polyimide matrix, or by adjusting the concentration of protons so that electron exchange within the polyimide is limited by the formation of an enhanced stability radical anion complex. In the former case, large hydrated cations slow the reduction process. In the case of protons, they may actually block further polyimide reduction depending upon other cell parameters (e.g., magnitude of applied cathodic potential).

We have seen how the nature of the aqueous electrolyte can affect the production of radical anion PMDA-ODA. Now, for cases where radical anion formation has occurred, let us examine the disposition of radical anion Kapton® and its impact on adhesion at the metal/polyimide interface.

Table I. Comparison of average (15-20 individual peels) adhesion force of 208.3 μm wide electrodes to Kapton® obtained following: (1) fabrication of test sample; (2) temperature and humidity conditioning (T&H) at 85 °C and 80% relative humidity for 100 hours; (3) cathodic bias at -1.5 V for 2 hours in an aqueous 2 M KCl solution; and (4) cathodic bias followed by T&H as described in (2) and (3) above.

Test Conditions	Relative Adhesion
control	1.00
T&H only	0.47
bias in KCl only	0.74
bias in KCl, T&H	0.04

Adhesion Degradation Mechanism

To examine the effect of radical anion production on adhesion at the metal/polyimide interface, peel tests were conducted on the samples shown in Table I. The control sample was measured following its fabrication and its peel value (g/mm) taken to be unity. Peel values of the remaining samples are normalized to this benchmark value. It is obvious from the results in Table I that accelerated aging (T&H) at 85 °C and 80% relative humidity for 100 hours results in approximately a 53% decrease of metal to polyimide adhesion. Bias alone, in the absence of T&H, resulted in a 26% drop in the average adhesion force measured. However, for parts that were subjected to bias conditions and then subjected to 100 hours of T&H conditioning, the peel value decreased (relative to the control sample peel) by 96%. This catastrophic loss of adhesion, while due in part to the *in-situ* generation of the radical anion species at the interface (and a necessarily decreased level of metal to polyimide charge transfer), was rapidly accelerated by the effects of increased humidity (a solvent) and elevated temperature (thermal activation energy). From these observations we conclude that the generation of radical anion Kapton® at the metal/ polymer interface is the first step in a chemical reaction sequence leading to decreased bonding. In the following we attempt to explain the observed phenomena in terms of a mechanistic chemical pathway.

As discussed previously, radical anion polyimide can be generated (from either the thermally or chemically imidized forms) by either contact with a cathode in an electrolyte solution or by contact with an appropriate reducing agent in a supporting electrolyte. Under controlled atmospheric conditions (dry N_2 and less than 10 ppm detectable O_2) in aprotic solvents the radical anion form is stable indefinitely. Similarly, the radical anion paired with a counterion is relatively stable in the absence of solvent despite the presence of oxygen. This is presumed to be the result of the dependence of the reduction potential of oxygen on its environment and/or due to the limited mobility of unsolvated cations in the dry film. UV-visible spectra obtained over an extended period showed no decline in absorbance in the 560-730 nm region indicating that the radical anion concentration was unchanged. Upon introduction of O_2 (in the presence of nonaqueous electrolyte and absence of water) the characteristic absorbance of the radical anion is observed to vanish rapidly, returning to that of virgin film. ATR-FTIR spectra of comparative samples (virgin vs. reduced and O_2 treated) showed recovery of the virgin polymer film following oxygen treatment indicating that oxygen, in the presence of electrolyte, is capable of scavenging an electron from the polyimide radical anion. Radical-induced polymer dimerization or cleavage is discounted based on these ATR-FTIR and UV-visible spectral results. The FTIR results offer strong support in light of its relatively shallow sampling depth into the polyimide (approx. 1-2 μm).

Direct reaction of radical anion polyimide with water has been discounted based on CV results. Only minimal shifts in reduction and oxidation waves are observed when CV scans of samples generated under dry nonaqueous O_2-free solution conditions are compared with those made with the additions of 1, 2, and 10% water to the nonaqueous electrolyte. This retention of reversible redox behavior indicates that little, if any, electron transfer from reduced polymer to water, or proton transfer directly to radical polyimide, is occurring. From this evidence we conclude that radical anion polyimide reacts readily (when there is sufficient solvent in the film) with oxygen according to Equation (1):

$$2O_2 + 2\text{PMDA-ODA}^{\cdot -} \rightarrow 2O_2^{\cdot -} + 2\text{PMDA-ODA} \quad (1)$$

The ability to produce the superoxide radical in solution by electron transfer from a reducing agent has been shown elsewhere.[15] The probability that superoxide will behave as a one-electron oxidant forming the peroxide dianion is very small[16]. The reactivity of superoxide in both aqueous and nonaqueous solutions has been studied by pulse radiolysis techniques[18-20] and has shown that the tendency of superoxide to disproportionate by abstraction of protons from both solvents and substrates (containing acidic hydrogens) is its most characteristic reaction pathway.[21] Proton abstraction from water has been shown to readily occur at pH 14 terminating with complete disproportionation of superoxide.[22] It has been estimated that superoxide has the ability to promote proton transfer from substrates (and solvents) to an extent equivalent to that of the conjugate base of an acid with pKa= 24.[21] The driving force for this reaction is the formation of the hydroperoxy radical shown in Equation (2) by proton abstraction from water (aqueous solution conditions) via Equation (3):

$$2O_2^{\cdot -} + 2H^+ \rightarrow 2HO_2^{\cdot} \quad (2)$$

$$2H_2O \rightarrow 2H^+ + 2OH^- \quad (3)$$

$$2HO_2^{\cdot} \rightarrow H_2O_2 + O_2 \quad (4)$$

The hydroperoxy radical is known to react according to Equation (4).[23] Summation of Equations (1) to (4) yields Equation (5) which gives the overall equation for the reaction occurring at the metal/polyimide interface.

$$2\text{PMDA-ODA}^{\cdot -} + 2O_2 + 2H_2O \rightarrow 2\text{PMDA-ODA} + H_2O_2 + O_2 + 2OH^- \quad (5)$$

The overall reaction is consistent with both the empirically observed behavior of reduced PMDA-ODA and the literature account of known reaction pathways. Here neutral Kapton® is regenerated by reaction with oxygen, forming superoxide. The disproportionation of superoxide yields hydrogen peroxide, oxygen, and hydroxide at the metal/polyimide interface. Stabilizing potassium cations, previously paired with reduced PMDA-ODA, are made available by radical anion reoxidation in stoichiometric amounts, maintaining overall electrical neutrality in the film. In addition, this chemical process liberates 1 mole of oxygen for every 2 moles consumed, thereby recycling oxygen for superoxide formation.

Adhesion degradation mechanisms would therefore be those that result from the action of H_2O_2 or OH^-. While H_2O_2 is known to react with tertiary amines to produce amine oxides[24] (good leaving groups), a nearby alkyl carbon chain, required for alkene formation by a Cope Elimination pathway, is absent. H_2O_2 and OH^- exist as a function of the equilibrium shown in Equation (6):

$$H_2O_2 + OH^- \leftrightarrow HO_2^- + H_2O \quad (6)$$

Therefore, a plausible polyimide reaction mechanism capable of affecting polymer hydrolysis is expected to result from the reaction of either hydroperoxy anion or hydroxide anion (both nucleophiles) following a route of attack at the PMDA carbonyl carbon. The presence of a cathodic circuit bias more negative than -1.0 V, and the demonstrated ability of radical anion PMDA-ODA to act as a one-electron reducing agent (reversible redox behavior) influence this equilibrium in the following way. The most significant electrochemical reaction capable of eliminating the hydroperoxy anion would do so by combining two electrons, two protons, and hydrogen peroxide to form two moles of water. This reaction occurs at +1.56 V vs. SCE. However, because few free protons are available in this system (neutral pH), the reaction is probably of little significance. More importantly, hydroperoxy anions can be combined with water and two electrons at +0.63 V vs. SCE to yield three moles of hydroxide anions. Not only is this reaction strongly favored, but it also gives rise to additional hydroxide at the interface (in addition to that produced by proton abstraction from water by superoxide). This reaction most probably also results in the removal of any hydroperoxy anions produced by reaction of water and oxygen with two electrons at -0.31 V vs SCE. Although the formation of hydroperoxy anions is possible by the preceding reaction, the formation is not as favored for electrochemical considerations as is the reduction of the anion to hydroxide. Overall, the formation of hydroperoxy anion is not favored in this electrochemical system. If kinetic limitations exist (diffusion of ions), then additional hydroxide may be produced by hydrolytic cleavage of water via either reduction and reaction with oxygen at +0.16 V vs. SCE yielding two moles of OH^- for every mole of water reduced; and/or water may be reduced directly to H_2 and OH^- at -1.07 V vs. SCE. These reactions may serve as additional sources of hydroxide. The reduction potentials quoted are, of course, strongly influenced by pH. From these reactions we believe that in all likelihood a high concentration of hydroxide is most probably the active species at the metal/polymer interface. The reactive nature of this nucleophile with PMDA-ODA is discussed below.

Base catalyzed hydrolysis of PMDA-ODA by alkali metal hydroxides (esp. KOH) is described in the literature[12] and proprietary formulations are in current use for etching of PMDA-ODA films.[25-26] These procedures detail the use of highly concentrated KOH at 80 to 95 °C for 5-10 min. as adequate for accomplishing complete hydrolysis of 50.8 μm thick films. According to the preceding discussion, hydroxide anion/potassium cation salts are produced and trapped locally at the metal/polyimide interface during electrochemical reduction. That comparatively less adhesion degradation is observed immediately following bias compared with post T&H exposure is consistent with the observed slow rate of hydrolysis of the PMDA-ODA by hydroxide at room temperature. However, T&H conditioning effectively reproduces the conditions of elevated temperature and solvent concentration required for efficient polyimide hydrolysis. The results presented in Table I are consistent with this effect. Again, the correlation between known polyimide reduction profiles and observed adhesion loss profiles supports the hypothesis that polyimide reduction ultimately gives rise to metal/polymer adhesion degradation. That a substantial increase in adhesion degradation is observed following T&H conditioning is consistent with the known rates of polyimide hydrolysis by hydroxide nucleophile (with potassium counterions) over this temperature range.

CONCLUSIONS

Radical anion formation is observed visually (by its characteristic green color) and confirmed spectroscopically at the combined metal electrode, polyimide, and solution interface for even relatively thick chemically imidized Kapton® films (12 μm and 50.8 μm) which have received a post-imidization thermal treatment. Reduction proceeds into the bulk polymer at rates consistent with solvated cation mobility in the swollen polymer matrix as determined by hydrated cation radial size and equivalent conductance data. The nature of the cation (acidity) and solution pH also strongly affect the rate of polyimide radical anion to cation complex formation. In the case of protons, a largely covalent polyimide radical-alcohol is postulated. When the cation is weakly acidic such as ammonium, a complex which is partially covalent in nature is suggested; and when alkali metal cations are used, a complex which is largely ionic in nature results. The relative stability of these complexes may act to determine the rate of polyimide reduction by limiting polyimide mediated electron exchange to neutral neighboring polyimide species. In the presence of oxygen, however, the radical anion rapidly undergoes electron transfer to oxygen forming superoxide. This highly reactive species readily abstracts protons from water yielding hydroxide anions and hydroperoxy radicals. These hydroperoxy radicals most likely undergo electrochemical reduction with water at the electrode surface and yield additional hydroxide anions. In addition, direct reduction of water and oxygen at the cathode is possible in oxygen containing systems. The net effect is a relatively high concentration of hydroxide anions paired with cations at the metal/polymer interface. While initial adhesion degradation occurs, this alkali hydroxide salt can remain in the polymer film until such time as sufficient solvent is again present for ion solvation and hydrolysis. That the rate of hydrolysis is also strongly dependent upon thermal input is shown by the adhesion data presented. Polyimide hydrolysis by nucleophilic attack of hydroxide at the PMDA carbonyl carbon is well understood and provides rationale for understanding the adhesion degradation occurring. As some manufactured microelectronic components and assemblies that incorporate PMDA-ODA polyimide as dielectric material also place metal conductors in direct contact with it, the phenomena described herein may present very real reliability concerns with respect to both adhesion loss and mobile ion contamination in either the short or long term depending upon the physical environment of the application.

ACKNOWLEDGEMENTS

The authors wish to thank Martin Goldberg (IBM Yorktown) for the PMDA-NBA model compound synthesis, Terrance O'Tool (IBM East Fishkill) for his contributions to the spectroscopic analysis, and Walter Pawlowski (IBM Endicott) for helpful discussions regarding the base catalyzed hydrolysis of PMDA-ODA.

REFERENCES

1. Kapton® and Pyralin® are registered trademarks of of E.I. Du Pont de Nemours & Co., Thermid® is a registered trademark of the National Starch and Chemical Corp., XU-218® is a registered trademark of Ciba-Geigy Corp., and Sure-Seal™ is a trademark of Aldrich Chemical Co.
2. For a review, see K.L. Mittal, Editor, "Polyimides: Synthesis, Characterization, and Applications," Vols. 1 and 2, Plenum Press, New York, (1984).
3. "Polymer Materials For Electronic Applications," ACS Symposium Series No. 184, Amer. Chem. Soc., Washington, D.C. (1982).
4. (a) R.C. Haushalter and L.J. Krause, Thin Solid Films, 102, 161 (1983); (b) 2, pp. 735-750.
5. S. Mazur, P.S. Lugg, and C. Yarnitzky, J. Electrochem. Soc., 134, 346 (1987).

6. (a) L.J. Krause and J.L. Bales, J. Electrochem. Soc., 135, 1137 (1988); (b) L.J. Krause, P.S. Lugg, and T.A. Speckhard, J. Electrochem. Soc., 136(5), 1379 (1989).
7. A. Viehbeck, C.A. Kovac, S.L. Buchwalter, M.J. Goldberg, and S.L. Tisdale, in "Metallization of Polymers," E. Šacher, J.-J. Pireaux, S.P. Kowalczyk editors, pp. 394-414, ACS Symposium Series no. 440, American Chemical Society (1990).
8. A. Viehbeck, M.J. Goldberg, and C.A. Kovac, J. Electrochem. Soc., 137, 1460 (1990).
9. Condensation of the aromatic anhydride and *n*-butylamine in N,N-dimethylacetamide.
10. Olin-Hunt High Speed Circuit Etch, Cat no. 838383, W. Patterson, NJ..
11. R. Paonessa, unpublished results, IBM Endicott (1989).
12. Du Pont Kapton® Bulletin # E-41395.
13. K. Itaya, T. Ataka, and S. Toshima, J. Am. Chem. Soc., 104, 4767 (1982).
14. R.C. Weast, editor, "Handbook of Physics and Chemistry," 63rd addition, CRC Press, Boca Raton, Florida, 1982-83.
15. D.T. Sawyer, D.T. Richens, E.J. Nanni, and M.D. Stallings, Dev. Biochem., 11(A), 1 (1980).
16. D.T. Sawyer, M.J. Gibian, M.M. Morrison, and E.T. Seo, J. Am. Chem. Soc., 100, 627 (1978).
17. J.A. Fec and J.S. Valentine, in "Superoxide and Superoxide Dismutase," Michelson, McCord, and Fridovich Editors, pp. 16-60, Academic Press, New York, (1977).
18. B.H. Bielski, J. Photochem. Photobiol., 28, 645 (1978).
19. Y.A. Han, D. Meisel, and G. Czapski, Israel J. Chem., 12, 891 (1974).
20. D. Behan, G. Czapski, J. Rabani, L.M. Dorfman, and H.A. Schwartz, J. Phys. Chem., 74, 3209 (1970).
21. D.T. Sawyer and J.S. Valentine, Acc. Chem. Res., 14, 393 (1981).
22. G.F. Carter, J.L. Margrove, and D.H. Templeton, Acta. Crystallogr., 5, 85 (1952).
23. G. Czapski and L.M. Dorfman, J. Phys. Chem., 68(5), 1169 (1964).
24. W.H. Reusch, "An Introduction to Organic Chemistry," pp. 339-343, Holden-Day, San Franscisco, California (1977).
25. US Patent #4846929, July 11, 1989.
26. US Patent #4857143, August 15, 1989.

PART IV. ADHESION ASPECTS OF METALLIZED PLASTICS

STICKY POLYMERS THROUGH ION BEAM PROCESSING

P. Anders Ingemarsson

Asea Brown Boveri Ltd.
Corporate Research
CH-5405 Baden-Dättwil
Switzerland

Ion beams can be used to enhance the adhesion of metal films to polymer substrates. Typical doses range from 10^{11} to $10^{16}/cm^2$ depending on materials, ion species, ion energies, etc. Recent progress in the understanding of radiation induced chemical and structural modifications, observed by various analytical techniques, is discussed.

INTRODUCTION

Polymer surfaces are generally associated with low sticking coefficients and poor adhesion. Depending on the function of the material, these features can be advantageous (e.g. in reducing wear or friction) or disadvantageous (e.g. in thin film applications). Accordingly, the metallization of polymers often implies that some additional processing step must be employed in order to increase the surface energy, thereby producing interfaces of sufficient strength and integrity[1]. The wide spectrum of potential applications for various metal-polymer combinations has encouraged numerous scientists and engineers in their search for effective methods to promote bonding and to obtain a better understanding of its controlling factors.

Besides mechanical[2], wet chemical[3], and flame surface pre-treatments[4], different types of particle bombardment techniques are used for the improvement of thin film adhesion. Such techniques can be more or less complicated, both with respect to the experimental equipment required and the physical mechanisms involved. In a plasma treatment, for example, the surface is exposed to an energetically distributed flux of atoms, ions, electrons, molecules as well as electromagnetic radiation[5]. Consequently, the energy transferred to the substrate is dissipated through a variety of processes, each associated with different impinging species. The corresponding surface alteration is therefore a result of the cumulative effect of these processes.

Metallized Plastics 2, Edited by K.L. Mittal
Plenum Press, New York, 1991

When studying the impact of a well-defined monoenergetic ion beam, the complexity of the particle-solid interactions is considerably reduced. These interactions are different in nature depending on the specific energy (energy per atomic mass unit) of the penetrating ions. At the energies most commonly used in materials science, two primary interactions can be distinguished: In the keV/amu region, energy is transferred mainly through elastic collisions with the target atoms (nuclear stopping) whereas above ~0.1 MeV/amu, ionization and excitation processes prevail (electronic stopping). Sketched in Figure 1 is the energy loss as a function of specific energy for the different modes of interaction.

Characteristic of ion beams in the electronic stopping region are large ion ranges and negligible sputtering of metal surfaces. These energies are thus well-suited also for post-deposition treatments of thin film structures. When such ions penetrate a metal-polymer interface, two phenomena are perceivable; while passing through the metal film, the fast electronic relaxation of ionizations and excitations will disperse any potential defects that could lead to atomic displacements and, accordingly, no damage will remain after the passage of the ions (the impact of nuclear stopping events can be neglected); reaching the polymer substrate, however, electronic disturbances caused by the impinging ions will last long enough to allow for atomic displacements around the ion tracks.

For several reasons, it is of special interest to compare the modifications of pre- and post-irradiated interfaces. Firstly, the influence of a present overlayer can be investigated; pre-bombardment allows for the erosion of contaminants and preferential sputtering, hence changing the surface composition, whilst post-bombardment should be expected to enhance the possibility of migration of one constituent into the other. Secondly, the poor adherence of metals to polymers has been attributed to the presence of Weak Boundary Layers (WBL) in the polymer surface[6,7], the idea being that the actual fracture does not occur at the film-substrate interface but rather within the substrate itself; if the impact of the ion beam would result in a toughening of these layers, such an effect would be observed following both pre- and post-irradiation treatment, respectively.

The intent of this report is to shed some light on the present status and recent progress in the understanding of ion beam induced chemical and structural modifications at metal-polymer interfaces. Rather than going through the particulars of numerous experiments, however, we will acquaint

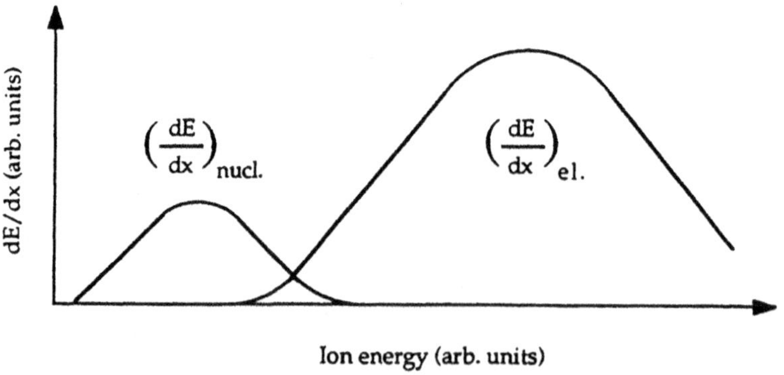

Figure 1. Schematics of energy loss versus ion energy, showing the nuclear stopping and electronic stopping regions.

ourselves with some general circumstances regarding the choice of ion species, energy, dose, pre-irradiation versus post-irradiation effects, etc. Furthermore, results will be presented indicating what chemical interactions are involved at the interface, what interface structures may develop subsequent to the ion bombardment, and also the importance of determining the locus of an adhesion failure. As examples, PTFE (polytetrafluoroethylene) and PVC (polyvinyl chloride) interfaces will be discussed in view of observations made by a number of analytical tools. The emphasis will be on the use of MeV ion beams since most of the author's work was carried out in this regime.

METAL-PTFE INTERFACES

An atomically clean Teflon surface only reluctantly forms bonds with adjacent atomic species. This manifests itself in a high degree of inertness to contamination and also in low sticking coefficients at the initial stage of thin film evaporations[8]. Figure 2 shows an example of how the amount of iron atoms deposited onto Si and PTFE surfaces, as measured by Rutherford Backscattering Spectrometry (RBS), relates to that simultaneously deposited onto a quartz crystal. As is clearly seen, the sticking coefficient for iron evaporated onto an untreated Teflon surface increases with increasing film thickness. Thus, in order to achieve a desired thickness of 25 Å, the crystal should record a thickness of 100 Å. Also shown in Figure 2 is the corresponding curve for iron evaporated onto silicon. In this case, the dependence is quite linear even at small thicknesses implying that the sticking coefficient of iron to silicon is approximately equal to that of iron to iron. At thicknesses above a certain limit, the lines appear to be parallel indicating that the substrate surface is completely covered by the iron film.

Metal films deposited onto Teflon surfaces can easily be peeled off by tape. Depending on what film material is being used, however, a wide

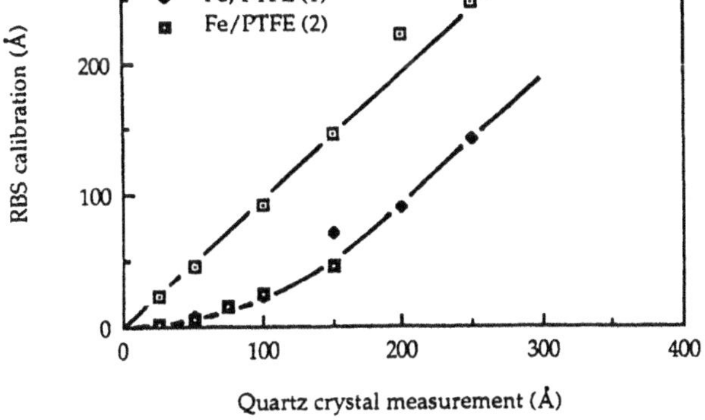

Figure 2. Sticking behaviour of Fe evaporated onto Si and PTFE, respectively; taken from Reference 8.

range of adhesion strengths can be attained (see Figure 3). This phenomenon has been attributed to differences in cross-link densities in the polymer surface layers, the idea being that the more reactive metals give rise to enhanced fluorine abstraction[9]. This interpretation, based on the WBL theory, was later supported by observations made in an X-ray Photoelectron Spectroscopy (XPS) study by Font and Montes[10], in which monolayers of iron were sequentially deposited onto Teflon surfaces. For each deposited monolayer, an increase in the signals corresponding to iron-fluorine bonds was recorded. Furthermore, the onset of these components was accompanied by cross-linking of the Teflon surface layers.

Ion irradiation of metal-PTFE interfaces at doses of between 10^{11} and 10^{16} ions/cm^2 results in a substantial adhesion improvement. The dose required to achieve a certain degree of adhesion strength, the so called threshold dose (D_{th}), depends on the stopping power of the incident ion beam. Table I lists ion species, ion energies and the corresponding doses for gold thin films on Teflon reported in the literature[8,11-15]. The dependence of D_{th} on the stopping cross section is plotted in Figure 4. The threshold dose, as measured by the Scotch Tape test (i.e. no film removed by the tape), for Au on Teflon scales as the electronic stopping cross section to the -1.1 power.

Liu and Dong[16] recently observed D_{th} to vary with the applied film material. In their study, an ion beam of 6 MeV fluorine was used to irradiate thin films of Ni, Al, Cu, Pd, Sn and In deposited onto Teflon

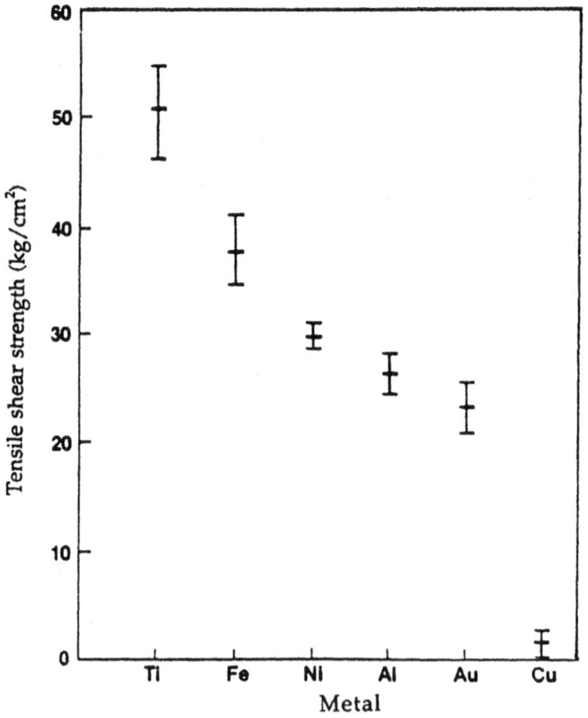

Figure 3. Tensile shear strengths of metal films evaporated onto PTFE; taken from Reference 9.

Table I. Particle beams and corresponding threshold doses required for adhesion enhancement of Au films on Teflon.

Beam particle	Threshold dose	Reference
1 MeV H^+	$1 \times 10^{14}/cm^2$	11
2 MeV H^+	$5 \times 10^{14}/cm^2$	12
1 MeV He^+	$\sim 3 \times 10^{13}/cm^2$	12
1.5 MeV He^+	$1 \times 10^{13}/cm^2$	11
5 MeV F^{2+}	$3 \times 10^{12}/cm^2$	12
12 MeV S^{3+}	$1 \times 10^{12}/cm^2$	13
16 MeV S^{3+}	$1 \times 10^{12}/cm^2$	8
20 MeV Cl^{4+}	$1 \times 10^{12}/cm^2$	14
0.25-3.5 keV e^-	$< 1 \times 10^{16}/cm^2$	15
0.2-3.0 MeV γ	$< 1 \times 10^7$ rads	15

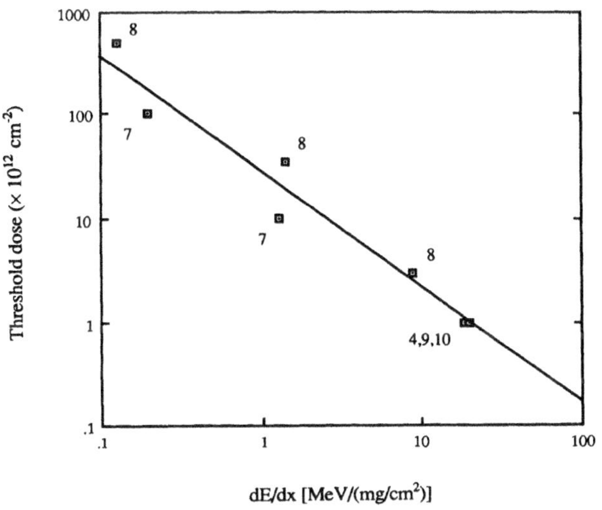

Figure 4. Dependence of threshold dose required for adhesion enhancement on stopping cross section; based on data listed in Table I.

substrates. Their results are depicted in Figure 5. Comparing these findings with the data presented in Figure 3, we see, for example, that although the adhesion of an as-deposited film is much better for Ni than for Cu, the required threshold dose is higher. The variation in D_{th} was found to be proportional to the ratio of the film material elasticity modulus to the electronic stopping power, and was attributed to changes in the internal stresses in the film and substrate. One might also consider differences in the adhesion between the applied tape and the film, the different film materials being associated with various rates of radiation induced cracking of hydrocarbons on the surface.

It is noteworthy that the ion beam treatment is just as effective when performed prior to thin film deposition[8] (for low energy ions actually *more* effective[17]), despite atmospheric exposure of the Teflon surface between the two experimental steps. The pre-bombardment effect can be investigated by a variety of techniques due to the absence of the film

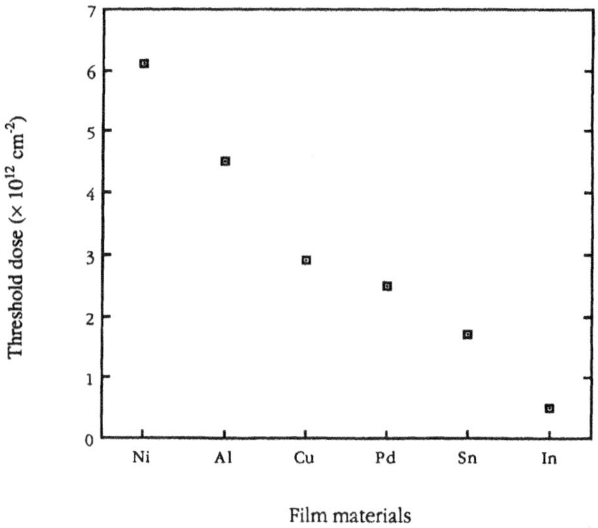

Figure 5. Dependence of threshold dose required for adhesion enhancement on film material; data taken from Reference 16.

material. By means of a simple mass spectrometer, extensive desorption yields of molecular species have been observed during irradiation[14]. Further information about the residual chemical state of the substrate surface may be provided by XPS, showing signals which can be attributed to cross-linking and branching of the polymer chains[14]. Such a cross-linking effect would result in a tougher surface layer structure and hence increase the interfacial strength.

The existence of weak boundary layers in the Teflon can easily be established by XPS analysis showing fluorine to be present at the surface of a removed film[14]. However, the WBL theory does not explain the bonding between film and substrate atoms, nor does the development of Fe-F bonds observed by Font and Montes[10] since the fluorine atoms can only be bonded either to iron atoms in the film or to carbon atoms in the substrate. Accordingly, in studying radiation induced interface effects, one has to look for analytical techniques capable of revealing chemical bonds between film atoms and carbon atoms in the substrate or, possibly, ternary compounds.

Evaporation of ^{57}Fe thin films onto pre-bombarded surfaces enables the use of Conversion Electron Mössbauer Spectroscopy (CEMS) to examine the interface chemistry and its response to ion irradiation[8]. Figure 6a shows a typical Mössbauer spectrum for an as-deposited 25 Å film of α-iron. When the film is evaporated onto a pre-bombarded surface, doublets indicative of iron-carbon (D1) as well as iron-fluorine (D2) compounds are seen to develop as a function of dose (see Figure 6b-c).

The CEMS technique can successfully be exploited also in studies of post-irradiation effects. In this case, since the iron film is present during the actual treatment, the formation of chemical compounds is even more pronounced, though presenting another distribution of the different species; as is evident from Figure 7, the formation of iron-fluorine species appears to be dominant. Furthermore, *two* iron-fluorine doublets (D2 and D3) can be observed.

It is clear that the development of D1 and D3 can be associated with pre- and post-irradiation, respectively, whereas the intensity of D2 shows a similar variation with dose for the two cases. Based on the intensity

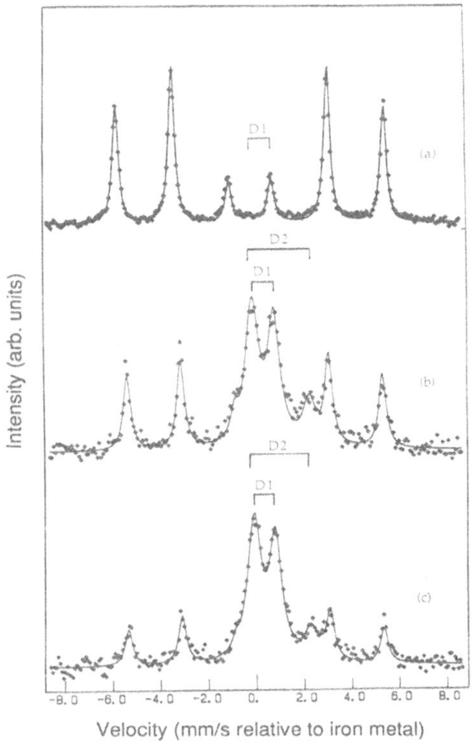

Figure 6. Mössbauer spectra for ^{57}Fe films deposited onto PTFE surfaces pre-bombarded with 16 MeV S^{3+} ion beams; (a) unirradiated, and irradiated at doses of (b) 10^{13} ions/cm^2 and (c) 10^{14} ions/cm^2; the doublets D1 and D2 correspond to Fe-C and Fe-F compounds, respectively; taken from Reference 8.

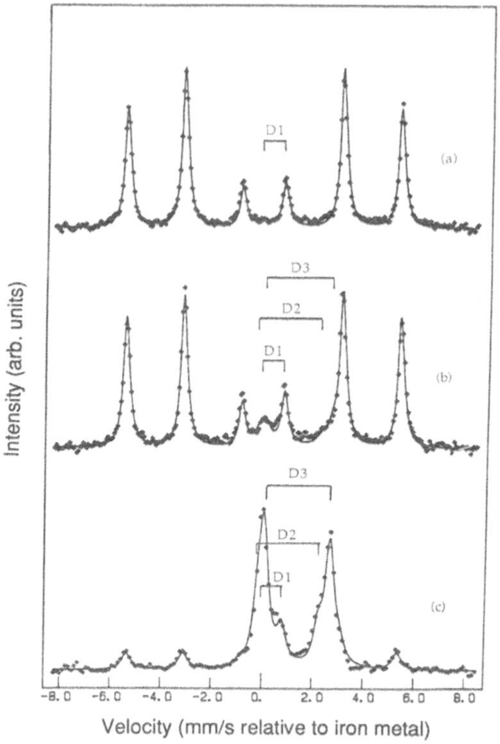

Figure 7. Mössbauer spectra recorded from ^{57}Fe films on Teflon PTFE (a) before, and after ion irradiation with 16 MeV S^{3+} to doses of (b) 10^{13} and (c) 10^{14} ions/cm^2; doublets indicate the presence of Fe-C (D1) and Fe-F (D2,D3) compounds; taken from Reference 8.

difference for D3 in the two cases, this component could be attributed to iron-fluorine bonds formed within the film. The D2 component, on the other hand, shows only a small difference between pre- and post-bombarded samples and could, accordingly, be related to interface compound formation. The centroid shift as well as the quadrupole splitting of this doublet are both consistent with a reduced iron-fluorine interaction. This situation might arise at the interface (due to the presence of carbon in the substrate) rather than within the film. Moreover, the relative intensity of D2 never exceeds ~20% which corresponds to a film thickness of 5 Å, i.e. 1 - 2 monolayers. The D2 doublet might correspond to the Fe-F signal observed by Font and Montes[10], though the interaction with the carbon at the interface may not be resolved by the XPS technique.

As concerns the D1 component, it should be noted that irradiation of Teflon surfaces has been found to result in a decrease in the relative abundance of fluorine[14]. Consequently, iron films deposited onto pre-irradiated Teflon substrates should encounter a comparatively carbon-rich surrounding, thus increasing the probability of forming iron-carbon compounds. When post-irradiated, however, the iron atoms in the film will preferentially react with the fluorine released from the substrate.

The fact that the changes in the Mössbauer spectra of pre-bombarded surfaces do not resemble those of post-bombarded samples indicates that

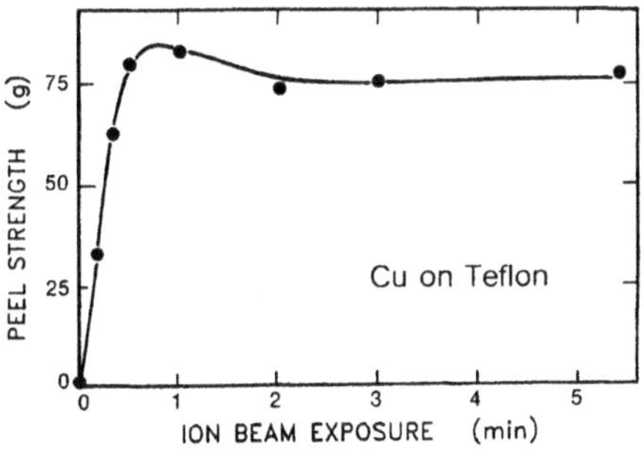

Figure 8. Adhesion between Cu and Teflon, expressed in terms of peel strength, as a function of sputtering time of Teflon prior to Cu deposition; taken from Reference 17.

the adhesion enhancement is not related to the observed bond formation in a simple way. This is in agreement with the XPS results[14] indicating that the locus of the weakest structure of a metal-Teflon interface is within the Teflon itself. If this is so, strengthening of the Teflon surface following ion impact would occur irrespective of when the irradiation is carried out (prior to, or after film deposition) and despite atmospheric exposure of a pre-bombarded surface. On the other hand, the D2 component, observed in the Mössbauer spectra at the same extent for pre- and post-irradiation and attributed to interface interactions between iron-fluorine species and carbon atoms in the substrate matrix, might disguise a chemical term necessary to explain the strong film-substrate adhesion.

A general behaviour, observed at metal-Teflon interfaces subjected to high-energy beam treatments, is that a maximum in adhesion strength is reached at a certain particle dose above which the adhesion starts to deteriorate[15,18]. Such an adhesion deterioration is further accompanied by a discolouration, i.e. carbonization, of the substrate surface. However, Chang *et al.*[17] found that low-energy pre-sputtering with 500 eV Ar^+ and subsequent evaporation of Cu thin films leads to an adhesion improvement which, above a certain dose, levels off at a maximum increase of ~50 times the original strength. The adhesion effect, which is quantitatively illustrated in Figure 8, was ascribed to enhanced surface morphology and chemical bonding. Apparently, such a 'soft' surface processing does not cause severe degradation of the Teflon as is the case at higher ion energies.

Also ageing effects have been observed. For example, post-irradiation with 1 MeV He^+ at $10^{13}/cm^2$ did not enhance bonding when tested immediately after irradiation, whereas five days later, the bonding was found to be comparable to that of samples exposed to slightly higher doses[12]. The authors mean that this effect could be attributed to alterations of the film surface which, when employing the Scotch Tape test, would lead to variations in the adhesion of the tape to the Au film.

METAL-PVC INTERFACES

When Fe-PVC interfaces are subjected to MeV ion irradiation, the resulting effects are more dramatic and occur at even lower doses than in the case of Teflon substrates. This implies that analyses with lower depth resolution, such as Elastic Recoil Detection Analysis (ERDA) and Rutherford Backscattering Spectrometry (RBS), can be used as a complement to the CEMS technique. Special care must be exercised to the ion beam analysis, however, since the impinging ions gradually modify the materials during the actual measurements.

Irradiation of Fe-PVC thin film structures with 48 MeV Br^{8+} ions leads to an enrichment of chlorine in the interface region at increasing ion doses, as revealed by both RBS and ERDA[19]. Figure 9 shows a typical ERDA energy spectrum for Cl in PVC coated with Fe, corresponding to an ion dose of $\sim 10^{15}$ ions/cm^2. The enrichment of Cl near the substrate surface is evidently seen on the right hand side of the spectrum. From this analysis, it is not possible to determine how the chlorine appears, i.e. whether it is bonded to film atoms or simply exists at a higher concentration in the outermost surface layers of the substrate.

Figure 10 displays typical Mössbauer spectra obtained from ^{57}Fe films on PVC[19]. Shown in Figure 10a is a spectrum recorded from a 100 Å thick unirradiated film, whilst 10b and 10c show spectra obtained for irradiated films of 100 Å and 50 Å thickness, respectively. A quadrupole doublet, attributed to Fe-Cl compounds, clearly develops at the expense of the original magnetic six-line pattern. In the 50 Å sample, complete conversion is attained after bombardment at a dose of 5×10^{12} ions/cm^2, whereas in the 100 Å sample, the intensity of the doublet is only 40%.

The CEMS technique thus provides the necessary information in order to conclude that the Cl enrichment observed in the ERDA spectrum is associated with a transformation of the Fe film into an ionic Fe-Cl compound. In addition, the films exhibited dramatic changes in their optical and electrical properties, from being metallic and conductive to transparent and insulating.

CONCLUDING REMARKS

Comparing the two systems discussed above, one finds that only the metal-PTFE interface maintains its layered structure when subjected to ion irradiation. The metal-PVC interface is already at very low doses com-

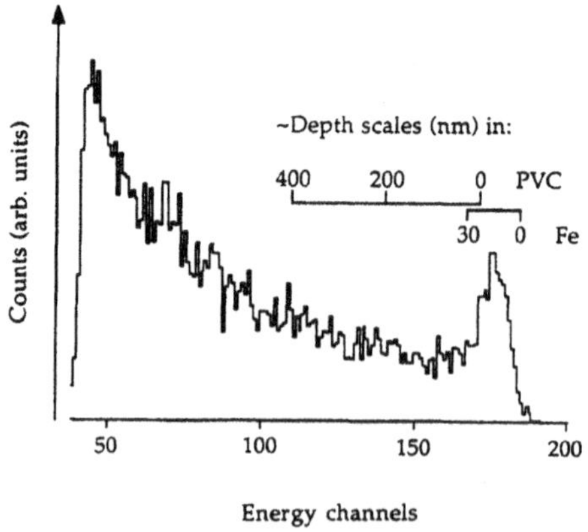

Figure 9. Recoil energy spectrum of Cl in PVC after irradiation with 48 MeV Br to 10^{15} ions/cm^2; the deduction of the depth scales is based on the initial composition of the materials; taken from Reference 19.

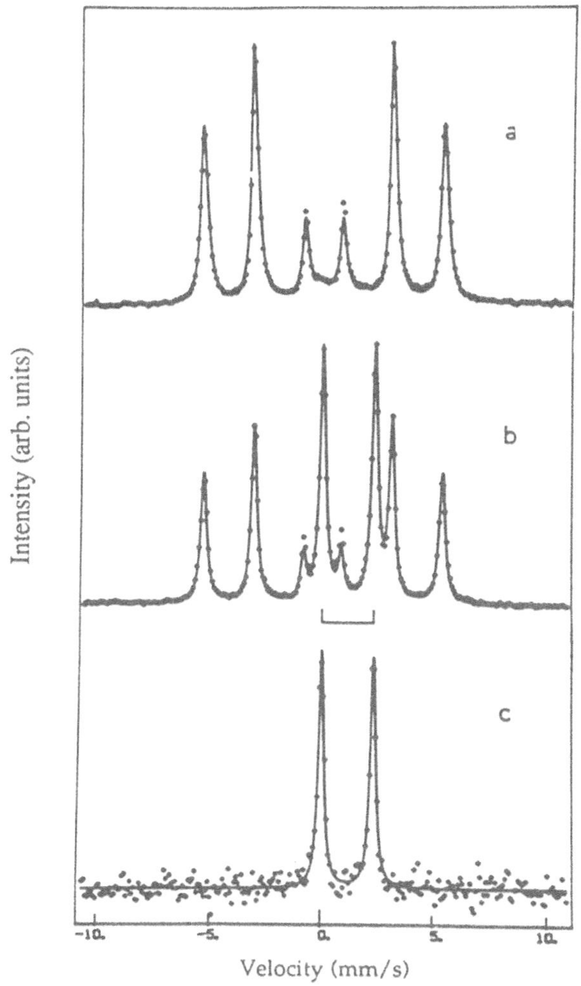

Figure 10. Mössbauer spectra recorded from a) unirradiated and b) irradiated 100 Å ^{57}Fe, and c) irradiated 50 Å ^{57}Fe films on PVC (irradiation dose: 5×10^{12} 16 MeV S ions/cm^2); taken from Reference 19.

pletely converted into a compound and, hence, it is not meaningful to discuss differences in adhesion. Nevertheless, the PVC system exhibits a number of interesting effects which are relatively distinct and can, therefore, be studied using a variety of complementary techniques.

The metal-PTFE interface has drawn enormous interest, partly because of its simple structure and partly because of its extremely poor adhesion. Regarding the mechanisms of bonding and, in a further step, the mechanisms of enhanced bonding, it seems that several effects – supporting as well as counteracting – are involved. In the above sections, we have addressed the reactivity and elasticity of the film material, cross-linking of weak boundary layers, chemical interactions between film and substrate atoms, surface morphology, and also ion beam parameters such as energy, stopping power, dose, etc. For further discussions on mechanisms for ion beam induced bonding of metal-PTFE and other interfaces, see References 20-24.

Numerous theories and explanations can be supported - rather few can be excluded. In lieu of trying to provide a patent solution to the many problems involved, I have tried to stress ambiguities and coherences so as to encourage new and unconventional routes in the designing of further experiments. Which theories are plausible and which should not even cross one's mind may be left to individual judgement.

ACKNOWLEDGEMENTS

I am indebted to Tore Ericsson, Ulrik Gelius, Michael Keane, Göran Possnert, Bo Sundqvist, Harry Whitlow, Roger Wäppling and Mikael Östling for their contribution to many of the results presented herein, and for sharing their knowledge and expertise during the course of my work at Uppsala University.

REFERENCES

1. See, for example, K. L. Mittal and J. R. Susko, editors, "Metallized Plastics 1: Fundamental and Applied Aspects", Plenum Press, New York, 1989.

2. W. Goldie, "Metallic Coatings of Plastics", Vol. 1, Electrochem. Publ., Middlesex, 1968, pp. 23.

3. D. Briggs, V. J. I. Zichy, D. M. Brewis, J. Comyn, R. H. Dahm, M. A. Green and M. B. Konieczko, Surf. Interface Anal., 2, 107 (1980).

4. D. Briggs, in "Surface Analysis and Pretreatments of Plastics and Metals", Applied Science, London, 1982, pp. 210.

5. See, for example, J. R. Hollahan and A. T. Bell, editors, "Techniques and Applications of Plasma Chemistry", John Wiley, New York, 1974.

6. J. J. Bikerman, Adhesives Age, 2, 23 (1959).

7. R. H. Hansen and H. Schonhorn, J. Polym. Sci., B4, 203 (1966).

8. P. A. Ingemarsson, T. Ericsson, G. Possnert and R. Wäppling, J. Adhesion Sci. Technol., 3, 503 (1989).

9. S. L. Vogel and H. Schonhorn, J. Appl. Polym. Sci., 23, 495 (1979).

10. D. Font and H. Montes, in "3èmes Journées sur L'Adhérence sur le Thème L'Interface Métal-Polymere", p. 16, Societé Français du Vide, Paris, 1987.

11. B. T. Werner, T. Vreeland, M. H. Mendenhall, Y. Qiu and T. A. Tombrello, Thin Solid Films, 104, 163 (1983)

12. J. E. Griffith, Y. Qiu and T. A. Tombrello, Nucl. Instrum. Meth., 198, 607 (1982).

13. S. Jacobson, B. Jönsson and B. Sundqvist, Thin Solid Films, 107, 89 (1983).

14. P. Anders Ingemarsson, Michael P. Keane and Ulrik Gelius, J. Appl. Phys., 66, 3548 (1989).

15. C. J. Sofield, C. J. Woods, C. Wild, J. C. Riviere and L. S. Welch, Mater. Res. Soc. Symp. Proc., 25, 197 (1984).

16. Zhengmin Liu and Jianjun Dong, Nucl. Instrum. Meth., B44, 445 (1990).

17. Chin-An Chang, J. E. E. Baglin, A. G. Schrott and K. C. Lin, Appl. Phys. Lett., 51, 103 (1987).

18. J. E. E. Baglin, G. J. Clark and J. Bøttiger, Mater. Res. Soc. Symp. Proc., 25, 179 (1984).

19. P. A. Ingemarsson, T. Ericsson, G. Possnert and R. Wäppling, Hyperfine Interactions, 46, 549 (1989).

20. T. A. Tombrello, Mater. Res. Soc. Symp. Proc., 25, 173 (1984).

21. T. A. Tombrello, Mater. Res. Soc. Symp. Proc., 119, 95 (1988).

22. J. E. E. Baglin, in "Fundamentals of Adhesion," L.-H. Lee, Editor, Plenum Press, New York, 1991.

23. J. E. E. Baglin, Nucl. Instrum. Meth., B39, 764 (1989).

24. P. Anders Ingemarsson, Nucl. Instrum. Meth., B44, 437 (1990).

MOLECULAR DYNAMIC PHENOMENA AT POLYMER SURFACES AND THEIR RELEVANCE TO POLYMER ADHESION BEHAVIOR

M. Morra, E. Occhiello and F. Garbassi

Istituto Guido Donegani S.p.A.
Via G. Fauser 4, 28100 Novara, ITALY

Hydrophobic recovery of a number of plasma-treated polymers (polyethylene, polypropylene, polyethyleneterephthalate, polycarbonate and polystyrene) has been studied. For high molecular weight polymers, it occurs through segmental motions which bury polar groups away from the surface. The extent of crosslinking, along with mobility of segments, limits the contact angles which can be reached upon recovery. For low molecular weight polymers, macromolecular motions overcome segmental motions, and thus the recovery is determined by a mechanism involving diffusion through crosslinked network.

Evaporated metal (Al) adhesion on plasma treated polymers depends dramatically upon hydrophobic recovery. The linear dependence of adhesion on the number of polar groups at the polymer surface points to an essentially chemical origin.

INTRODUCTION

It is often necessary to improve the hydrophilicity of polymer surfaces prior to their use for technological purposes, such as printing, metallization, adhesion, etc.[1]. Several methods, including chemical, flame, corona and cold plasma treatments, have been devised for this purpose[1].

Increasing the surface free energy of a polymer surface involves imposing a stress, since there is a thermodynamic drive to lower the interfacial energy with the medium. Stresses imposed by chemical modification of the surface or by contacting with high energy media are met by strains, resulting in surface restructuring, first observed by Langmuir[2] in 1938 and reviewed by Lavielle[3], Andrade[4] and ourselves[5].

After treatments leading to increases of surface free energy and decreases of contact angles, frequently time-dependent increases of contact angles and decreases of surface free energy are observed. This phenomenon is known as hydrophobic recovery and it has been observed for both grafted and plasma treated polymer surfaces[6-14].

In this work high molecular weight polyethylene (PE), polypropylene (PP), polyethyleneterephthalate (PET) and Bisphenol A polycarbonate (PC) along with two very different molecular weight fractions of

Metallized Plastics 2, Edited by K.L. Mittal
Plenum Press, New York, 1991

polystyrene (PS) were used. Their surfaces were treated using a single and reproducible oxygen plasma. Hydrophobic recovery of treated surfaces was then studied, together with its effect on metallizing behavior. To limit the number of parameters, crystallinity effects were limited in the case of HDPE, PP and PET by quenching the samples to have as little formation of crystallites as possible.

EXPERIMENTAL

HDPE (Himont), PP (Himont), PET (Montefibre) and PC (Bayer) were injection molded to yield 60x60x2.5 mm thick plaques. To minimize crystallinity, all samples were quenched with liquid nitrogen.

PS narrow molecular weight fractions (2700 and 1020000, $M_w/M_n < 1.1$) were purchased from Aldrich, dissolved in dichloromethane (10% w/v solutions) and spin-coated on clean glass slides. Solvent removal was completed by annealing the samples under vacuum.

Oxygen plasma treatments were performed using a parallel plate reactor. The upper (powered) electrode was 15 cm in diameter, the samples were located on the water-cooled grounded electrode, 30 cm in diameter, and the interelectrode distance was 15 cm. The plasma parameters were: excitation frequency 13.56 MHz, power 100 W, pressure 2 Pa, gas flow 8 cc(STP)/min, and treatment time 20 s. Oxygen from lecture bottles, supplied by Carlo Erba, was used.

PE, PP, PET and PC treated samples were aged at room temperature, relative humidity was kept constant at 35%. PS samples were aged at temperatures between 293 K and 413 K, as reported in ref. 13.

Water advancing and receding contact angles were determined by the sessile drop technique[1], using a Ramé-Hart contact angle goniometer. We kept the capillary pipette of the microsyringe immersed in the drop during the entire measurement and used carefully deionized and twice-distilled water (surface tension = 72.8 mN/m).

XPS (X-ray Photoelectron Spectroscopy) spectra were obtained on a PHI model 548 XPS-AES spectrometer, using the Mg Kα radiation (1253.6 eV) from a 400 W source. The analysis chamber pressure was maintained near 2×10^{-7} Pa, without baking. The electron take-off angle was 48°.

Signal averaging to improve signal-to-noise ratios was possible due to the connection of the spectrometer to a PDP 11/50 computer. Further data processing (smoothing, Shirley or linear background subtraction, integration, Gaussian/Lorentzian curve fitting) was carried out by using in-house software on a Sperry 1100/72 mainframe computer, details are reported elsewhere[15]. The sensitivity factors recommended by the manufacturer and verified by us were used to obtain surface compositions.

Aluminium evaporation was carried out in the above-described plasma reactor, maintaining a 10^{-3} Pa pressure and resistively heating the aluminium in a tungsten boat. The thickness of the deposited layer was determined by a quartz microbalance and was kept constant at 50 nm.

Pull tests were performed using a Sebastian II instrument, supplied by Quad Group. Epoxy coated pull studs were provided by the manufacturer. They were adhered to the metallized samples and cured at 363 K for 1 hour, pull angle was 90° and pull rate cannot be controlled with this instrument.

RESULTS

Polymeric samples were oxygen plasma treated and then allowed to undergo aging at room temperature. Periodically, samples were analysed to see the effect of aging (hydrophobic recovery) on surface chemistry and

Table I. XPS Surface Compositions (Atomic Percentages) for Selected Aging Conditions of PE, PP, PET and PC. Typical Error Margins are ± 3%.

Aging condition	C	O
PE : Untreated	97	3
PE : Just treated	75	25
PE: Aged 24 hours	77	23
PE : Aged 375 hours	77	23
PP : Untreated	98	2
PP : Just treated	84	16
PP : Aged 24 hours	85	15
PP : Aged 375 hours	85	15
PET: Untreated	72	28
PET: Just treated	68	32
PET: Aged 375 hours	69	31
PC : Untreated	84	16
PC : Just treated	79	21
PC : Aged 24 hours	80	20
PC : Aged 400 hours	80	20

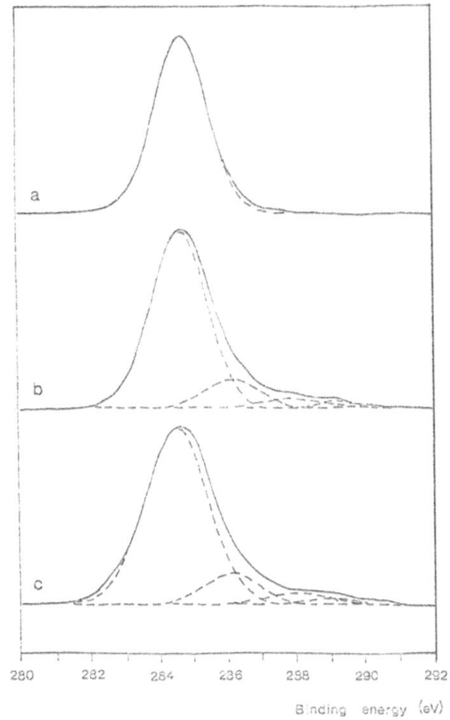

Figure 1. XPS C-1s line-fitted peaks relative to untreated PP (a), just treated PP (b), and treated PP aged for 375 hours at 373 K.

wettability and then were metallized to assess the effect on metal-polymer adhesion.

XPS data

Surface compositions of untreated and oxygen plasma treated samples are presented in Table I. The first observation is that plasma treated specimens have a higher oxygen concentration than untreated specimens. This effect is in good agreement with the extensive literature on reactions of polymer surfaces with plasma species, mainly oxygen atoms[1,16-17].

A more interesting result is the fact that aging does not affect in a significant way the composition obtained by XPS, i.e. any alteration leading to worsening of adhesion occurs within the XPS-observed layer. Also the distribution of XPS components created by plasma treatment is essentially unaffected by hydrophobic recovery, as shown in Fig. 1 for PP. Furthermore, no effect of varying the aging temperature was observed[8-10, 13-14]. Any contribution to the hydrophobic recovery by diffusion effects is, therefore, ruled out.

In the case of PS, the effects of varying molecular weight and aging temperature were studied and the results are shown in Table II. For high molecular weight specimens, the result is similar to that in Table I, i.e. the composition of the surface layer is essentially unaffected by aging at any temperature. On the other hand, for low molecular weight PS and high aging temperatures, outdiffusion of bulk polystyrene occurs, thus providing another contribution to hydrophobic recovery.

Contact Angle Measurements

In Fig. 2 the behavior of water advancing and receding contact angles as a function of aging time at room temperature is presented for plasma treated PE, PP, PET and PC. In all cases an increase of contact angles with aging is observed. Only in the case of PP the advancing angle corresponding to that for untreated PP is obtained; whereas for PET, PC and PE a limiting advancing angle rather lower than that of the untreated polymer is reached. As to receding angles, in all cases values appreciably lower than those of untreated specimens are obtained.

In Fig. 3 water advancing and receding angles of oxygen plasma treated PS are plotted vs. aging for different molecular weights and aging temperatures. For aging at room temperature, only marginal differences were observed. As in the case of other polymers (PE, PET, PC) limiting advancing angles and receding angles quite below those relative to untreated PS are observed.

Table II. XPS Surface Compositions (Atomic Percentages) for Oxygen Plasma Treated PS as a Function of Molecular Weight, Aging, and Temperature. Typical Error Margins are ± 1-2%.

Aging condition	Mol. weight	C	O
Untreated	1020000	98	2
Just treated	1020000	80	20
Aged 375 hours (293 K)	1020000	81	19
Aged 48 hours (413 K)	1020000	81	19
Untreated	2700	97	3
Just treated	2700	79	21
Aged 375 hours (293 K)	2700	80	20
Aged 48 hours (413 K)	2700	94	6

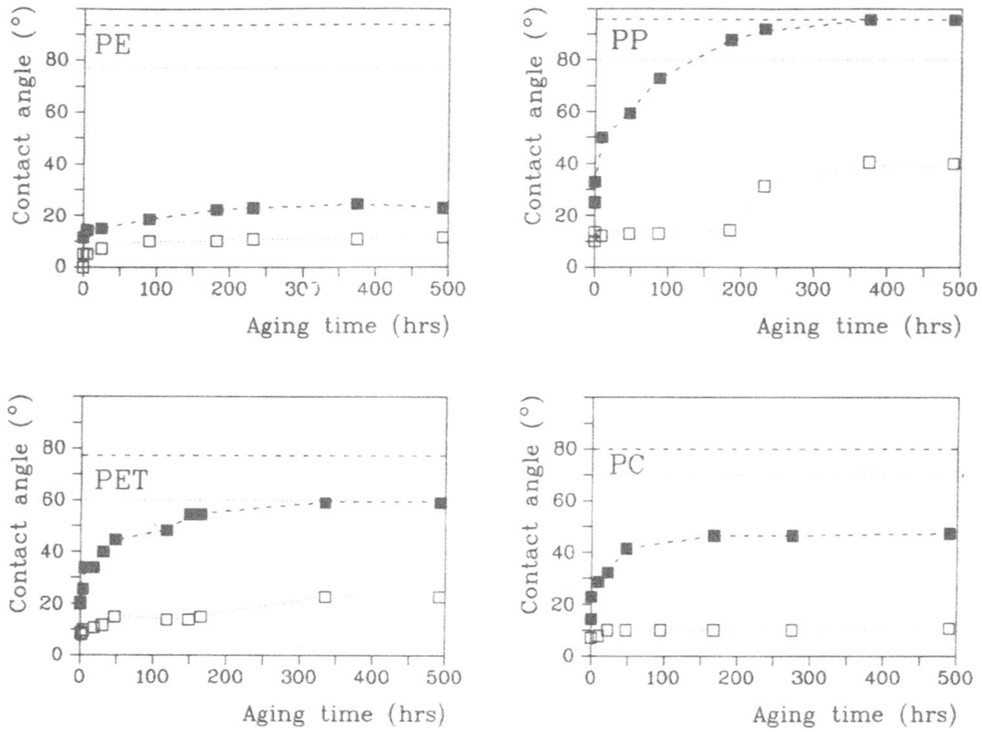

Figure 2 Advancing (■) and receding (□) water contact angles on oxygen plasma treated polymer surfaces aged at 293 K. Dashed and dotted lines represent advancing and receding angles on the untreated polymer.

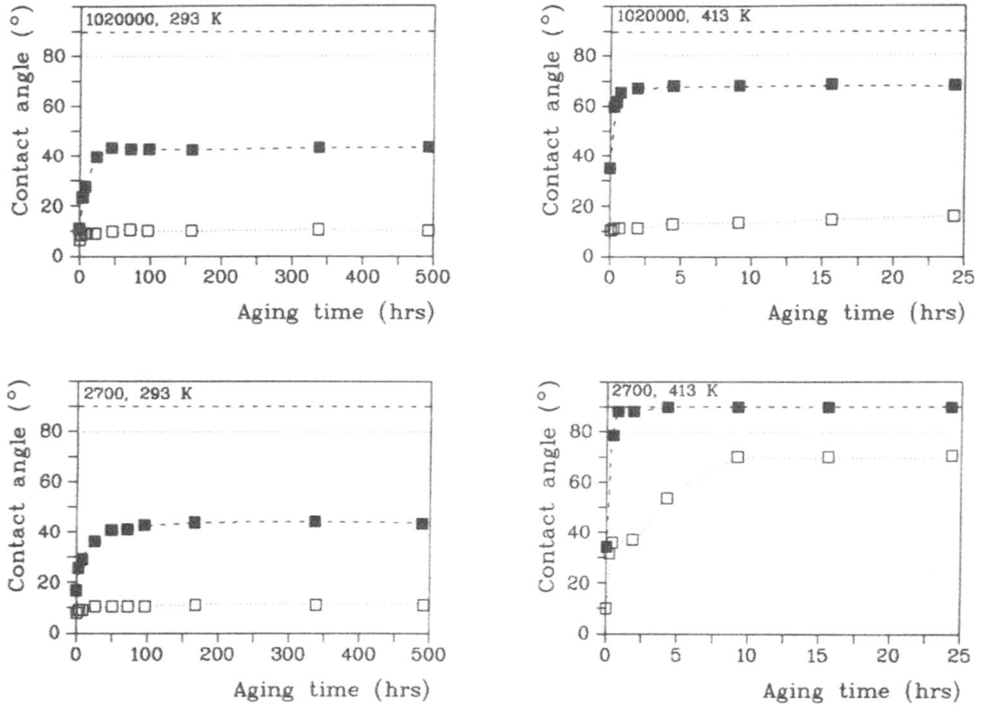

Figure 3 **Advancing (■) and receding (□) water contact angles on oxygen plasma treated PS. Dashed and dotted horizontal lines have the same meaning as in Fig. 2.**

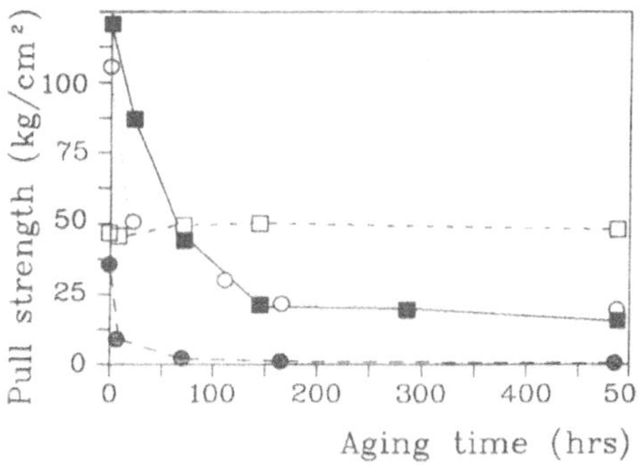

Figure 4 Pull strength of aluminium coating (50 nm thick) on plasma treated PE (□), PP (●), PET (○) and PC (■) as a function of aging at room temperature.

At higher aging temperatures, high molecular weight PS shows an increase in limiting advancing and receding angles, but these still remain quite different from the value of untreated PS. On the other hand, for the low molecular weight fraction, values close to those of untreated PS are obtained, pointing to efficient outdiffusion, in agreement with XPS data.

Metal Adhesion to Polymers

Results of pull testing of aluminium-coated polymer specimens as a function of aging before metallization are presented in Fig. 4. In the case of PE no dependence of adhesion on aging was observed, the pull strength remained essentially constant. On the other hand, for PP, PET, and PC a clear decrease in adhesion with aging was observed. In all cases a steep initial decrease followed by a plateau is observed, in good agreement with the trend in contact angles (Fig. 2).

DISCUSSION

Dynamics of Modified Polymer Surfaces

After plasma modification of polymer surfaces, to use Lavielle's words[3], "there is an asymmetry of the force field at the interface". Introducing an analogy with viscoelastic stress-strain relationships[18-19], for plasma-treated surfaces the stress is the gradient in surface free energy between treated and untreated surfaces, the time-dependent strain is the change in contact angles with time.

Looking at the strain vs. time curves reported in polymer dynamics textbooks[20-21], for instance in the shear-creep compliance curve, three broad regions are normally found. The first is related to local movements in the glassy state, the second is the rubbery plateau or rubbery state and the third is flow.

In our case, for high-molecular-weight samples, only the first two regions are observed in a contact angle vs. time plot. As shown by XPS (Tables I and II), all motions are confined within the plasma-modified layer. Initially contact angles increase with time, due to segmental motions leading to burial of polar groups away from the surface. A plateau is then reached and using molecular modelling we have shown that the relative mobility of different clusters of oxidized species is an important factor in determining the position of the limiting plateau[10].

The value of limiting contact angles depends also on crosslink density. Elsewhere we have shown that the PS surface is definitely crosslinked by plasma treatment[13] and that the mechanical properties of treated PE and PP/epoxy adhesive joints are related to different extents of plasma-induced crosslinking[11]. The fact that the efficiency of hydrophobic recovery, as shown in Figs. 2 and 3, follows the PP > PET, PC, PS > PE sequence is indeed in remarkable agreement with the corresponding crosslink yields[22].

In the case of PS, at sufficiently low molecular weight (below M_c, critical molecular weight below which effects are not present) and high temperatures (Table II, Fig. 3), segmental motions are overcome by macromolecular motions in determining the extent of hydrophobic recovery. The latter is controlled in this case by the mobility of polymer chains through a crosslinked network: the plasma modified layer. In turn, the mobility of polymer chains entangled with the surface layer depends essentially on molecular weight[13].

Effects on Adhesion

Autohesion and adhesion between thermoplastic polymer surfaces and in particular the time evolution of joint strength have been interpreted in terms of reptation diffusion coefficients[23-24]. But in the case of metallized polymers[25] the availability of oxygen containing groups at the interface is the "conditio sine qua non" for adhesion.

By comparing Fig. 2 with Fig. 4 it is evident that high adhesion strengths correspond to low water contact angles[26]. Since low contact angles have been shown by SSIMS[11] to correspond to high amounts of oxygen at the surface, metal/polymer adhesion seems to be driven by an essentially chemical mechanism.

Hydrophobic recovery induces a decrease in the number of oxygen containing functional groups available at the surface and, therefore, in the ability to interact with the metal layer. Thus a correspondence between hydrophobic recovery and the loss of adhesion of the metal layer is expected. Actually the strongest decrease in adhesion has been observed for PP, where hydrophobic recovery is most efficient; while no detectable effect was observed for PE, where hydrophobic recovery is nearly absent. Intermediate behaviors, for both burial of polar groups away from the surface and the effect of aging on adhesion, were detected for PET and PC.

CONCLUSION

Segmental and macromolecular motions intervene in hydrophobic recovery of plasma treated polymers. If the treatment condition is constant, the relative importance of the two mechanisms depends on substrate and temperature. For HDPE, PP, PET, PC and high molecular weight PS segmental motions dominated recovery. In the case of low molecular weight PS, at high temperatures, outdiffusion of non-modified PS became important. An analogy with viscoelastic stress-strain relationships is possible.

Metal adhesion was found to be related to the extent of hydrophobic recovery, thus being sensitive to the amount of oxygen containing groups present at the surface.

REFERENCES

1) S. Wu, "Polymer Interface and Adhesion", Marcel Dekker, New York, 1982
2) I. Langmuir, Science, 87, 493 (1938)
3) L. Lavielle, in "Polymer Surface Dynamics", J. D. Andrade, Ed., p. 45, Plenum Press, New York, 1988
4) J. D. Andrade, D. E. Gregonis and L. M. Smith, in "Surface and Interfacial Aspects of Biomedical Polymers", Vol. 1, p. 15, J. D. Andrade, Ed., Plenum Press, New York, 1985
5) M. Morra, E. Occhiello and F. Garbassi, Adv. Colloid Interface Sci., 32, 79 (1990)
6) A. Baszkin, M. Nishino and L. Ter-Minassian-Saraga, J. Colloid Interface Sci., 54, 317 (1976)
7) H. Yasuda, H. K. Sharma and T. Yasuda, J. Polym. Sci. Phys. Ed., 19, 1285 (1981)
8) Y. Hsieh and E. Y. Chen, Ind. Eng. Chem. Prod. Res. Dev., 24, 246 (1985)
9) M. Morra, E. Occhiello and F. Garbassi, J. Colloid Interface Sci., 132, 504 (1989)
10) F. Garbassi, E. Occhiello, M. Morra, L. Barino and R. Scordamaglia, Surface Interface Anal., 14, 595 (1989)
11) E. Occhiello, M. Morra, G. Morini, F. Garbassi and P. Humphrey, J. Appl. Polym. Sci., in press
12) M. Morra, E. Occhiello, R. Marola, F. Garbassi, P. Humphrey and D. Johnson, J. Colloid Interface Sci., 137, 11 (1990)
13) M. Morra, E. Occhiello, P. Cinquina and F. Garbassi, Polymer Preprints, 31, 308 (1990)
14) M. Morra, E. Occhiello and F. Garbassi, Angew. Makromol. Chem., in press
15) F. Garbassi, Surface Interface Anal., 2, 165 (1980)
16) M. Hudis, in "Techniques and Applications of Plasma Chemistry", p.113, J. R. Hollahan and A. T. Bell, Eds., Wiley, New York, 1974
17) H. V. Boenig,"Plasma Science and Technology", Cornell University Press, Ithaca, New York, 1982
18) A. V. Tobolsky, "Properties and Structure of Polymers", Wiley, New York, 1960
19) J. D. Ferry, "Viscoelastic Properties of Polymers", Wiley, New York, 1980
20) P.-G. de Gennes, "Scaling Concepts in Polymer Physics", Cornell University Press, Ithaca, NY, 1979
21) M. Doi and S. F. Edwards, "The Theory of Polymer Dynamics", Oxford University Press, London, 1986
22) V. D. McGinniss, in "Encyclopaedia of Polymer Science and Technology", Vol. 4, p. 418, Wiley, 1986
23) P.-G. de Gennes, C. R. Acad. Sci. Paris, 291, 219 (1980)
24) S. Prager and M. Tirrell, J. Chem. Phys., 75, 5194 (1981)
25) J. M. Burkstrand, in "Physicochemical Aspects of Polymer Surfaces", K. L. Mittal, Ed., Vol. 2, p. 1093, Plenum Press, New York, 1983
26) K. L. Mittal, in "Adhesion Science and Technology", L. H. Lee, Ed., Part A, p. 129, Plenum Press, New York, 1975

ENHANCED METAL/POLYMER ADHESION BY ION ASSISTED DEPOSITION

R.A. Moody, T.G. Tetreault, and J.K. Hirvonen

Spire Corporation

Patriot's Park, Bedford, MA 01730

Significant improvements in the adhesion of several metals to polymers have been achieved through the ion beam assisted deposition (IBAD) technique. Films of silver, gold, and copper were deposited via IBAD onto poly(tetrafluoroethylene) (PTFE) substrates. These films are substantially more adherent than similar non-IBAD coatings. Film characterization was performed with conventional pull tests, selective abrasion tests, optical microscopy, ESCA, RBS, and static SIMS. Evidence is given for the "smearing" of the normally sharp metal/polymer interface by the ion beam recoil events. Further analysis, however, suggests that the improved adhesion may result from an increase in the shear strength of the PTFE surface layers due to the promotion of crosslinking.

I. INTRODUCTION

There is a wide variety of products whose integrity, durability, and stability depend on the quality of adhesion of thin film coatings. This is especially true in the electronics industry, where sets of materials with vastly different electronic states (metals, ceramics, semiconductors, and polymers) are juxtaposed in multilayer configurations.

In general, good adhesion occurs when the interfacial energy of two materials is at a minimum. This state is usually (but not always) achieved with mutually reactive materials or when molecular or atomic interdiffusion occurs. The interfacial energy is typically controlled by any combination of the following four mechanisms: a) the creation of primary chemical (covalent) bonds, b) molecular or atomic interdiffusion, c) mechanical interactions (surface morphology), and d) the creation of weak secondary (van der Waals, hydrogen, etc.) bonds.

Conversely, the interface between nonreactive species is expected to produce, at best, weak adhesion via the creation of secondary bonds. This is due to the drastically different electronic states of the two species. Minimizing the interfacial energy of ordinarily immiscible materials would, in theory, produce significant improvements in adhesion. This would demand actively energizing the interface to externally induce one or more of the above three (a-c) mechanisms.

Conventional techniques, including chemical and physical deposition, are unable to produce the necessary conditions for achieving significant improvements in the adhesion between metals and polymers. Multiple ion beam techniques have been reported which have produced enhancements in the interfacial adhesion of metals and insulators.[1-5] Although mechanisms for the ion beam induced enhancements are not well understood, the basic concepts leading to the improved adhesion with ion beams consist of:

i) The ion irradiation serving primarily to disrupt the electronic structure of chemical/metallic bonding of interfacial atoms, producing new bonding configurations and developing an interfacial layer where atoms of both film and substrate species are stably bonded,

ii) Ion beam mixing, wherein the individual atoms of the surface coating are struck by energetic ions and are driven into the substrate by elastic recoil to develop an intermixed interfacial layer.

Earlier work by our group[6,7] has clearly illustrated the qualitative enhancements in adhesion possible with the IBAD process on many metal/polymer couples and offered a possible mechanism for coating failure. The thrust of the present effort was to expand the previous work by further characterizing the nature of the adhesion enhancement and by identifying the mechanisms for coating failure through quantitative means. The metals of interest for this study were those commonly used in electronic and biomedical applications, and the polymer of choice, due to its low modifier surface energy, was poly(tetrafluoroethylene), PTFE. The underlying thought being that if the metal were to adhere well to PTFE, then it might adhere even better to other polymers.

Ion beam assisted deposition (IBAD) is a physical vapor deposition process with concurrent ion implantation. In this coating procedure, a material is formed on a substrate with a low energy, high rate deposition technique, such as electron beam evaporation or magnetron sputtering. A low energy (hundreds of volts) ion beam is simultaneously used to bombard the coating during deposition. In this way, the ions provide recoil events which enhance the effect of "merging" the evaporated film with the substrate. The resulting coatings have a graded, "interdiffused" interfacial region or interphase (see Figure 1).

Housing the ion source and evaporator in the same chamber allows substrate pretreatment without having to break vacuum. This ion beam pretreatment (sputter cleaning and/or texturing) is an established technique for enhancement of metal/polymer adhesion.[3,8] In addition, the potential for promoting crosslinking between chain segments in this interface region and those of the underlying polymeric substrate should provide even greater resistance to delamination than conventional coating processes.

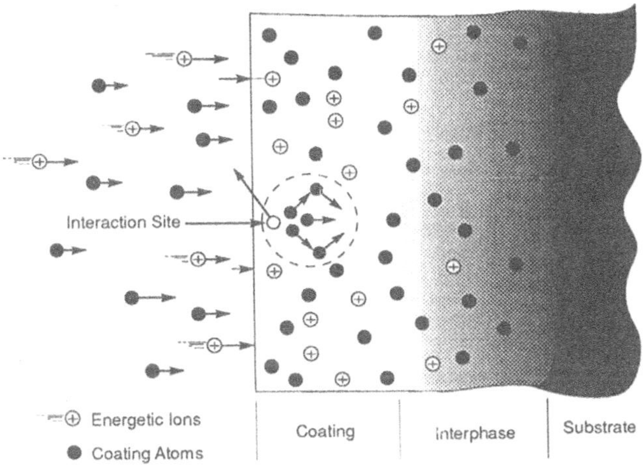

Figure 1. Schematic Representation of the IBAD Process.

II. EXPERIMENTAL

The IBAD facility, shown in Figure 2, was used to perform all the depositions. It consists of an endstation mated to a low energy, high current Kaufman ion source. The endstation is subdivided into two separate chambers: an upper, cryopumped chamber which houses the sample holder; and a lower, diffusion pumped chamber which contains an electron beam evaporator. Silver, gold, and copper were deposited via the IBAD process onto PTFE test coupons which had been solvent cleaned using iterative ultrasonic cleaning with distilled water, propanol, and acetone. The evaporation of each metal was set at a nominal rate of 1 to 10 Å/s as controlled by a quartz crystal deposition rate controller positioned in the same plane as the sample. For 30 seconds before and during the evaporation, 400 eV Ar^+ ions bombarded the substrate at a flux of 10 to 50 $\mu A/cm^2$. Coating thicknesses between 1000 Å and 1 μm were used for the adhesion investigations. During the IBAD process, a thin tubing was used to shield portions of the substrate from both the vertically incident evaporant and the horizontally incident ion bombardment. This provided an intrinsic set of control samples that could be used for direct comparison: a section of IBAD coating, a section of metal evaporation only, and a section of ion bombardment only.

The coating adhesion, termed the practical adhesion, is the force/area necessary to remove the coating from the substrate. The practical adhesion is the sum of the basic or fundamental adhesion and other factors such as, coating stresses, structural faults, etc. Basic adhesion is the sum of all intermolecular interactions at the interface and can only be theoretically calculated.[9] Our discussion will therefore be limited to practical adhesion and enhancements therein. When measuring the practical adhesion, the failure can occur in either

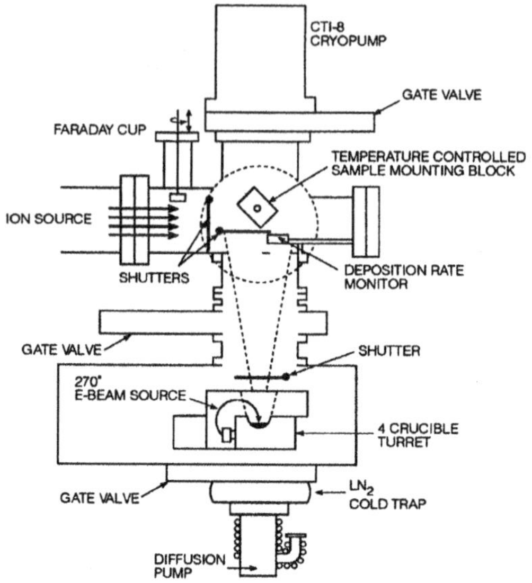

Figure 2. Functional Diagram of Spire's IBAD Facility.

the interfacial region (interphase) or within the bulk of one of the materials. The quantity measured by the former is the interphasial tensile strength (ITS) of the region, while the latter quantity is known as the cohesive strength of the material.

The interphasial tensile strength (ITS) of the IBAD and non-IBAD depositions was determined both qualitatively and quantitatively. Adhesive tape pull tests were performed as preliminary qualitative screening tests. The tests were performed by firmly affixing identical pieces of 3M Scotch tape to the surfaces of the coatings and slowly pulling them off normal to the sample surfaces. The extent of improved adhesion was determined by estimating the film coverage remaining after tape removal using optical microscopy. Quantitative ITS was performed using an Instron 1000 equipped with a 10 lb. beam and a load cell. Siding nails to be used as bonding studs were applied to the metal/polymer system with Devcon five-minute epoxy and allowed to properly cure. The test coupon was placed in the mounting jig (Figure 3) and the stud was attached to the Instron. Tensile failure of the sample was recorded at a crosshead rate of 0.5 inch/min. The areas under test were estimated and compared using photographic means.

Enhancement of the abrasion resistance of the ion beam assisted coatings was evaluated by scratching the surface of the metallized polymers with a probe of a modified Moh's hardness apparatus. A series of scratches using tips of different hardnesses was made across the IBAD/non-IBAD interface of the metallized polymers. Optical microscopy was used to determine the extent of interfacial failure.

The premise of this investigation, as stated, is to demonstrate the effectiveness of the ion beam in promoting improved adhesion between a

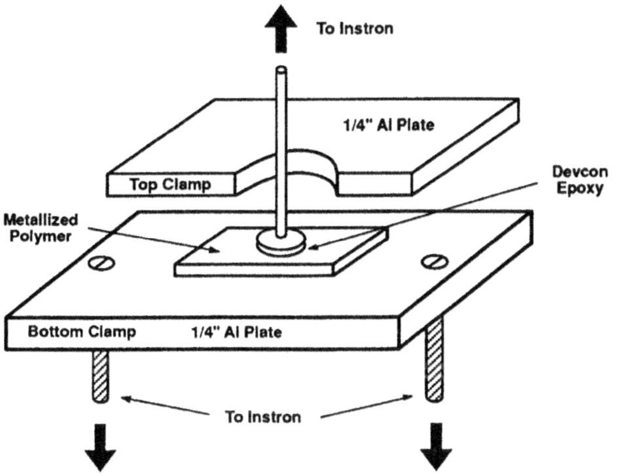

Figure 3. Schematic of Tensile Pull Test Apparatus.

metal film and a polymeric substrate. To this end, both qualitative and quantitative adhesion tests were performed to the point of interfacial failure. Since presumably this failure will occur at the weakest point within the metal/polymer interphase, elemental surface analysis of the failure boundary will identify its location and determine whether the failure is interphasial or cohesive in type. We therefore need analysis techniques which have good elemental specificity as well as sensitivity. Additionally, since we are interested in detecting evidence of bond structure alterations such as crosslinking, we need a technique sensitive to fractionation, i.e., to the mass of fractionation products. The requirements of the former are satisfied by Rutherford backscattering spectroscopy (RBS) and electron spectroscopy for chemical analysis (ESCA) while those of the latter by static secondary ion mass spectroscopy (SIMS).

RBS spectra were obtained on the University of Connecticut (Storrs) Van de Graaf system using 2.0 MeV He^+ ions. Given the Z_{ave} of PTFE is eight, then the sensitivities for Au, Ag, and Cu are 1×10^{12} atoms/cm^2, 3×10^{12} atoms/cm^2, and 8×10^{12} atoms/cm^2 respectively, i.e., in the range of 10^{-3} to 10^{-2} monolayers.

ESCA, also called X-ray photoelectron spectroscopy (XPS), is generally used for the determination of elemental composition at the near surface region. The detection of trace elements is dependent on the cross section of the desired element and the background of the other species present. For metals in a polymer background, sensitivity is expected in the parts per thousand range. For our studies, an SSL Model SSX100 Small Spot XPS Spectrometer with a spot size of 1000 μm was utilized to investigate the removed surface coating (herein called failed coating) and the substrate from both the IBAD and non-IBAD coatings.

Static SIMS surface analysis is performed by mass spectroscopy of the sputtered ions from a sample surface. The sputtering beam is composed of low energy, heavy, inert ions at current densities which are several orders of magnitude lower than those used in dynamic SIMS. This technique analyzes monolayers of the material and detects all of the compositional elements, including hydrogen, with a much higher sensitivity than ESCA. It is, therefore, more suited for determining the exact locus and chemical composition of the material's surface. Static SIMS will also yield information regarding polymeric structural features, such as unsaturation, branching, crosslinking, and even surface orientation of polymeric molecules or segments (which ESCA cannot). The surface structural changes due to ion bombardment during IBAD can be clearly seen with this analytical technique. Xe^+ static SIMS measurements were carried on a Physical Electronics (PHI) 6300 spectrometer using a 60° incident, 4.5 keV, 60 pA Xe^+ ion beam rastered across an area of 1 mm^2. The secondary ions were filtered using a 90° energy filter before entering a quadrupole equipped with 16 mm rods in a vacuum of 1×10^{-10} torr. In addition to the Xe source data, measurements were obtained using the SF_6 source instrument at Idaho National Engineering Laboratory (Idaho Falls, ID.) at 10 keV, 1 pA neutrals or 4 pA particles ($SF_6{}^0$ + $SF_6{}^-$) with a beam diameter of ca. 2 mm^2 in a vacuum of 1×10^{-8} torr. Static SIMS measurements were performed on fresh surfaces of the failed coatings and substrates from the Cu/PTFE and Au/PTFE samples.

III. RESULTS AND DISCUSSION

The typical abrasion resistance of the metallized films is shown in Figure 4. For each sample, the integrity of the non-IBAD coating was severely affected; while, for the ion assisted surface, the coating integrity was retained. The samples which exhibited excellent adhesion (no delamination) in the IBAD regions during the tape pulls (see Figure 5) were further analyzed with Instron pull tests. The results of the Instron ITS studies of silver, gold, and copper on PTFE are shown in Table I. The IBAD processing resulted in improved adhesion in each case. The IBAD Cu/PTFE exhibited the greatest improvement (300%) over non-IBAD Cu/PTFE, while Ag/PTFE showed the least improvement (16%). Results for the Au/PTFE were intermediate (180%).

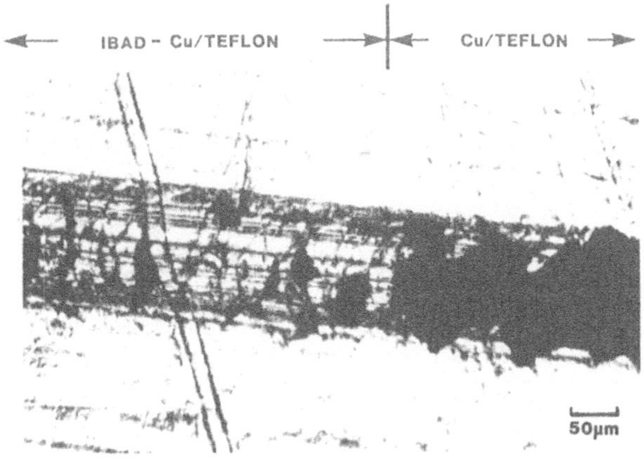

Figure 4. Optical Micrograph of a Scratch made with a 7.5 Moh's Tip. Severe damage is seen in the non-IBAD region (right side).

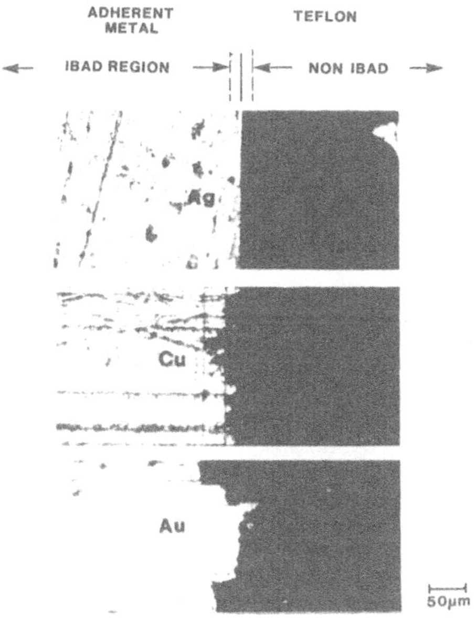

Figure 5. Results of Tape Pulls on Au/PTFE, Ag/PTFE, and Cu/PTFE across the IBAD/Non-IBAD Interface.

Table I. Interphasial Tensile Strength Comparison of IBAD and Non-IBAD Metallizations.

COATING SYSTEM	INTERPHASIAL TENSILE STRENGTH lbs/in^2 *(g/mm^2)*		% IMPROVEMENT
	non-IBAD	IBAD	
Ag/Teflon	20.30 ± 7.60 *(28.80 ± 10.79)*	23.60 ± 12.20 *(33.50 ± 17.32)*	16
Au/Teflon	16.80 ± 5.74 *(23.80 ± 8.15)*	46.50 ± 9.00 *(66.00 ± 12.77)*	180
Cu/Teflon	13.30 ± 7.23 *(18.90 ± 10.26)*	53.60 ± 12.50 *(76.10 ± 17.74)*	300

The improvement in adhesion due to ion beam processing appears to be species dependent. This is evident from a comparison of the absolute ITS values. It is interesting to note the ordering reversal due to IBAD, observed in the Ag and Cu systems.

RBS data taken on both IBAD and non-IBAD regions of a Au/PTFE sample (substrate side) are shown in Figure 6. Following the coating removal, the PTFE substrate in the non-IBAD region was essentially atomically clean of gold indicating that the failure occurred at a distinct Au/PTFE interface or completely within the PTFE. The IBAD region, however, showed a significant amount of gold still remaining (6×10^{14} Au/cm^2) within the substrate providing evidence for the possible smearing of the interface between the metal and the polymer due to atomic mixing initiated by ion recoil events. The asymmetric nature of the Au peak, the left edge corresponding to deeper regions within the material, provides indirect evidence for this nonuniform (graded) depth profile of Au in the PTFE substrate. In addition, the total Au peak width (ca. 50 keV) is considerably larger than detector resolution, also suggesting depth penetration to allow proper identification above signal noise. The ESCA compositional profile of a freshly pulled metallized layer from the IBAD region (Figure 7) further illustrates this probable mixing of Au and PTFE species.

The presence of PTFE on the failed coating for the IBAD Au layer was also observed with ESCA. This suggests that the delamination may be the result of cohesive failure within the polymer rather than interfacial failure at the metal/polymer interphase. This implies that the binding between the metal and the polymer was stronger than the shear strength of the surface layers within the polymer and that this difference was a result of the ion beam treatment and the metal species.

The more sensitive surface probing technique, SIMS, provided data which was in complete agreement with the ESCA analysis. Complete coverage of PTFE was observed on each of the failed coatings (from both IBAD and non-IBAD regions), while no metal signal was observed on any

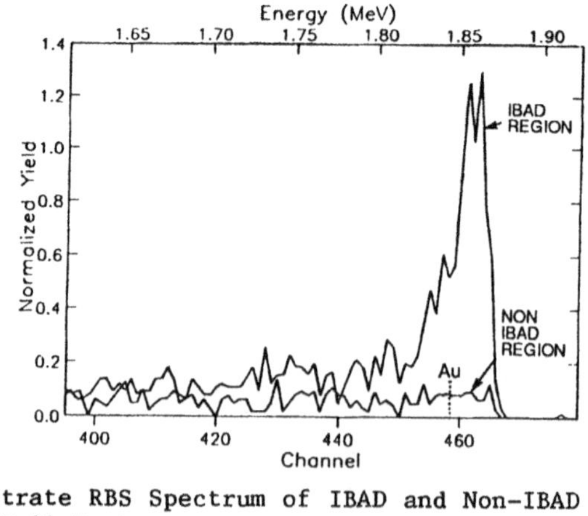

Figure 6. Substrate RBS Spectrum of IBAD and Non-IBAD Regions Following Pull Test.

of the polymer substrates. This clearly demonstrates that the failure was completely cohesive within the polymer. This also means that, with or without the ion beam, the metal forms a bond with the PTFE which is stronger than the shear strength of the near surface region of the pristine PTFE. The SIMS spectra (0-250 amu) of the IBAD and non-IBAD regions of the failed coating are shown in Figures 8a and 8b. For contrast, Figure 9 shows the 0-250 amu spectrum for a control PTFE sample. Note that the three figures are very similar, with the exception of hydrocarbon contamination. The evidence for the enhanced PTFE shear strength and, therefore, for the observed adhesion enhancement for the IBAD metallized polymers is shown by contrasting Figures 10a and 10b with Figure 11. These 150-350 amu SIMS spectra

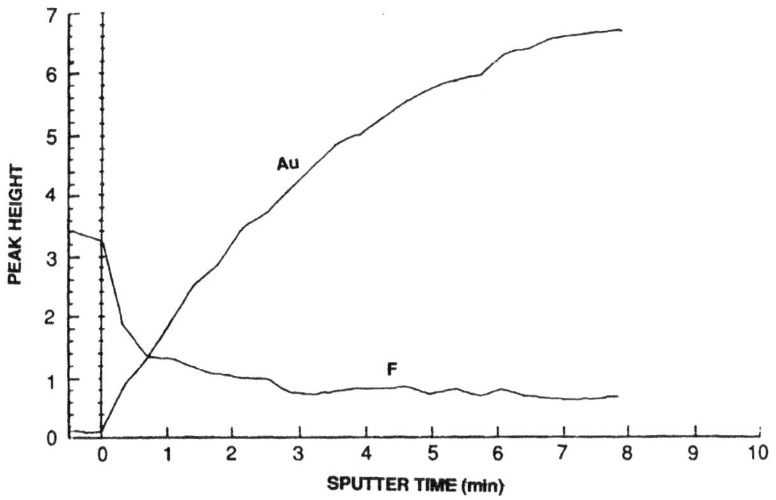

Figure 7. Sputter ESCA Compositional Profile of the IBAD Au/PTFE Interface.

were obtained with the SF_6 source. It should be noted that the 0-250 amu spectra obtained from the SF_6 source were virtually identical to those obtained with the Xe^+ source; that is, no trace (ions or fragments) of the SF_6 primary ion was observed. The observed increases in the yields of the higher mass fragments for the IBAD region (note the change in vertical scales) imply that the fragments were more branched than those from the non-IBAD region and the control PTFE. Similar increases in branching were correlated with the increased crosslinking in aliphatic hydrocarbons and have been observed with static SIMS.[10] Although this branching has not been previously seen in PTFE, it strongly implies that the ion irradiation during IBAD processing produced crosslinks in the surface region of the PTFE. This crosslinking is in turn responsible for the increase in the PTFE shear strength and, ultimately, the increased practical adhesion of the IBAD metal coating.

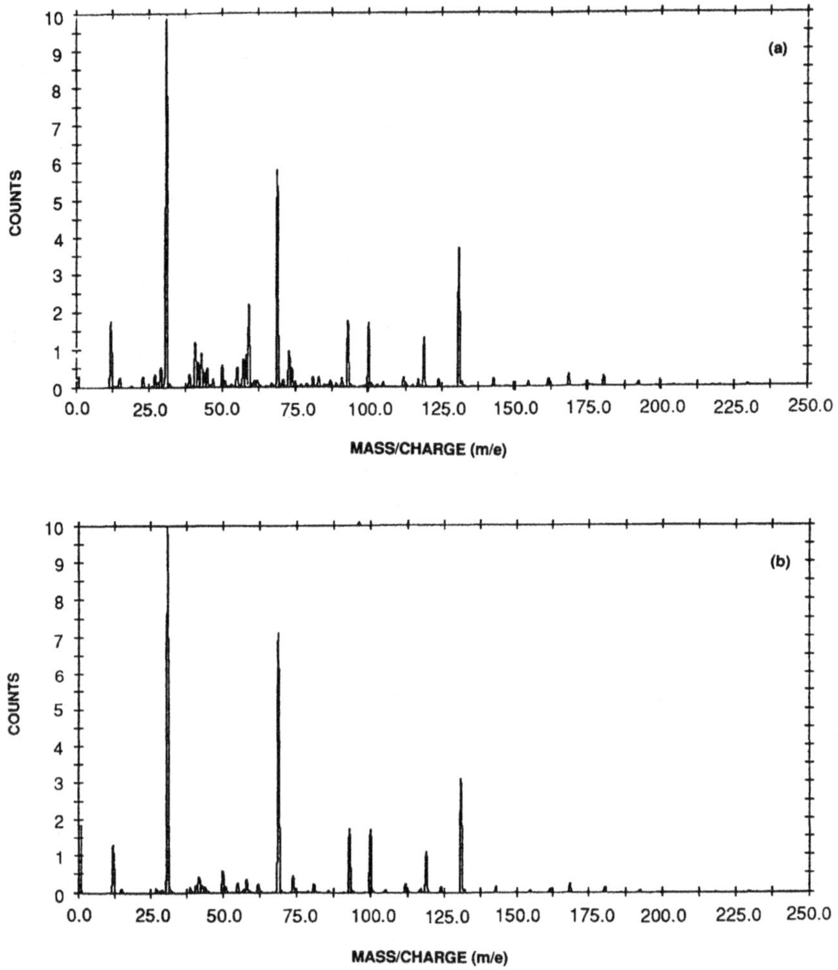

Figure 8. Xe^+ Static SIMS Spectra for Au/PTFE Metal Failed Coating a) Non-IBAD Region and b) IBAD Region.[11]

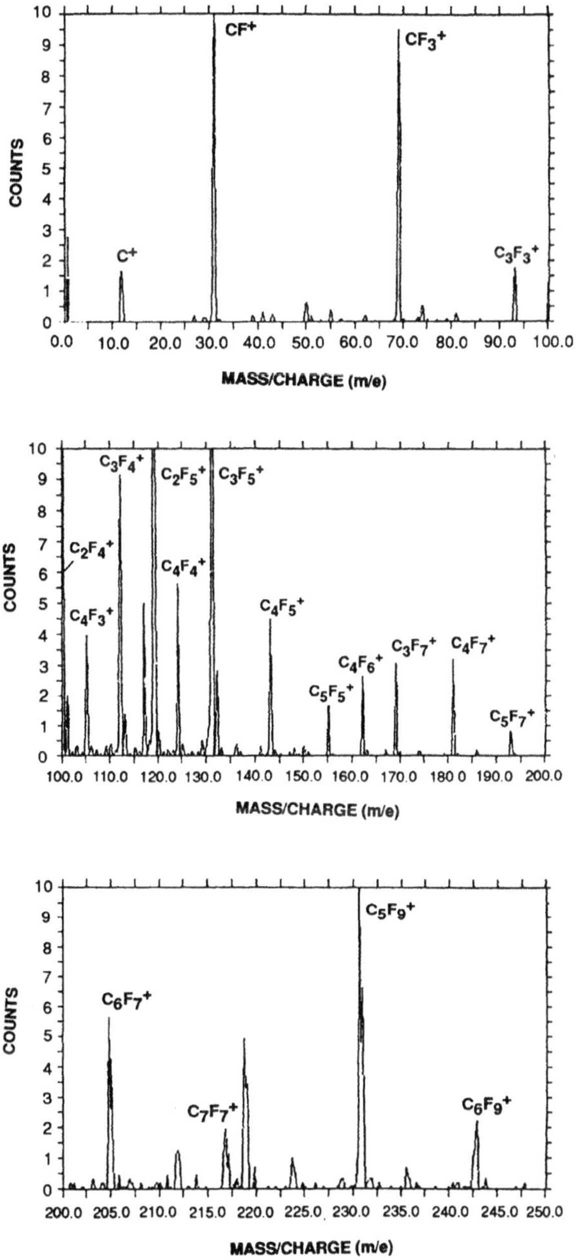

Figure 9. Xe^+ Static SIMS Spectra for Control PTFE Sample.[11]

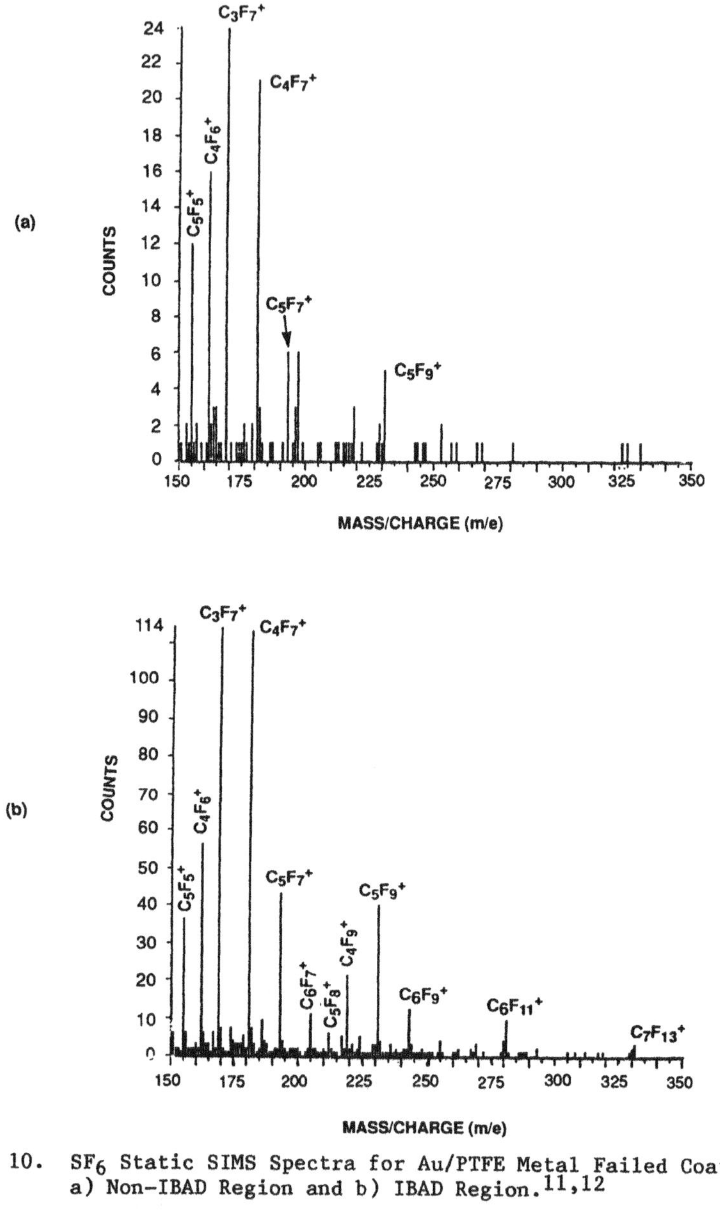

Figure 10. SF_6 Static SIMS Spectra for Au/PTFE Metal Failed Coating a) Non-IBAD Region and b) IBAD Region.[11,12]

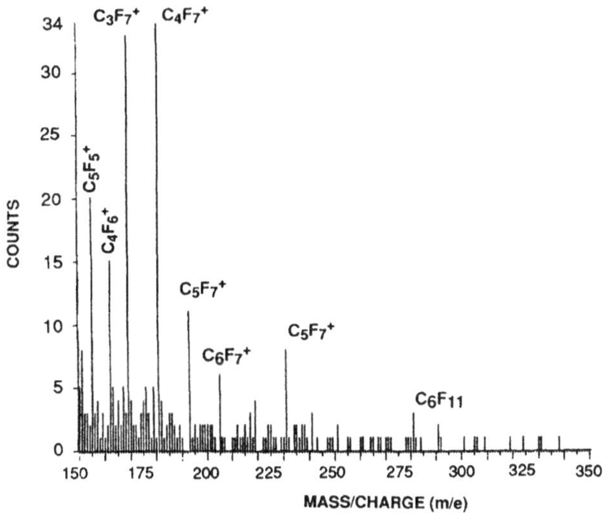

Figure 11. SF_6 Static SIMS Spectrum for PTFE.[11]

IV. SUMMARY

Significant improvements in the adhesion of several metals to polymers have been achieved through the ion beam assisted deposition (IBAD) technique. These films are substantially more adherent than similar non-IBAD coatings. Earlier work by our group suggested that the enhanced metal/polymer adhesion was the result of the "smearing" of the normally sharp metal/polymer interface through ion beam recoil events. The extreme surface sensitivity of the static SIMS analysis has suggested that the improved adhesion may actually be the result of increases in the cohesive strength of the surface layers in the PTFE substrates through the promotion of crosslinking. The radiation damage in polymers associated with ion irradiation can be very destructive to the polymeric structure depending on the conditions. This damage, initiated through chain scissions and consequently free radical formation, gas evolution, and dangling bond creation, may result in crosslinking in some polymer systems, while in others it may result in chain reorientation and/or carbonization, i.e., film degradation. Regardless of the type of chain reorganization which may occur, ion irradiated polymers will eventually approach graphitization at intense beam conditions.[13-20] The results of the static SIMS analyses of the ion beam modified polymer systems investigated by our group have indicated that PTFE is prone to crosslinking at moderate beam conditions[19,20] and, as seen by others, surface texturing at severe ion beam treatments.[20,21] Thus the possible enhanced adhesion through IBAD is contingent upon the degradation process that the polymeric substrate will follow as determined by the ion beam parameters used. This mechanism also helps explain the lack of improved adhesion observed in certain polymer systems.[6] The relatively minor adhesion improvement seen with silver is not as yet understood and further investigation is underway.

V. ACKNOWLEDGEMENT

This work was partially supported by DOE SBIR Phase II (Contract # DE-AC02-88ER80613).

VI. REFERENCES

1. J.E.E. Baglin, Mater. Res. Soc. Symp. Proc., 47, 3 (1985).
2. J.E. Griffith, Yuanxum Qiu, and T.A. Tombrello, Nucl. Instrum. Methods, 198, 607 (1982).
3. Chin-An Chang, J.E.E. Baglin, A.G. Schrott, and K.C. Lin, Appl. Phys. Lett., 51(2), 13 July, 1987.
4. C.J. Sofield, C.J. Woods, C. Wild, J.C. Riviere, and L.S. Welch, Mater. Res. Soc. Symp. Proc., 25, 197 (1984).
5. J.E.E. Baglin and G.J. Clark, Nucl. Instrum. Methods, B7/8, 881 (1985).
6. I.H. Loh and J.K. Hirvonen, Mater. Res. Soc. Symp. Proc., 108, 241 (1987).
7. I.H. Loh, R.A. Moody, and J.C. Huang, Membrane Sci., 50, 31-49 (1990).
8. Chin-An Chang, Appl. Phys. Lett., 51(16), 19 October, 1987.
9. K.L. Mittal, Polym. Eng. Sci., 17, 467 (1977).
10. W.J. van Ooij and R.H.G. Brinkhuis, Surface Interface Anal., 11, 430 (1988).
11. W.J. van Ooij and M.M. Nindi, Service Report on SIMS of Metal/PTFE, Polycarbonate, and Polyimide Samples, 1989.*
12. W.J. van Ooij, Mater. Res. Soc. Symp. Proc., 153, 149 (1989).
13. S. Noda and T. Hioki, Carbon, 22, No. 4/5, 359 (1984).
14. M.S. Dresselhaus, B. Wasserman, and G.E. Wnek, Mater. Res. Soc. Symp. Proc., 27, 413 (1984).
15. B. Wasserman, Ph.D. Thesis, Massachusetts Institute of Technology, 1985.
16. T. Venkatesan, L. Calcagno, B.S. Elman, and G. Foti, in "Ion Beam Modification of Insulators," P. Mazzoldi and G.W. Arnold, editors, p.301, Elsevier, New York, 1987.
17. B.S. Elman, G.F. Blackburn, M.K. Thakur, D.J. Sandman, L.A. Samuelson, and D.G. Kenneson, Nucl. Instrum. Methods, B19/20, 872 (1987).
18. I.H. Loh, R.W. Oliver, and P. Sioshansi, Nucl. Instrum. Methods, B34, 337 (1988).
19. R.A. Moody, P.H. Lu, and I.H. Loh, Technical Reports for Phase II DOE Contract # DE-AC02-87ER80450 entitled "Electrically Conductive Polymers by Ion Implantation," 1989.*
20. Unpublished Spire results of ion beam processes on polycarbonate, PTFE, and polyimide, 1989-1990.
21. L. Torrisi, L. Calcagno, and A.M. Foti, Nucl. Instrum. Methods, B32, 142 (1987).

---

* Available from Publications Dept., Spire Corporation.

AN IMPROVED PROCESS TO FACILITATE THE CHEMICAL BONDING OF ELECTROLESS COPPER TO POLYETHERIMIDE SURFACES

B.R. Karas[1], D.F. Foust[1], W. V. Dumas[2], and E.J. Lamby[1]

[1]GE Corporate Research and Development, Schenectady, NY 12301
[2]GE Armament and Electrical Systems Department, Lakeside Avenue Burlington, VT 05401

An improved, chromium-free, non-swell and etch process for treating Ultem® polyetherimide is described. A peel strength between electroless copper and the polymer of 150-200 g/mm is obtained while producing little surface roughness. The results of ultrasonic aluminum ball bond tests to gilded polyetherimide surfaces show an order of magnitude decrease in wires failing to bond relative to previous systems. Results from scanning and transmission electron microscopic, X-ray photoelectron spectroscopic, and Auger analyses of the system are discussed and indicate that a well-defined interface exists. Classes of adhesion promoters and the effects of process parameters on the innate properties of the polymer are described. Cohesive failure occurs in the polymer layer during 90°peel testing. The applicability of the system to a manufacturing environment is discussed via simulated production techniques.

INTRODUCTION

The advent of high temperature plastics such as polyimides has led to the use of polymers in a variety of electronic applications. General Electric's Ultem® polyetherimide is a unique type of polyimide which is capable of being injection molded. Moreover, a wide variety of fillers can be added to enhance the thermal and physical properties of the plastic. These properties have led to the use of Ultem® polyetherimide as a circuit board dielectric material. The advantages of an Ultem® substrate over the traditional copper-clad-epoxy laminate include a reduced dielectric constant, increased glass transition temperature, a Z-axis expansion more closely matched to copper, and a reduced coefficient of thermal expansion[1-4]. A unique advantage is the capability to mold 3-dimensional shapes with this material and thus incorporate functional and structural components as integral parts of the circuit board[1-5]. Moreover, through holes can be formed during the molding process, thereby eliminating the need for drilling and deburring.

The manufacture of a 3-dimensional, molded circuit device such as the sample shown in Figure 1 differs greatly from the traditional printed circuit board. In the manufacture of a typical printed circuit board, the metal traces are formed from the laminated copper. In producing a molded circuit board, the metal must be deposited onto the substrate surface.

® Ultem is a registered trademark of GE

Metallized Plastics 2, Edited by K.L. Mittal
Plenum Press, New York, 1991

Figure 1. Macroview of demonstration 3-dimensional molded circuit board made from Ultem® 2312.

In recent years, a large number of techniques have been developed to obtain metal traces on the resins used in molded circuit boards. These include photosensitive catalysts[6], laser imaging of catalysts[7], pressed circuitry, two-shot molding[8], conductive inks[9], laser induced decomposition of the plastic to produce conductive carbon[10], and laser patterning of metals[11].

The adhesion between the deposited metal layer and plastic rather than the method of metallization is critical. Several theories have been developed to explain the metal/polymer interactions and these can be categorized as being physical or chemical or mixed physical-chemical in nature[12,13].

A common method of obtaining adhesion between an electrolessly deposited metal layer and a plastic substrate utilizes an organic compound to soften or swell the resin and a strong oxidizing agent, like chromic acid/sulfuric acid, to etch the resin. As the chemical resistance of the resins increases, more aggressive chemicals are required to promote metal adhesion. Figure 2 presents a cross-section of a copper-plated sample of a glass-filled polymer treated by a "swell and etch" process. The cracks in the resin result from excess chemical attack and extend deep, 300μm, into the plastic bulk. These cracks have a deleterious effect on the mechanical and electrical properties of the resin (Table I). In addition, a rough metal layer can result from the irregular polymer surface created during etching . The physical attack of the swell and etch chemistries on the polymer is generic and is not related to a specific resin.

The metal/polymer bond produced by a swell and etch process is almost completely mechanical in nature. The desire to improve the surface finish and eliminate the detrimental effects on the resin properties indicates the need for an adhesion system with a chemical component. Recently, a number of inventions dealing with chemical treatments of polyimides to improve the bonding of copper have been described[14-18]. These processes are void of organic swellants and chromic acid.

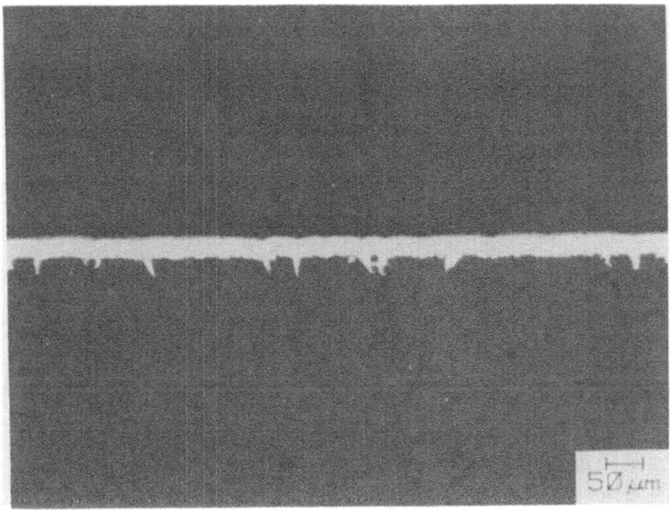

Figure 2. Cross-section of a glass-filled polymer treated by typical swell-and-etch process showing cracks into polymer interior.

Foust and Dumas described a process for treating Ultem® polyetherimide which yields an improved surface vis-a-vis swell and etch treatments and a mixed mechanical/ chemical bond between the copper and plastic[17-18]. Their approach typically produces a peel strength of 150-200 g/mm and employs a chemical adhesion promoter, e.g., hydroxylamine. Although the surface quality improvement is significant relative to swell and etch techniques, further improvement is required for certain applications such as wire bonding. An approach to obtaining a larger chemical and a smaller mechanical component of adhesion, while minimizing surface roughness and maintaining metal/plastic peel strength, is described in this work.

EXPERIMENTAL

The plastic substrates utilized in this study were based upon Ultem® polyetherimide available from General Electric. Samples of unfilled resin (Ultem® 1000) were employed to assess changes in the plastic during chemical treatment. Because of their unique thermal and mechanical properties filled Ultem®grades were also utilized. Table II compares the properties of unfilled and glass-filled Ultem® resins. Peel strength data are reported for filled grades of Ultem® resins.

Table I. Comparison of Polyetherimide Properties Following Adhesion Treatment.

Property	Traditional Swell & Etch	This Work
Flexural Strength	30% reduction	No change
Weight Loss	10-15% reduction	No change
Electrical	No change	No change
Electrical (48°C/50% RH)	25% reduction	No change

Table II. Ultem® Polyetherimide Properties

Sample	Filler	Deflection Temperature 1.82 MPa(°C)	Flexural Modulus (MPa)	Coefficient of Themal Expansion (m/m/°C)
Ultem® 1000	None	200	3,300	$5.6x10^{-5}$
Ultem® 2312	30% Glass	210	9,000	$2.0x10^{-5}$

Initial metallization of the substrates was accomplished via electroless deposition techniques. The metal thickness was then increased electrolytically in an acid copper bath. Typically, 30-35 µm of copper was applied[19].

Adhesion measurements were performed by applying 3.2 mm (0.125 inch) wide tape to a copper-plated substrate, etching the undesired copper in concentrated nitric acid, water rinsing, and removing the tape. One end of the strip was connected to an Ametek digital force gauge connected to a computer processor. Force values required to lift the metal strips from the substrate were recorded and converted to g/mm peel values by the computer. The peel was conducted at 90°to the substrate. Multiple measurements (50 point) were taken for each strip and the average value reported.

Scanning electron microscopic analysis was performed on a Hitachi S-800 field emission microscope. X-ray photoelectron spectroscopic analysis was performed on a Surface Science Instruments SSX-100 system employing a monochromatic Al Kα source, focussed to 300 or 600 µm. Spectra were obtained using a fixed pass energy of 150 eV for survey scans and 50 eV for detail scans of individual core levels. All binding energies were corrected for sample charging by assigning the C-H component of the C 1s line to 284.6 eV. A Physical Electronics, Inc. Auger spectrometer model 545 was used for the Auger depth profiles. The electron beam diameter was typically around 10 µm with a beam voltage of 3 kV. The sputtering was done using a PHI differential ion gun and xenon ions at 2.5 kV and an emission current of about 25 mA. The beam was rastered over a 2 mm square area and using these conditions the sputtering rate for copper is about 100 angstroms per min. The atomic concentrations were calculated using peak to peak heights multiplied by published sensitivity factors, and the results normalized to 100 percent.

The Auger line scans were made using the Perkin Elmer, Physical Electronics Division scanning Auger microprobe Model 600. The electron beam voltage was 10 kV with an emission current of about 20 nA. Analysis was made at 200 points per line and cycled for 30 min. A low energy background is subtracted from each point of N(E) data. These line scans represent only Auger signal minus background and no quantitative significance is suggested. These line scans do show variations in elemental Auger intensity across the scan line.

RESULTS AND DISCUSSION

A schematic of the previously described Foust and Dumas process[17,18] to introduce chemical adhesion into the copper/Ultem® resin bond is shown in Figure 3.

Hydroxylamine is the adhesion promoter and typically yields 150-200 g/mm peel strength. The plastic surface is cleaned of grease, fingerprints and mold release, normalized via a sulfuric acid etch, and then a chemical adhesion promoter applied. The pieces are subsequently catalyzed with a palladium colloid, electrolessly plated with copper, heat treated, electrolytically plated, and heat treated a second time. A cross-section of metallized Ultem® 2312 treated as described by Foust and Dumas is shown in Figure 4. Note the improved surface and decreased resin damage relative to the swell and etch example displayed in Figure 2 . Further surface refinement was necessary, however, for certain applications such as wire bonding.

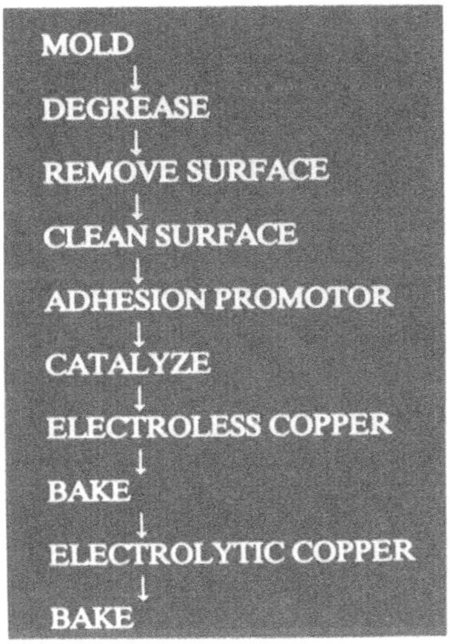

Figure 3. Overview of process described by Dumas and Foust[17].

The processing step responsible for rough metal deposits and the rough plastic surface is the removal of the outer plastic layer, i.e., the sulfuric acid. The key, then, is to find improved adhesion promoters which will increase the copper to polymer chemical adhesion component while allowing a decrease in the mechanical component and thus yield a smoother surface topography. A comparison of thiourea and hydroxylamine as adhesion promoters as a function of post-electrolytic heat treatment is shown in Figure 5. The sulfuric acid dwell time was 3 minutes. The thiourea-treated specimens not only achieved a greater peel strength, but did so more rapidly than the hydroxylamine samples.
The effect of decreasing the etch time on adhesion for Ultem® 2312 is shown in Figure 6 for both hydroxylamine and thiourea adhesion promoters. The disparity between the two

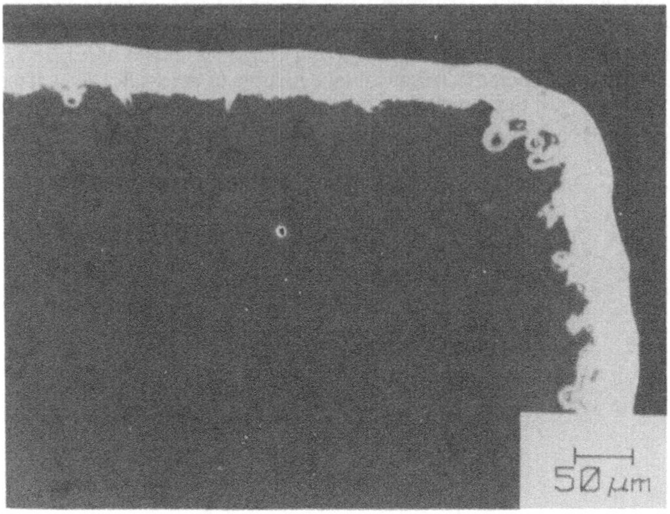

Figure 4. Cross section of Ultem® 2312 processed as described by Dumas and Foust[17].

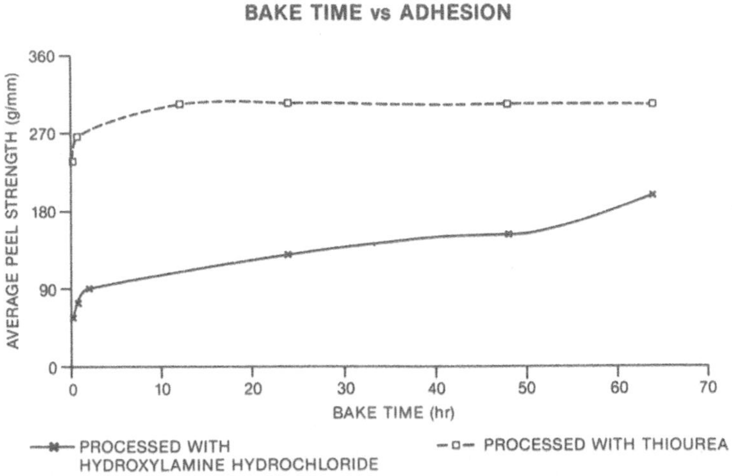

Figure 5. Adhesion vs bake time for Ultem 2312 for thiourea (❑) and hydroxylamine (x) adhesion promoters. The acid dwell time was 3 minutes.

chemical bonding agents is most pronounced at the lower etch times. Adhesion values greater than 180 g/mm are possible with minimal surface damage. Tests on unfilled Ultem® resin indicate that 1 µm of surface is removed. The improvements in surface quality are most striking. Figure 7 presents scanning electron micrographs (SEMs) of metallized Ultem® 2312 treated by the Foust and Dumas process, 3 minute sulfuric acid dwell, and the improved process described herein, 0.5 minute acid immersion. The loss of the large-scale surface roughness due to the filler and the overall improved appearance is clearly visible in the right-hand SEM in the figure. Figure 8 displays a cross-section of a 0.5 minute sulfuric acid, thiourea-treated, glass-filled Ultem® specimen containing gold, nickel and copper metallization. The increase in surface smoothness, desirable for wire bond applications, can be seen by referencing Figure 4.

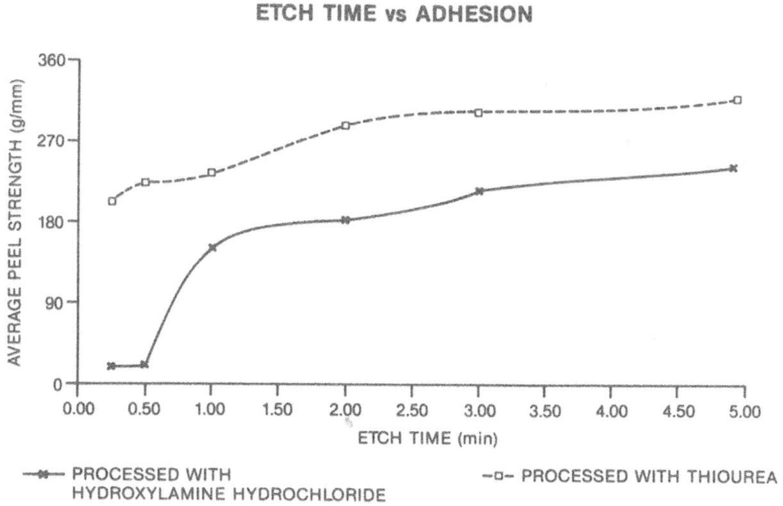

Figure 6. The effect of sulfuric acid etch time on adhesion of copper to Ultem®2312. The adhesion promoters were thiourea (❑) and hydroxylamine (x).

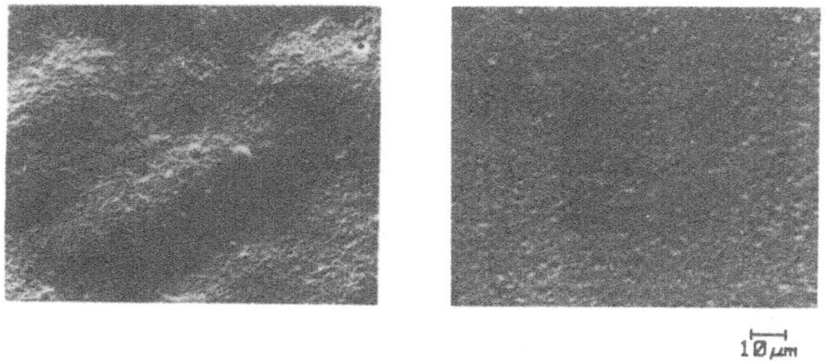

Figure 7. Scanning electron micrographs of Ultem® 2312 showing surface topography for 3 minute (left) and 0.5 minute (right) sulfuric acid immersion.

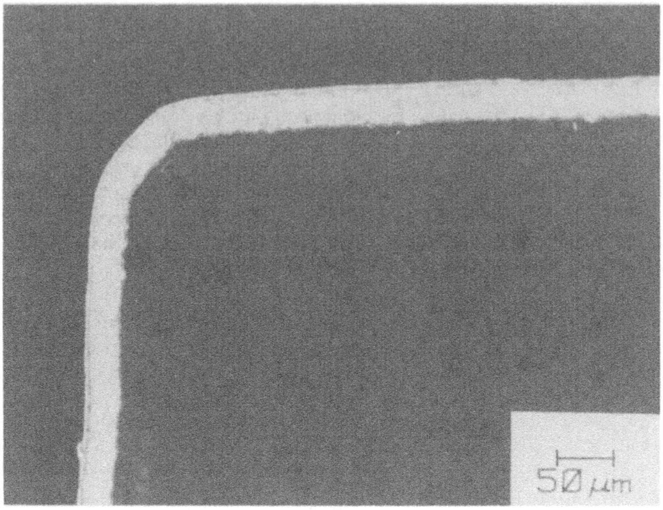

Figure 8. Cross-section of 0.5 minute sulfuric acid dwell, thiourea-treated Ultem® 2312.

The effect of this surface improvement is illustrated by functional wire bond test results. Figure 9 presents an overview of a plastic pin grid array possessing 0.18 mm traces and 0.13 mm spaces. Wire bonding is a common way to make electrical connection between the metal trace and the integrated circuit. A cross-section of a trace displaying the metallization scheme and the polymer and metal surfaces is shown in Figure 10. The results of 25μm aluminum wires ultrasonically ball bonded to the metallized plastic surfaces are shown in Table III. An order of magnitude decrease in the number of wires failing to bond and an increase in the bond strength of those that did bond was observed. The decline in bond failure was attributed to the smoother surface of the thiourea-treated plastic.

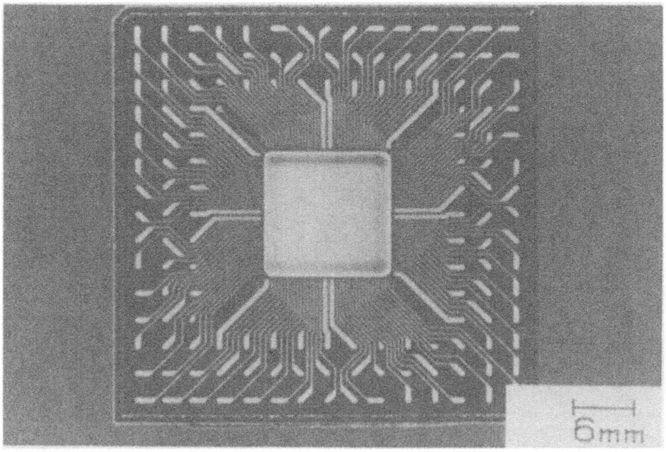

Figure 9. Macroview of metallized plastic pin grid array. The traces are 0.18 mm and the spaces 0.13 mm.

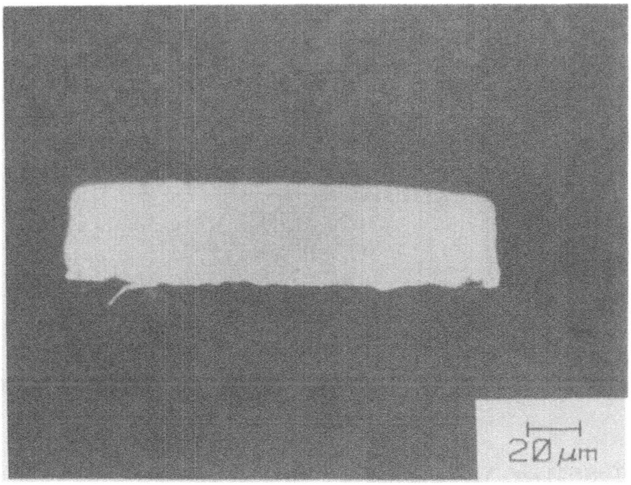

Figure 10. Cross section of trace on plastic pin grid array.

In the sections which preceded this one, we described the functional and surface improvements achieved via a new adhesion promoter. The sections below delineate the chemical treatment employed[19], interfacial analysis, and applicability of the process to a manufacturing environment.

Process Description

Initially, the molded samples (Figure 11) were cleaned in Freon®* to remove grease, fingerprints and mold release, cf Figure 12. There is no significant change in either the appearance or chemical composition of the unfilled Ultem® specimen. Ultem®1000 was used as a model system to study the effects of the pretreatment steps on the surface morphology and chemical composition. While other cleaners such as Micro®function satisfactorily [18,20], Freon-type cleaners are preferred because of their volatility, ensuring complete and rapid drying prior to further processing.

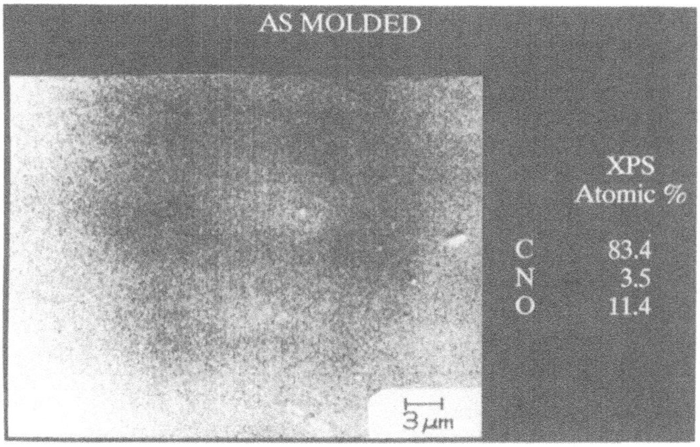

Figure 11. SEM and XPS-analysis of relative atomic concentration of as-molded Ultem® 1000.

*® Freon is a registered trademark of E.I. duPont deNemours & Co.

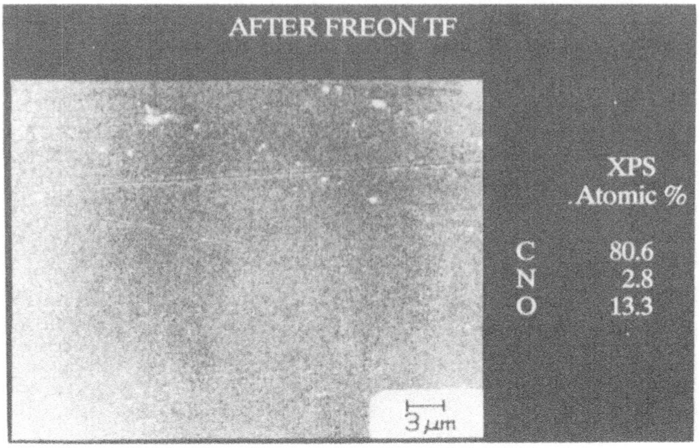

Figure 12. SEM and XPS-analysis of relative atomic concentrations of Freon-treated Ultem® 1000.

The next step is to "normalize" the resin surface. This is necessary to eliminate any inconsistencies in the plastic surface. The variations may be chemical or physical in nature. Although a select number of reagents can interact with Ultem® polyetherimide, concentrated sulfuric acid was chosen for environmental and safety reasons, and because it is a solvent for polyetherimide. Following the brief, 0.5 minute sulfuric acid immersion and a water rinse, a non-adherent white film is present on the resin surface. Figure 13 presents an SEM and X-ray photoelectron spectroscopic (XPS) data which show the film and the oxidized nature of the surface. Furthermore, the binding energy of the S_{2p} signal indicates that S is present in an oxidized state.

The white-film is modified by treatment with a strong alcoholic base. The base physically and chemically modifies the residue, cf Figure 14. The film has taken on a crusty, less granular appearance relative to the film after the sulfuric acid rinse. X-ray photoelectron spectroscopy reveals that the Ultem®polyethermide ring is hydrolyzed to the potassium salt of the carboxylic acid. This behavior is consistent with previously published results [18,20].

Table III. Wire Bonding Results for Al to Gilded Ultem® 2312

	Hydroxylamine Based Adhesion Promoter	Thiourea Based Adhesion Promoter
Average Bond Strength (g)	9.1	11.5
Bond Strength Range (g)	6-13	7-16
Wires failing to bond (%)	5	0.5

Figure 13. SEM and XPS-analysis of relative atomic concentrations of sulfuric acid-treated Ultem® 1000.

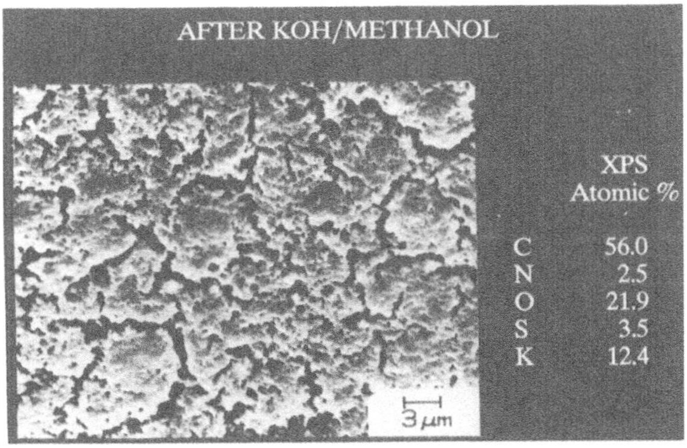

Figure 14. SEM and XPS-analysis of relative atomic concentrations of KOH/methanol-treated Ultem® 1000.

The modified white-residue is removed in a 1:1 mixture of alcohol and an aggressive solvent. Figure 15 shows the surface and chemical analysis of the polyetherimide after treatment with a 50/50 mixture of methanol and N,N—dimethylformamide. Comparison of this photograph to Figure 11 reveals that little physical change has occurred to the surface. However, the chemical nature of the surface has been altered and the resin is now water wettable.

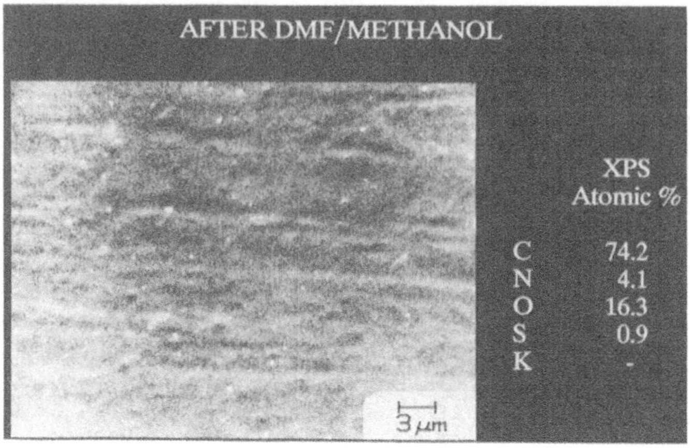

Figure15. SEM and XPS-analysis of relative atomic concentrations of DMF/methanol-treated Ultem® 1000.

Following a water rinse, the "conditioned" Ultem® polyetherimide sample was immersed in an aqueous solution containing thiourea; this species acts as an adhesion promoter. Figure 16 depicts the surface following this treatment. The surface morphology is unchanged relative to the initial surface. The polymer surface has been enriched in oxygen and, based on the C and S binding energies, contains C = S linkages.

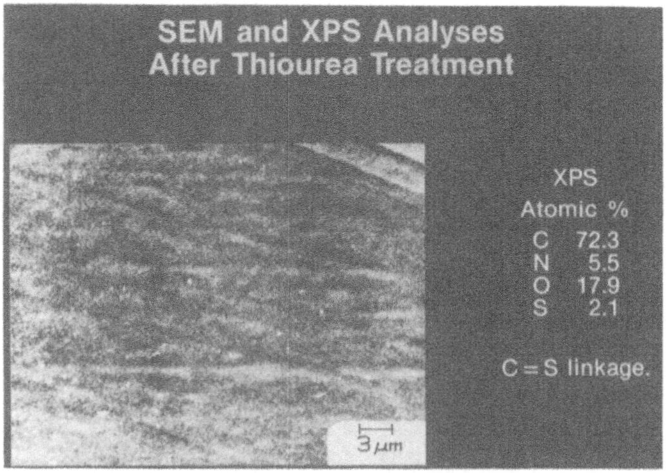

Figure 16. SEM and XPS-analysis of relative atomic concentrations of thiourea-treated Ultem® 1000.

Figure 17 presents an Auger depth-profile of a copper-plated Ultem® 1000 specimen. Sulfur is detected only at the copper/polyetherimide interface. The presence of copper, oxygen, carbon, and sulfur at the interface is shown in the line scan presented in Figure 18. Transmission electron microscopic analysis (Figure 19) shows a well- defined interface with an overall roughness of less than a few tenths of a micron. No diffusion of the copper into the polyetherimide is detected.

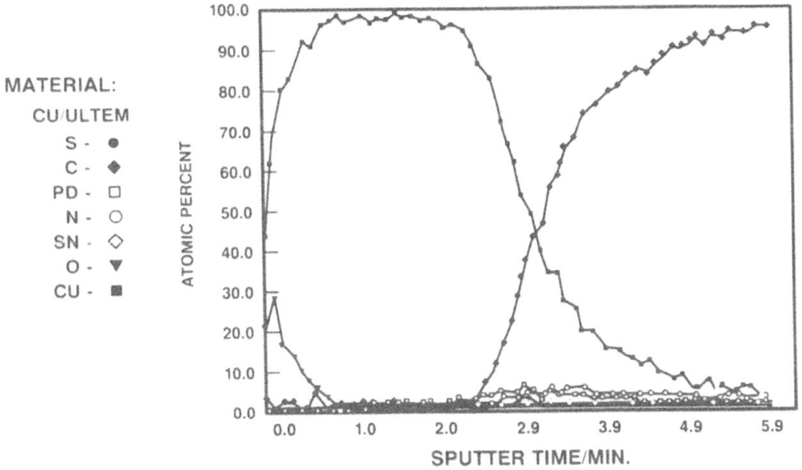

Figure 17. Auger depth profile of copper-plated Ultem® 1000.

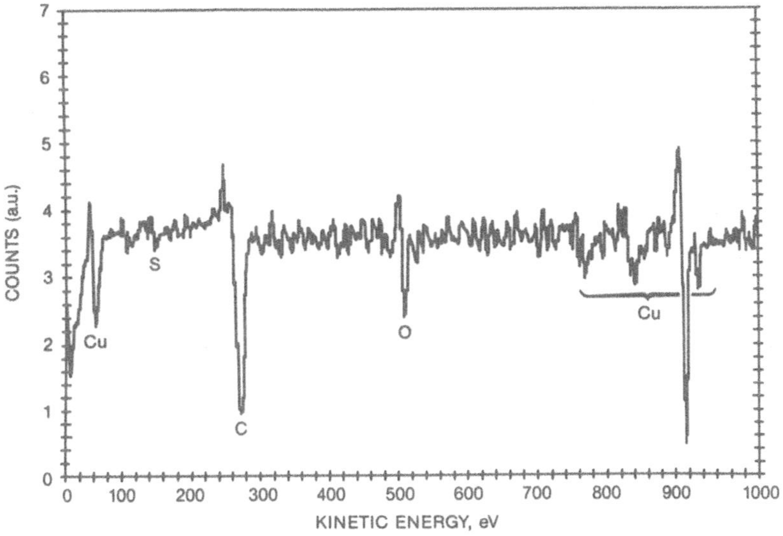

Figure 18. Auger line scan of copper/Ultem® 1000 interface.

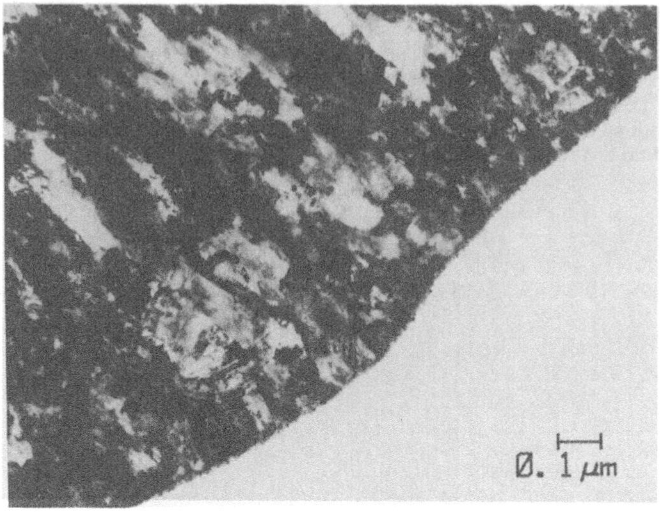

Figure 19. Transmission electron micrograph of copper/Ultem® 1000 interface.

The adhesion-promoted surface is catalyzed with a commercial tin/palladium colloid, accelerated to remove the tin sheath from the catalyst, and electrolessly plated to 1.0-1.5 μm of copper. Following a 2 hour heat treatment at 75°C the metal thickness is built up electrolytically to 37.5 μm of copper and the sample heat treated again.

The peel strength of the copper to the plastic was then determined. A broad range of adhesion promoters, which are capable of chelating copper have been identified and yield peel strengths of 150-200 g/mm, Table IV. In contrast, utilization of hydroxylamine and the 0.5 minute dwell in sulfuric acid described herein produced 18 g/mm peel values. The improvement in adhesion promoting ability of thiourea vs hydroxylamine is once again apparent.

The copper/polymer bond fails cohesively in the plastic layer during 90° peel testing, this is shown in the scanning electron micrographs which reveal resin on the metal-side of the peel. Figure 20 presents SEMs of the failed resin and metal regions for metallized glass-filled polyetherimide. Note the presence of glass on the back of the metal trace. Examination of these areas by XPS shows the presence of metallic copper and amide-type nitrogen on the metal side of the peel, and copper (II) and imide-type nitrogen

Table IV. Adhesion Promoting Agents for Ultem® 2312

Chemical Compound	Peel Strength (g/mm)
Diacetamide	130
1,1,3,3-tetramethylurea	135
Formamide	142
Thiosemicarbazide hydrochloride	157
2, 4-dithiobiuret	158
Semicarbazide hydrochloride	160
Acetyl thiourea	162
Thiourea	166
Acetamide	175
Urea	176
Thiocyanuric acid	189
None	114

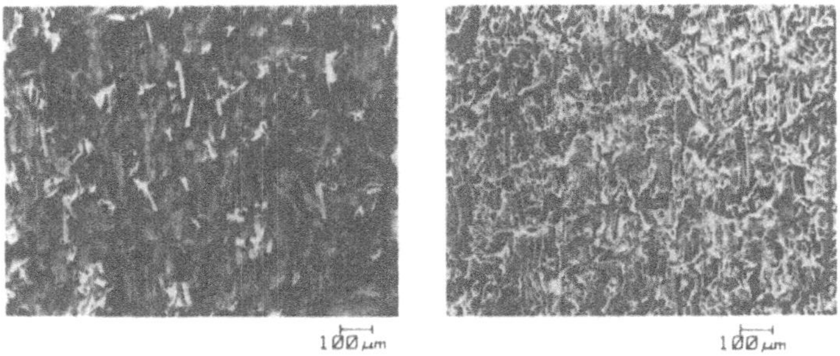

Figure 20. SEMs of copper and Ultem® 2312 surfaces following peel.

on the polymer side of the peel. Ion beam sputtering of the resin layer on the back of the metal during XPS analysis indicates that the polymer layer is 3000-4000Å thick.

The applicability of the system described in this paper to the manufacture of large quantities of metallized molded circuit boards was evaluated by simulating production techniques. Figure 21 presents the low, average and high adhesion values obtained as a function of throughput. The adhesion consistency is clearly evident in the figure. During the course of the simulated production study, no detectable consumption of pre-catalyzation chemistry was detected.

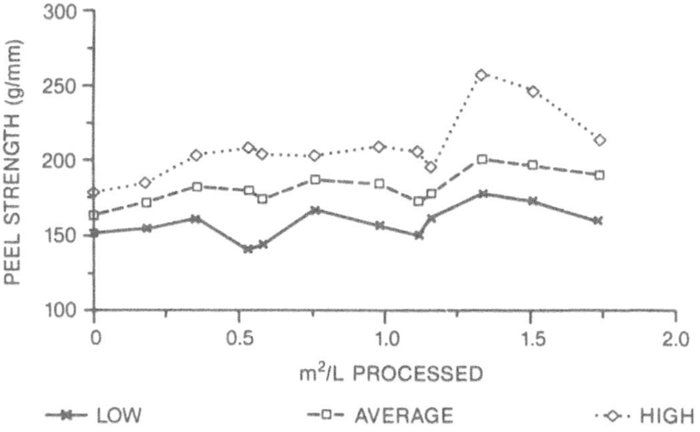

Figure 21. Adhesion as function of m^2/L of Ultem® 2312 processed.

SUMMARY

A method for obtaining 150-200 g/mm peel strengths between copper and Ultem® polyetherimide has been described. The adhesion values are obtained with minimal surface degradation. The metallized surface is amenable to wire bonding and is improved relative to other non-swell and etch techniques. The improved surface is a result of a decreased chemical etch allowed by the use of superior adhesion promoters. Analysis of the bond reveals a well-defined copper/polymer interface, and that failure during peel occurs cohesively in the polymer layer.

ACKNOWLEDGEMENTS

The authors wish to thank the members of the Light Microscopy Group for cross-sectional analyses, M. C. Burrell and J.J. Chera for XPS, L. King for SEM, M. McConnell for Auger, and E. Koch for TEM analyses.

REFERENCES

1. J. Travis and J. Ganjei, Proc. 41st Reinforced Plastics/Composite Institute, 1986, Session 8-A, p 1.

2. J.M. Mihelcic, Printed Circuit Design, 41 (Feb. 1988).

3. ICI Technical Bulletin 400-186.

4. J.D. Gandre, Printed Circuit Design, 3, No. 6 (June 1986).

5. J. Rose, Connection Technol. 5, No. 1, 21-24 (January 1989).

6. D.C. Frisch, and W. Weber, US Patent 4,594,311 (1986).

7. H. Cole, Y.S. Liu, J.W. Rose, R. Guida, L.M. Levinson, and H.R.Phillip, Proceedings of Symposium on Laser Processes for Microelectronic Applications, Electrochemical Society, #88-10, 1988, p. 187.

8. D.C. Frisch, J. Row, and J. Somers, paper presented at the National Electronic Packaging Conference, Nepcon West, Los Angeles, 1987.

9. L. Firestone, Printed Circuit Fabrication 11, No. 11, 107 (November 1988).

10. A.M. Lyons, F.T. Mendenhall, Jr., M. Robbins, N.R. Quick, and C.W. Wilkins, Jr., US Patent 4,691,091 (1987).

11. H.S. Cole, Y.S. Liu, J.W. Rose, and R.Guida, Appl. Phys. Lett., 53, 2111(1988).

12. K. L. Mittal, J. Vac. Sci. Technol., 13, 19 (1976).

13. W.P. Townsend, in "Handbook of Adhesives", 2nd Edition, I. Skeist, ed., p. 846, van Nostand Reinhold, New York (1977).

14. H. Malkow and W.Strache, German Patent DE 3612822 (1987).

15. H. Mahlkow, M. Romer, G. Rosskomp, H.-M. Seidenzsinner, L.Stein, and W. Strache, German Patent DE 3708214 (1988).

16. J. Grapentin, H. Mahlkon, and J. Skupsch, US Patent 4,517,254 (1985).

17. W.V. Dumas and D.F. Foust, US Patent 4,775,449 (1988).

18. D.F. Foust and W.V. Dumas, in "Metallization of Polymers", E. Sacher, J. -J. Pireaux, and S. P. Kowalczyk, eds., ACS Symposium Series No. 440, Chapter 35, (1990).

19. D.F. Foust, W.V. Dumas, E.J. Lamby, and B.R. Karas, US Patent 4,842,946 (1989).

20. M.C. Burrell, B. R. Karas, D. F. Foust, W. V. Dumas, E. J. Lamby, W. T. Grubb, and J.J. Chera, in "Metallized Plastics 1: Fundamental and Applied Aspects," K. L. Mittal and J. R. Susko, eds., pp. 223-233, Plenum Press, New York (1989).

ADHERENT ELECTROLESS COPPER LAYERS ON POLYSTYRENE

Charles E. Baumgartner

GE Corporate Research & Development

Schenectady, NY 12301

Molded polystyrene and polystyrene containing blends such as Noryl®resin* can be chemically modified using chlorosulfonic acid to produce a sulfonic acid derivatized surface which is capable of being electrolessly copper plated using standard catalyzation and plating procedures. Following metallization, adhesion develops across the copper/polystyrene interface via a chemical bond that forms through the sulfonic acid linkage concomitant with the loss of interfacial water. The rate at which adhesion develops after plating increases with bake temperature and, for a given temperature, increases with decreasing relative humidity. No adhesion develops in ambients with a relative humidity approaching 100%.

Once formed, the adherent copper/polystyrene interface loses adhesion with prolonged heating at about 60°C in a high humidity environment. Adhesion loss is initially reversible as the interface is rehydrated, then becomes irreversible as the polystyrene surface is desulfonated.

INTRODUCTION

An electroless copper layer deposited onto molded virgin polystyrene possesses little to no interfacial adhesion. Adhesion is developed in the commercial plating of polystyrene and polystyrene containing blends by first etching the plastic with chromic acid as part of a pretreatment process, followed by catalyzation and metal deposition[1,2]. With this treatment, etch-generated crevices on the plastic surface provide regions for the mechanical attachment of the metal deposit to the polymer. However, these etch cracks can also roughen the plastic surface complicating aesthetic surface finishing operations and can increase the risk of fracture, especially in regions of high residual molding stress.

One approach in obtaining metal/polymer adhesion without the necessity of mechanically etching the surface is to alter the chemistry of the polymer surface in such a way as to promote chemical bonding across the metal/polymer interface[3]. Several authors have reported an increase in the adhesion of deposited metals to polystyrene by prior exposure of the polymer to a sulfonating agent[4-8]. This work explores the chemistry associated with the development of adhesion between a deposit of electroless copper and either polystyrene or GE's Noryl® resin, a blend of polystyrene and polyphenylene oxide, following exposure of the polymer to chlorosulfonic acid ($ClSO_3H$) dissolved in anhydrous trichlorotriflouroethane (Freon 113).

---

*Trademark of GE Company

Metallized Plastics 2, Edited by K.L. Mittal
Plenum Press, New York, 1991

EXPERIMENTAL

Compression molded plaques of polystyrene, obtained from Borg-Warner, and Noryl® resin, a blend of polystyrene and polyphenylene oxide manufactured by GE, were ultrasonically cleaned in 1,1,2 trichlorotrifluoroethane (Freon 113), exposed to $ClSO_3H$ as a 0.1 vol% solution in Freon 113, and H_2O rinsed as a pretreatment to electroless copper plating. The treated samples were catalyzed using a 5 min soak in 0.1 M $SnCl_2$ dissolved in 0.01 M HCl and 2 min soak in 0.01 M $PdCl_2$ dissolved in 0.01 M HCl with an intermediate H_2O rinse. Copper was then deposited from a commercial electroless copper plating bath (Shipley #251) to a minimum thickness of 5μm. Following plating, the samples were heated in air under various relative humidity conditions using a controlled humidity oven. Copper adhesion was measured on these samples by electrolytically depositing 0.04 mm of copper over the electroless deposit, striping the copper with tape, and back etching the copper with HNO_3 to leave 3.2 mm wide copper strips. 90° peel measurements were recorded on these copper strips using a computer controlled peel system attached to the sample through a force gage at a linear peel rate of 5 cm/min. Surface compositional studies were performed using x-ray photoelectron spectroscopy (XPS) and secondary ion mass spectrometry(SIMS).

RESULTS AND DISCUSSION

Exposing a polymer to a sulfonating agent such as SO_3 or $ClSO_3H$ as a pretreatment step prior to metallization has, for certain classes of polymers, been shown to improve adhesion across the metal/polymer interface[4-8] Since the polymer surface is typically not chemically roughened during the sulfonation process, the improved adhesion results from a chemical interaction between the added sulfonic acid functional groups and the metallic deposit with only a minor added contribution from mechanical adhesion. One of the processes described in the patent literature is the sulfonation of the polymer surface using a dilute solution of $ClSO_3H$ in a non-reactive organic solvent[8]. Exposure of the polymer to this solution for a "short" period of time leads to the formation of a polyarylsulfonyl chloride which is hydrolyzed by H_2O to form the sulfonic acid derivative. This reaction occurs with little alteration to the polymer surface morphology; continued exposure, however, can often lead to polymer degradation. The precise exposure conditions which differentiate between a modified and a degraded polymer are obviously dependent on the polymer type and solution reaction conditions.

A variation of this process was utilized here to pretreat polystyrene prior to electroless plating. Noryl® resin was also explored under the scope of this study and the data considered together with that obtained for polystyrene because the exterior surface of molded Noryl® resin was determined by XPS to be almost exclusively polystyrene with very little surface exposure of polyphenylene oxide. Figure 1 shows a flowchart of the modification and plating protocol. A layer of electroless copper deposited onto virgin polystyrene following catalyzation, without any prior surface modification, exhibits insufficient adhesion to the polymer surface to withstand a tape test. However, if the polystyrene had been pretreated with dilute $ClSO_3H$, such as is described in the literature[8], adhesion between the deposited metal and the modified polymer develops soon after withdrawal from the final plating rinse solution. For example, polystyrene samples were treated using 0.1 vol% $ClSO_3H$ in anhydrous Freon 113 prior to catalyzation and plating. Following deposition of a 2 μm thick electroless copper film,metal/polymer adhesion sufficient to survive a cross hatched tape test develops in about 30 min under ambient conditions or about 5 min at 60°C. Thicker copper layers require a higher temperature and/or longer time to attain the same level of adhesion.

A more quantitative adhesion measurement than a pass-or-fail tape test is to measure the force required to peel the metal from the polymer surface. A 5 to 7μm thick electroless copper layer was deposited on the samples After removal from the plating bath, the samples were heated under controlled conditions to develop adhesion across the copper/polymer interface. A layer of electrolytically deposited copper was then applied over the top, copper peel strips delineated by covering with an adhesive tape, and the remaining copper dissolved in HNO_3. The level of metal/polymer adhesion was measured by peeling these copper strips

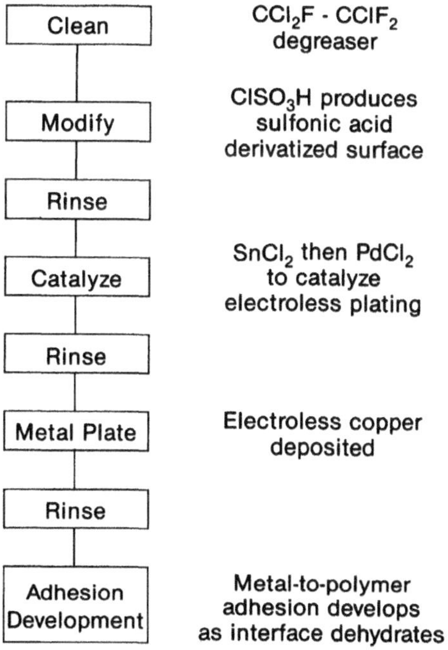

Figure 1. Polymer modification and plating process.

90° to the sample surface. The maximum adhesion obtained following plating depends upon the extent of surface functionalization, i.e., the $ClSO_3H$ concentration and exposure time prior to metallization. These data are seen in Figure 2 in which the extent of surface sulfonation is determined using SIMS as a function of exposure time to a 0.1 vol% $ClSO_3H$ solution and compared to the final copper/polymer adhesion measured via a 90° peel test. A characteristic maximum is seen in the adhesion vs treatment time curve as adhesion initially increases due to the presence of an increasing number of bonding sites and then decreases again at longer exposure times due to polymer degradation. Continued exposure to $ClSO_3H$ for much longer times would ultimately result in a barely measurable copper adhesion.

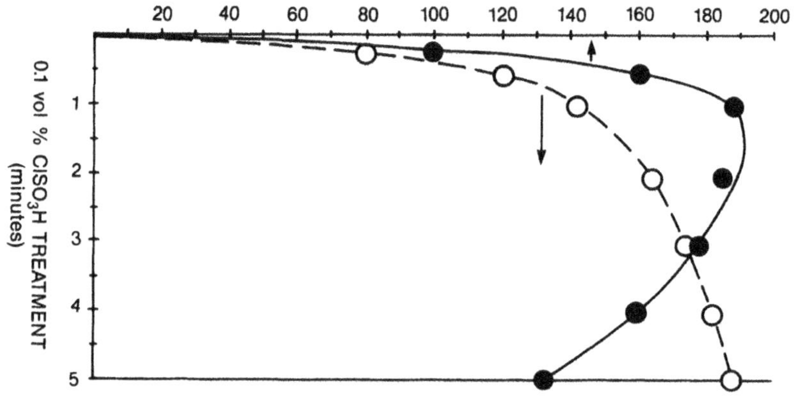

Figure 2. Copper adhesion on polystyrene as a function of exposure time to $ClSO_3H$.

Figures 3 and 4 show microscopic photographs of the plated metal/polymer interface following mounting and cross-sectional polishing. The sample was molded Noryl® resin treated with $ClSO_3H$ for 60 sec. In Figure 3, the copper/Noryl® interface can be seen to be very smooth and free of etch pits. Further magnification of this interface however, such as can be seen in the transmission electron micrographs seen in Figure 4, shows a very slight degree of roughening on the polymer surface characterized by the interpenetration of the deposited copper into the polymer. The extent of this interpenetration increases with $ClSO_3H$ exposure time and the resulting boundary can be described as a composite interphase. The absence of etching on the $ClSO_3H$ treated surface can be further seen in the scanning electron micrographs, Figure 5, which show the surface of the sample before and following $ClSO_3H$ treatment. The two surfaces are identical in texture. In contrast, the bottom photograph seen in Figure 5 shows the surface of a Noryl® sample after removal of the copper strip in a peel test. The remaining surface shows that failure was cohesive and ductile within a region of the polymer under the metal/polymer interface. In addition, examination of the "backside" of the copper peel strip shows that it is covered with a layer of polymer (photo not shown). The thickness of the polymer remaining on the back of the peel strip varies greatly ranging from less than 0.05μm to over 1μm. Longer $ClSO_3H$ exposure times typically yield a thicker polymer layer remaining on the copper peel strip regardless of the measured peel value.

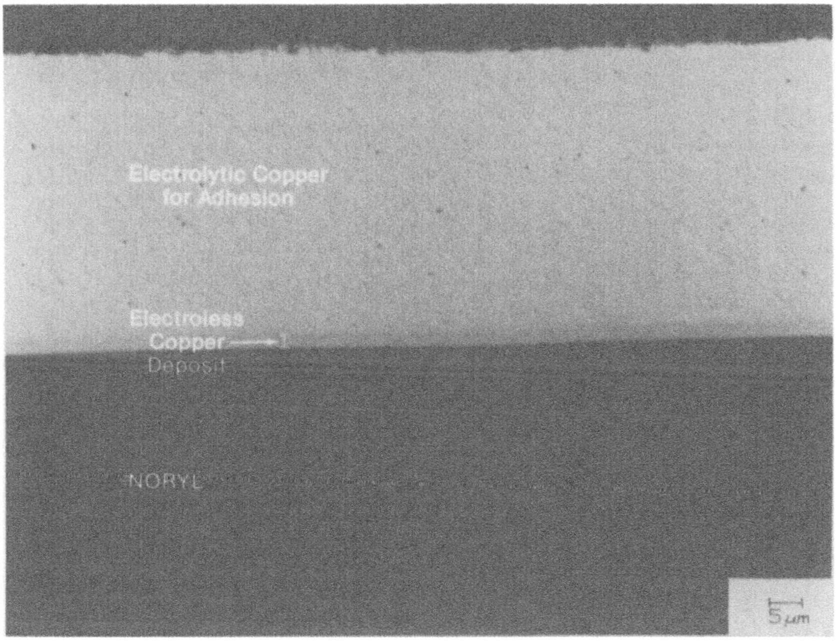

Figure 3. Photograph of mounted and polished cross-sections of $ClSO_3H$ treated molded Noryl® resin plated with electroless/electrolytic copper.

The samples shown in the preceding example were all baked for 48 hrs at 73°C following the deposition of the electroless copper layer. The rate at which adhesion develops is critically dependent upon the selection of proper bake conditions, with the relative humidity being the variable of primary dependence. This dependence can be seen in Figure 6 where the adhesion of copper plated Noryl®resin samples was explored as a function of bake time at 73° C and at constant relative humidities from 19 to 88%. All samples were heated for the prescribed conditions, then electroplated with 0.04 mm electrolytic copper, striped, and peel tested. The measured adhesion level would have been higher in many cases if the sample was reheated following electrolytic copper deposition, however these data are not included here. The figure shows two interesting points. First, the rate of adhesion development is clearly dependent on the bake conditions. However, the maximum adhesion obtained is independent of bake conditions providing the sample is heated for a sufficient time. The one exception to this was the sample baked at a relative humidity of 88% in which no adhesion development occurred. These data suggest that adhesion development involves water elimination from the interface as part of the rate controlling step.

The chemistry of this interfacial interaction was explored using XPS analysis of samples following various stages in the process. Following $ClSO_3H$ exposure, a sulfur 2p signal was observed on polystyrene with a binding energy of 163eV, characteristic of an organosulfonic acid. The presence of Sn and Pd were also observed following their respective catalyzation steps. Deposition of a Cu layer followed by its removal without an intermediate bake step shows small traces of copper remaining on the polymer surface but

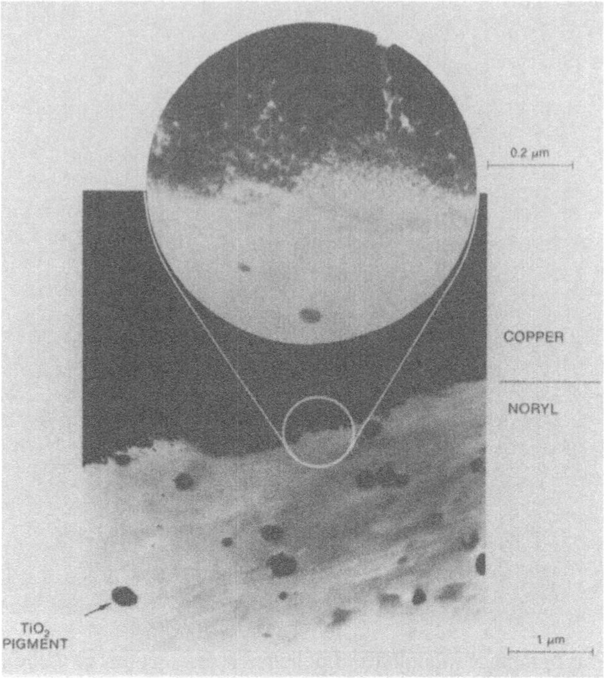

Figure 4. Transmission electron micrograph of $ClSO_3H$ treated molded Noryl® electroless copper/resin interface.

predominantly the spectrum resembles that of the sulfonated polystyrene. In this case the binding energy of the S 2p signal remains at 163eV. As the plated sample is heated, failure during peel changes from interfacial to cohesive within the body of the polymer as was seen above. XPS of a peeled sample shows only a polystyrene material with little residual S remaining on the sample. XPS analysis of the polymer remaining on the backside of the copper peel strip does show the presence of the sulfonated polymer; however, in this case the binding energy of the S 2p signal shifted to 168eV, a level more characteristic of an inorganic sulfate. Since this increase in S 2p binding energy accompanies the development of interfacial adhesion it is likely that the added sulfonic acid groups are directly involved in bond formation. The precise chemistry involved in this sequence has not yet been determined, however, a Bronsted-type acid/base interaction between he nascent hydroxide on the deposited copper and the organosulfonic acid would not be unreasonable.

Aromatic sulfonic acids are often capable of undergoing the reverse or desulfonation reaction when exposed to water at high temperatures. Polystyrene sulfonated with $ClSO_3H$ is also capable of being desulfonated in this fashion. Figure 7 shows the decrease in surface S concentrations of sulfonated polystyrene following heating in water, based on SIMS analyses. The SIMS data indicate that the degree of sulfonation was unaffected by soaking in 30°C water while 75% of the S is removed in 30 min if the water is heated to 90° C.

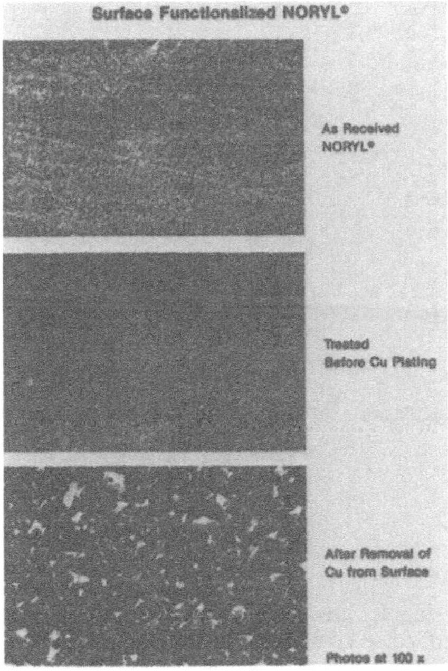

Figure 5. Scanning electron micrographs of the surface of virgin, $ClSO_3H$ treated, and copper peeled molded Noryl® resin.

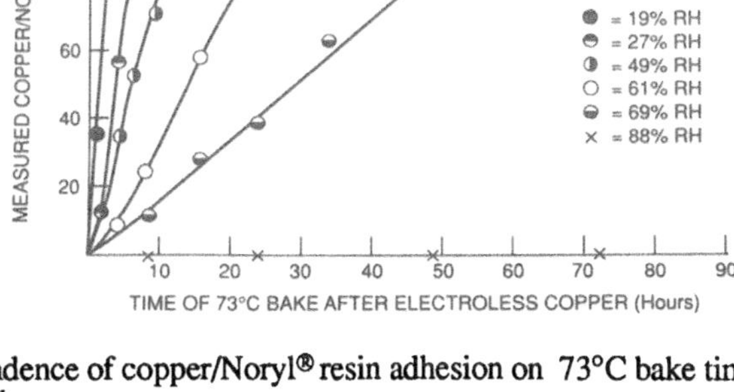

Figure 6. Dependence of copper/Noryl® resin adhesion on 73°C bake time at different relative humidities.

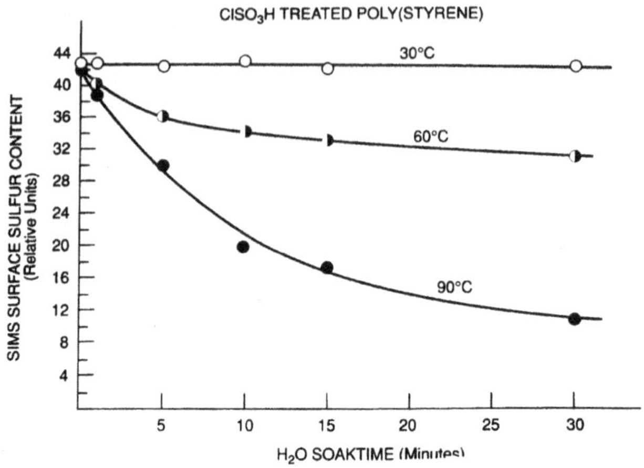

Figure 7. Sulfur concentration for $ClSO_3H$ treated polystyrene as a function of soaktime in H_2O.

Copper plated samples are also affected by heating in a moist atmosphere. Figure 8 shows adhesion data collected on samples following heating in a 100% RH ambient at temperatures between 30 to 90°C. Samples stored in this high humidity ambient at 30°C do not show any decline in adhesion following 16 hrs. However, if the temperature is increased to 45° C, a slight decrease in adhesion is seen. The rates of adhesion decline appear to be diffusionally limited, if the data from Figure 8 are replotted as a function of $(time)^{1/2}$, Figure 9. Removal of the sample from the high humidity oven, followed by rebaking in a low humidity ambient, restores the adhesion to the original level. In other words, the loss of adhesion accompanying rehydration of the metal/polymer interface is reversible. This feature was not observed in all cases however. Samples which were heated for longer than an hour at 90° C or 8 hrs at 60° C, denoted by an asterisk in the figure, were unable to recover most of the decline in measured adhesion. Examination of the surface of these samples following removal of the copper peel strip using SIMS showed the S concentration to be significantly lower than that observed immediately following sulfonation in a manner identical to that seen for a high temperature water soak of a non-plated sample. This arises from desulfonation of the polystyrene surface following rehydration of the interface. An extreme example of this behavior is seen in the copper plated sample exposed to a 90°C/100% RH ambient for 2 hrs. This sample completely lost all interfacial adhesion so that the Cu film delaminated from the surface as the sample was withdrawn from the oven.

SUMMARY

It is clear from this study that copper/polymer adhesion develops in this case through the added sulfonic acid linkage as this modified interface is dehydrated. The level of interfacial adhesion which develops is dependent upon the extent to which the surface was modified prior to metallization and can readily exceed the cohesive failure limit of the polymer. An acid-base interaction between the organosulfonic acid and copper surface is a likely chemical route by which adhesion develops.

Once an adherent copper/polymer interface has formed, exposure to a warm, high humidity ambient results in a decrease in the level of interfacial adhesion The adhesion loss is initially reversible, as the rehydrated interface can again be dehydrated, and then becomes irreversible as the polystyrene surface is desulfonated.

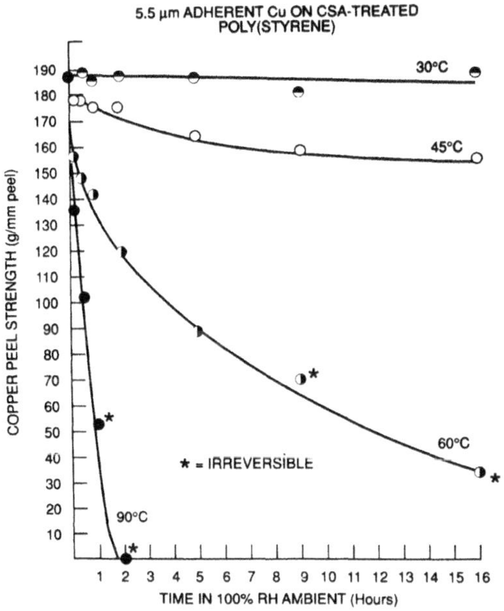

Figure 8. Copper adhesion decrease on baked, $ClSO_3H$ treated Noryl® resin as a function of heating at different temperatures in a 100% relative humidity ambient.

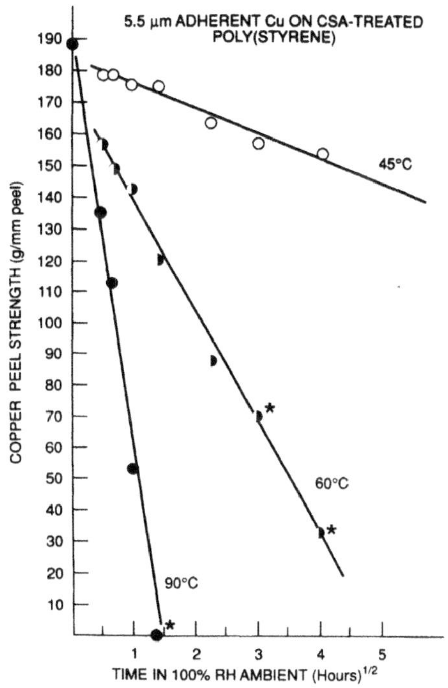

Figure 9. Data from Figure 8 replotted as $(\text{time})^{1/2}$

REFERENCES

[1] K. Wiebusch, H. Hendus, and E. Zahn, "Metallic Coatings for Plastics" Hanser, Munich (1966).

[2] R. R. Smith, "Metal Plating of ABS and Other Plastics", Plast. Inst. Conf. London (1966).

[3] D. F. Foust, W. V. Dumas, E. J. Lamby, and B. R. Karas, US Patent 4,842,946,June 1989.

[4] R. Huss, US Patent 4,308,301, December 1981.

[5] J. Eichhorn and J. M. Steinmetz, US Patent 2,945,942, July 1960.

[6] J. E. McCaskie and C. Tsiamis, US Patent 4,520,046, May 1986.

[7] J. Roubal and J. Korpiun, US Patent 4,039,714, August 1977.

[8] R. Rathsack, US Patent 3,607,350, September 1971.

CORRELATION BETWEEN SURFACE CHEMISTRY OF POLYCARBONATE AND ITS ADHESION BEHAVIOR TO ELECTROLESSLY PLATED COPPER

Ileana Isern-Flecha and Bo Boatwright

IBM Corporation
Entry Systems Division
1000 N. W. 51st Street
Boca Raton, FL 33432

A fundamental requirement for improving adhesion of electrolessly plated copper (Cu) on polycarbonate (PC) components is to understand the chemical and physical changes that occur on the PC substrate surface during the multi-step electroless Cu plating process. X-ray photoelectron spectroscopy has been used to study the surface chemistry of polycarbonate in the pre-copper plating steps, i.e., pre-etching, etching, and tin/palladium seeding reactions, of a commercial plating process. Surface imperfections, chemical changes in the polymer surface, and delamination between Cu and PC resulting from non-optimum conditions in the pre-etch and etch steps were observed. Furthermore, the relative concentrations of palladium and tin on the surface, which are known to be critical to the electroless plating ability, were also correlated with pre-etch and etch conditions.

INTRODUCTION

The use of polymeric structural components has become an established practice in the computer industry. In these applications, there is a need for electromagnetic and radio-frequency shielding, which is normally met by incorporating a metallic inner-shield, or by applying a conductive coating to the polymer surface[1,2]. Due to its high conductivity, electroless copper is an excellent coating for polymeric (mostly, polycarbonate and

Metallized Plastics 2, Edited by K.L. Mittal
Plenum Press, New York, 1991

ABS) structural components, overplated by electroless nickel to improve its corrosion resistance[2,3]. Various studies have been published on corrosion of electroless deposits on ABS[3] and polycarbonate[4], and on adhesion of electroless deposits on ABS[5]. However, a limited amount of information is available on the chemical changes that occur on their surfaces during the multi-step electroless plating process. Moreover, the effects non-optimum conditions have on delamination of Cu-PC are not known. The purpose of this study was to determine the correlation between surface chemistry on polycarbonate and adhesion failure for electrolessly plated copper. Specifically, X-ray photelectron spectroscopy, XPS (or electron spectroscopy for chemical analysis, ESCA), was used to study changes that occur on a polycarbonate surface at various stages of a commercial electroless plating process. Emphasis was placed on the effects of pre-etching, etching and tin/palladium seeding steps of the cycle. Furthermore, the ESCA data were complemented with that from secondary ion mass spectrometry (SIMS) and Scanning Auger Microscopy (SAM) analyses.

EXPERIMENTAL

Polycarbonate samples (LEXAN* BE 2130, General Electric) were plated by ETHONE INC., using their ENSHIELD* electroless plating process. Processing variables for this process (e.g., bath concentration, temperature and conditioning time) were optimized for BE 2130 non-filled polycarbonate[6]. Samples were supplied at the various steps of the electroless plating cycle; intermediate rinses (with de-ionized water) were required after each step, and in this study they were considered part of their preceding major step. Substrate plaques were cut from a molded structural component, and prepared by immersion (in sequence) in one or more of the following solutions:

1. Cleaner or pre-etcher (Enplate* Pre-Etch 3489)
2. Etcher (chromic acid, sulfuric acid and Enplate Wetter 820 LF)
3. Neutralizer (Enplate Neutralizer 835)
4. Sensitizer (Enplate Promoter 3923)
5. Activator (Enplate Activator 850)
6. Accelerator (Enplate Accelerator 860)
7. Electroless copper (Enplate Cu-872)
8. Initiator (Enplate Initiator 852)
9. Electroless nickel (Enplate NI-426)

---

* ENPLATE and ENSHIELD are registered Trademarks of ENTHONE, INC. LEXAN is a registered Trademark of the General Electric Company.

It was previously established that processing variables in the pre-etch and etch step dictate the quality of electroless plating (i.e., adhesion) in the Enshield electroless plating process[6]. Therefore, in order to study their effect on adhesion, some samples were pre-etched and etched in solutions with non-optimum concentrations (i.e., dilutions of Enplate solutions), and analyzed after treatment or after being subjected to subsequent steps of the electroless plating process. Specifically, samples were subjected to pre-etch solutions with concentrations of 65%, 70% (optimum concentration), and 75% (V/V), and processed up to and including the accelerator step; samples pre-etched under these conditions were etched in solutions with Cr^{+3} concentrations of 1.8 g/l (optimum) or 4.1 g/l (non-optimum).

ESCA analyses were conducted on a Perkin-Elmer (PHI) 5400 small spot ESCA unit equipped with a hemispherical analyzer and a position sensitive detector. Unmonochromatized magnesium (1253.6 eV) or aluminum (1486.6 eV) anodes were used as excitation sources, and the aperture lenses were set to provide an analysis area with a 600-um diameter. The analyzer was operated at a pass energy of 89.4 eV and 71.5 eV for low resolution survey and multiplex scans, respectively, and at a pass energy of 44.7 eV for high resolution scans. Peaks were referenced to the Au (4f 7/2) or Ag (3d 5/2) lines from a piece of metal in contact with the sample; special provisions for charge compensation were not necessary in these experiments. Metallized polymer samples were sputter-etched using a differentially pumped ion gun (3-keV Ar ions, 15 mPa), with a raster rate of 40% (covering an area of ≈0.4 cm^2); sputter rate was estimated at ≈17 Å/min, based on a SiO2 standard. The base pressure in the analysis chamber was in the low-10^{-9} torr, and it increased to the high- 10^{-9} torr with each sample. ESCA data were normally obtained with a 45° photoelectron take-off angle, but these were also obtained at 15° and 90° angles, thus providing chemical information at different sampling depths (e.g., 10 Å and 100 Å). Chemical state information was obtained by comparing the binding energies determined experimentally with values reported in the literature. Quantification was based on sensitivity factors available from PHI ESCA software[7]; although the results are not accurate, trends in elemental composition as a result of processing steps can be established. All ESCA experiments were done at least in duplicate.

In addition, electron micrographs of metallized samples were obtained on a Perkin-Elmer (PHI) 610 scanning Auger microprobe, using a 5-keV incident electron beam and 1-2 nA beam current. To avoid charging, samples were wrapped in aluminum foil, exposing a plated surface area of ≈2 mm X 2 mm. Furthermore, samples of the raw material and polymer samples at a pre-etched and etched stage were subjected to static-SIMS analysis at Perkin-Elmer Physical Electronics Laboratory (Eden Prairie, Minnesota). Experiments were performed using a 4-keV xenon ion beam, 1 nA ion beam current, rastered over an area of ≈4 mm X 4 mm.

RESULTS AND DISCUSSION

In this study, primary attention is given to the carbon (C) and oxygen (O) peaks in the ESCA spectra of non-metallized samples (i.e., samples subjected to various stages of the electroless plating process up to and including the accelerator step). For the as-received material, the relative atomic concentrations for C and O are 79.6 and 18.5%, respectively; small amounts of silicon were also detected (1.9% relative atomic concentration), see Table I. In analogy with a recent XPS study on as-received polycarbonate[8], the C 1s signal contains resolvable components at ≈284.6 eV, 286.1 eV and 290.4 eV (figure 1a), which are attributed to C-C or C-H, C-O, and carbonate groups, respectively. The C 1s signal can also be resolved to include the relatively small shake-up peak due to photoemission from the aromatic rings at 291.7 eV. The O 1s signal was resolvable into various peaks characteristic of polycarbonate, (e.g., C=O and C-O groups at 532.3 eV and 534.0 eV respectively), and peaks of lower intensity attributed to the compound containing silicon. However, with few exceptions, only the resolvable components of the C 1s signal are discussed in detail in this paper; mostly, peak width, peak energy, spectral line shapes and elemental relative atomic concentrations were used to establish trends in surface chemistry of the polycarbonate samples at the different stages of the electroless plating process. Angle-resolved ESCA experiments demonstrated that the silicon content was higher on the outermost layers (i.e., at sampling depths (10 Å), and not at the depth normally sampled in the standard ESCA experiments ($\leq$ 50 Å). Furthermore, the binding energy of the Si 2p peak (102.9 eV) was in agreement with literature values for polymethylsiloxane[9].

The negative ion SIMS spectra of the as-received material were dominated by peaks characteristic of polycarbonate[10] (i.e., m/z 25, 211, m/z 133 and m/z 73, with 100%, 80%, 25% and 25% relative intensity, respectively); less intense (<10% relative intensity) peaks at m/z 149 and m/z 75 which are characteristic of polydimethylsiloxane[10] were also present. Moreover, the

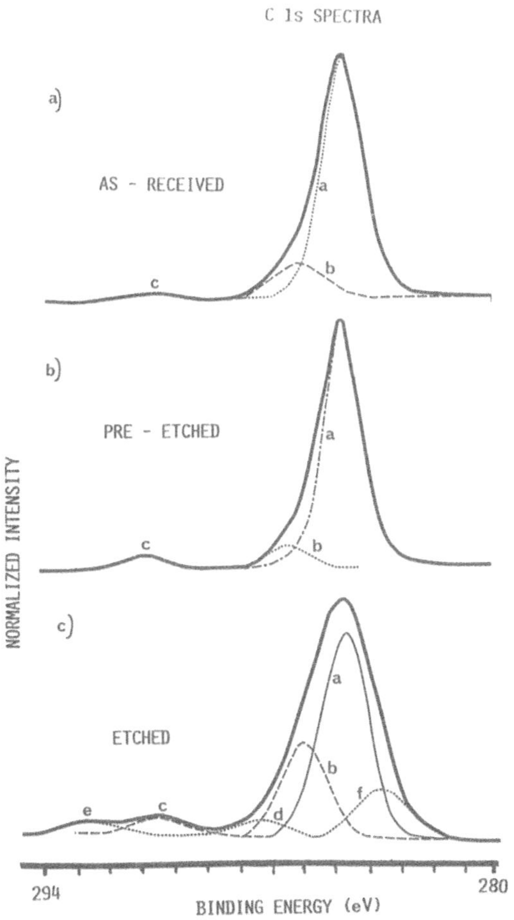

Figure 1. C 1s spectra of polycarbonate samples in as-received (a), pre-etched (b), and etched conditions (c); a, b, c, d, e and f peaks were attributed to C-C/C-H, C-O, carbonate, ketone/aldehyde, fluorocarbon, and unknown components, respectively.

positive ion SIMS spectra of the as-received material was dominated by peaks attributed to polydimethylsiloxane (e.g., base peak at m/z 73, and peaks of lower intensity at m/z 147, m/z 207 and m/z 221); slightly less abundant low molecular weight components (<m/z 50) characteristic of polycarbonate were also present. The difference in intensity of peaks attributed to polydimethylsiloxane and polycarbonate between the negative and positive SIMS spectra are typical of the SIMS ionization process[11,12], and will not be discussed further in this paper. In spite of these

Table Ia. Surface Composition at Different Stages of the Plating Process.

Sample Condition (a)	Relative Atomic Concentration (%)					
	C	O	Si	F	Cr	C/O
AS RECEIVED	79.6	18.5	1.9	nd(b)	-	4.3
65% V/V PRE-ETCHER	79.8	16.3	3.8	-	-	4.9
70% V/V PRE-ETCHER	84.0	13.6	2.2	0.5	-	6.2
75% V/V PRE-ETCHER	84.9	12.7	1.6	0.9	-	6.7
65% PE + ETCHER 1 (c)	78.6	18.3	0.6	2.2	0.9	4.3
70% PE + ETCHER 1	76.1	18.7	nd	4.1	1.0	4.1
75% PE + ETCHER 1	73.4	19.0	nd	6.6	1.0	3.9
65% PE - ETCHER 2 (d)	82.6	13.5	nd	3.3	0.6	6.1
70% PE - ETCHER 2	72.8	19.2	nd	7.2	0.7	3.8
75% PE - ETCHER 2	71.6	20.1	nd	7.7	0.7	3.6
65% PE - NEUTRALIZER	77.3	19.5	nd	2.4	0.8	4.0
70% PE - NEUTRALIZER	73.7	19.9	nd	5.5	0.8	3.7
75% PE - NEUTRALIZER	73.3	20.0	nd	5.7	1.1	3.7
65% PE - PROMOTER	76.3	19.5	nd	3.5	0.7	3.9
70% PE - PROMOTER	73.3	20.2	nd	5.6	0.8	3.6
75% PE - PROMOTER	73.3	20.5	nd	5.5	0.8	3.6

(a) Sample condition reflects processing up to and including the step specified and subsequent rinse(s).

(b) There might be trace amounts of the element, but below the limit of detection of the instrument (≈0.5% relative atomic concentration).

(c) 65% PE and ETCHER 1 denote 65% V/V pre-etcher and 1.8 g/l Cr^{+3} etcher solution, respectively.

(d) ETCHER 2 denotes 4.1 g/l Cr^{+3} etch solution.

Table Ib. Surface Composition at Different Stages of the Plating Process.

Sample Condition	Relative Atomic Concentration (%)					
	C	O	F	Sn	Pd	C/O
65% PE - ACTIVATOR	56.8	30.4	1.5	9.8	1.5	1.9
70% PE - ACTIVATOR	54.7	31.1	2.4	10.5	1.3	1.8
75% PE - ACTIVATOR	53.1	30.9	4.2	10.7	1.1	1.7
65% PE - ACCELERATOR	69.3	24.4	3.6	0.9	1.8	2.8
70% PE - ACCELERATOR	72.4	21.1	4.0	0.8	1.7	3.4
75% PE - ACCELERATOR	69.9	24.7	2.9	0.8	1.8	2.8

differences, the data can be interpreted as polydimethylsiloxane being more abundant than polycarbonate in the outermost surface (≈5Å). The sampling depth of SIMS analysis (≈5 Å) is below that of ESCA[13], which can account for the difference in abundance of both compounds for results obtained with the two techniques. Polycarbonate pellets of the same material used to mold the parts were surface-analyzed and the spectra were characteristic of polycarbonate[10]. Therefore, the presence of polydimethylsiloxane might be

Table II. Peak Width (FWHM) and Energy of C 1s and O 1s XPS Signals at Different Stages of Processing.

	FWHM (PEAK BINDING ENERGY) (eV) C 1s	O 1s
AS RECEIVED	1.7 (284.5)	2.8 (533.0)
PRE-ETCHER	1.7 (284.5)*	2.1 (533.0)
ETCHER	2.4 (284.5)	3.6 (533.0)
NEUTRALIZER	2.4 (284.6)	3.8 (533.0)
PROMOTER	2.4.(284.6)	3.7 (533.0)
ACTIVATOR	2.2 (285.5)	3.3 (533.4)
ACCELERATOR	2.5.(284.6)	3.8 (533.2)

* Slight increase in FWHM in direct proportion to pre-etcher solution concentration.

Table III. Effect of Pre-Etch Concentration On Relative Abundance of C-C/C-H, and C-O Peaks (From XPS C 1s Signal).

SAMPLE CONDITION	C-C/C-H (%)	C-O (%)
AS RECEIVED	83.1	16.9
PRE-ETCHED IN 65% V/V	84.6	15.4
PRE-ETCHED IN 70% V/V	87.8	12.2
PRE-ETCHED IN 75% V/V	89.4	10.6

attributed to residues of oil/grease introduced during the molding process or during shipping of the molded part, (e.g., contamination from the container or wrapping material).

There was no major change in the C 1s and O 1s peak energy or peak width (i.e., full width at half maximum, FWHM) as a result of pre-etching the polycarbonate samples (Table II). However, the curve fitted data for the C 1s signal demonstrated that the pre-etching step produced a decrease in the abundance of the C-O peak, relative to that of the C-C/C-H groups for the C 1s signal (figure Ib) and a slight increase in the carbonate peak (290.4 eV). These changes were in direct proportion to the pre-etch solution concentration (Table III), and they correlate with changes in the relative atomic concentration for carbon and oxygen (Table I). The ESCA data also showed a slight increase in the relative atomic concentration of silicon on the polymer surface with the pre-etch step (≈4% relative atomic

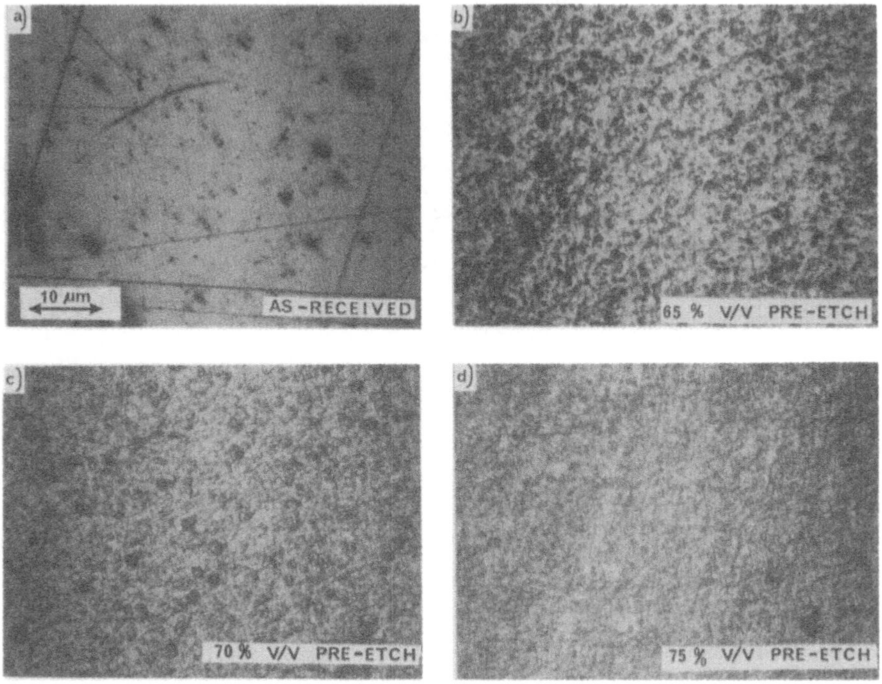

Figure 2. Photographs (200X) of polycarbonate samples in as-received condition, and conditioned in pre-etch solutions of different concentrations.

concentration) for the 65% V/V pre-etch solution , but it decreased as the pre-etch concentration was increased (Table I). Evidently, the pre-etch solution served as a partial extractant of polydimethylsiloxane, i.e., it caused migration of polydimethylsiloxane to the outermost polymer surface. The highest pre-etch concentration produced the most efficient extracting/cleaning ability of the polymer surface, and apparently part of the polydimethylsiloxane went into solution.

It is noteworthy that as a result of pre-etching, small amounts ($\leq$ 1% relative atomic concentration) of fluorine (F) were detected on the polycarbonate surface, but only when pre-etch solution concentrations of 70% and 75% V/V were used (Table I). Furthermore, a slight increase in the F content was observed for samples processed with the higher pre-etch solution concentration, i.e., 75% V/V. The presence of fluorine on the polycarbonate surface will be explained in detail when the effect of the etch step is discussed.

Interpretation of the ESCA data was further supported by results from SIMS analyses. First, in the positive SIMS spectra, the relative intensity of peaks characteristic of polydimethylsiloxane (e.g., m/z 73, m/z 147, m/z 207, and m/z 221) was much higher for samples conditioned in the pre-etched solution, than for the as-received material. For the latter, peaks characteristic of polycarbonate fragment ions, e.g., $C_2H_x^+$, $C_3H_x^+$, and $C_4H_x^+$, dominated the spectra. Thus, the pre-etch solution served as an extractant of the polydimethylsiloxane surface contaminant. Second, the negative ion SIMS spectra of the pre-etched samples showed an increase in the relative intensity of peaks in the lower mass range, e.g., < m/z 50, m/z 93, and m/z 133; while that of the peak at m/z 211 decreased considerably (for peak assignment refer to above discussion on spectra of as-received material). The relative increase of fragment ions in the lower mass-to-charge range is attributed to a bond-breaking effect of the pre-etch solution on the polycarbonate surface molecules. These results correlate with the role of the pre-etching step, i.e., to swell the polymer surface and break bonds.

In addition to surface analyses, samples were visually inspected at magnifications of 200X, and a comparison of the effect of the pre-etching step on the polycarbonate surface is presented in figures 2a-2d. It can readily be noticed that non-uniform modification of the surface occurs when the sample was conditioned in 65% V/V pre-etch solution; the stronger pre-etch solutions (70% and 75%) produced a more uniform attack of the surface. These results are in agreement with surface analyses data, but more importantly, they suggest non-complete polymer surface modification with the low pre-etch solution concentration.

It has been previously established[6] that the best adhesion for Cu/PC occurs with the higher pre-etch concentrations (i.e., 70%-75% V/V). In this study, it has been proven that the pre-etch concentration dictates physical (including surface roughness) and chemical properties of the polymer surface, and thus its wetting ability. Furthermore, the pre-etch solution facilitated the removal of dimethylsiloxane in the etch step, a contaminant which can interfere with adhesion of electroless Cu plating to the polycarbonate substrate.

By etching the polycarbonate samples in the chromic acid/sulfuric acid solution, various changes were observed in the elements detected on the surface by ESCA analysis. First, it was deduced (Table I) that silicon disappeared from the polymer surface, specially when the samples were previously conditioned in the pre-etched solution of higher concentration. Second, small amounts (approximately 1% relative atomic concentration) of chromium were detected on the surface of etched polymer samples. The Cr 2p 3/2 peak was observed at ≈578.3 eV; based on literature values[7], it was attributed to hexavalent Cr, i.e., CrO_3, as theoretically expected for the Enplate electroless plating process[6]. Third, the fluorine content increased significantly as a result of etching the polymer surface (Table I), and in analogy with the pre-etch step, the higher content was observed for samples which were previously conditioned in the pre-etch solution of higher concentration (75% V/V). Moreover, angle-resolved ESCA experiments proved that the fluorine content was higher on the outermost surface layers.

As a result of the etch step, other changes were also observed in the ESCA spectra. For example, considerable peak broadening was observed in the C 1s and O 1s signal (Table II), thus revealing the presence of additional functional groups on the polymer surface. By resolving the C 1s envelope, it can be established that a new component of relatively high abundance was introduced to the polymer surface (estimated binding energy of 283.7 eV) with the etch step (figure 1c). The relatively low binding energy of this peak does not match literature values of the C 1s peak of different organic (standards) compounds, and therefore its nature is unknown. Possible peak assignment includes species where polycarbonate groups are bound to Cr, as it has been reported for polyethylene[14] and polypropylene[15]. It is also possible that this lower binding energy C 1s peak is due to low molecular weight scission products which do not contain O or F. Other changes produced in the C 1s signal as a result of the etch step include an increase in the relative concentration of the carbonate groups, the presence of ketone/aldehyde groups (≈287.6 eV), and of the peak(s) at ≈292 eV (figure 1c); the latter is attributed to fluorocarbon components, in agreement with an increase in the relative atomic concentration of fluorine in etched samples (Table I). Clearly, the increase in the groups containing O, as derived from the C 1s signal, is supported by a decrease in the C/O relative atomic concentration ratio (Table I), and with peak broadening of the O 1s signal (Table II). Furthermore, the increase in oxidized species on the surface can imply a predicted hydrophilicity or increase in surface polarity of the etched polymer surface, which should be at a level that will favor good wetting and adhesion characteristics[14,16]

Etching with a more concentrated solution, i.e., 4.1 g/l Cr^{+3} vs. 1.8 g/l Cr^{+3}, produced a higher increase in the surface content of fluorine (Table I), and a slight decrease in the relative atomic concentration ratio for C/O (Table I). The extent of these changes was also in direct proportion to the concentration of the pre-etch solution. In addition, slightly lower amounts of Cr (Table I) were detected on the surfaces of the polymer samples etched in the solution of higher concentration. Furthermore, a slight increase in the relative abundance of peaks due to carbonate and fluorocarbon groups was evident from the C 1s signal of samples etched in the more concentrated solution. The increase in the O and F relative atomic concentration correlates well with a higher relative abundance of the resolvable C 1s peaks attributed to polycarbonate oxidation products and to its fluorocarbon component, respectively.

In agreement with the ESCA results, SIMS data of etched samples indicate that the fluorine peaks are due to a fluorocarbon, that its content is directly proportional to the concentration of the etch solution, and that it predominates on the outermost surface. Moreover, the SIMS spectra of samples conditioned in the etched solution showed an increase (compared to both the as-received material and samples conditioned in the pre-etch solution) in the relative abundance of peaks attributed to low mass fragment ions characteristic of polycarbonate; this behavior is attributed to a relatively strong bond-breaking effect of the etch solution on the polymer surface.

Evidently, the fluorocarbon, a minor constituent of the polymer, migrated to its surface as a result of conditioning in the pre-etch and etch solutions, the effect being much more pronounced in the latter. In addition, the surface F content increased in direct proportion to the pre-etch and etch solution concentrations; the highest F content was detected for samples conditioned in 75% V/V pre-etch solution with subsequent treatment in a 4.1 g/l Cr^{+3} etch solution. Interestingly, ESCA analyses were performed on many polycarbonate samples which exhibited delamination of electroless Ni/Cu plating, and in most cases, a higher than expected fluorine content was detected on the exposed polymer surface and underplating. This behavior is easily noticed by comparing the ESCA spectra of such samples (figure 3a) with that of a polymer sample which was etched in a highly concentrated chromic acid solution (figure 3b). Harsh etching or pre-etching conditions, i.e., higher than normal solution concentrations, could have produced the higher F content in the plated sample which exhibited delamination. It is also possible that longer than

XPS SPECTRA

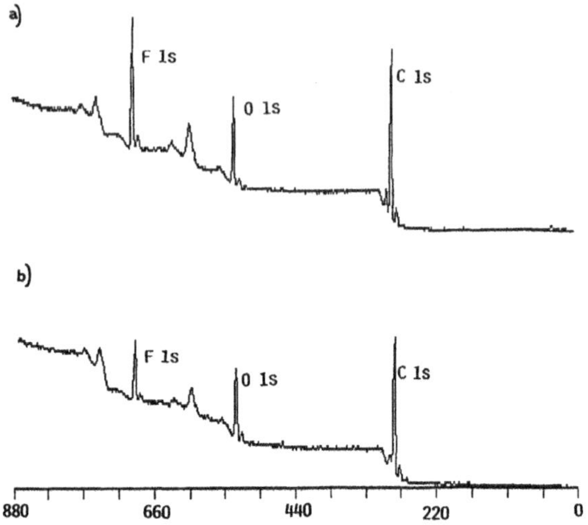

Figure 3. XPS spectra of polycarbonate sample which exhibited plating delamination (a), and of sample which was pre-etched (75% V/V solution) and etched in a 4.1 g/l Cr^{+3} solution (b).

normal conditioning time in the pre-etch or etch solutions, or higher than normal solution temperatures can have the same effect as a higher solution concentration, thus favoring surface migration of the fluorocarbon additive in polycarbonate. It should be noted that electroless plating delamination problems were not observed in samples which were conditioned in an etch solution with optimum concentration (1.8 g/l Cr^{+3}), even when F was detected on the surface at up to ≈5% relative atomic concentration. These results suggest that small amounts of fluorine do not interfere with adhesion.

It has been proven that chromic acid treatment not only oxidizes the polymer surface, but it also causes loss of material into the etch solution[14,16]. Therefore, the removal of polydimethylsiloxane from the polymer surface should be expected for polycarbonate samples conditioned in the etch solution. Moreover, migration to the surface of minor constituents/additives in polymers has been reported for many polymers[17], and it might occur naturally, or under special conditions, e.g., heating, molding. Therefore, it is not unusual to observe the same behavior, i.e., migration of the fluorocarbon additive of polycarbonate (as a result of the etch step) or of polydimethylsiloxane contaminant (as a result of the pre-etch step) in the experiments reported here. Furthermore, it should be

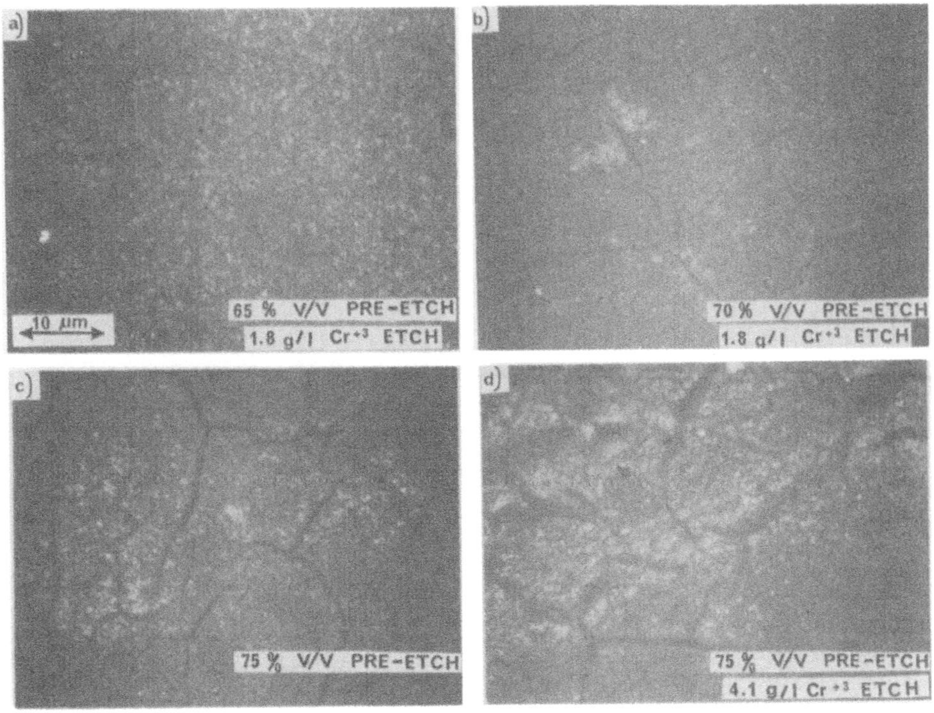

Figure 4. Photographs (200X) of polycarbonate samples which were pre-etched and etched.

expected that harsh etch, pre-etch or molding conditions can cause excessive migration of fluorocarbon to the polymer surface, thus interfering with adhesion.

Visual examination (figures 4a-4d) of the surfaces of etched samples showed evidence of further modification of the polymer, and which was also dependent on the pre-etch solution concentrations. The etched sample which was previously conditioned in the 65% V/V pre-etch solution exhibited a uniform surface; those which were previously pre-etched in the 70% and 75% V/V solutions exhibited micro-cracks ("mud-cracks") on the surface; the higher number of micro-cracks, of larger size and which extended deeper into the polymer surface, was observed for the etched samples which were previously conditioned in the 75% V/V pre-etch solution. It is noteworthy that in other studies from this laboratory[4,18], electroless Ni/Cu plated polycarbonate samples which exhibited plating delamination problems and/or a higher susceptibility for corrosion were characterized by a plated surface with a higher content of micro-cracks (and of larger size) than that observed in electrolessly-plated parts of acceptable plating quality.

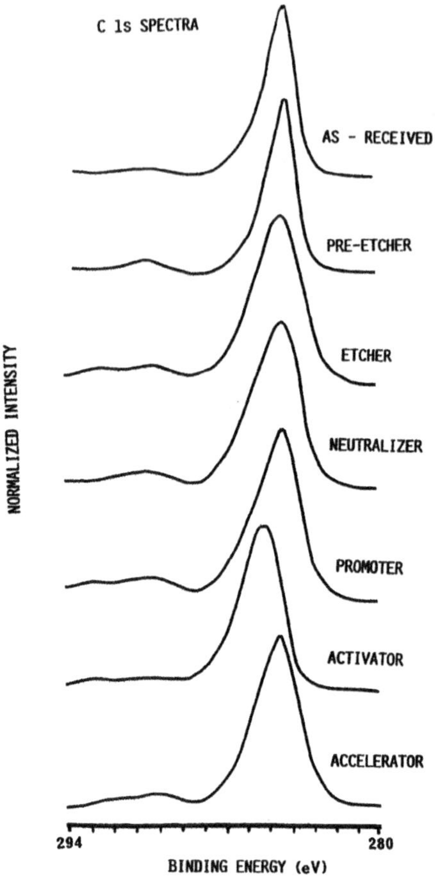

Figure 5. C 1s spectra of polycarbonate samples at different stages of processing.

No major changes were observed in peak energy, or peak width (Table II) of the carbon, oxygen or fluorine XPS 1s signal as a result of subjecting the polycarbonate samples to the neutralizer and promoter solutions. However, a slight decrease was observed in the relative atomic concentration of carbon, while a slight increase was observed for oxygen (Table I). By examining the C 1s peak (figure 5) of neutralized samples, it is evident that the reduction in the carbon content, and the increase in the oxygen content, is due to a small increase of carbonate groups on the polymer surface. Moreover, samples conditioned in the neutralizer solution exhibited a slight decrease in the surface Cr content (to ≈0.8% relative atomic concentration, Table I), and a change in its oxidation state. Specifically, the peak binding energy of ≈ 576.8 eV for the Cr 2p 3/2 peak of neutralized samples matches literature values[9] for Cr in a trivalent state; therefore, Cr suffered a reduction from the hexavalent state to the trivalent state (e.g., Cr_2O_3). These results are in agreement with the effect the neutralizer solution should have on an etched polymer surface[6], i.e., neutralization and removal of excess/residual etchant on the polymer surface, and with studies[14,16] in which Cr^{+3} was detected on the surface of etched polyethylene and polypropylene samples. With the promoter step, and in analogy with the neutralizer step, a slight decrease was observed in the relative atomic concentration of Cr (Table I). Since the promoter solution is slightly acidic (it also contains organic cationic surfactants), it should favor removal of Cr still bound to the polymer surface[14]. (Based on the experiments reported here, and in agreement with other studies (14), it remains unknown if the amount of surface Cr influences metal adhesion). No changes were evident in peak shape for the C 1s (figure 5) or O 1s signal of samples conditioned in the promoter solution; a minor decrease in the relative atomic concentration ratio of C/O (Table I) was detected, compared with results from the neutralizer step. The role of the promoter solution is to increase the adsorption of tin/palladium during the activating step, and only minor changes were observed on the polycarbonate surface as a result of this step. Although the low magnitude of these changes might influence subsequent steps of the electroless plating process, it is also possible that other changes were produced on the polycarbonate surface at the promoter step, which were not detected by ESCA (based on its resolving power).

As expected, samples immersed in the activator solution exhibited considerable changes in peak energy, shape and width for the C 1s and O 1s signals, and more importantly, tin and smaller amounts (<2% relative atomic concentration) of palladium and chlorine were also detected on the surface. By examining figure 5 it can be noticed that when the polymer sample was conditioned in the activating solution, the C 1s peak maximum shifted to a higher binding energy, ≈285.5 eV, (compared to the previous step), and the peak became narrower (e.g., FWHM changed from 2.4 eV to 2.2 eV). The C 1s signal (figure 5) can be resolved into its various components, to show a reduction in the C-O and carbonate peaks relative to the C-C/C-H peak, and no change in the peaks due to the fluorocarbon. Despite the decrease in the relative abundance of the C 1s peaks due to oxidized hydrocarbon, the relative atomic concentration of oxygen increased in the surface-activated samples, and an increase was also observed in the C/O content ratio (Table I). Therefore, the additional oxygen detected is attributed to the "seeding species", i.e., oxidized tin (Sn) and palladium, discussed below. While no shifts were observed in peak binding energy (Table II) for the O

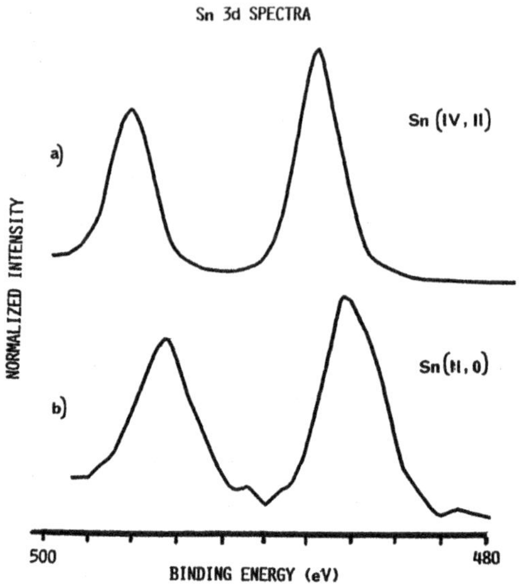

Figure 6. Sn 3d spectra of polycarbonate samples at activating (a) and accelerating (b) stages of the plating process.

1s signal, the peak width decreased, i.e., comparable to all previous steps (with exception of pre-etched samples and untreated polycarbonate); these results too are indicative of a change in the functional groups being present on the surface.

As a result of the activating step, the amount (Table I) of tin deposited on the polymer surface is large (≈10% relative atomic concentration) compared to that of palladium (≈1% relative atomic concentration), which correlates with published results for seeding processes based on colloidal tin/palladium[19-21]. The peak height of the Sn 3d 5/2 signal (figure 6a) is observed at ≈487.5 eV; this value is relatively high compared to literature values for SnO, SnO_2, or $Sn(OH)_4$, but it is in closer agreement with values for oxidized tin compounds[9,22]. It is also possible that tin is bonded to some polycarbonate groups, thus explaining the shift in binding energy for the C 1s peak.

Moreover, the relatively wide (≈2 eV FWHM) tin peak (Table III) is characteristic of an oxidized state. The peak binding energy for Sn (II) and Sn (IV) species is almost identical (within ≈+/- 0.3 eV), which makes further differentiation impossible. An analogous problem exists for the O 1s peak, where the binding energy for the oxidized Sn (II) and Sn (IV) species is within (≈0.1-0.5 eV[9,22]).

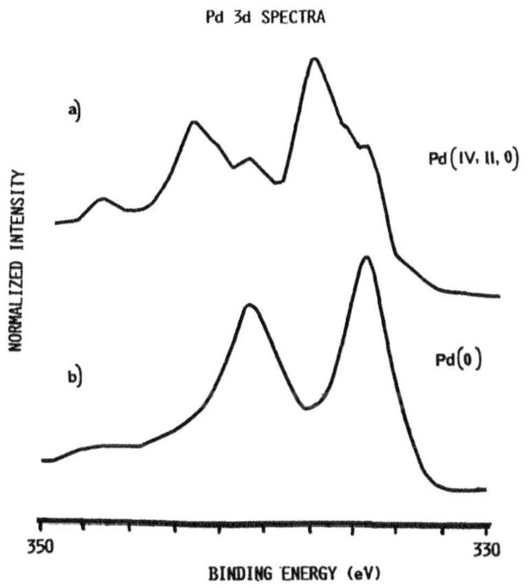

Figure 7. Pd 3d spectra of polycarbonate samples at activating (a) and accelerating (b) stages of the plating process.

It is also possible that tin is bonded to some polycarbonate groups, thus explaining not only the relatively high peak binding energy of the C 1s signal, but also that of the Sn 3d 5/2 peak as well. The Pd 3d signal is very broad (figure 7a), and it can be resolved into at least three components, e.g., at ≈338.0 eV, 336.5, 335.5 eV. Based on literature values[22,23], they are attributed to $PdCl_2$ or PdO_2, PdO and elemental Pd. The relatively small amounts of chlorine (<1% relative atomic concentration) detected in the surface-activated samples might be due to $PdCl_2$ or to residues from the acidic Sn/Pd seeding solution components, as claimed in a recent study[19]. Judging from the relatively low concentration of chlorine, it can be concluded that if $SnCl_2$ is present, it can only exist as a trace component compared with other tin species. It should also be noted that no correlation could be established between the extent of oxidation of Sn and Pd vs. the concentration of the pre-etch solution.

By examining Table I, it can be deduced that the highest surface concentration of Sn was detected in surface-activated samples which were previously conditioned in the pre-etch solution of higher concentration, i.e., 75% V/V. In contrast, the highest surface-concentration of Pd was detected in samples which were previously conditioned in the pre-etch solution of lower concentration, i.e., 65% V/V. Moreover, the Sn/Pd content ratio was ≈6.5, 8.1 and 9.7 for samples pre-etched in solutions with concentrations of 65%, 70% and 75% V/V, respectively. From these results it can be established that after conditioning polycarbonate samples in the activator solution, the Sn/Pd content ratio increased in direct proportion to the concentration of the pre-etch solution.

The accelerator step lowered the peak binding energy of the C 1s signal, while no changes were observed in the peak energy for the O 1s signal (Table I). A slight increase (≈0.3 eV) was observed in peak width for both the C 1s and O 1s signals, indicative of changes in the functional groups being present on the polycarbonate surface, or in their relative concentration. With the accelerator step, the relative atomic concentrations of C and Pd increased, while those of O and Sn decreased (Table I). In addition, by analyzing the data in Table I, a correlation can not be established between surface content of Sn, Pd or Sn/Pd ratio vs. pre-etch solution concentration, for samples conditioned in the accelerator solution. Therefore, it is concluded that delamination of the electroless copper plating from the polycarbonate substrate due to non-optimum pre-etch conditions is not necessarily related to the amounts of Pd or Sn adsorbed on the polycarbonate surface in the accelerator step. However, it is clear from the binding energy (≈335.4 eV) of the Pd 3d 5/2 peak (figure 7b) that elemental palladium is present on the polymer surface at this stage of the electroless plating process. In contrast, the tin peak is now broader than in the previous step (figure 6), and indicative of tin species with different oxidation states, e.g., SnO and elemental Sn. Despite the mixture of Sn species, the accelerator step served its purpose, i.e., to dissolve tin from the "seeder" Sn/Pd colloidal species[6,19], thus exposing elemental palladium which initiates the deposition of copper in the electroless copper bath.

Not surprisingly, no differences were observed in surface chemistry of copper-plated samples which were processed under different pre-etch or etch conditions. However, visual examination at high magnification (200X) and electron micrographs (figures 8a-8f) showed differences in the extent of micro-crack formation on the surface of plated samples, which were dependent on the concentration of the pre-etch and etch solutions. In agreement with visual observations on the surface of polymer samples at the etch stage (figures 2a-2d); these imperfections were favored for samples conditioned in pre-etch and etch solutions of higher concentrations. It can also be noticed (figures 8a-8f) that plated samples which were conditioned in the lower concentration (65% V/V) pre-etch solution

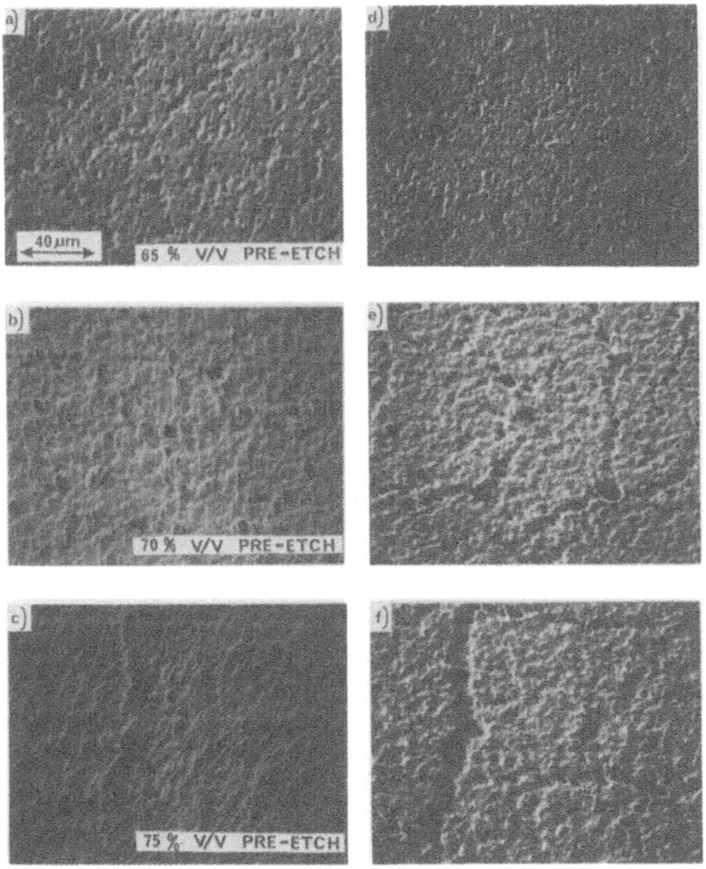

Figure 8. Electron micrographs of Cu-plated polycarbonate samples. Samples were prepared by conditioning in pre-etch solutions with concentrations of 65%, 70% and 75% V/V, (a, b, and c, respectively), and etched in solutions with concentrations of 1.8 g/l Cr^{+3}, or conditioned in pre-etch solutions with concentrations of 65%, 70% and 75% V/V (d, e, and f, respectively) and etched in solutions with concentrations of 4.1 g/l Cr^{+3}.

exhibited a smoother surface (low surface roughness), where only micropores were present (no microcracks). This was true for samples conditioned in "optimum" (1.8 g/l Cr^{+3}) and "non-optimum" (4.1 g/l Cr^{+3}) etcher solution concentrations. These results suggest that good adhesion of Cu/PC requires optimum micro-etching or surface roughness, i.e., the presence of some microcracks does not interfere with adhesion and it might even be desirable, but too large or too many microcracks favor delamination at the Cu/PC interface.

In this investigation, emphasis was placed on determining possible causes of delamination at the Cu/PC interface, basically because delamination at the Ni/Cu interface seldom occurs[6]. However, evidence has been obtained on possible causes of delamination at Ni/PC interface, some of them having its origins at the Cu/PC interface. First, it should be mentioned that in most cases, copper oxide and/or copper chloride were detected on the exposed Ni and Cu surfaces where delamination occurred[18]. The role of the initiator solution (refer to EXPERIMENTAL section for sequence of steps in electroless plating process) is to remove small amounts of copper oxide that are formed during rinsing, drying and transfer operations, and to produce a smoother copper film. It is then possible that when delamination at the Ni/Cu interface occurs, the conditions in the initiator step were inadequate (e.g., a solution of lower concentration or shorter conditioning time will not remove all oxides from the copper surface). In analogy, copper chloride could have been formed as a result of inadequate rinses after conditioning in the initiator solution (3% Pd in diluted hydrochloric acid solution). Second, it should be mentioned that in an earlier study from this laboratory[4], evidence was obtained for propagation of microcracks and other surface imperfections all the way from the electroless Ni and Cu plating and into the polymer. Therefore, the pre-etch and etch solutions concentrations were also critical to the quality of electroless Ni plating.

CONCLUSIONS

Based on this study, various possible explanations were offered as to why the pre-etch and etch step are the two most critical steps in providing good adhesion of electroless copper over polycarbonate. The pre-etch solution not only serves its main purpose of swelling the polymer surface and break bonds, but it acts as a mild cleaner and extractant of surface contaminants (e.g., polydimethylsiloxane) and polymer additives (e.g., fluorocarbon). It was also established that the surface roughness and microcracking produced in the etch step, and the adsorption of Sn and Pd at the activating step, depend on the concentration of the pre-etcher solution. However, larger amounts of palladium at the surface (in the activator step) were not necessarily in proportion to the amount adsorbed in the accelerator step, nor with the quality of electroless plating. In the etch step, the expected trends observed include etching or removal of material from the polymer surface, bond-breaking and surface polarity changes, i.e., from a hydrophobic to a hydrophilic surface. Furthermore, and in analogy with the pre-etch step, it was proven that the etch solution acts as a strong cleaner or extractant of surface contaminants and polymer additives, and that its concentration influences the formation of surface microcracks. Based on the results from samples which exhibited electroless plating delamination problems at the Cu/PC interface, small amounts of some polymer additives (i.e., fluorocarbon) or surface imperfections do not interfere with good adhesion of Cu on PC. However, in order to maintain their presence within acceptable limits, pre-etch and etch step conditions (e.g., concentration of solutions) should be cautiously monitored.

ACKNOWLEDGEMENT

The authors are very grateful to Richard Bellemare (Enthone, Inc.), who prepared all samples used in this study, and for many helpful discussions. The authors also acknowledge Ed Bastenbeck and Richard Brander (Enthone, Inc.), and Frank Laberti and Richard Pikul (General Electric), for valuable discussions, and John Moulder (Perkin-Elmer, PHI Div.) who performed SIMS analyses.

REFERENCES

1. E. M. Hassan, Plast. Technol, 27(7), 67 (1981).
2. L. K. English, Mater. Eng., 106(3), 37 (1989).
3. E. B. Saubestre, J. Hajdu, and G. Leibowitz, Electrochem. Technol., 6, 341 (1968).
4. I. Isern-Flecha, J. Vac. Sci. Technol. A, 7(3), 1702 (1989).
5. E. B. Saubestre, L. J. Durney, J. Hajdu, and E. Bastenbeck, paper presented at the American Electroplaters Society 52nd. Annual Convention, New York, N.Y., July 1965.
6. E. Bastenbeck and R. Bellemare (Enthone, Inc.), personal communication, 1988.
7. ESCA 5000 SERIES Systems Software, Version 2.0/Rev. B, (Physical Electronics Division, Perkin-Elmer, Eden Prairie, MN) 1988.
8. M. C. Burrell and J. J. Chera, Appl. Surface Sci.,35, 110 (1988-89).
9. 5000 Series ESCA Systems Manual, Version 2.0, Physical Electronics Division, Perkin-Elmer, Eden Prairie, MN) 1987.

10. Spectra were compared with those obtained from standards (SIMS spectra library) at the Perkin-Elmer Physical Electronics Laboratory (Eden Prairie, MN).
11. A. Benninghoven, J. Giber, J. Laslo, M. Riedel and H. Werner, editors "Secondary Ion Mass Spectrometry" Springer-Verlag, New York, 1982.
12. T. Ogama, H. Kurokawa and I. Karino, Applications Surface Sci., 22/23, 82 (1985).
13. D. Briggs, Br. Polymer J., 21, 3 (1989).
14. D. Briggs, V.J.I. Zichy, D. M. Brewis, J. Comyn, R. H. Dahm, M. A. Green and M. B. Konieczko, Surface Interface Anal., 2, 107 (1980).
15. G. M. Whitesides, R. Rasmussen and E. R. Stedronsky, J. Amer. Chem. Soc., 99, 4736 (1977).
16. D. M. Brewis and D. Briggs, Polymer, 22, 7 (1981).
17. D. Briggs, in "Practical Surface Analysis by Auger and X-ray Photoelectron Spectroscopy," D. Briggs and M. P. Seah, editors, Ch. 9, John Wiley & Sons, New York, 1985.
18. I. Isern-Flecha, unpublished results, 1988.
19. D. J. Auerbach, C. R. Brundle and D. C. Miller, IBM J. Res. Dev., 32, 669 (1988).
20. R. L. Cohen, J. F. D'Amico and K. W. West, J. Electrochem. Soc., 118, 2042 (1971).
21. R. L. Cohen and K. W. West, J. Electrochem Soc., 119, 433 (1972).
22. K. Kishi and S. Ikeda, Bull. Chem. Soc. Japan, 46, 342 (1976).
23. K. S. Kim, A. F. Gossmann and N. Winograd, Anal. Chem., 46, 197 (1974).

WATER-INDUCED DEGRADATION OF METAL/POLYIMIDE INTERFACES

B.K.Furman, D.Neugroschl and S.Purushothaman

IBM T.J.Watson Research Center
Yorktown Heights, NY 10598, USA

The effects of deposition process conditions and post-deposition annealing on the adhesion of electron beam evaporated Cr, Ta, Nb and Cr/Cu alloys to fully cured polyimide have been evaluated by 90° peel testing and surface analysis. Auger analysis of the peel failure loci indicates a cohesive failure within the polyimide 5-20 nm from the metal interface for as-deposited and peeled samples which show peel strengths in the range from 60-80 g/mm depending on process conditions. After annealing in forming gas, degradation of peel strengths is always observed. Two modes of degradation occur following thermal processing. One mode results in the reduction of peel strengths by 25-35% to 45-55 g/mm. This is characterized, through Auger analysis, as a cohesive failure in the polyimide. A more dramatic failure results in peel strengths of 0-10 g/mm and is characterized as an interfacial failure at the metal oxide/polyimide interface. Results indicate that polyimide surface pre-treatment by RF Ar plasma creates a region of modified polyimide and metallic impurities which retard the degradation process.

This paper proposes that the water absorbed in the polyimide is responsible for both modes of degradation. In the case of interfacial failure, water was observed to react with the metal during the elevated temperature annealing resulting in the formation of an oxygen rich region at the interface and loss of adhesion. Secondary Ion Mass Spectrometry (SIMS) results using isotopically enriched water are presented in support of this hypothesis.

INTRODUCTION

Most studies of metal to polymer bonding have entailed observing the deposition of metals onto idealized surfaces of polyimide while still in a UHV system. [1,2] Other studies have shown the effect of cleaning processes, commonly used prior to metal deposition, on polyimide surfaces. [3-5] The energetic ions from a plasma or ion gun used to remove contaminant layers induced from prior processing and/or storage can cause significant alterations in the polyimide. It has also been demonstrated [6] that a fully cured polyimide will absorb and desorb H_2O with subsequent annealing. Although these three focuses of study are useful in developing processes for fabrication they will not necessarily anticipate the real

Metallized Plastics 2, Edited by K.L. Mittal
Plenum Press, New York, 1991

process situation where all three converge and one must deal with metal adhesion to a pre-processed polyimide and maintain good adhesion through post deposition processing.

This paper will address metal/polyimide adhesion as a function of these realistic processing conditions. Using thin film peel strips evaporated onto in situ sputter cleaned polyimide samples under high vacuum process conditions (10^{-7} Torr to 10^{-6} Torr), we have performed peel strength measurements to characterize adhesion and surface analysis of the peel failure surfaces to understand the failure mechanisms. Based on peel strength measurements and their associated failure surfaces, a model is proposed to explain adhesion degradation from as deposited to those having undergone subsequent annealing steps. The model proposes that water absorbed by the polyimide during processing or from the ambient can react with the damaged region of the polyimide or the metal layer resulting in adhesion degradation. The model was critically tested using isotopically tagged water exposure of the polyimide and SIMS, to track the source of the oxygen involved in the observed failures. An in situ annealing experiment that precluded water absorption in the polyimide was also conducted. The results of these experiments support the proposed adhesion degradation mechanism.

EXPERIMENTAL DETAILS

Substrates used in this study were 57 mm silicon wafers which were spin coated with a 5 μm layer of commercial polyimide (PMDA-ODA based) which was fully cured in a flowing nitrogen ambient to a maximum temperature of 400°C. A 200 nm edge coating of copper was applied on top of the polyimide to act as a peel initiating release layer prior to metal deposition, the polyimide surface was sputter cleaned in situ in an RF Ar^+ plasma to remove surface contaminants and to normalize the surface conditions on all the samples studied. After RF sputter etching the substrates were heated to 90°C or 150°C and peel strips of 20 nm Ta,Nb,Cr/8 μm Cu thickness were then electron beam evaporated without breaking vacuum on these wafers, through Cu coated stainless steel masks with 1.6mm wide slits in them. The wafers were stored in dry nitrogen after deposition until they were removed for peel testing or annealing. 90° peel testing was done in a mini-peel tester at a peel rate of 0.5 - 8.5 mm/min and peel forces were measured using a Sensotec compact load cell that was calibrated to a sensitivity of 50 g/volt. Peel strengths in g/mm were obtained from the peel force traces over peel lengths of at least 30 mm and by averaging the results from at least two to three peels per condition. The wafers were then put through various annealing cycles in forming gas (90% N_2 and 10% H_2), as described in Figure 1, to evaluate the effect of such exposures.

Surface analysis was carried out on the peeled metal strips as well as the substrate side of the peel failure to characterize the failure loci. Auger analyses were performed using a Physical Electronics (PHI) Model 595 Scanning Auger Spectrometer. A 3 keV electron beam with a 50 nA beam current was used for the analysis of fracture surfaces and elemental depth profiles. A 2 keV Ar^+ ion beam was used for depth profiling. For SIMS analysis, a CAMECA IMS 3F Ion Microscope was used. Depth profiles were obtained by monitoring negative secondary ions of H, D, O^{18}, O^{16} while sputtering with 10 keV Cs^+ ions. Typical beam currents were 0.5 μA rastered over an area of 250 x 250 μm resulting in an average sputtering rate of 2.5 nm/sec.

SIMS analysis was also used to quantify the cleaning process and to trace the source of the oxygen observed at fracture surfaces of low strength failures. To quantify the cleaning process, analysis of the

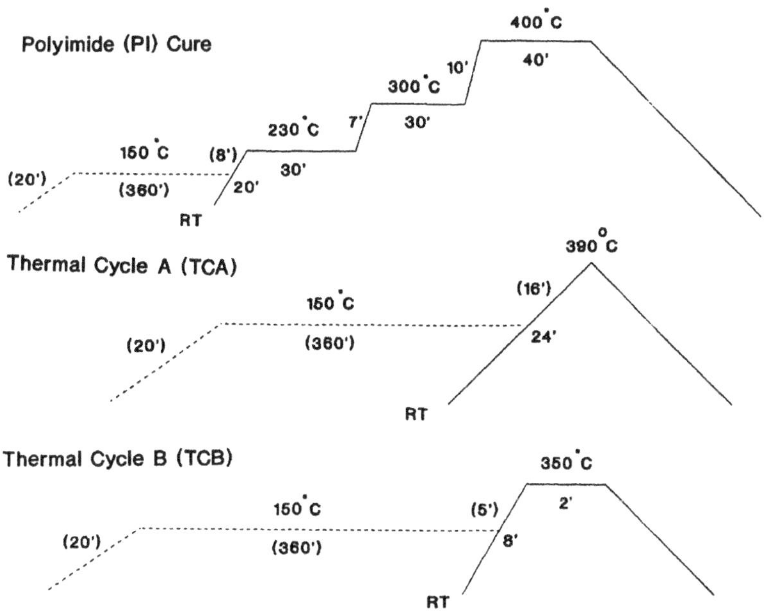

Figure 1. Profiles of the annealing cycles used. The dotted lines represent the initial step added for the extended heat treatment cycles (6 hour dwell at 150°C). The numbers in parentheses are the times for the corresponding legs of the extended heat cycle profile. All cycles were done under forming gas and allowed to cool below 50°C before exposing to air. Thermal cycle B (TCB) was repeated 15 times with exposure to air in between each cycle. Peel tests were done after 5, 10 and 15 cycles (i.e. TCB #1, 2 and 3 in Table I).

polyimide surface was done before and after RF sputter cleaning to monitor the contaminant levels. Oxygen tracing experiments were done by exposing the polyimide to water tagged with deuterium or O^{18} either before or after the metal deposition, followed by annealing and depth profile analysis. Thus, one can clearly trace the ingress and migration of absorbed water by monitoring the distributions of O^{18} by SIMS and distinguish these from the oxygen associated with the polyimide substrate as these would be O^{16}. Further, the O^{16} peak would also serve as an internal standard for surface analysis of the isotopic mass peak, thus helping in eliminating artifacts associated with depth profiling by sputter etching.

RESULTS AND DISCUSSION

Table I shows the effect of process variables on the peel strength of several metal/polyimide systems. Parameters investigated include the metal deposited onto the polyimide ,the coating on the evaporation chamber prior to RF cleaning (pre-coat) and the power and time for RF prior to metal deposition. Peel strengths and failure loci are reported for as-deposited and thermally processed samples.

Table I. Summary of metal/polyimide interface degradation.

METAL*	PRE-COAT**	RF CLEAN		AS DEPOSITED		PI CURE		TCA		TCB #1		TCB #2		TCB #3	
		Power	Time	Strength	Failure	Strength	Failure	Strength	Failure	Strength	Failure	Strength	Failure	Strength	Failure
Cr (200)+	Cu	200 W	30min	66 g/mm	PI-PI	47 g/mm	PI-PI	47 g/mm	PI-PI	44 g/mm	PI-PI	44 g/mm	PI-PI	44 g/mm	PI-PI
Cr (200)	Cr	200	30	77	PI-PI	50	PI-PI	53	PI-PI	42-0	PI/CrO	43- 0	PI/CrO	0	CrO
Cr (200)+	Cr	200	30	71	PI-PI	53	PI-PI	54	PI-PI	53-42	PI-PI	56-15	PI- mix	51-0	PI/mix
Ta (200)	Ta	200	30	74	PI	51	PI	0	mix						
Nb (200)	Nb	200	30	69	PI-PI	0	mix								
Nb(200)+	Nb	200	30	66	PI-PI	50	PI-PI	0-41							

* The number in parentheses is the thickness of the metal in Angstroms.

** pre-coat represents the metal coating deposited on the evaporator walls prior to RF cleaning.

+ signifies that extended heat cycles were used (see Figure 1).

In all cases as-deposited samples were observed to fracture cohesively within the polyimide 10-20 nm below the metal layer. The strength and depth of the fracture showed little difference regardless of metal deposited. It was noted that poor adhesion was observed if no in-situ cleaning preceded metal deposition and that differences in cohesive peel strengths were observed when RF sputter parameters were changed. SIMS surface surveys of as-received and RF sputtered polyimide in which the chamber walls had been coated with Cu are shown in Figure 2. These results identified that our in-situ "clean" was indeed more complex than the simple removal of surface polyimide . In fact, metal ions coated onto the walls of the evaporator were being implanted into the near surface region of the polyimide. The level and species of metal were observed to influence the initial peel strength and subsequent thermal degradation rate. The most pronounced effect was the differences observed in Cr/Polyimide thermal degradation rates depending on metallic impurity incorporated into the near surface region of the polyimide. It was observed that a Cr precoating resulted in severe adhesion degradation after a few thermal cycles while the Cu precoat samples degraded only slightly and then stabilized. This effect also accounted for spatial adhesion differences when Cr was intentionally incorporated. Regions adjacent to the Cu release layer were observed to degrade much less than non-adjacent areas. This was subsequently explained by the fact that Cu was backsputtered from the release layer into adjacent regions of the polyimide thus these regions behaved similar to Cu presputtered samples. To establish consistency within our experiments the evaporation chamber was pre-coated with 200 nm of Cu prior to RF sputter cleaning of polyimide samples. Auger analysis has also quantified the level of Cu as 5 % Atomic approximately 5-10 nm deep in the polyimide.

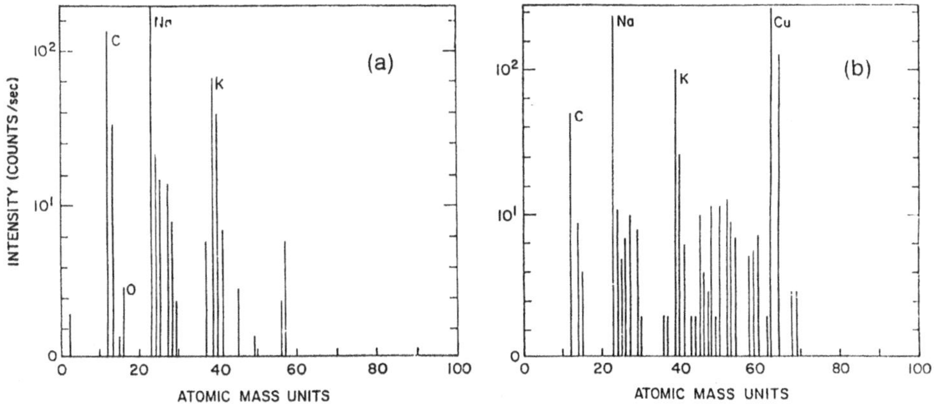

Figure 2. SIMS Surface surveys of (a) as-recieved polyimide and (b) after RF sputter cleaning in Ar ion plasma (200W, 30 min)

This Cu appears innocuous to the cohesive strength of the Cr/polyimide interfacial region in that similar levels of Cr or low level metal impurity levels with Ar ion beam treatment all result in similar strength when peeled. It was noted that the dose of Ar ion bombardment and resulting damage to the polyimide did affect as-deposited strength. It has been observed that Ar ion bombardment significantly modifies the composition of the polyimide surface which is subsequently observed as a weakening of the polyimide below the metal layer. In general, low dose bombardment (little damage) results in higher as-deposited peel strengths than the 200W, 30 minute treatment. It has also been observed that only the near surface region of the polyimide is affected by extended RF due to the fact this modified region is being removed by sputtering and little surface roughening occurs. In this study we have chosen to in-situ clean for longer times (200W for 30 minutes) to avoid the significant run to run variations that can be observed with low dose exposure as well as the enhanced thermal degradation.

As-deposited Cu/Cr thin film peel strips yielded peel strengths of 66-77 g/mm, depending on deposition conditions, for the 200W 30 minute RF preclean. Peel strengths as high as 90 g/mm have been obtained for light sputter cleaned samples (50W, 30min) but with inconsistent results. In addition, if no RF was done prior to Cr deposition, poor adhesion resulted (0-5 g/mm). Figure 3 shows Auger spectra of the resulting peel fracture surfaces. On the peel surface of the sample peeled in the as-deposited condition only C, O and N, characteristic of a cohesive failure in the polyimide, are observed. In most samples, either Cu or Cr introduced during RF cleaning can also be detected. Subsequent Auger depth profiling, shown in Figure 4(a), indicates that the peel failure occurs within the polyimide approximately 5-10 nm below the Cr/polyimide interface regardless of pre-coated metal. This type of fracture was observed with Nb and Ta as well.

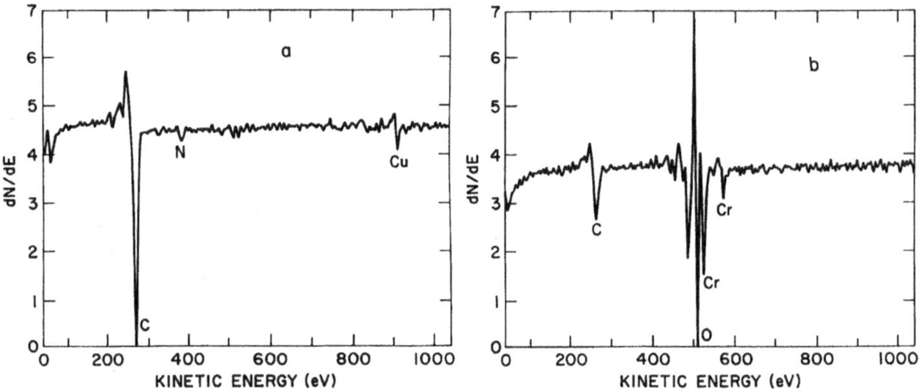

Figure 3. Auger spectra of Cr/Cu strip surfaces peeled from polyimide: (a) as-deposited (200W, 30 min RF) and (b) after 350°C forming gas annealing (50W, 10 min RF)

Annealing in forming gas or nitrogen after exposure to laboratory ambient routinely results in degradation of peel strengths. Two modes of degradation occur following these cycles. One mode results in the reduction of peel strengths by 25-35% to 45-55 g/mm. This is characterized, through Auger analysis, as a cohesive failure in the polyimide, similar to that observed in as-deposited samples. The fact that failure occurs in the same location of the polyimide except at a lower strength indicates that either the polyimide has been weakened by the thermal process or the mechanical properties of the thick Cu layer are changing. To separate the thermal effects both on the Cu and on the polyimide from stresses introduced from the coefficient of thermal expansion (CTE) mismatch of the metal/polyimide, similar samples were identically annealed without exposure to ambient. Subsequent mechanical testing showed little or no degradation in peel strengths. These results indicate that the degradation in polyimide strength may be due to chemical reaction with air and or water vapor absorbed by the polyimide when exposed to ambient. Similar degradation has been observed when samples are subjected to extended temperature and humidity exposure.

A more dramatic failure results in peel strengths of 0-10 g/mm and is characterized as an interfacial failure at the metal oxide/polyimide interface, or a mixed mode failure where some polyimide is detected on the metal side of the peel. In the case of Cr layers, Auger analysis of the peel side fracture surface of samples that had low peel strengths after annealing shows high levels of Cr and O indicative of an interfacial failure between the oxidized Cr layer and the polyimide layer (see Figure 3(b)). Auger analysis of the polyimide side of this fracture shows no Cr present, confirming interfacial failure has occurred. Some carbon is also detected which may be due to small islands of cohesive polyimide separation. Auger depth profiling into the Cr film on the peel strips (Figure 4(b)) shows some polyimide present and the Cr layer partially oxidized. If one examines the Auger carbon peak shapes in several areas of the as deposited and annealed peel strips an interesting trend is observed. The peak shape within the polyimide is characteristic of an organic carbon, as expected, while at the polyimide/Cr interface a much different carbon peak shape is observed. This shape has been previously reported as a

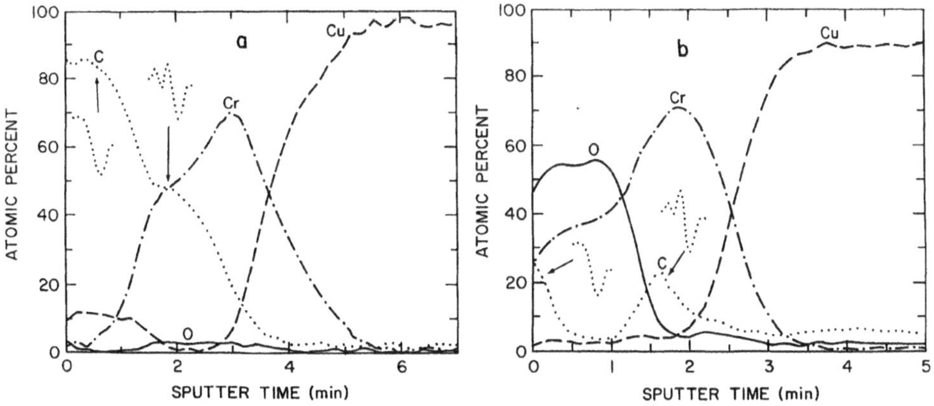

Figure 4. Auger depth profiles of strip surfaces peeled from polyimide: (a) as-deposited and (b) after 350°C forming gas annealing (50W, 10 min)

carbidic carbon peak. This result is in agreement with our analysis of the RF treated surface which is enriched in graphitic carbon prior to Cr deposition and the results of Kovac et.al. [8], where Cr-carbide bonds are reported at high coverages of Cr (greater than one monolayer). The Auger results support the mechanism that the Cr/polyimide interface consists primarily of Cr-C type bonds.

The unexpected result is that after annealing and Cr oxidation, this carbidic carbon peak is detected between the oxidized and unoxidized Cr regions. It is proposed that this Cr-C peak represents the original Cr-polyimide interface as shown in the as-deposited samples. As Cr oxidizes, oxygen is unable to diffuse into the Cr due to its low solubility and Cr oxide is forced to grow outward from the Cr-C interface due to the low diffusion rates of Cr into Cr oxide In doing so the Cr-C is trapped between the two films and the polyimide bonds are broken by the growth of chromium oxide which results in poor adhesion.

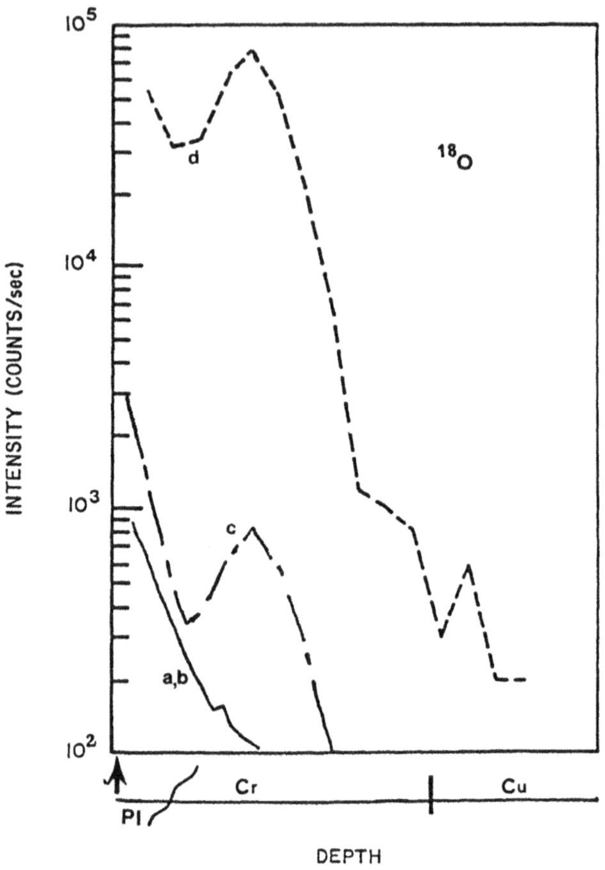

Figure 5. SIMS O^{18} depth profiles of Cu/Cr peel strip surfaces deposited on polyimide, (a) and peeled, (b) exposed to H_2O and peeled (c) exposed to H_2O, annealed in forming gas at 350°C (200W, 30 min RF) and peeled and (d) exposed to H_2O, annealed in forming gas at 350°C (50W, 10 min RF) and peeled.

SIMS depth profiles of strips peeled from the polyimide, similar to those shown in Figure 4 except exposed to H_2O^{18}, are shown in Figure 5. As deposited and peeled, figure 5(a) and exposed to tagged water and peeled, figure 5(b) show little or no O^{18} enrichment. Samples RF sputter cleaned for 30 minutes at 200W with Cu precoated chambers are shown in figure 5(c). This O^{18} profile shows some oxygen from the water which has penetrated the Cr but not to a sufficient level to result in interfacial failure. In the case of a 50W, 10min RF clean followed by H_2O exposure and anneal, a significant level of O^{18} is observed, figure 5(d). This sample has poor adhesion resulting from interfacial failure of the Cr oxide to polyimide interface. The fact that the oxide formed is predominantly O^{18} indicates water absorbed within the polyimide is the principal source of oxygen. These results identify that the high oxygen observed by Auger analysis at the fracture surface was originally present as water absorbed within the polyimide and not from the polyimide itself or ambient effects. A schematic representation of this experiment is shown in Figure 6.

In the case of Cr, however, surface pretreatment has been shown to significantly retard the rate at which water absorbed within the polyimide can react with the Cr/polyimide interface. SIMS results with isotopically tagged water are shown in Figure 7 for thin Cu/Cr deposited on polyimide. Prior to Cr deposition, the polyimide was in-situ sputter cleaned for 30 minutes at 200 W for all samples except the one shown in Figure7(d) which saw a low dose clean of 10 minutes at 50W. Figure 7(a) represents the naturally occurring O^{18} distribution in the polyimide. Figure 7(b) is observed after the sample is soaked for 1 hour in H_2O or exposed to H_2O vapor for 48 hours. We observe a significant (200x) increase in the O^{18} level in the region of Cr/polyimide interface. These results indicate that this region may be capable of gettering water from the bulk of the polyimide. Since no adhesion degradation occurs after the soaking only, it is unlikely that significant chemical reaction has occurred at this time. In addition, similar distributions are obtained for D_2O, supporting the hypothesis that only absorption and segregation has occurred.

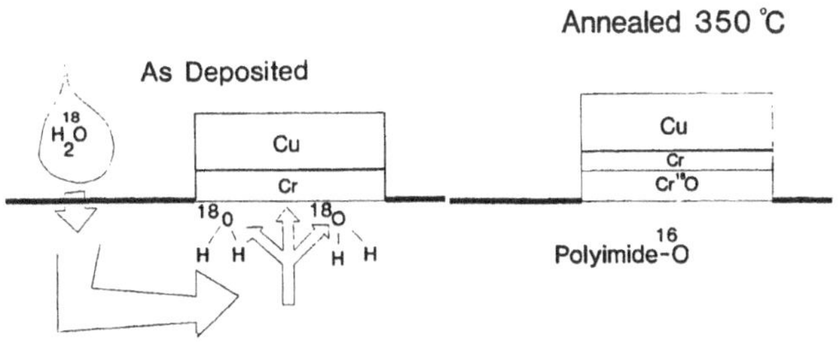

Figure 6. Schematic representation of isotopic water absorption and redistribution after anneal.

If we in-situ sputter clean, at 200W for 30 minutes, as described earlier, expose to H_2O and anneal, we see no significant penetration or reaction of the O^{18} into the Cr. However, there is a significant level remaining within the polyimide in the region whose cohesive strength has degraded 25-35%. These results indicate that the water has reacted with the polyimide resulting in mechanical weakening of this region. This is further supported by the fact that identical samples subjected to "dry" thermal processing show no degradation. In contrast, if the polyimide surface is exposed to a short sputter clean at low power (10 min , 50W) prior to Cr deposition, significant Cr/polyimide adhesion degradation occurs. Figure 7(d) shows the O^{18} distribution in a similarly treated sample. In this case 50x levels of O^{18} are observed in the Cr layer compared to the 30 minute 200W sample. Auger analysis confirmed that this corresponds to significant levels of oxygen (50-60 % At.) within the Cr and accounts for the catastrophic failure at the Cr/polyimide interface.

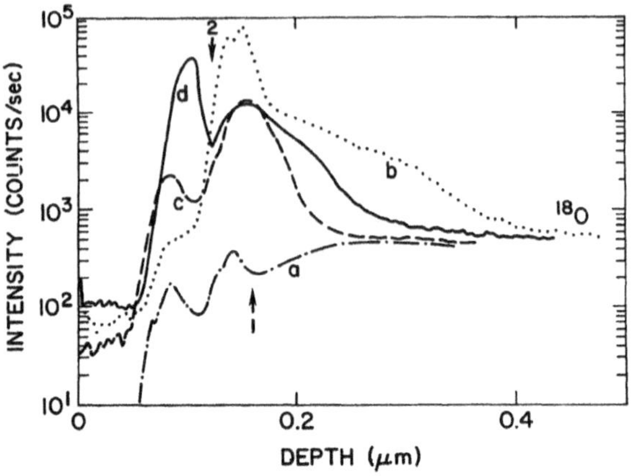

Figure 7. SIMS depth profiles of Cr/thin Cu films on polyimide: (a) as-deposited, showing O^{18} from ambient humidity exposure (normalized for isotopic abundance); (b) exposed to H_2O and (c) exposed to H_2O, annealed in forming gas at 350°C (200W, 30 min RF), and (d) exposed to H_2O and annealed in forming gas at 350°C (50W, 10 min RF). Arrows indicate region of fracture for peeled samples: a,b,c fail at point 1 at high peel strength 30-70 g/mm (thick Cu backing) while d fails at Point 2 at 0-10 g/mm.

From the above results it is clear that the observed degradation of adhesion of Cr to polyimide is a result of the reaction of the water absorbed in polyimide with Cr during the elevated temperature anneals. It appears that the absorbed water segregates under the Cr/polyimide interface. On annealing, the water reacts with the Cr at the interface most probably forming an oxide.

In order to verify this claim and to definitively establish the role of absorbed water in the polyimide, the following experiment was performed. Cu/Cr peel strips were deposited on polyimide coated substrates as before, but the vacuum chamber was purged with N_2 or forming gas after completing the deposition and the samples were annealed at 350°C, in situ. Since the parts were never exposed to ambient air prior to annealing, water absorption in the polyimide was precluded. Peel strengths and peel failure loci chemistry in these in situ annealed samples were similar to those of samples peel tested in the as-deposited condition. This conclusively demonstrates that the presence of absorbed water in the polyimide prior to elevated temperature annealing is essential for the thermal degradation of metal/polyimide interface adhesion.

The exact role of RF treatment on the kinetics of Cr/polyimide adhesion degradation is still a topic of research. Two potential contributors are the role of trace metals such as Cu introduced during RF cleaning , and modification of the polyimide during ion bombardment. Experiments to identify their joint and separate effects on water diffusion across the metal /polyimide interface and their ability to getter and thus minimize the amount of water available at the metal /interface are in progress.

CONCLUSIONS

In the case of chromium, very good adhesion to sputter cleaned polyimide in the as-deposited condition was observed. However, on subsequent thermal exposure adhesion degradation is observed. This degradation is a result of absorbed water reacting with the polyimide to result in weakening of the polyimide where it has been damaged by ion bombardment and subsequently the metal layer (Cr,Ta) during annealing to form an oxygen rich metal layer at the Cr/polyimide interface. Similar results have been previously observed for Ti and Zr. [9,10] In the case of chromium, where the reaction kinetics appear slower than Ti, Zr and Ta ,we have found that surface pretreatment of the polyimide prior to Cr deposition can significantly alter the rate of metal/ polyimide degradation. Details as to the role of metallic impurities and a modified polyimide layer in retarding water reaction with the metal/polyimide interface are still under study.

ACKNOWLEDGEMENTS

The authors would like to thank C. Parks for his help with SIMS measurements and R. Davis with data reduction of some Auger depth profiles. We also thank H. Arter for her preparation of some of the polyimide coated wafers and E.Casellani and S.Renick for deposition of metal films.

REFERENCES

1. P.S. Ho, P.O. Hahn, J.W. Bartha, G.W. Rubloff, F.K. LeGoues, and B.D. Silverman , J. Vac. Sci. Technol.,A3(3), 739-745 (1985)
2. R.M. Tromp, F.K. LeGoues, and P.S. Ho , J. Vac. Sci. Technol.,A3(3), 782-785 (1985)
3. W.E. Vanderlinde, P.J. Mills, E.J. Kramer, and A.L. Ruoff, J. Vac. Sci. Technol.,B3(5), 1362-1364 (1985)
4. L.J. Matienzo, F. Emmi, D.C. Van Hart, and T.P. Gall , J. Vac. Sci. Technol., A7(3), 1784-1789 (1989)
5. B.J. Bachman and M.J. Vasile , Proc. 38th Electronic Components Conference ,p. 444 (1988)
6. P.O. Hahn, G.W. Rubloff, and P.S. Ho , J. Vac. Sci. Technol., A2(2), 756-760 (1984)
7. C.C. Chang, "*Characterization of Solid Surfaces*", p.536,Plenum Press,New York (1974)
8. C.A.Kovac, J.L.Jordan, M.J.Goldberg , J.G.Clabes, A.Viebeck and R.A.Pollak , IBM J. Res. Develop., 32(5) ,13 (1988)
9. K. Seshan, S.N.S. Reddy, and S.K. Ray Proc. 37th Electronic Components Conference, p. 604 (1987)
10. B.K.Furman, S.Purushothaman, E.Castellani and S.Renick, Proc. of Symp. on Multilevel Metalization,Interconnection and Contact Technol., ECS ,p.142 (1986)

THIN FILM ADHESION MEASUREMENT USING EXCIMER LASER ABLATION AND TENSILE EXTENSION TESTS

W.-C. Lee and V. W. Lindberg*

Materials Science and Engineering Program
Rochester Institute of Technology
Rochester, New York 14623

P. H. Wojciechowski and F. J. Duarte

Research Laboratories
Eastman Kodak Company
Rochester, New York, 14652-3701

* Address correspondence to Dr. Lindberg

We propose and demonstrate two tests for determining the adhesion of thin films deposited on organic and inorganic substrates. The tests are (i) ablation by direct irradiation of the film using an excimer laser, and (ii) film fracture by tensile extension of the film-substrate system. For the laser test, the film can be any material that sufficiently absorbs UV laser radiation (e.g. metal or polymer). For the tensile test, the film can be any material that is less ductile than the substrate. The substrate can be rigid or flexible. In the laser test, pulses of various energy densities are directed onto different sites at a local region of the sample. At the threshold of damage, a visible crack appears and the corresponding energy density is used as a quantitative measure of adhesion. Results for metallic thin films on untreated and plasma-treated substrates correlate well with other adhesion tests (e.g. tape peel). The damage mechanism appears to be a thermal wave induced by absorption of the laser beam. In the tensile test, correlation is observed between the degree of adhesion and the nature of film fracture.

INTRODUCTION

Of all the properties of thin films the one that is universally important and most difficult to measure is adhesion. In most cases we wish to have good adhesion between the thin film and the substrate, and in a few we want to have weak adhesion so that the film can be removed from the substrate; seldom do we want intermittent adhesion.

Fundamental adhesion can be considered as the force between the atoms of the thin film and the substrate at the interface. Careful measurements of the energies during nucleation can lead to this value.

Generally, however, we wish to determine the practical adhesion[1] as measured by a variety of tests. The practical adhesion is a measure of how much extra force or energy is required to remove a film from a substrate. Practical adhesion will differ from fundamental adhesion (defined as the summation of all interfacial intermolecular interactions) due to several factors. The growing film will add stress to the film thus reducing the external force required to remove the film. Imperfections in the film or substrate may provide initiation sites

Metallized Plastics 2, Edited by K.L. Mittal
Plenum Press, New York, 1991

for adhesion failure. Porous films may allow gas and water vapor to reach the interface where they can deleteriously affect adhesion.

Many common tests of adhesion[2] are variants on either the tape test or the scratch test. In its simplest form the tape test consists of applying an adhesive coated tape to the thin film and peeling it off. The film passes the test if no material is removed. A refinement, the peel test, uses a tensile tester to measure the force required to peel a film off a substrate. The simplest form of a scratch test is an eraser test where an eraser is rubbed across the film and the number of strokes needed to abrade through the film is used as a measure of adhesion. The refined version of the test loads a stylus with some weight and pulls it across the film. The weight required to scratch through the film is used as a measure of adhesion.

Both these tests, and most other tests, involve considerable sample preparation. The peel test requires a tape to be applied to the film. Variables which will change the force-per-unit-width required to remove the film include the thickness and elasticity of the tape, film, and substrate; the angle of removal of the tape from the film; and the rate of pulling of the tape. The peel test is also limited to those films which have moderate or poor adhesion and can thus be removed.

The scratch test is affected by the size and history of the stylus, and by the elastic properties of the film and the substrate.[3] For a soft substrate, such as a plastic, the test may be overly influenced by the flow of the substrate. For ductile coatings, the thin film may flow under the stylus leaving an ultra-thin layer of material which is only observed by surface analysis techniques.

Both tests remain in heavy use because they do work to some extent. A better test of adhesion, one which is quick, repeatable, sensitive, and able to quantify the adhesion of strongly adhering films, is desireable.

Anderholm and Goodman[4] in 1970 suggested a method of measuring adhesion using a laser pulse or other generator of tensile stress. The pulse was incident on the back of the substrate and created a compressive stress wave in the material. The stress wave passed into the thin film and reflected from the free surface, subjecting the bond between film and substrate to a tensile stress wave. If the tensile stress was sufficiently large, interfacial failure was observed.

Vossen[5] used this "Laser Spallation Test" in 1978 to study the adhesion of thin films to a substrate. A 1.06-μm laser pulse was directed onto an absorbing layer at the back of the substrate which could be the substrate or a coating on the substrate. A portion of the absorbing layer was rapidly heated and evaporated explosively, sending a compressive shock wave through the substrate. The thin film on the front of the substrate was patterned into dots which had a size smaller than the laser beam size. The laser power needed to effect removal of the thin film was used as a measure of adhesion. Results of adhesion of various metals on SiO_2 were reported. Vossen also stated that soft substrates such as polymers cannot be used since they excessively dampen the shock wave.

This paper reports on a variation of the laser ablation test in which the laser pulse is incident on the thin film material rather than the substrate, thus overcoming the problem of polymer substrates. Preliminary results were reported by Wojciechowski *et al.*[6] in 1988. The further work reported here confirms the earlier results and also clarifies, to some extent, the mechanism of the adhesion failure in this test.

A second method of adhesion measurement is also discussed. Flexible substrates are stretched in an Instron tensile tester and the onset of cracks is observed. The results of these two tests are compared to results of tape tests.

EXPERIMENTAL DETAILS

The system under study in this report was silver evaporated onto poly(ethylene terephthalate) (PET). All samples were coated in a roll coater in which the web was transported across the deposition region at various speeds. The base pressure of the system was below 1 x 10^{-5} Torr, and was this high due to the outgassing from the unrolling web. The source of the silver was an electron-beam-heated crucible holding 99.999% pure silver. The e-gun source was run at constant power, and hence a constant evaporation rate. The deposition thickness was controlled by adjusting the speed of transport of the web. Variations in coating rate and thickness were about 12%. Prior to deposition, the web passed over a plasma treatment box in which an oxygen glow discharge could be maintained.

A XeCl excimer laser (Tachisto Inc., Model 401XR) was used for the laser ablation portion of this work. An electrical discharge in the gas resulted in a laser pulse of wavelength 308 nm. An electromagnetic shutter limited the time of the emitted laser pulse to 20 ns. The pulse was shaped into a rectangle of exit width 12 mm and length 25 mm. The energy density along the width of the pulse was Gaussian, while along the length of the pulse it was roughly constant with Gaussian edges.

The pulse was focused by a quartz lens of focal length 27 cm, and impinged on the sample. The size of the pulse and hence the radiative energy density (energy per unit area) directed onto the sample was controlled by changing the distance between the sample and the lens. A portion of the beam, ~2%, was split out and sent to a thermal sensor (Molectron Detector, Inc., Model J25) which integrated over the area of the pulse to determine the energy. A Tektronix 7854 oscilloscope monitored the output voltage signal of the thermal sensor. This monitoring was performed during all tests. In order to estimate the energy density we need to know the area. The size of the laser beam at various distances from the lens was measured by exposing a piece of photograpic paper. Measurement of the effective beam width by this technique is imprecise. We did not consider variations of the energy density across the area of the beam.

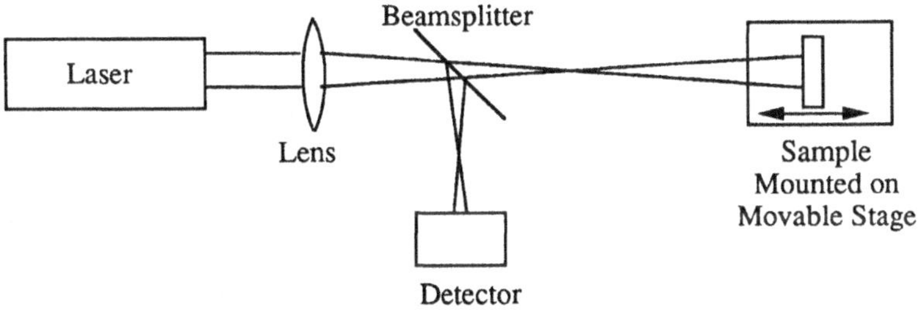

Laser: Tachisto Inc., Model 401 XR Excimer Laser.
Lens: Quartz lens with focal length of 27 cm at 308 nm.
Beamsplitter: ~2% reflected. This varied as described in text.
Detector: Molectron Detector, Inc., Model J25 Thermal Sensor.
Sample Stage: Movable in three axes by micrometer control.

Figure 1. Schematic of the Excimer Laser Test Apparatus.

The specimen to be tested was mounted on a stage which could be moved in three orthogonal directions, towards the laser and laterally. A laser shot was made on a portion of the sample, then the sample was moved to a new location and the procedure was repeated. Laser energies were recorded for each pulse. A schematic of the experimental set-up is shown in Figure 1.

The test specimen had several areas where the laser had impinged, with different energy density for each area. The areas were directly observed. In all cases a slight discoloration showed the area of the pulse. If the energy density was sufficiently high, a crack was evident in the center of the damaged area. The threshold of crack production was chosen as a measure of adhesion. Above threshold a very noticeable flash of visible light accompanied the laser pulse. Subsequent to the test the damaged areas were examined in a microscope and photomicrographs were taken, as shown in Figure 2. Even before the crack threshold was reached we could observe a slight bump on the back side of the substrate.

For some of the samples, a tensile extension test was also performed. Strips of the coated substrate (15 mm x 50 mm) were stretched in an Instron tensile testing apparatus while the film was viewed during real-time elongation with a video recorder at 40X magnification. In addition, photomicrographs at 150X were taken of samples stretched to various end-strain values. Figure 3 shows typical results. The value of the strain at which cracks appeared was used as a measure of the adhesion.

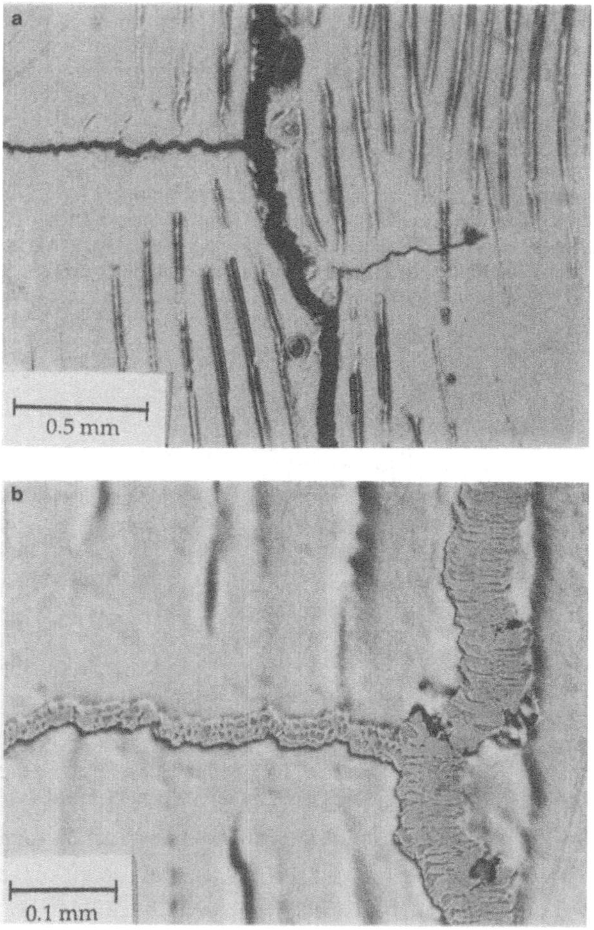

Figure 2. Photomicrographs of threshold laser damage of 650-nm thick Ag deposited on 7-mil (178 μm) Kodak Estar (PET).

Results of the laser and tensile extension tests were compared to standard tape tests. Scotch Magic tape (3M Corp.) was smoothed onto the film and burnished lightly. The tape was removed at a 90° angle to the substrate with a rapid manual pull. The percentage of the film that was removed was recorded. An average was taken for several locations on the substrate.

THEORETICAL CONSIDERATIONS

Laser Ablation Test

A model for the ablative damage of the metallized polymeric web system is now considered. Based on our observations, as well as those of others, we speculate that the mechanism of film failure due to laser heating follows the four-phase process depicted in Figure 4. Although our discussion can apply to most metallic-thin-film-on-polymer systems, we will focus our discussion on the system studied most in this investigation—thin Ag films on PET substrates.

In the first phase, the 20-ns laser pulse is absorbed by the Ag thin film. The absorption length of 308-nm wavelength radiation for Ag is approximately 25 nm. Thus 98% of the radiation is absorbed in the first 100 nm of film thickness. The films in this study have thicknesses in the range of 200 nm to 4000 nm. Therefore, virtually all of the UV laser light is absorbed by the film. Upon laser irradiation of the material, the electromagnetic energy is converted into electronic excitation. We calculate that for the entire volume of Ag directly under the footprint of the 50-mJ laser beam, the average energy is of the order of 1 eV/atom. It is important to note that although this is a significant amount of energy per atom, it is not shared equally among them. After a very short period of degradation and collisions of the electrons in the absorbed region of the film, the laser energy is converted into other kinds of energies. A large fraction of the laser energy is converted into thermal excitation of atoms and molecules. This leads to melting and rapid evaporation at the surface.

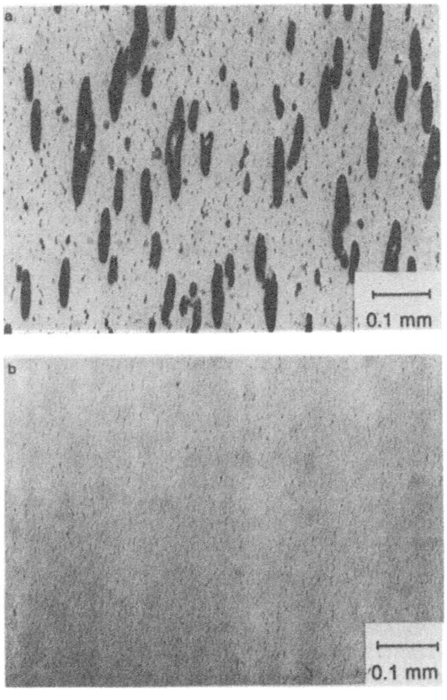

Figure 3. Photomicrographs of 410-nm thick Ag deposited on 4-mil (101 μm) DuPont Mylar (PET) strained in tension to 18%. (a) untreated, and (b) oxygen-glow-discharge treated prior to deposition.

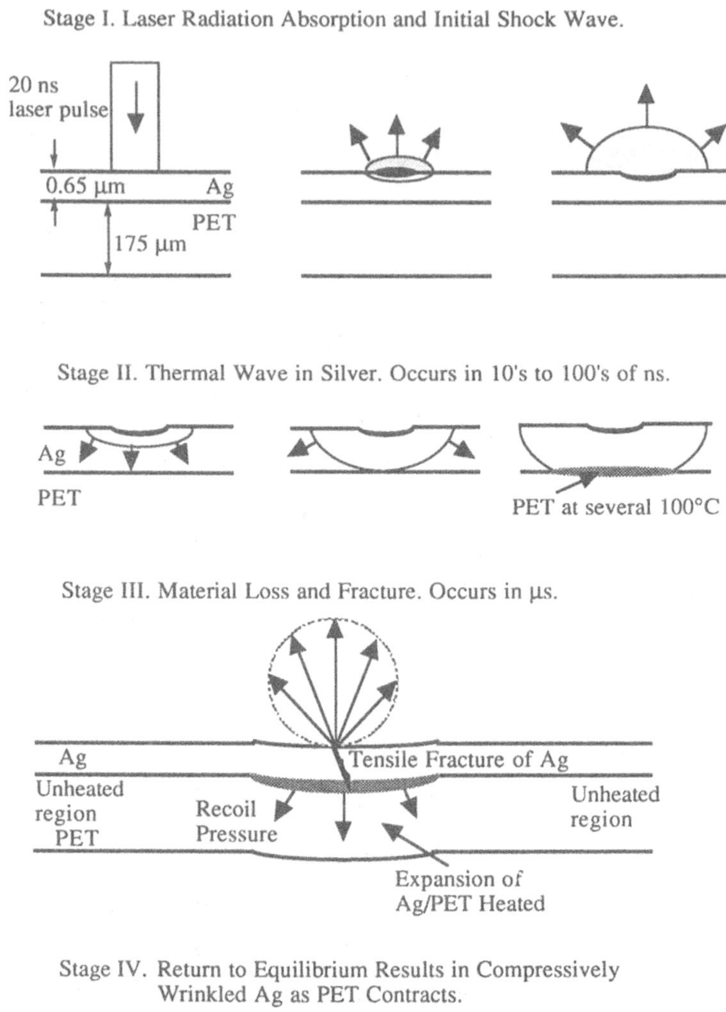

Figure 4. Proposed mechanism of laser ablation failure of metallized plastic.

In this initial stage, rapid evaporation of atoms from the surface of the metal as well as adiabatic heating of the ambient air at the interface causes a spherical shock wave to propagate from the surface. This effect is seen in the remarkable photographs of laser-heated polymeric materials obtained by Braren.[7] In addition, the adiabatic process of laser radiative heat transfer to the material may result in the formation of a layer of plasma at the metal surface. The loss of surface material and the reaction of the shock wave against the new surface result in a shallow crater of material that is densified and compressed against the substrate.

The second phase of our model is characterized by propagation of the resulting thermal wave into the thin film layer toward the substrate. Laser heating of the metallic thin film results in a temperature distribution within the film. The heat conduction equation, modified to include the absorbed laser energy Q(x,y,z,t), can be solved to obtain the temperature profile T(x,y,z,t), where the coordinates are the usual ones of position and time.

The problem can be simplified to a one-dimensional case by assuming a uniform photon distribution in the x-y plane. The temperature profile, T(z,t), is a function of time and direction perpendicular to the plane of the film. The result is a complementary error function solution for T(z,t). From this order-of-magnitude analysis, estimates can be made for critical values of temperature such as the temperature at the interface.

As indicated previously, in the film-substrate systems studied here, the thickness, d, of the metal film ranged from 200 to 4000 nm, whereas the absorption depth for UV radiation in silver is about 25 nm. Thus the heat source for the temperature field within the metallic film is essentially the surface of the film. This greatly simplifies the error function solution which has the form

$$T(z,t) = \frac{2\,I_0}{K\sqrt{Dt}}\ \mathrm{ierfc}\left\{\frac{z}{2\sqrt{Dt}}\right\}(1 - R)$$

where z is the spatial coordinate perpendicular to the film plane, t is time, I_0 is the average incident radiative intensity, K is the thermal conductivity, D is the thermal diffusivity, and R is the reflectance of the thin film surface. For silver (thermal diffusivity $D = 17 \times 10^{-5}\ m^2/s$) subject to a 20-ns laser pulse, the thermal diffusion length, $\lambda = \sqrt{(2Dt)}$, is approximately 2000 nm. From the above we determine that the temperature at the Ag-PET interface is several hundred °C. Because the softening temperature of the PET (Tg ~ 90°C) is much lower than that of the Ag, the polymer at the interface can be molten while the metallic film at the interface remains solid. When this occurs there will be a thin fluid layer of polymer at the interface. (As one reviewer noted, this mechanism may not occur when the polymer surface is fluid.)

In addition, during this heating phase the locally-heated film-substrate region is under the influence of thermal stresses. This is in part due to the mismatch in the linear thermal expansion coefficient (TEC) of the Ag film (TEC = 18 μm/m°C) adhering to the PET (TEC = 38 μm/m°C). This mismatch may lead to stresses that are large enough to cause tensile cracking of the Ag layer.

The third phase of our model is characterized by decomposition, fracture, and loss of material. The evaporated species escaping from the hot surface have a Maxwellian distribution corresponding to the surface temperature, but their velocities are directed away from the surface. Thus the particles have a velocity component perpendicular to the surface that is much larger than the other two components which are parallel to the surface. After a few mean free paths, typically micrometers, the velocity distribution is restored to isotropic. This region above the evaporating surface is referred to as the Knudsen layer.

Particles in the Knudsen layer that experience large angle collisions are scattered back to the surface with a very high velocity. It is these scattered particles, a fraction of which are condensed back onto the surface, that result in a very high ablative pressure at the surface. Anisimov[8] has calculated the velocity of the vaporization front, the surface temperature of the metal, and the recoil momentum for a laser-heated metal surface in vacuum.

The rapid thermal expansion of this local film-substrate region results in a compressive force against the surrounding rigid, nonheated region. This stress, combined with the recoil pressure of the exiting material, results in a pronounced permanent dimpling of the metal-polymer composite. Additional cracking of the Ag film may occur in this phase. This is plausible in light of the observation by Braren[7] of the material plume that exits from regions of laser-heated polymeric material. She has measured a relatively large time delay (10 μs - 100 μs) between the duration of the laser pulse (~10 ns) and the time of maximum material ejection.

The fourth phase involves the cooling and return to equilibrium of the metal-polymer system. Immediately after the high-velocity loss of material, the system returns (only partially) to its original planar state. Upon cooling, the system (which is now nearly thermally equilibrated in the z-direction) undergoes stress, again due to the mismatch of thermal expansion coefficients of the two materials. In this cooling process, however, as the system temperature decreases, compressive stresses will build up in the metal layer. These stresses can become large enough to cause wrinkles as seen in Figure 2.

Tensile Extension Test

Consider a two-layer system consisting of a thin, low-elongation film (e.g., metal or dielectric) adhering to a flexible polymeric substrate. When an externally applied, uniaxial tensile force acts on the system, the film and the substrate are strained equally until cracks form in the film owing to the larger value of failure strain of the substrate. (Such cracks can also result from stresses such as those due to thermal mismatch of different layers or intrinsic (growth) stresses.) Eventually, a distribution of cracks in the thin film develops and has, for a given set of conditions, a mean length designated as L. Upon first cracking of the film, any future cracking is due to the transfer of stress from the strained substrate, across the film-substrate interface, and into the film layer. A review and analytical treatment of this problem is given by Wojciechowski and Mendolia.[9] Obviously, the nature of this stress transfer across the interface, and hence the nature of the resulting cracked film layer, is determined to a large degree by the nature of the interfacial adhesion. Mittal[2] reported a qualitative relationship between interfacial adhesion and cracking phenomena. He noted that as the adhesion of a system decreases, the crack patterns formed during loading of the substrate (in a scratch test) become increasingly coarse. Moreover, once the film fails in the cracking (cohesive) mode, it can then fail in the interfacial mode by peeling away from the substrate as shown in Figure 5(a). Chow and co-workers[10,11] present a mathematically rigorous treatment of the onset of this peeling failure for brittle films on low-modulus polymeric substrates. They use Griffith's energy criterion and continuum-mechanics boundary value analysis to develop a relationship between thin-film adhesion and the externally applied, critical tensile force required to initiate interfacial (peeling) failure.

The mechanisms of cracking and peeling failure can be explained in terms of the stresses that act both within and on the cracked film segment. Figure 5(b) shows a section of cracked thin film adhering to the substrate prior to peeling ($a = 0$). We consider the elemental volume of thin-film material shown in Figure 5(b). The normal stresses of interest in this element are σ_{yy} and σ_{zz}. The closed form analysis of Aleck[12] and the finite element analysis of Kouyumdjiev[13] both give the normal stress distribution in the region of the crack edge as illustrated in Figure 5(c). In addition, their analyses yield the shear stress distribution τ_{zy} at the film-substrate interface, which is included in Figure 5(c). This result is also obtained by the boundary-value analysis of Chow[10] and the shear-lag analysis of Wojciechowski and Mendolia.[9]

Consider the stress functions illustrated in Figure 5(c). The normal stress σ_{yy} is identically zero at the edge of the crack and increases monotonically to some maximum at the center of the crack element. This build-up of normal stress in the crack element is due to the accumulation of shear stress integrated along the interface from edge to edge of the crack element. This is given by the equilibrium condition expressed as

$$\sigma_{yy} = (1/d) \int \tau_{zy} \, dy$$

where the interfacial shear stress τ_{zy} must be determined analytically. As the value of the normal stress σ_{yy} in the film builds up due to the accumulation of shear stress integrated along the interface, the failure stress σ_{yy}^f will be reached. This is the condition for cohesive cracking of the film layer. As the adhesion strength at the film-substrate interface increases for a given tensile pull force F, less interfacial slip will occur resulting in an increase in the shear stress at the interface. This, in turn, will cause a more rapid build-up in the y-direction

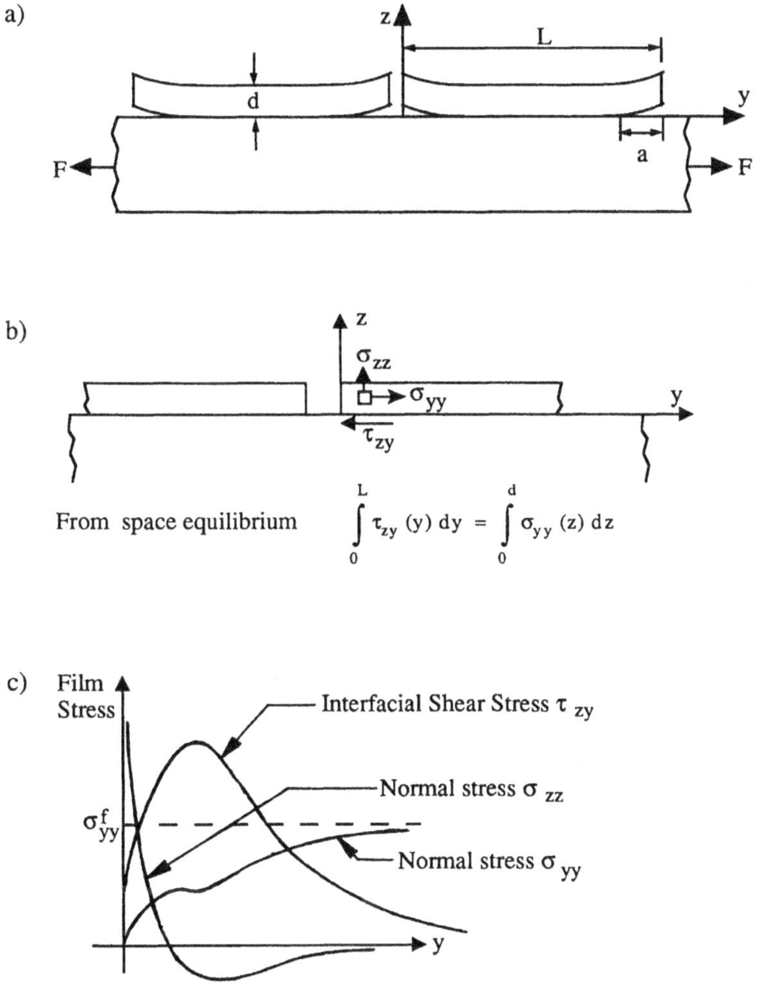

Figure 5. Mechanical behavior of a cracked metal film deposited on a plastic substrate, and distribution of stresses associated with the cracked film element.

of σ_{yy}, which will eventually lead to smaller lengths L of crack segments as the system is extended further. As this is occurring, the normal stress σ_{zz} likewise is increasing until it exceeds the interfacial bonding force per unit area normal to the interface. This is the condition for interfacial failure of the film which results in the commonly observed phenomenon of sharp lift-off of the edges of crack segments. Examples include cracked dried mud flats and cracked paint sections undergoing peeling from a rigid substrate. Often, particularly in the case of metal-polymeric systems, the interfacial bond strength is stronger than the cohesive strength of the polymeric layers below the interface. Consequently, both the film and a layer of adhering polymer undergo lift-off which is observable using high-power microscopy and/or XPS analysis at the locus of failure.

These failure effects are observed in Figure 3(a) for the strained Ag-on-Mylar sample that did not undergo oxygen glow discharge treatment prior to e-beam deposition of Ag. Two sizes of cracks are observed: small and large. The small cracks uniformly populate the surface and result in a milky appearance of the Ag film to the unaided eye. The large cracks are essentially the result of coalescence of the smaller ones where greater stress relief occurred due to the applied elongating strain. In these larger cracks, the lift-up at the crack edges is clearly observable. The tape test applied to the unstrained non-treated (by glow discharge) sample resulted in nearly 100% failure. Once the sample was stretched, the cracked Ag film would rub off under a gently applied fingernail scratch.

When the oxygen-glow-discharge treated sample underwent the same value of tensile strain (18%), no cracks were observed visually. Upon closer microscopic examination, the film appeared to have developed a texture of very small, micrometer-sized features. However, the degree of specular reflectivity of the sample remained virtually unchanged, although not quantified in this study. The tape test of the unstrained sample resulted in no failure. Once stretched, the film remained adherent and would not rub off under firmly applied fingernail scratch.

RESULTS

Table I shows the results of silver evaporated onto PET at a rate of 43 ± 5 nm/s to a thickness of 650 ± 70 nm. PET samples from four different manufacturers were used, some of which had proprietary surface modification or surface additives intended to promote adhesion.

Table I. Silver Evaporated onto PET From Different Manufacturers at 43±5 nm/sec to a Thickness of 650 nm.

Source of PET	Threshold Energy Density for Laser Ablation, mJ/cm^2	Tape Test Amount Removed
Dupont 700D	13	90%
Kodak Estar	43	40%
Hostaphan	47	60%
Melinex	54	10%

The threshold energy densities for laser-ablation-induced cracks correlate well with the tape test results. In all cases the uncertainty in the tape test was ±10% (e.g. (60±10)% removed) while the uncertainty in the threshold energy density was estimated at less than 10% (e.g. 47±10%(47) mJ/cm^2).

The Melinex had the smallest amount (10%) of Ag removal. This corresponded to the highest laser threshold energy of 54 mJ/cm^2. The Estar and Hostaphan were next with 40% and 60% Ag removal by the tape test, respectively. The corresponding laser energy thresholds were 43 mJ/cm^2 and 47 mJ/cm^2, respectively. Adhesion of silver to Dupont 700D is weakest with 90% removed in the tape test, and a threshold of 13 mJ/cm^2. These results confirm the earlier results[6] and show that the laser ablation test provides a broad range of threshold values for a corresponding range of adhesion.

Table II shows the results of tests on silver evaporated onto Melinex PET. Some of the samples were pretreated in an oxygen glow discharge plasma at 15 mTorr and 30 mTorr prior to deposition of silver.

All samples passed the tape test with less than 10% removal of silver. In all cases, however, we observed failure with the laser ablation test. The threshold energy density differentiates between conditions of no glow, 172 mJ/cm^2, glow at 15 mTorr, 255 mJ/cm^2, and 30 mTorr, 191 mJ/cm^2. Energy densities in this test appear different than those in the previous table since the laser had been realigned and the beam splitter had been adjusted. A

Table II. Effect of Oxygen Glow Discharge on Adhesion. Silver evaporated at 43±5 nm/s to a thickness of 650 ±70 nm onto Melinex PET.

Glow Condition	Threshold Energy Density for Laser Ablation, mJ/cm^2 ± 8	Tape Test Amount Removed ±10%
No glow	383	1%
Glow at 15 mTorr	541	10%
Glow at 30 mTorr	413	7%

new splitting ratio was not determined for this arrangement, and so the numbers can be used in only a relative sense. We speculate that the treatment at 30 mTorr may have overtreated the polymer surface resulting in a weak boundary layer and reduced adhesion. Further work is needed to clarify this.

Table III shows results of tests on various thicknesses of silver on PET. Thin coatings of silver, less than 500 nm, pass the tape test while thick coatings, 1000 and 4000 nm, fail to some extent. A mechanism for this effect is given by Kim and Kim.[14] As the silver thickness increased, the threshold energy density increased until at 4000 nm thickness of silver no failure occurred in the laser ablation test. The two tests at first seem to give contradictory results, that the tape test indicates that the adhesion decreases while the laser ablation test indicates that the adhesion improves.

Table III. Silver Evaporated onto PET. Rates and Thicknesses Vary. Notice that the threshold energy density increases as the silver thickness increases. The reasons for this are discussed in the text.

Thickness of Silver, nm.	Threshold Energy Density for Laser Ablation, mJ/cm^2	Tape Test Amount Removed
200	14	0%
500	24	10%
1000	47	40%
4000	Not reached	100%

The resolution of this contradiction is provided by consideration of the thermal diffusion in the silver. As mentioned earlier, a simplified model of the temperature profile provided an estimate of a one-dimensional thermal diffusion length of 2000 nm. Assuming that the three-dimensional length is similar we should expect the thermal energy to diffuse away from the original laser site, and when the silver becomes thick enough the thermal energy reaching the interface will be insufficient to cause adhesion failure.

CONCLUSIONS

Based on the results to date we can make some preliminary conclusions. Further work should elaborate and clarify these points.

First and foremost is that the laser ablation test does work as a measure of practical adhesion. The work presented here shows correlation between the tape test and the laser ablation test, and shows two advantages versus the tape test. First, it gives a broader measurement range of adhesion, working for systems that totally fail the tape test to ones that totally pass the tape test. Second, the laser ablation test apparently shows small differences in adhesion that the tape test cannot.

The results reported here were for silver on PET, but the test also works for other substrates. Some parts of the mechanism proposed earlier for polymeric substrates must be different for the glass substrate and bear further investigation.

The major drawback of the test is that it requires an expensive piece of hardware — the excimer laser. If this is available the test is relatively easy and quick, and requires no special sample preparation. The energy per pulse must be carefully measured to make conclusions in this test. The effective area of the laser pulse bears further study. We used the area of the pulse as determined by exposing a piece of photographic paper. We are currently considering whether this is the proper area to use to determine the energy density. Finally we note that the test is very sensitive to the thickness of the metal film. The adhesion of non-metallic thin films should be amenable to measurement with laser ablation provided that the materials absorb the UV light. It may be interesting to repeat the work of Vossen[5] using excimer-produced UV light directly on the back of a polymer substrate.

The uniaxial stress test shows promise as an adhesion test for low-elongation materials deposited on high-elongation substrates. The test is quick and requires no special sample preparation. It should be especially useful with microscopic inspection of the film during stress.

ACKNOWLEDGEMENT

The authors gratefully acknowledge the support for this work provided by Eastman Kodak Company and, in particular, for the graduate stipend awarded to W.-C. Lee for his part of this research.

REFERENCES

1. K. L. Mittal, J. Adhesion Sci. Technol. 1, 247 (1987).
2. K. L. Mittal, Electrocomponent Sci. Technol. 3, 21-42 (1976).
3. J. Valli, J. Vac. Sci. Technol. A4, 3007 (1986).
4. N .C. Anderholm and A. Goodman, U. S. Patent 3,605,486 (1971).
5. J. L. Vossen, in "Adhesion Measurement of Thin Films, Thick Films, and Bulk Coatings", STP 640, K. L. Mittal, Editor, pp. 122-133, Amer. Soc. Testing and Materials, Philadelphia, PA (1978).
6. P. H. Wojciechowski, F. J. Duarte, and A. L. Hrycin, "Proc. Int'l Conf. on Lasers '87", F. J. Duarte, Editor, pp. 1097-1104, STS Press, McLean, VA (1988).
7. B. Braren, Laser surface modification of polymers and its implications for metallization, presented at this symposium.
8. S. I. Anisimov, Soviet Physics JETP 27, 182 (1968).
9. P. H. Wojciechowski and M. S. Mendolia, J. Vac. Sci. Technol. A7, 1282-88 (1989).
10. T. S. Chow, in "Adhesion Science and Technology", L. H. Lee, Editor, Part B, p. 687, Plenum Press, New York (1975).
11. R. C. Penwall, K. S. Liang, and T. S. Chow, Thin Solid Films 60, 133-139 (1979).
12. B. J. Aleck, J. Appl. Mech. 16, 118 (1949).
13. C. N. Kouyumdjiev, Surface Technol. 26, 45 (1985).
14. K.-S. Kim and J. Kim, ASME J. Eng. Mater. Tech. 110, 266-273 (1988).

ABOUT THE CONTRIBUTORS

SURYADEVARA BABU is currently Professor of Chemical Engineering at Clarkson University. He received his Ph.D. in 1971 from the State University of New York at Stony Brook. He spent 1987-88 with Bell Communications Research in Red Bank, NJ on sabbatical leave and was with IBM at Endicott as a visiting professor/summer faculty member during 1983-86. Prior to that he was on the faculty at the Indian Institute of Technology in Kanpur and was a visitor at the Niels Bohr Institute in Copenhagen and the International Center for Theoretical Physics in Trieste. His current research interests include optical microlithography, plasma and laser processing, and molecular thermodynamics.

CHARLES BAUMGARTNER is a Senior Scientist with GE Corporate Research and Development in Schenectady, NY. Before joining GE in 1980, he received his Ph.D. from Rensselaer Polytechnic Institute. His interest in surface and interfacial chemistry has prompted diverse studies in battery and fuel cell electrodes, piezoelectric ceramic fabrication, and metallization for GE. His recent activities in the realm of plastic metallization are concerned with understanding the relationship between plastics surface chemistry and adhesion.

HOWARD K. BIRNBAUM is Professor in the Department of Materials Science and Engineering, University of Illinois at Urbana-Champaign. He received his Ph.D. degree from University of Illinois. He has carried out research in the areas of diffusion, plastic deformation, phase transitions, hydride structures and transitions, acoustic attenuation, and elastic constants. His recent research is on hydrogen effects on deformation and fracture and on the chemistry and mechanical properties of interfaces and grain boundaries.

BO BOATWRIGHT is a senior technician in the Materials Engineering Laboratory at IBM Corp., Boca Raton, FL. He has coauthored various papers and invention disclosures dealing with new methods for stripping/etching metals, applications of conductive adhesives and corrosion of connectors. He is currently involved in the evaluation and development of conductive adhesives and metallized polymers.

ANDREW B. BOCARSLY is Associate Professor of Chemistry at Princeton University where he has been active in research since 1980. In addition to interests in the photolithographic deposition of transition metals, he maintains research interests in the area of electrochemistry and photochemistry of organometallic complexes.

PETER BODO is Assistant Professor of the Laboratory of Surface Physics and Chemistry at Linkoping University, Linkoping, Sweden. He received a Ph.D. in Materials Science at Linkoping University in 1986. During 1988-1989 he held a Post-Doc position at IBM, T.J. Watson Research Center,

Yorktown Heights, New York. His research interests are focused at surface science of organic materials, especially interfaces between small organic molecules and inorganic surfaces.

MADELEINE BOU is currently preparing thesis at Ecole Centrale de Lyon, France. She graduated as Engineer from Lyon University. Her specific interest is in the application of spectroscopic tools and electron microscopy to adhesion of metal/polymer systems.

ALEXANDER MARIAN BRADSHAW is currently director of the Department of Surface Physics at the Fritz-Haber-Institut of the Max-Planck-Society in Berlin, Germany. He studied chemistry at Queen Mary College, University of London and received his Ph.D. in 1968. Subsequently, he did postdoctoral research at Ludwig-Maximilians-Universität and Technische Universität München, Germany.

JEAN-LUC BREDAS is Professor at the Universite de Mons-Hainaut (Chair of Chemistry of Novel Materials) and Head of the Service de Chimie des Materiaux Nouveaux. His main field of research is electronic properties and nonlinear optical properties of conjugated molecules and polymers. He has a Ph.D. degree.

L. PAIVIKKI BUCHWALTER is currently at IBM T.J. Watson Research Center, where she has recently transferred from IBM East Fishkill. She received her Diploma Chemical Engineer degree from Helsinki University of Technology in 1976. Her research and development efforts have primarily focused on polymer/inorganic adhesion and surface and interface characterization using X-ray photoelectron spectroscopy, and she has authored a number of publications in these areas.

STEPHEN L. BUCHWALTER has been with IBM T.J. Watson Research Center since 1985 where he has been involved in processing and characterization of polyimides, epoxy resins, and siloxanes for packaging applications, and became manager of VLSI Packaging Materials in 1987. He received his Ph.D. in physical organic chemistry from Harvard University in 1974 and did postdoctoral work at the University of Chicago before beginning his industrial research in 1976. His research publications have been primarily in the areas of physical organic and polymer chemistry, while patents have been on organic coating compositions and processes and novel materials and processes for electronic packaging.

DOUGLAS L. CHAMBERS currently General Manager of XYTORR Corporation. He has worked in vacuum deposition, sputtering, ion-plating, chemical vapor deposition, plasma spray coating processes, material development for solar energy, superconductivity, controlled thermonuclear reaction, and plasma gasification of coal. He is currently a member of the AVS' Vacuum Metallurgy Division's Executive Committee and Past President of the Society of Vacuum Coaters. He has authored numerous reports and papers. He is the inventor of more than a dozen patents involving coating processes which have won him numerous honors and awards.

RANDY CAMERON has been working for Lockheed Corp. since 1986. He received his Ph.D. from Princeton University in 1986, having completed a dissertation in the area of photochemistry of transition metal organometallic complexes.

MOHAMMED CHTAÏB is Research Associate at Laboratoire Interdépartemental de Spectroscopie Electronique, Facultes Universitaires Notre-Dame de la Paix, Namur (Belgium). His main field of research is surface modifications of polymers (metallization, laser treatment). He has a Ph.D. degree.

JEANINE A. CONKLIN is currently a graduate student in chemistry at the University of California at Los Agneles. She received her B.S. degree in chemistry from the Pennsylvania State University in 1990. She represented the Penn State chemistry department at the National Conference on Undergraduate Research in 1990. While at Penn State she was actively involved with the Nittany Chemical Society and participated in several chemistry magic shows.

PER DANNETUN is a graduate student, Laboratory of Surface Physics and Chemistry, Physics Department, Linkoping University (Sweden). His main fields of research are electronic structure of conjugated polymers, and surface studies of biologically-active molecules.

JOHN H. DAS is currently completing his Ph.D. research on thermal stability of metals in polymers at the State University of New York at Binghamton. He received his MSEE degree from the South Dakota School Mines and Technology, and during his thesis research he developed a CO_2 waveguide laser with transverse RF excitation. His principal areas of research are microelectronics devices and materials, which also include quantum electronics and computer simulations of dynamic processes.

MOSES DAVID received his Ph.D. in Chemical Engineering from Clarkson University in September 1990. He received his B. Tech. degree in 1985 from Indian Institute of Technology and M. Eng. degree from McGill University in 1987. His research interests are in the area of novel laser and plasma based processes for depositing thin films and powders.

G. DIETLER has been working since 1988 on adhesion problems at the University of Fribourg, Switzerland. He received his Ph.D. in Biophysics at the ETH Zurich in 1985, and did postdoctoral research in Santa Barbara, where he studied fractal aggregates and phase transitions in surfactants solutions.

KARL DOBLHOFER is presently a group leader in electrochemistry at the Fritz-Haber-Institut der Max-Planck-Gesellschaft in Berlin. He received his Ph.D. degree in Electrochemistry in 1969 from Colorado State University, Fort Collins, and had a postdoctoral appointment with A.A. Pilla at Electric Storage Batteries (ESB), Inc., in Yardley, Pennsylvania. His current interests include the study of organic coatings with electrochemical and IR-spectroscopic techniques.

F.J. DUARTE is presently a Senior Research Scientist at Eastman Kodak Company. He is the author of some 40 research papers and is co-editor of Dye Laser Principles (Academic Press, New York, 1990). He received his Ph.D. in Physics in 1983 from Macquarie University, Sydney, Australia.

WILLIAM V. DUMAS is a Chemist at General Electric Armament Systems Department in Burlington, VT. He received his Bachelor's degree in Chemistry from Union College, Schenectady, in 1984. He is involved in the metallization of high performance plastics and metals.

RICHARD EISFELLER is currently a staff chemist, at the Davidson Technology Center in Dover, NH. He received a BS in chemistry from Illinois Institute of Technology in Chicago. He has had more than 20 years experience in the development of vacuum deposited thin films and the study of the adhesion of these films. He has authored several patents and publications in these areas.

THOMAS FLOTTMANN is with Akzo Research Laboratories Obernburg, Germany, which he joined in 1987. He received his Ph.D. in Physics from the University of Heidelberg in 1982 and carried out postdoctoral research at the Institut Laue Langevin in Grenoble. Since 1989 he has been the leader of the thin film technology group at Akzo. His current research interests include metallization for electronics packaging, adhesion of polymers to metals, and ion beam assisted processing for electronics applications.

DONALD F. FOUST is a Staff Chemist at General Electric Co. Corporate R&D Center in Schenectady, NY. He received his Ph.D. degree in Inorganic Chemistry from the University of Massachusetts, Amherst, in 1981. He is involved in the metallization of high performance plastics and has interest in the surface modification of polymers.

BRUCE K. FURMAN has been a Research Staff Member since 1985 at the IBM T.J. Watson Research Center and is currently working on metallization and interconnection for high performance packaging. He received his Ph.D. in Chemistry from Cornell University in 1980. After working at Charles Evans and Associates as a consultant on surface analysis and materials characterization, he joined IBM EF Product Assurance in Poughkeepsie, NY.

FABIO GARBASSI has worked since 1968 at Istituto Guido Donegeni, Novara, Italy, the corporate research center of Enimont. He received his university degree in chemistry at the University of Trieste. He has authored over 90 scientific publications, holds 15 patents, and is the co-editor of the book High Energy Density Technologies in Materials Science. His experience encompasses x-ray diffraction studies of inorganic and organic structures, heterogeneous catalysis, chemico-physical characterization of surfaces, adhesion studies, and modification of polymer surfaces. In 1984 he was appointed Senior Scientist by Montedison for his studies in surface science.

MICHAEL GRUNZE has been with the Fritz-Haber-Institute of the Max-Planck-Gesellschaft in Berlin for seven years and was Professor of Physics at the University of Maine and Director of the Laboratory for Surface Science and Technology before he returned to Germany in 1987 as Chair Professor in Applied Physical Chemistry at the University of Heidelberg. He received his Ph.D. in Physical Chemistry in 1974 and his Habilitation in Physical Chemistry in 1980 from the Freie Universitat in Berlin, Germany. He has published over 90 papers in scientific journals, has edited three books, and has been an invited speaker at many international conferences. His research interest presently focuses around the chemical and physical interactions at organic/inorganic interfaces.

CHRISTA HAAG is Processing Engineer at IBM in Sindelfingen, Germany. Prior to that she did postdoctoral research in the adhesion of metals to PTFE and other polymers by plasma deposited thin metallic films at the University of Tuebingen. She received her Ph.D. in Chemistry at the University of Tuebingen in 1987.

CHRISTIAN HAHN is a graduate student at the University of Maine working in the Laboratory for Surface Science and Technology. He received Diploma-Physiker in 1990. His current research interest is in polymer adhesion.

J.K. HIRVONEN is a Senior Scientist with the Advanced Films Division of Spire Corporation. He has more than 18 years of experience involving beam effects on materials. Prior to joining Spire Coporation, he was a founder, V.P., and Director of Research for Zymet. From 1971-1982 he was a technical staff member at the Naval Research Laboratory where he headed a seven member technical group (1976-1982) conducting basic and applied R&D on the application of ion implantation. He received his Ph.D. in Physics, and has coauthored over 60 papers and has edited two books on ion implantation. He has organized topical symposia and has presented papers at numerous international conferences.

KARSTEN HORN is presently staff scientist at Fritz-Haber-Institut of the Max-Planck-Society in Berlin, Germany. He studied physics at Rheinisch-Westfälische Technische Hochschule in Aachen and Queen Mary College, University of London. He received his Ph.D. in 1976.

R. ANDERS INGEMARSSON is presently working on electroceramic materials at Asea Brown Boveri in Baden-Dattwil, Switzerland. He received his Ph.D. degree from Uppsala University, Sweden in 1989.

ILEANA ISERN-FLECHA is an advisory scientist in the Materials Engineering Laboratory at IBM Corp., Boca Raton, FL. She received her Ph.D. in analytical chemistry from Purdue University (1986). She has authored or coauthored various papers covering a variety of topics in mass spectrometry, thermal analysis, surface analysis and corrosion. Her current research interests include applications of surface analysis to studies on mechanisms of corrosion and polymer metallization.

HYO-SOO JEONG is a graduate research assistant in the Microelectronics Sciences Laboratories, Columbia University. He received his MS degree in Materials Science from Columbia University in 1986, and is currently working toward the Ph.D. His research topics include low energy ion beam modification of high performance polymers used in electronics applications and direct measurement of the adhesion strength between these polymers and thin films used for microelectronics packaging.

JOHN S. JUDGE is currently president of an independent consulting firm, John Judge & Assoc., Inc. He has had over 25 years experience in materials research and development in the electronic industry--primarily in magnetic recording and memory materials. He has worked for IBM, Control Data, Data General, Raytheon and Alcoa. He has a Ph.D. degree in physical chemistry from Syracuse University, and served as chairman of the Boston Section of the Electrochemical Society.

BRADLEY R. KARAS is a Staff Chemist at the General Electric Co. Corporate R&D Center in Schenectady, NY. He received his Ph.D. in Inorganic Chemistry from the University of Wisconsin-Madison in 1981. His research activities involve metallization of plastics and novel imaging techniques for 3-dimensional circuit boards.

ANDREAS KILLINGER is a Ph.D. student in Prof. Grunze's surface science group at the Physikalisch-Chemisches Institut, University of Heidelberg. He carried out Diploma research work at Dalhousie University, Canada. His main research deals with surface specific spectroscopy of organic and inorganic adsorbates on metal surfaces.

J.E. KLEMBERG-SAPIEHA has been research scientist in the Department of Engineering Physics, Ecole Polytechnique, Montreal since 1978. Her major activity is in plasma processing for deposition of films and treatment of materials, that has resulted into more than 40 publications.

OTTO KNOTEK has held the Chair of Materials Science and has directed the Materials Science Institute in Aachen since 1970. He received his doctorate in 1950 from the Technical University in Graz, Austria. In 1957 he acquired his professorial qualification (Habilitation) at the Technical University Aachen. He has authored some 300 papers and other publications.

LARRY J. KRAUSE is currently a Senior Research Specialist in the Industrial and Electronics Sector Research Laboratory of 3M in St. Paul. He received his Ph.D. in Chemistry in 1981 from the University of Illinois at Chicago. After a postdoctoral appointment at Argonne National

Laboratory, he joined 3M in 1982 in their central laboratories in St. Paul. His current research interests include physical polymer chemistry and physical electrochemistry.

O. KÜTTEL is a researcher at the University of Fribourg, Switzerland. He received his education in plasma physics. During 1989-1990 he spent his fellowship in the Engineering Physics Department, Ecole Polytechnique, Montreal, with the main emphasis on plasma processing.

ROBERT H. LACOMBE is with IBM Corporation, E. Fishkill Facility, in Hopewell Junction, NY.

EDWARD J. LAMBY is a Chemist at the Corporate R&D Center of General Electric at Schenectady, NY. He received his M.S. degree from Union College, Schenectady, NY, in 1971. He is involved in the metallization of high performance plastics and photopatterning processes.

ROBERTO LAZZARONI is Chargé de Recherches of the Belgian National Fund for Scientific Research (FNRS), presently working at the Service de Chimie des Materiaux Nouveaux, Université de Mons-Hainaut (Belgium). His main field of research is electronic structure of organic molecules and conjugated polymers. He has a Ph.D. degree.

WEN-CHIEH (JERRY) LEE is continuing his studies toward a Ph.D. at The State University of New Jersey at Rutgers. He received his MS in Materials Science and Technology from Rochester Institute of Technology.

THIERRY LEMOGNE is C.N.R.S. engineer at Ecole Centrale de Lyon, France. His interests are tribology, adhesion, XPS, metals, ceramics and polymers, and has published more than 5 papers.

VERN LINDBERG has taught at the Rochester Institute of Technology for the past 11 years and is currently an Associate Professor in the Physics Department and in the Center for Materials Science and Technology. He received his Ph.D. from Case Western Reserve University. His interests are in adhesion of metals to polymers, surface modification of polymers, and optical coatings. He has been active in the Upstate Chapter of the American Vacuum Society, and will serve as president for the next two years.

FRANK LÖFFLER has worked as a research scientist at the Technical University Aachen since 1986 and since 1989 has directed The Department of PVD Technology at the Institute of Materials Science. He received his doctorate in the field of ceramic hard coatings in 1989, and is the author of some 20 papers and other publications.

MICHAEL LÖGDLUND is a Graduate student, Laboratory of Surface Physics and Chemistry, Physics Department, Linkoping University (Sweden). His main fields of research are electronic structure of conjugated polymers and surface studies of biologically-active molecules.

WOLFGANG LOHMANN is with the Akzo research laboratories in Obernburg, Germany which he joined in 1987 and was appointed leader of the physical surface modification laboratory. Since 1989, he has been manager of the Department Physics and Analytics. He received his Ph.D. in Materials Science from the Technical University in Aachen. During 1979-1986, he was affiliated with the Nuclear Research Center Julich, Germany, and during this period he spent one year at The Los Alamos National Laboratory. His current research interests are in surface-sensitive characterization methods.

DAVID MAGER, Executive Vice President - Vapor Technologies, Inc., is one of the company's founders. He has 15 years experience in high technology product development and program management with inventions, and experience in the PVD, power generation, electronics, food and biotechnology industries. He holds two U.S. patents and has B.S. degree in Biology from State University of New York, Stony Brook.

GIOVANNI MARLETTA has been since 1984 Assistant Professor of Chemical Physics at the University of Catania, Italy, where he received his Dr.Sc. in Chemistry in 1980. Since 1988 he has been a Member of the National Committee for Chemistry and the National Committee for Informatics of the C.N.R. (Italy). His research interests cover surface and interface chemistry by electron spectroscopies with application to microelectronics, photovoltaics and metallurgy; and ion- and electron-beam effects on inorganic salts and polymers. He has bee invited speaker in many international conferences, and has authored more than 30 papers.

JEAN MICHEL MARTIN is presently professor at the Ecole Centrale de Lyon engineer school. He graduated as chemical engineer from the ESCIL school in Lyon in 1972 and was awarded a Ph.D. degree in 1978. He has published over 40 papers, and his interests cover mainly tribochemistry and the application of microscopic analysis in the field of mechanics.

L. MARTINU has been with Ecole Polytechnique, Montreal since 1988 and is a senior researcher with the main interests in plasma-surface interactions, plasma chemistry and thin film physics, that have resulted into more than 40 publications. He received his education at The Charles University, Prague.

R. MAURON is a student who worked on surface analysis for his Masters degree at The University of Fribourg, Switzerland.

HEINRICH MEYER has been since 1985 research director of the Electroplating Division of Schering AG in Berlin, and has been qualified for university lecturing in Inorganic Chemistry since 1987. He received his Ph.D. from the Ludwig Maximilian University in Munich. In 1981 he joined W.C. Heraeus GmbH, Hanau (Germany) as head of research and development of the Chemical Division. His main areas of research are electrochemistry, catalysis, reaction mechanisms, and relationship between properties and structure of solid materials.

KASHMIRI LAL MITTAL * is presently employed at the IBM US Technical Education in Thornwood, N.Y. He received his M.Sc. (First Class First) in 1966 from Indian Institute of Technology, New Delhi, and Ph.D. in Colloid Chemistry in 1970 from the University of Southern California. He has organized and chaired a number of very successful international symposia and in addition to this volume, he has edited 34 more books as follows: Adsorption at Interfaces, and Colloidal Dispersions and Micellar Behavior (1975); Micellization, Solubilization, and Microemulsions, Volumes 1 & 2 (1977); Adhesion Measurement of Thin Films, Thick Films and Bulk Coatings (1978); Surface Contamination: Genesis, Detection, and Control, Volumes 1 & 2(1979); Solution Chemistry of Surfactants, Volumes 1 & 2 (1979); Solution Behavior of Surfactants: Theoretical and Applied Aspects, Volumes 1 & 2 (1982); Adhesion Aspects of Polymeric Coatings, (1983); Physicochemical Aspects of Polymer Surfaces, Volumes 1 & 2 (1983);

* As the editor of this volume.

Surfactants in Solution, Volumes 1, 2 & 3 (1984), Adhesive Joints: Formation, Characteristics, and Testing (1984), Polyimides: Synthesis, Characterization and Applications, Volumes 1 & 2 (1984); Surfactants in Solution, Volumes 4, 5 & 6 (1986); Surface and Colloid Science in Computer Technology (1987); Particles on Surfaces 1: Detection, Adhesion and Removal, (1988); Particles in Gases and Liquids 1: Detection, Characterization and Control (1989); Particles on Surfaces 2: Detection, Adhesion and Removal (1989); Surfactants in Solution, Volumes 7-10 (1989); Metallized Plastics 1: Fundamental and Applied Aspects (1989); Polymers in Information Storage Technology (1989) and Particles in Gases and Liquids 2: Detection, Characterization and Control (1990). Also he is Editor of the Series, Treatise on Clean Surface Technology, the premier volume appeared in 1987. In addition to these books he has published about 60 papers in the areas of surface and colloid chemistry, adhesion, polymers, etc. He has given many invited talks on the multifarious facets of surface science, particularly adhesion, on the invitation of various societies and organizations in many countries all over the world, and is always a sought-after speaker. He is a Fellow of the American Institute of Chemists and Indian Chemical Society, is listed in American Men and Women of Science, Who's Who in the East, Men of Achievement and many other reference works. He is or has been a member of the Editorial Boards of a number of scientific and technical journals, and is the Editor of the Journal of Adhesion Science and Technology, which made its debut in 1987. In 1990 he received the Charles B. Dudley Award of the American Society for Testing and Materials.

RICHARD A. MOODY is a Staff Scientist within the Polymer Development Group of the Materials Modification Division of Spire Corporation, and is currently working on a Ph.D. in Polymer Physics at the University of Lowell. His current research interests include the modification of the electrical, mechanical, optical and chemical properties of polymers using ion beam and laser processes. At the University of Lowell, he is active in several research programs including the synthesis and characterization of novel electroactive polymers and the characterization of Langmuir-Blodgett thin films. He has published over 15 papers and has several pending patents.

MARCO MORRA has been since 1987 at Istituto Guido Donegani, the corporate research center of Enimont, in Novara, Italy. He graduated with honors at the University of Torino in 1986. He has authored over 12 scientific publications and holds four patents. His experience includes chemico-physical characterization of surfaces and technologies for modification of polymer and glass surfaces.

JAMES MORRIS is Professor and Chair of Electrical Engineering in the Thomas J. Watson School at The State University of New York at Binghamton. He has degrees in physics and electrical engineering and has held academic appointments in both disciplines. He is active in many professional societies and was the first director of SUNY-Binghamton's Institute for Research in Electronics Packaging. He is editor of the new VNR monograph series: Electronics Packaging Forum. His primary research interests cover a wide variety of thin film topics, especially the nucleation, growth and electrical properties of discontinuous metal films.

MANISH NANDI is currently a research chemist with ARCO Chemical Co. in Newton Square, PA. He obtained his Ph.D. degree in Chemistry from the Pennsylvania State University working on the chemistry of polymers, ceramics and composite materials.

K.R. NARENDRNATH is Manager of Coating Operations and Analysis at Vapor Technologies, Inc. He obtained his Ph.D. in Metallurgy and Materials Science from Polytechnic University, Brooklyn, New York. His areas of interest include coating deposition for industrial and aerospace applications, advanced materials processing, property evaluation and correlation to structure, bulk and surface analyses through SEM/EDS, WDS, X-ray diffraction, XPS/ESCA, TEM, substrate surface preparation and adhesion, and failure analysis.

DANIEL NEUGROSCHL is a graduate student at the Henry Krumb School of Mines, Columbia University. This work was done during a temporary assignment at the IBM T.J. Watson Research Center. He holds M.S. in Materials Science from Columbia University.

S. NOWAK has been working since 1988 on surface treatment and analysis of polymers at the University of Fribourg, Switzerland where he received his Ph.D. in Physics in 1986. He did postdoctoral research in Eindhoven on spectroscopy of thermal plasmas.

ERNESTO OCCHIELLO has been since 1985 at Istituto Guido Donegani, the Corporate research center of Enimont, in Novara, Italy. He graduated in Chemistry with honors at the University of Torino in 1984. In 1986, on a sabbatical, he worked at the IBM Almaden Research Center, San Jose, CA studying plasma etching and deposition processes. He has authored over 25 scientific publications and holds 8 patents. His experience includes NMR spectroscopy, chemico-physical characterization of surfaces, applications of high energy density technologies (laser, plasma) in materials science, and synthesis and modification of polymers.

AARON ORTEGA-VILLAMIL is presently Associate Professor at the National University of Colombia, Bogota. He studied physics at the University of Colombia, at State University of New York at Buffalo, and at Technische Universitat Berlin. He received his Ph.D. in 1980.

RAGHUNATH PADIYATH is currently a Ph.D. student in the Department of Chemical Engineering at Clarkson University. He received his B. Tech. degree in 1988 from Calicut University, India and his M.S. degree from Clarkson University in 1990. He is interested in thin films and metallization processes for VLSI applications.

SALVATORE PIGNATARO is Professor at the University of Catania, Italy which he joined as Chairman of Physical Chemistry in 1975. He is currently also general manager of a Consortium between Catania University and several industries and public agencies. He received his degree in chemistry in 1962 from the University of Catania. He has been or is on the editorial boards of some journals, and has been very active as member of various scientific or steering committees of Italian National Research Council or Italian Chemical Society. He has published more than 120 papers in the areas of mass spectrometry, photoelectron spectroscopy, interaction of ion beams with matter, and physical chemistry of surfaces.

MIRO M. PLECHATY is presently employed at the IBM T.J. Watson Research Center in Yorktown Hts., NY. He is a graduate of the Technical University in Prague, Czechoslovakia (M.Sc.) and the Union Institute in Cincinnati, Ohio (Ph.D.). His interest is in the field of trace analytical chemistry and has published in this area.

SAMPATH PURUSHOTHAMAN is Manager/Research Staff Member in the High Performance Packaging Technology department at the IBM T.J. Watson Research Center, Yorktown Hts., NY. He holds D.E.Sc. degree in Materials Science from Columbia University. Prior to joining IBM, he was a Senior

Research Associate and Adjunct Professor at the Henry Krumb School of Mines, Columbia University. He holds several patents in microelectronic packaging and has published over 30 technical papers.

WILLIAM R. SALANECK is Professor of the Laboratory of Surface Physics and Chemistry at Linkoping University, Linkoping, Sweden. Previously, he studied at the University of Pennsylvania, Philadelphia, receiving a Ph.D. in Solid State Physics in 1968. He subsequently held several scientific positions at the Xerox Webster Research Center in Webster, New York, before moving to Linkoping in 1983. His principal research interests include the surface science of organic molecular solids, polymeric materials, especially conjugated polymers, and polymer interfaces.

MARK J. SCHADT is a Staff Scientist at IBM Systems Technology Division in Endicott, NY. His graduate research was in the area of photochemistry. He joined IBM in 1985 at the Thomas J. Watson Research Center in Yorktown Hts., NY and worked on the development of photoresist processes for submicron applications. Since 1988 he has been working on the metallization of dielectric materials for Thin Film packaging applications.

GISELA SCHAMMLER is currently scientist/group leader in the Research Center for Microperipheric Technologies at the Technical University of Berlin, where she received her Ph.D. in chemistry in 1989. She was for two years an associated researcher at Schering AG, Berlin, electroplating division. Her professional experience includes polymeric materials and glasses, electrodeposition and electroless plating of polymers. She has several publications in the field of electroplating of polymers.

L. SCHLAPBACH has been since 1988 professor of experimental physics at the University of Fribourg, Switzerland. He received his Ph.D. from the ETH Zurich in 1973 for work done on magnetic properties of liquid alloys. He is editor and coauthor of Topics in Applied Physics, Vols. 63 and 64 on Hydrogen in Intermetallic Compounds I and II, Springer, 1988 and 1990.

RALF SCHULZ has been working since 1986 in the Research Department of Schering Electroplating in Berlin. He studied chemistry at the Technical University of Berlin and received his Ph.D. degree. He has carried out basic and applied research on the metallizing of polymers, especially the investigation of the metal/polymer interphase, and his current responsibilities are quality assurance and materials science of plated metal coatings.

FRANK SCHWAB received his B.A. degree from Princeton University in 1986 and is currently completing a degree in medicine.

AYUSMAN SEN is currently a Professor of Chemistry at the Pennsylvania State University, which he joined in 1979. He holds a Ph.D. in Chemistry from the University of Chicago, and carried out research at the California Institute of Technology. His awards include an Alfred P. Sloan Research Fellowship and the Paul J. Flory Award from IBM.

KABUL S. SENGUPTA is currently pursuing his Ph.D. degree as a Research Assistant with Professor Howard K. Birnbaum in the Department of Materials Science and Engineering, University of Illinois at Urbana-Champaign. His interests include metal-polymer and metal-ceramic interfaces using microscopy and surface analysis techniques.

KRISHNA SESHAN is with IBM Corporation, E. Fishkill Facility, in Hopewell Junction, NY.

PHILIPPE SNAUWAERT is a Graduate student, Laboratoire Interdepartemental de Spectroscopie Electronique, Facultes Universitaires Notre-Dame de la Paix, Namur (Belgium). His main field of research is synthesis and characterization of conducting polymers (polyanilines).

TOM SPECKHARD is currently a Chemical Engineering Specialist in the Commercial Graphics Division Laboratory of 3M in St. Paul working in electrostatic printing. Since joining 3M he has worked in the Electronic and Information Technologies Sector Laboratory and the Industrial and Electronic Sector Laboratory with projects in the areas of antistatic materials and metallization of polyimide. He received his Ph.D. in Chemical Engineering from the University of Wisconsin in 1985.

JURGEN SPRINGER has been Professor of Macromolecular Chemistry at the Technical University of Berlin since 1972. He worked on polymers in the Fritz-Haber-Institute of the Max-Planck-Gesellschaft and in 1963 he was awarded the doctorate degree in Natural Sciences (Dr. rer. nat.) from the Free University of Berlin. From 1963 to 1972 he was a research scientist at the Fritz-Haber-Institute. His research has been concerned with solution properties as well as permeability and adhesion phenomena of polymers, and has written numerous papers in these areas.

THOMAS STRUNSKUS is a graduate student at the University of Maine working in the Laboratory for Surface Science and Technology. He received Diploma-Chemiker in 1988. His current research interest is in polymer adhesion.

HARALD SUHR has been since 1960 at the Department of Organic Chemistry, University of Tübingen, Germany. He received his Ph.D. in Physical Chemistry from the University of Frankfurt in 1956. His main area of research is plasma chemistry of organic and organometallic compounds.

GUY T. SUSI is currently Manager of Manufacturing for XYTORR Corporation. He has over 25 years experience in setting up, checking out, and directing the manufacturing operations of large coating systems.

KENNETH A. TAYLOR is currently a consultant to XYTORR Corporation responsible for the design and development of production-scale vacuum coating systems. He has more than twenty years of experience in design and manufacturing of vacuum coating systems utilizing many different Physical Vapor Deposition processes. He has published many technical papers pertaining to thin film coatings and the mechanics of systems needed to produce such coatings.

T.G. TETREAULT is a Staff Scientist within the Advanced Films Section of the Surface Modification Division of Spire Corporation. His interests are in the development and tailoring of IBAD coatings to specific application needs which may require certain optical, mechanical, or tribological properties. He received B.S. degree in Physics from the University of Vermont in 1980. He has published several papers and has coauthored a chapter in the book Synthesis and Properties of Boron Nitride.

RICHARD R. THOMAS is currently a Research Staff Member in the VLSI Packaging Materials group at IBM T.J. Watson Research Center, Yorktown Hts., NY. He received Ph.D. degree in Inorganic Chemistry from the Pennsylvania State University in 1985. His research interests include adhesion, polymer modification and electroless plating of metals on polymeric substrates.

CHRISTINA THÜMMLER is a Ph.D. student in Prof. Grunze's surface science group at the Physikalisch-Chemisches Institut, University of Heidelberg.

She carried out Diploma research work at Fritz-Haber-Institut der Max-Planck-Gesellschaft, Berlin. Her main research deals with organic and inorganic adsorbate/metal interfaces.

KAJSA UVDAL is a graduate student of the Laboratory of Surface Physics and Chemistry at Linkoping University, Linkoping, Sweden. Her main research interest is surface science of organic molecular solids, especially conjugated polymers, and interfaces between small organic molecules and inorganic surfaces.

ALFRED VIEHBECK is a Research Staff Member at IBM Research Division, T.J. Watson Research Center, Yorktown Hts., NY and is currently managing the Dielectric Materials and Processes Group in the Packaging Department. He received his Ph.D. degree in Physical Chemistry from Rice University in 1982. His research background includes electrochemistry and corrosion science. Since joining IBM in 1986, he has been involved in the development of new packaging processes with primary emphasis on metallization of polymeric materials.

LOUIS VOVELLE is presently working at Rhone-Poulenc in Saint Fons, France. He graduated as chemical engineer from the Paris chemistry school in Paris 6 and was awarded a Ph.D. degree in 1986. He has published over 20 papers.

CHONG T. WAN is currently Director of Research and Development at the XYTORR Corporation. He has worked in the areas of: (1) reactive sputtering of various nitrides and oxides, (2) monolithic and multilayer transparent conductive coatings, (3) magnetic thin films, (4) erosion and corrosion resistant coatings, and (5) titanium nitride coating for tool bits. He is the inventor of more than a dozen of patents on glass coating method and apparatus, glass coating materials and magnetron cathode.

M.R. WERTHEIMER has been Professor, Department of Engineering Physics, Ecole Polytechnic, Montreal since 1974. His principal research interests are plasma processing of materials, physics of dielectrics, and thin film science and technology. He has 180 publications and 10 patents to his credit. He is IEEE Fellow, and Killam Research Fellow (1990-1992).

ROBERT C. WHITE is currently Assistant Professor of Electrical Engineering at Columbia University. He received his Ph.D. in Chemical Physics from the University of Utah in 1986. He worked in the Physical Sciences Department, IBM T.J. Watson Research Center until 1988 when he went to Columbia. His current research efforts include surface physics and chemistry of microelectronics materials, polymers in electronics, interconnect and packaging technologies as well as sensors and transducers.

PAUL H. WOJCIECHOWSKI is currently Senior Staff Research Associate in the Physical Performance Laboratory at Eastman Kodak Company. Prior to joining Kodak in 1985, he served for 13 years as professor in the Department of Mechanical Engineering at Rochester Institute of Technology. He received his Ph.D. in Mechanical and Aerospace Sciences from the University of Rochester.

MEISHENG ZHOU recently received her Ph.D. from Princeton University and is currently carrying out postdoctoral research at the University of Rochester's Science and Technology Center. Her thesis work was in the area of photolithographic plating of platinum and copper.

INDEX

GPSR Compliance

The European Union's (EU) General Product Safety Regulation (GPSR) is a set of rules that requires consumer products to be safe and our obligations to ensure this.

If you have any concerns about our products, you can contact us on ProductSafety@springernature.com

In case Publisher is established outside the EU, the EU authorized representative is:

Springer Nature Customer Service Center GmbH
Europaplatz 3
69115 Heidelberg, Germany

Batch number: 10761599

Printed by Printforce, the Netherlands